T0211130

# Multi-scale Modeling of Structural Concrete

# Multi-scale Modeling of Structural Concrete

Koichi Maekawa, Tetsuya Ishida and Toshiharu Kishi

**CRC Press**
Taylor & Francis Group
Boca Raton London New York

CRC Press is an imprint of the
Taylor & Francis Group, an **informa** business

A TAYLOR & FRANCIS BOOK

CRC Press
Taylor & Francis Group
6000 Broken Sound Parkway NW, Suite 300
Boca Raton, FL 33487-2742

First issued in paperback 2019

© 2009 Koichi Maekawa, Tetsuya Ishida and Toshiharu Kishi
CRC Press is an imprint of Taylor & Francis Group, an Informa business

No claim to original U.S. Government works

ISBN-13: 978-0-415-46554-0 (hbk)
ISBN-13: 978-0-367-86622-8 (pbk)

Typeset in Sabon by Pindar NZ, Auckland, New Zealand

This book contains information obtained from authentic and highly regarded sources. Reasonable efforts have been made to publish reliable data and information, but the author and publisher cannot assume responsibility for the validity of all materials or the consequences of their use. The authors and publishers have attempted to trace the copyright holders of all material reproduced in this publication and apologize to copyright holders if permission to publish in this form has not been obtained. If any copyright material has not been acknowledged please write and let us know so we may rectify in any future reprint.

Except as permitted under U.S. Copyright Law, no part of this book may be reprinted, reproduced, transmitted, or utilized in any form by any electronic, mechanical, or other means, now known or hereafter invented, including photocopying, microfilming, and recording, or in any information storage or retrieval system, without written permission from the publishers.

**Trademark Notice:** Product or corporate names may be trademarks or registered trademarks, and are used only for identification and explanation without intent to infringe.

*British Library Cataloguing in Publication Data*
A catalogue record for this book is available from the British Library

*Library of Congress Cataloging-in-Publication Data*
Maekawa, Koichi, 1938–
Multi-scale modelling of structural concrete / Koichi Maekawa,
Tetsuya Ishida, and Toshiharu Kishi. – 1st ed.
    p. cm.
 Includes bibliographical references and index.
 1. Concrete construction—Mathematical models. 2. Structural
analysis (Engineering)—Mathematical models. 3. Concrete—
Evaluation—Mathematical models. I. Ishida, Tetsuya. II. Kishi,
Toshiharu, 1955- III. Title.
 TA681.5.M34 2008
 624.1'834—dc22                              2008022441

Visit the Taylor & Francis Web site at
http://www.taylorandfrancis.com
and the CRC Press Web site at
http://www.crcpress.com

# Contents

# Preface

This book presents both micro-scale chemo-physics and semi-macroscopic mechanics of structural concrete, including crack-like damages and corrosion. The authors aim to present the mutual linkage of chemo-physics and mechanistic events which develop over the control volumes of different scales, and the way to simulate the macroscopic behaviors of structural concrete under combined external loads and ambient conditions. The heat hydration and temperature, carbonation and $CO_2$ penetration, chloride migration and corrosion of steel in concrete, calcium leaching and dynamics of micro-pore structures, moisture hygro thermodynamics and creep and shrinkage, low and high cycle fatigue are computationally simulated and this strong mutual interaction is taken into account.

In 1999, the authors' first proposal to evaluate concrete performance with regard to durability was put together in a book entitled *Modeling of Concrete Performance: Hydration, Microstructure and Mass Transport* and the computer code named *DuCOM* (Durability COncrete Model) was partially released to the construction industries. Here, thermo-hygro physics, which concurrently advances inside multi-scale micro-pores with mutual linkage, were tried to be systematically integrated for macroscopic material properties and durability performances. Along these lines, further development on carbonation, chloride ion penetration, steel corrosion and calcium leaching has been combined, based upon the multi-scale platform. The research is still going on, but some of the integrated knowledge is being applied to engineering problems, such as life-cycle prediction of concrete structures and remaining life assessment of existing infrastructures. The authors decided that it was the right time to summarize the progress of the past decade in a consistent way. Thus, this book can be considered a revision of the previous publication.

Another point of importance is the strong coupling with structural mechanics on a large scale. The constitutive modeling of cracked concrete is described on the control volume of cm ~ meter scale including cracks. Another monograph entitled *Nonlinear Mechanics of Reinforced Concrete* was published mainly on structural safety and seismic performance assessment in 2003. Here, the multi-scale mechanical modeling ranges from $10^{-3}$ to $10^{-0}$ meter-scale volume including micro and macroscopic cracks. Conversely, the cementitious composite is idealized as an assembly of micro-pores with $10^{-9}$

to $10^{-3}$ meter scale. This book focuses on the further global linkage of micro-pore structure modeling with structural concrete having rather macroscopic damage and plasticity. Thus, this book constitutes a second revision of the previous work on nonlinear mechanics, which is represented by the computational code named COM3 (COncrete Model of 3-Dimension).

At the same time, full-scale integration of chemo-physics and mechanical events from $10^{-9}$ to $10^{-0}$ meter is intended by consistently combining the contents of two books, that is to say, integrated DuCOM-COM3. The authors recognize that this integration of multi-scale modeling may materialize the knowledge-base of concrete engineering, and that the computerized multi-scale behavioral simulation is made possible for the performance assessment of material and structural concrete. Currently, maintenance engineering of the existing infrastructure is in great demand, and a huge amount of professional knowledge, combined with experience and their integration are crucial for solving such a tough problem. Concrete engineering is rather system oriented. The authors expect *Multi-scale Modeling of Structural Concrete* to become a platform to which knowledge and experiences can be applied to meet the challenge of practice.

The authors sincerely appreciate Ph.D. scholars who have greatly contributed to the research on multi-scale modeling in Department of Civil Engineering at The University of Tokyo, and again express their sincere gratitude to Dr. R. P. Chaube (micro-pore modeling and hygro-equilibrium), Prof. K. Nakarai of Gunma University (calcium leaching and cemented soil), Dr. S. Asamoto of Saitama University (solidification and creep), Dr. K. Toongoenthong of Taisei Corporation (damaged structural modeling), Dr. K. F. El-Kashif of Cairo University (nonlinear creep), Dr. C. Li of Gunma University (carbonation), Dr. Y. B. Zhu of China Institute of Water Resources and Hydropower Research (solidification), Dr. R. Mabrouk of Cairo University (solidification), Dr. P. O. Iqbal (moisture and corrosion), Dr. R. R. Hussain of The University of Tokyo (corrosion rate), Dr. K. Y. Choi of Samsung Construction Corp. (time and temperature dependent tension stiffness), Dr. E. Gebreyouhannes of Addis Ababa University (fatigue simulation and shear transfer), Prof. M. Soltani Mohammadi of Tarbiat Modares University (joint interface fatigue), Dr. Y. Otabe of Sumitomo-Osaka Cement Corp. (hydration and strength model) and Ph.D. candidates, Mr. N. Chijiwa and Mrs. C. Fujiyama of The University of Tokyo and Mr. M. Azenha of University of Porto. The authors' gratitude is further addressed to three scholars, Dr. E. Gebreyouhannes, Dr. N. Bongochgetsakul and Dr. Lin Zhihai, for their editorial support and assistance. Without their efforts and enthusiasm, this book would never have existed.

Since 1990, the authors have incessantly upgraded the software DuCOM-COM3 as the computational platform of life-span simulation, and deeply thank Dr. N. Fukuura of Taisei Corporation and Dr. T. Mishima of Maeda Corporation for their effort to enhance the program efficiency and reliability by means of the massive parallel computation and super-high speed solver of huge matrixes. The first author fully shows his appreciation to Prof. X. An of Tsinghua University and Prof. H. W. Song of Yonsei University, who were former faculty and the authors' colleagues at The University of Tokyo, for their valuable advices and the scientific cooperation on durability design and maintenance, as well as an engineering application of the performance assessment software DuCOM-COM3 to Asian industries. The research for this book was financially supported by Grant-in-Aid for Scientific Research (S) No. 15106008 from Japan Society of the Promotion of Science (2003–2008) to the first author.

Koichi MAEKAWA,
Professor, Department of Civil Engineering,
University of Tokyo

Tetsuya ISHIDA,
Associate Professor, Department of Civil Engineering,
University of Tokyo

Toshiharu KISHI,
Associate Professor, Institute of Industrial Science,
University of Tokyo

# Abbreviations

| | |
|---|---|
| ASR | alkali aggregate reaction |
| ASTM | American Society of Testing and Materials |
| BET theory | per Brunauer, Emmet and Teller (1938) |
| BRC | belite-rich cement |
| BS | British Standard |
| CH | calcium hydroxide |
| CSE | copper–copper sulphate electrode |
| CSG | cemented soil and gravel |
| C–S–H | calcium silicate hydrate(s) |
| EDS | energy dispersive X-ray spectroscopy |
| ELG | epoxy-coated lightweight gravel |
| EPMA | electron probe micro analyzer |
| FA | fly ash |
| FEM | finite element method |
| Gy | gypsum |
| *hcp* | hardened cement matrix |
| HPC | high early-strength Portland cement |
| ITZ | interfacial transition zone |
| JCI | Japan Concrete Institute |
| KT theory | per Katz and Thompson (1987) |
| LG | lightweight gravel |
| LOI | loss on ignition |
| LPC | low-heat Portland cement |
| LS | limestone |
| MIP | mercury intrusion porosimetry |
| MPC | moderate-heat cement |
| NHE | normal hydrogen electrode |
| OG | ordinary gravel |
| OPC | ordinary Portland cement |
| PC | pre-stressed concrete |
| PRC | pre-stressed reinforced concrete |
| PSTs | pore structure theories |
| RC | reinforced concrete |
| REV | representative elementary volume |
| RH | relative humidity |

| | |
|---|---|
| R–R | Raleigh–Ritz distribution |
| SEM | scanning electron microscope |
| SF | silica fume |
| SG | blast furnace slag |
| SHE | standard hydrogen electrode |
| SP | super plasticizer |
| TGA | thermal gravimetric analysis |
| TG-DTA | thermogravimetry-differential thermal analysis |
| W/C | water-to-cement ratio |
| W/P | water-to-powder ratio |
| XRD | X-ray diffractometry |

# 1 Introduction

In this chapter, the authors present the outline of the multi-scale modeling of structural concrete as a systematic knowledge base of both cementitious composites and structural mechanics. An object-oriented computational scheme is proposed for life-span simulation of reinforced concrete and soil foundation. Conservation of moisture (Chapter 2, Chapter 3), carbon dioxide (Chapter 4), calcium (Chapter 5), oxygen, chloride (Chapter 6) and momentum is simultaneously solved with hydration, carbonation, steel corrosion, ion dissolution, damage evolution (Chapters 7–10) and their thermodynamic/mechanical equilibrium. As it is essential for verification of the holistic system to be based upon reality, efforts are also directed to the systematic examination and checking of modeling. The detailed modeling at each scale is presented in the subsequent chapters of this book.

To ensure the sustainable development of vast urban agglomerations where natural elements and people co-exist, infrastructure needs to retain its performance over the long term. In order to construct durable and reliable social environments, functionality, safety and their likely life-time will be estimated on the basis of scientific and engineering knowledge. For currently deteriorated infrastructure, a rational maintenance plan should be implemented in accordance with present situations and anticipated future requirements. It is indispensable to consider entire structural performances under predictable ambient and load conditions during the designated service life. The objective of this book is to develop a multi-scale modeling of structural concrete with soil foundation and to realize a so-called lifespan simulator capable of predicting overall structural behaviors as well as constituent materials.

Figure 1.1 shows a schematic representation of the targeted multi-scale modeling of material science and structural mechanics. Macroscopic characters of concrete composites are dependent upon micro-pore structures and thermodynamic states associated with durability. In turn, chemo-physical states of substances in nano~micro-pores are strongly associated in structural mechanics with damage induced by loads and weather actions. In order to consider how knowledge might best be applied in a practical sense, the two sub-fields of structural mechanics and material science have been investigated, in conjunction with a number of related engineering factors associated with

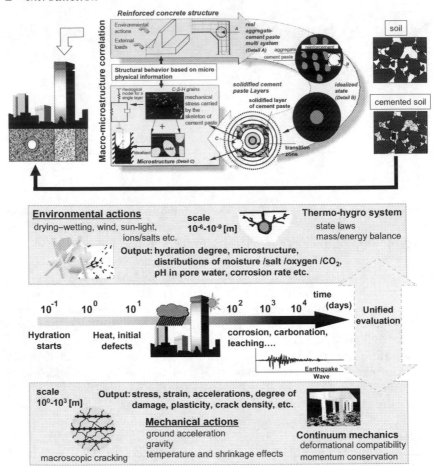

*Figure 1.1*  Multi-scale scheme and lifespan simulation for materials and structures.

the macroscopic features of composites, like uniaxial strength. Currently, much valuable knowledge is available to us. As representatives of academia, the authors have recognized that it makes sense for them to synthesize the considerable body of knowledge about concrete in a systematic way. This knowledge system should not be considered simply as so many archives or indexes, but as the raw material from which quantitative estimations can be drawn. Thus, a rich vein of overlaid multiple linkages might result from info-technological mining and cross-referencing.

In practice, multi-scale behavioral simulation of both constituent materials and structures is expected to serve engineers in charge of design, planning and policymaking. Indeed, addressing real maintenance issues requires an understanding of materials, structures, construction and management. Thus, we seek to present a computerized scheme for coming to grips holistically with such information by means of explicit mathematical formulae originating from past research. This book will propose one possible means of building up a systematic knowledge base from the amalgamation of insights drawn from material and structural engineering, including soil foundation (Figure 1.1). As a main point of discussion concerns system formation, some relevant models deriving from microphysical chemistry have necessarily been simplified. Doubtless, a system's functional integrity relies significantly on its constituent elements. Thus, individual modeling will be experimentally verified for continuous development as a quantitative knowledge resource for engineers. This book also aims to show the verification procedure.

## 1.1 Basic computational scheme

In principle, a dispersed object-oriented scheme on open memory space is implemented as shown in Figure 1.2. Physical chemistry and mechanical events in three dimensions are individually solved in a particular time step. The authors select kinematic chemo-physics and mechanical events with different geometrical scales of representative elementary volume (REV) where governing equations hold as indicated in Table 1.1,

Notably, these events are not independent but mutually interlinked. A complex figure of interaction is mathematically expressed in terms of state parameters commonly shared by each event. For example, Kelvin

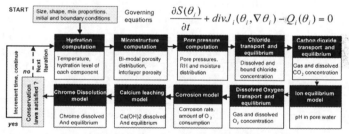

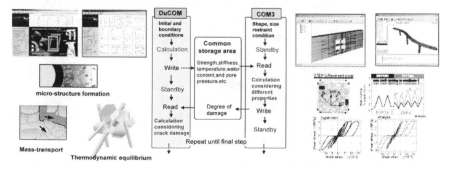

*Figure 1.2* Sub-structure platform of *Durability COncrete Model (DuCOM)* and integration of microphysics and macro-structural analysis.

*Table 1.1* Kinematic chemo-physics and mechanical events with different geometrical scales

| Chemo-physics and mechanical events | Governing principle | Scale |
| --- | --- | --- |
| Cement heat hydration and thermal conduction | thermal conservation | $10^{-6} \sim 10^{-4}$ m |
| Pore structure formation and moisture equilibrium/transport | water mass conservation | $10^{-10} \sim 10^{-6}$ m |
| Free/bound chloride equilibrium and chloride ion transport | chloride ion conservation | $10^{-8} \sim 10^{-6}$ m |
| Carbonation and dissolved carbon dioxide migration | $CO_2/Ca(OH)_2$ mass conservation | $10^{-9} \sim 10^{-6}$ m |
| Corrosion of steel and dissolved oxygen transport | $O_2$/proton conservation | $10^{-9} \sim 10^{-6}$ m |
| Calcium ion leaching from $Ca(OH)_2$ and transport | $Ca^{++}$ ion + $Ca(OH)_2$ conservation | $10^{-9} \sim 10^{-6}$ m |
| Chrome dissolution and migration | Cr ion mass conservation | $10^{-9} \sim 10^{-6}$ m |
| Macro-damage evolution and momentum conservation | static/dynamic equilibrium | $10^{-3} \sim 10^{-0}$ m |

temperature and pore water pressure are seen in the modeling of cement hydration rate, moisture equilibrium, the constitutive law of hardened cement paste, the conductivity of carbon dioxide, bound and free chloride equilibrium, etc. In order to simultaneously solve these multi-scaled chemo-physics/mechanics issues, each event is sequentially processed, the state parameters commonly referred to are revised and the computation is cycled till the whole conservation requirement is satisfied at each time step. Approximately 250 non-linear governing equations that need simultaneously to be solved and the independent variables of the same numbers are included in this book, along with their highlighted details. In other words, concrete performance of micro and macro scales is characterized by these sets of variables having multi-overlaid structures of cement composites. Individual physical chemistry events will be discussed subsequently.

Each sub-event is discretized in space by applying Galarkin's method of weighted residual function on a finite element scheme. This three-dimensional multi-scale coupled system is built on thermo-hydro physics coded by *Du-COM* for both early aged and aging concrete (Maekawa *et al.* 1999) and on non-linear mechanics that has been finite element mesh-coded by *COM3* (Maekawa *et al.* 2003b) for seismic performance assessment of reinforced concrete. The authors have selected an object-oriented scheme because of its stability of convergence and easy extensibility. In fact, soil foundation space, micro-organism activity and electric potential field were easily overlaid on this system for further extension and are being investigated.

## 1.2 Multi-scale modeling

### 1.2.1 *Pore-structure model of cement paste, $10^{-10} \sim 10^{-6}$ m scale*

Computational modeling of varying micro-pore structures of hardening cement paste media is a central issue of multi-scale analysis. Moisture migration, durability-related substances (this book discusses calcium, chloride, dissolved $CO_2$ and $O_2$) and diffusion of gaseous phases are highly influenced by micro-pore structures. Without information of these mass transport and equilibrated thermodynamic states, we cannot predict hydration of cement, or carbonation and corrosion of steel in concrete. Volumetric change caused by drying and the associated stresses are affected by the micro-pore moisture and the solidified skeleton mechanics of concrete. Here, the authors initiate multi-scale modeling with nm~μm scale hardening cement with calcium silicate hydrates (C–S–H) (Maekawa *et al.* 1999).

Suspended particles of cement are idealized as being spherical, and a representative unit cell consisting of a particle and water at the initial stage is defined as shown in Figure 1.3. For simplicity, the radius of individual cement particles is assumed to be constant and equal to an average representative radius. Cement hydration creates C–S–H grains inside and outside of the original spherical geometry. The inner product is thought to have no capillary distance between C–S–H grains. But, outside the original sphere, gaps are formed in between C–S–H grains. In this book, this gap space surrounded by C–S–H grains is defined as capillary pores (Chapter 3 and Chapter 7), as shown in Figure 1.3.

In order to identify statistically the micro-pores of nm~μm scales, the authors introduce the following hypotheses (H) and computational assumptions (A), based on cement solid science knowledge learned in the past several decades.

(H1) The particle size and geometry of C–S–H grains are constant. (Volume to surface area ratio = 19 nm for ordinary Portland cement.)

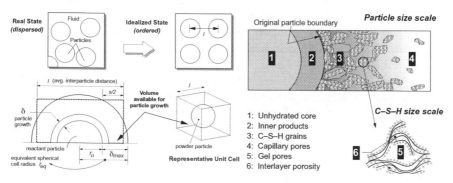

*Figure 1.3* Statistical modeling of micro-pore geometry and size for hardening cement paste.

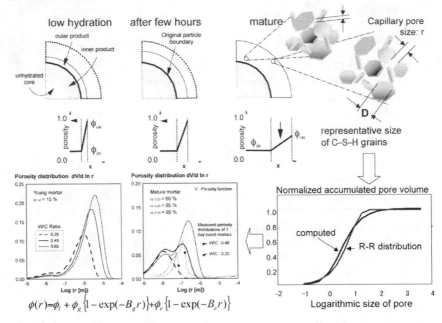

*Figure 1.4* Micro-pore development and statistical expression.

(H2) No capillary space is made inside the original sphere and the porosity of the outer product varies linearly (Figure 1.4).

(H3) C–S–H grains contain interlayer pores and gel pores, as shown in Figure 1.3.

When we direct our attention to the inside of C–S–H grains, the interlayer pore whose size is similar to that of a water molecule is identified. The authors further assume the following:

(A1) Total porosity of C–S–H grains is constant (0.28) and the size of interlayer in C–S–H grains is assumed 2.8A.

(A2) The specific surface area of interlayer is 510m²/g for ordinary Portland cement and that of gel pores is 40m²/g as constant.

From hypotheses (H1) and (H2), the capillary pore size distribution can be derived (Figure 1.4). As capillary porosity at location $x$ from the original boundary of the cement particle is proportional to the gap size $r$ (capillary pore size) between C–S–H grains and is assumed to linearly vary in space, we have,

$$\phi_c(x) = \frac{r}{D+r} = \left(\frac{x}{\delta_{max}}\right)\phi_{ou} \tag{1.1}$$

where $D$ is the size of C–S–H grains, $\delta_{max}$ is the distance between the extreme radius of the unit cell (REV) and the original boundary (Figure 1.3), and $\phi_{ou}$ is porosity at the extreme radius of the unit cell (Figure 1.3 and Figure 1.4). Accumulated volume of capillary pores formed up to distance $x$ is computed by integrating porosity with respect to location as

$$V_x = \int_0^x \phi_c \, 4\pi (x + x_o)^2 \, dx \tag{1.2}$$

Substitution of Equation 1.1 into Equation 1.2 yields

$$V_x = A\left(\frac{1}{4}x^4 + \frac{2}{3}x_o x^3 + \frac{1}{2}x_o^2 x^2\right) \tag{1.3}$$

Since location $x$ is non-linearly related to the capillary pore size by Equation 1.1, the accumulated volume of capillary pores can be implicitly obtained in terms of the pore size, as shown in Figure 1.4 (in case of $\delta_m = 1$, $X_o = 1$, $\phi_{ou} = 1$). Since the explicit expression of pore volume with respect to the pore size is convenient for computation, Raleigh-Ritz (R-R) distribution function nearly equivalent to the statistical pore size distribution derived from Equation 1.3 is accepted (Figure 1.4) and the equivalent expression of entire micro-pore structures are proposed as

$$\phi(r) = \phi_{lr} + \phi_{gl}\left(1 - \exp\left(-B_{gl} r\right)\right) + \phi_{cp}\left(1 - \exp\left(-B_{cp} r\right)\right) \tag{1.4}$$

where the same R-R distribution function is applied to the gel pores (the second term) and the size of interlayer pore (the first term) is assumed constant by assumption (A1). The porosity of each specific pore denoted by $\phi_{lr}$, $\phi_{gl}$, $\phi_{cp}$ in Equation 1.4 is computed based upon the hypothesis previously stated and the hydration degree. The sole parameter to specify the pore size distribution $B$ can be uniquely decided by the surface area of each pore system. Provided that the micro-pore is of tubular geometry of random connection, the specific surface area denoted by $S$ yields

$$S = 2\phi \int r^{-1} dV = 2\phi \int_{r_{min}}^{\infty} B \exp(-Br) \, d\ln r \tag{1.5}$$

Thus, Raleigh-Ritz parameter $B$ can be uniquely determined by $S$. As the specific surface area of gel is given by assumption (A2) as constant, $B_{gl}$ is not variable during the hydration process but the porosity is solely increasing as hypothesis (H1) assumes. The surface area of capillary pores derives from hypothesis (H1) and (H2) by integrating the local surface area of capillary pores and mass equilibrium (Figure 1.5). The surface area of capillary pores drastically varies in accordance with cement hydration and its mean size is

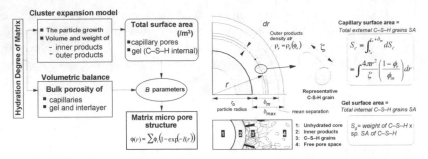

*Figure 1.5*  Computation of varying surface area of hydrates and capillary pores.

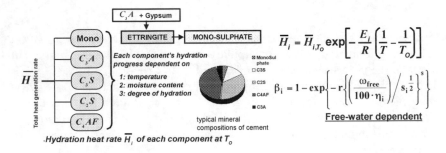

*Figure 1.6*  Multi-mineral component model of cement hydration and heat generation.

decreasing. The detailed derivation of porosity and surface area of each pore system from hydration degree is presented in Chapter 3.

The multi-component cement hydration model is used (Chapter 2). As cement consists of several minerals, the averaged activation energy is not constant in nature and the hydration pattern looks complex. In considering the engineering practice for different sorts of Portland cement and mixed cement with pozzolans, the authors computationally define referential hydration (heat) rate and activation energy of each mineral compound as shown in Figure 1.6. Although each mineral's activation is constant, apparently varying activation energy of cement, which is extracted from adiabatic temperature rise tests, can be simulated in terms of hydration degree. The effect of free water on hydration rate is modeled by hard shell concept of hydrated cluster.

## 1.2.2 Moisture transport and equilibrium, $10^{-10} \sim 10^{-6}$ m scale

Moisture mass balance must be strictly solved in both vapor and condensed water. The conservation equation is expressed with capacity, conductivity and sink terms on the referential volume, as shown in Figure 1.7 (Chapter 3). Pore pressure of condensed water is selected as a chief variable so that both saturated and unsaturated states can be covered with perfect consistency.

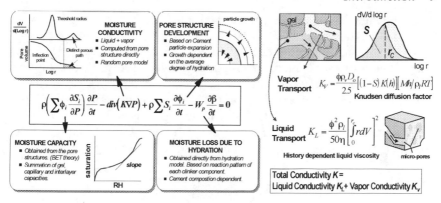

*Figure 1.7* Moisture conservation and transport based on statistical structural model of micro-pores.

Another key issue here is that material characteristic parameters are variable with respect to micro-pore development.

Knudsen diffusion theory was applied for vapor transport (Chapter 3) and the conductivity component is formulated based on vacant micro-pores where vapor can move as shown in Figure 1.7. For liquid transport, the random pore model described by percolation threshold was applied and its conductivity is computed based on the micro-pore distribution occupied by condensed water. Moisture transport in a porous body would be driven by temperature, pore pressure, and vapor pressure difference. Here, total flux of moisture $J$ is formulated as

$$J = -\left(D_p \nabla P_l + D_T \nabla T\right) \tag{1.6}$$

where $D_p$ and $D_T$ represent the macroscopic moisture conductivities with respect to pore pressure and temperature potential gradient, respectively. The conductivity $D_p$ in Equation 1.6 includes vapor diffusivity as well as liquid permeability, since vapor diffusivity and its gradient can be described together in the first term of the right-hand side (Figure 1.7). Under isothermal conditions (293[K]), vapor flux can be described as (Maekawa *et al.* 1999)

$$q_v = -\frac{\rho_v^{sat} \phi D_0}{\Omega} \int_{r_s}^{\infty} \frac{dV}{1 + N_k} \nabla h \quad N_k = \frac{l_m}{2\left(r - t_a\right)} \tag{1.7}$$

where $\rho_v^{sat}$ is the vapor density under saturated condition [kg/m³], $\phi$ is porosity, $D_0$ is vapor diffusivity [m²/s] in free atmosphere at 293 [K], $\Omega$ is the parameter representing the tortuosity of the pore (= $(\pi/2)^2$), $N_k$ is the Knudsen number, $V$ is pore volume [m³], $h$ is relative humidity, $l_m$ is the mean free path of a water molecule [m], and $t_a$ is the thickness of the adsorbed layer [m]. In the model, factors reducing the apparent diffusivity of vapor are taken into account, such as a complicated pore network, reduction of pore volume through which vapor movement can take place with increasing saturation and Knudsen diffusion.

When the vapor flux under various temperature conditions is solved, the gradient of relative humidity cannot be defined as its potential. In other words, relative humidity at different temperatures does not represent the driving force correctly, since the saturated vapor pressure depends on temperature, which leads to different relative humidity even though vapor density of the system is the same. Therefore, in order to generalize Equation 1.7 for arbitrary temperature, the authors tailored the equation as (Chapter 3)

$$q_v = -\frac{\phi D_0(T)}{\Omega} \int_{r_c}^{\infty} \frac{dV}{1+N_k} \nabla \rho_V = -D_v \nabla \rho_V \quad N_k = \frac{l_m}{2(r-t_a)} \tag{1.8}$$

The vapor flux described by Equation 1.8 is driven by the gradient of absolute vapor density $\rho_v$ of the system, instead of relative humidity. Here, absolute vapor density $\rho_v$ corresponds to the product of relative humidity $h$ and $\rho_v^{sat}$ in Equation 1.7. It has to be noted that vapor diffusivity has a strong dependence on temperature of the system. From a thermodynamic point of view, mass diffusivity under different temperature conditions can be derived by

$$\frac{D_0(T_1)}{D_0(T_2)} = \left(\frac{T_1}{T_2}\right)^{3/2} \left(\frac{\Omega_{D,T2}}{\Omega_{D,T1}}\right) \tag{1.9}$$

where $\Omega_{D,T1}$ and $\Omega_{D,T2}$ are collision integrals for molecular diffusion at temperature $T_1$ and $T_2$, respectively, which are functions of Boltzmann constant, temperature, and so on (Welty *et al.* 1969).

Next, the flux of liquid water $q_l$ can be described as (Maekawa *et al.* 1999)

$$q_l = -\frac{\rho_l \phi^2}{50\eta} \left(\int_0^{r_c} r dV\right)^2 \nabla P_l = -K_l \nabla P_l \tag{1.10}$$

where $\eta$ is the viscosity of fluid. It has been reported that the actual viscosity in the cementitious microstructure is far from the one of bulk water under ideal conditions. To account for such a phenomenon, Chaube *et al.* (1993) proposed the following model based on thermodynamics as (Maekawa *et al.* 1999)

$$\eta = \eta_i \exp\left(\frac{G_e}{RT}\right) \tag{1.11}$$

where $\eta_i$ is the referential viscosity under ideal conditions and $G_e$ is the free energy of flow activation in excess of that required for ideal flow conditions.

Considering a pore-water and microstructure interaction, Gibbs' free energy $G_e$ has been formulated as a function of characteristics of pore structure and moisture history. In the proposed model, the different value of viscosity $\eta_i$ is given according to temperature in a pore, whereas the authors use the same formulation to evaluate Gibbs' energy since the effect of temperature is implicitly considered by the original formula. Then, we have (Ishida and Maekawa 2002)

$$
\begin{aligned}
J &= -\left(D_p\nabla P_l + D_T\nabla T\right) \\
&= -\left\{D_V\left(\frac{\partial \rho_V}{\partial P_l}\nabla P_l + \frac{\partial \rho_V}{\partial T}\nabla T\right) + K_l\nabla P_l + K_T\nabla T\right\} \\
&= -\left(D_V\frac{\partial \rho_V}{\partial P_l} + K_l\right)\nabla P_l - \left(D_V\frac{\partial \rho_V}{\partial T} + K_T\right)\nabla T
\end{aligned}
\tag{1.12}
$$

In general, mass transport due to temperature gradient is known as the Soret effect or thermal diffusion that is represented by the last term of the right-hand side, i.e., $K_T\cdot\text{grad}(T)$. However, contribution of thermal diffusion to the total flux has not been cleared in concrete engineering. Moreover, this phenomenon normally plays a minor role in diffusion compared with moisture transfer driven by pore pressure and vapor pressure gradient (Welty *et al.* 1969). Then, as the first approximation, the thermal diffusion is neglected ($K_T \approx 0$).

The moisture capacity term enumerates moisture content with regard to the pore pressure. Quasi-thermo equilibrium is assumed at any time in nm~μm sized pores residing in REV. Vapor partial pressure can be obtained by equating Gibbs' free energy of vapor to that of condensed water as

$$
P_l = \frac{\rho_l RT}{M_w}\ln\frac{p_{vap}}{p^*}
\tag{1.13}
$$

where $R$ is the gas constant [J/mol.K], $p^*$ is saturated vapor pressure [Pa], $M_w$ is the molecular mass of water [kg/mol], and $\rho_l$ is the density of liquid water [kg/m³]. Under isothermal conditions, $p_{vap}/p^*$ corresponds to relative humidity $h$ inside pores.

By considering local thermodynamic and interface equilibriums, vapor and liquid interfaces would be formed in the pore structure due to pressure gradients caused by capillarity. When the interface is a part of an ideal spherical surface, the relation can be described by the Laplace equation as

$$
P_l = \frac{2\gamma}{r_s}
\tag{1.14}
$$

where $\gamma$ is the surface tension of liquid water [N/m] and $r_s$ is the radius of the pore in which the interface is created [m]. Based on the thermodynamic conditions represented by Equation 1.13 and Equation 1.14, a certain group of pores whose radii is smaller than the specific radius $r_s$ at which a liquid-vapor interface forms are completely filled with water, whereas larger pores remain empty or partially saturated.

If the porosity distribution of micro-pore structures is known, Equations 1.13 and 1.14 allow us to obtain the amount of condensed water at a given relative humidity or absolute vapor pressure. By combining the theory with the micro-pore structure distribution model, we evaluate the moisture profile under an isothermal condition (Figure 1.8). In the model, adsorbed phases on the pore wall described by a theory put forward by Brunauer, Emmet and Teller (1938) and trapped water due to the "ink-bottle effect" are taken into account as well as the condensed water (Chapter 3).

In Equation 1.13, liquid density, surface tension and saturated vapor pressure are temperature-dependent. For implicitly considering interdependency between temperature and moisture profile, liquid density and surface tension are given as non-linear functions in terms of temperature based on measured values in experiments (Welty *et al.* 1969). With regard to the saturated vapor pressure varying with temperature inside pores, the authors apply the Clausius-Clapeyron equation as

$$\frac{d \ln p}{dT} = \frac{\Delta H_{vap}}{RT^2} \tag{1.15}$$

where $\Delta H_{vap}$ is the heat of vaporization [kJ/mol]. Here, $\Delta H_{vap}$ is assumed to be constant (40.7 [kJ/mol]), since temperature variation is limited to the

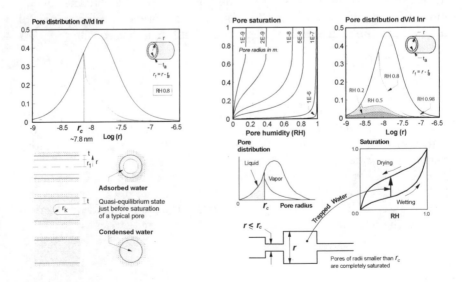

*Figure 1.8* Thermodynamic moisture equilibrium of condensed and adsorbed water in micro-gel and capillary pores.

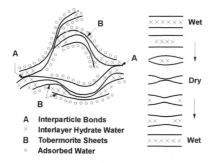
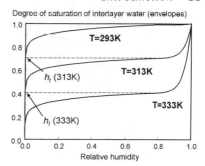

*Figure 1.9* Thermodynamic moisture equilibrium in micro-interlayer pores.

range from 273[K] to 373[K] in this book. By assuming the constant $\Delta H_{vap}$, integral of Equation 1.15 yields

$$p_{vap} = p^* \exp\left(\frac{P_l M_w}{\rho_l RT}\right) = p_0 \exp\left\{-\left(\frac{\Delta H_{vap}}{R}\right)\left(\frac{1}{T} - \frac{1}{T_0}\right)\right\} \exp\left(\frac{P_l M_w}{\rho_l RT}\right) \quad (1.16)$$

where $p_0$ and $T_0$ are the referential pressure and temperature, respectively.

From these discussions, the relationship of pore pressure, saturated vapor pressure and absolute vapor pressure is specified. By combining the pore structure model with these formulations, moisture in terms of both vapor and liquid inside gel and capillary pores can be related to pore pressure.

Thermodynamics is hardly applied to behaviors of moisture inherent in interlayer pores with the angstrom scale but individual molecular dynamics is the matter. Here, an empirically observed isotherm as shown in Figure 1.9 is used for evaluating total moisture capacity. Temperature is a key parameter, especially for low water-to-cement ratio concrete having a great deal of interlayer pores. The general modeling is presented in Chapter 3.

### 1.2.3 Chloride ion transport and equilibrium, $10^{-8} \sim 10^{-6}$ m scale

Chloride transport in cementitious materials under usual conditions is an advective-diffusive phenomenon. In modeling, the advective transport due to bulk movement of pore solution phase is considered, as well as ionic diffusion due to concentration gradients. Mass balance for free (movable) chlorides can be expressed as (Chapter 6 and Figure 1.10)

$$\frac{\partial}{\partial t}\left(\phi S C_{cl}\right) + div J_{Cl} - Q_{Cl} = 0, \quad J_{Cl} = -\frac{\phi S}{\Omega} D_{Cl} \nabla C_{Cl} + \phi S \mathbf{u} C_{Cl}, \quad \mathbf{u} = -\frac{K \nabla P}{\rho \phi S} \quad (1.17)$$

where $\phi$ is porosity of the porous medium, $S$ is the degree of saturation of the porous medium, $C_{Cl}$ is the concentration of ions in the pore solution phase [mol/l], $J_{Cl}$ is the flux vector of the ions [mol/m².s], $\mathbf{u}^T = [u_x \, u_y \, u_z]$ is the

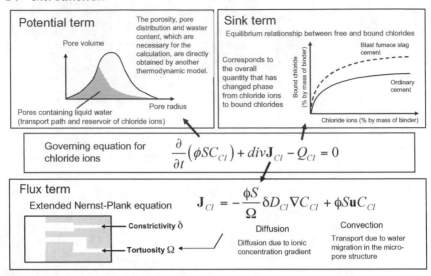

*Figure 1.10* Governing equation and constitutive models for chloride ion.

advective velocity of ions due to the bulk movement of pore solution phase [m/s], and $\Omega$ is the tortuosity of pore-structures for a three-dimensional pore network which is uniformly and randomly connected. Tortuosity is a parameter that expresses a reduction factor in terms of chloride penetration rate due to complex micro-pore structure. This parameter is determined by considering the geometric characteristics of pore structures (Chapter 6).

Material parameters shown in Equation 1.17, such as porosity, saturation and advective velocity, are directly obtained based on the thermo-hydro physics. The advective velocity u is obtained from the pore pressure gradient $\nabla P$ and liquid water conductivity K which is calculated by the moisture equilibrium and the transport model, according to water content, micro-pore structures, and moisture history. In the case of chloride ion transport in cementitious materials, S represents the degree of saturation in terms of free water only, as adsorbed and interlayer components of water are also present. It has to be noted here that the diffusion coefficient $D_{Cl}$ in the pore solution may be a function of ion concentration, since ionic interaction effects will be significant in the fine microstructures at increased concentrations, thereby reducing the apparent diffusive movement driven by the ion concentration gradient (Gjørv and Sakai 1995).

This mechanism, however, is not clearly understood, so the authors do not consider the dependence of ionic concentration on the diffusion process in this book. From several numerical sensitivity analyses, a constant value of $3.0 \times 10^{-11}$ [m²/s] is adopted for $D_{Cl}$. From these discussions, the total diffusivity of concrete is described as the product of $D_{Cl}$ and $\phi S/\Omega$, which depends on the achieved microstructures (the initial mix and curing condition-dependent variables) and moisture history.

As is well-known, chlorides in cementitious materials have free and bound components. The bound components exist in the form of chloro-aluminates and adsorbed phases on the pore walls, making them unavailable for free transport. The relationship between free and bound components of chlorides is expressed by the equilibrium model (Chapter 6), as shown in Figure 1.10. Here, the bound chlorides are classified into two phases: adsorbed and chemically combined components. Through these studies, it can be assumed that the amount of combined phases is approximately constant in terms of weight percent of hydrated gel products, whereas that of the adsorbed phase is strongly dependent on the constituent powder materials. For example, in the case of blast furnace slag (SG), the amount of adsorbed phases becomes larger compared to the case of ordinary Portland cement (OPC), which leads to a higher binding capacity of SG concrete and mortar. It means that the adsorbed component plays a major role in the chloride binding capacity of concrete.

By assuming local equilibrium conditions based on this relationship, the rate of binding or the change of free chloride to bound chloride per unit volume $Q_{Cl}$ can be obtained. From these discussions and formulations, the distribution of bound and free chloride ions can be obtained without any empirical formula and/or intentional fitting, once mix proportions, powder materials, curing and environmental conditions are applied to the analytical system.

## 1.2.4 CO₂ transport, equilibrium and carbonation, $10^{-9} \sim 10^{-6}$ m scale

For simulating carbonation in concrete, equilibrium of gaseous and dissolved carbon dioxide, their transport, ionic equilibriums and the carbonation reaction process are formulated on the basis of thermodynamics and chemical equilibrium theory (Chapter 4). Mass balance condition for dissolved and gaseous carbon dioxide in porous medium can be expressed as

$$\frac{\partial}{\partial t}\{\phi[(1-S)\cdot\rho_{gCO_2}+S\cdot\rho_{dCO_2}]\}+divJ_{CO_2}-Q_{CO_2}=0 \tag{1.18}$$

where $\rho_{gCO_2}$ is the density of $CO_2$ gas [kg/m³], $\rho_{dCO_2}$ is the density of dissolved $CO_2$ in pore water [kg/m³], and $J_{CO_2}$ is the total flux of dissolved and gaseous $CO_2$ [kg/m².s]. The local equilibrium of gaseous and dissolved carbon dioxide is represented by Henry's law, which states the relation of gas solubility in pore water and the partial gas pressure. $CO_2$ transport is considered in both phases of dissolved and gaseous carbon dioxide. By considering the effect of Knudsen diffusion, tortuosity and connectivity of pores on the diffusivity, Fick's first law of diffusion yields the flux of $CO_2$ as

$$J_{CO_2}=-\left(D_{dCO_2}\nabla\rho_{dCO_2}+D_{gCO_2}\nabla\rho_{gCO_2}\right),\ D_{dCO_2}=\frac{\phi D_0^d}{\Omega}\int_0^{r_c}dV\ \ D_{gCO_2}=\frac{\phi\cdot D_0^g}{\Omega}\int_{r_c}^{\infty}\frac{dV}{1+N_k} \tag{1.19}$$

where $D_{gCO2}$ is the diffusion coefficient of gaseous $CO_2$ in porous medium[$m^2$/s], $D_{dCO2}$ is the diffusion coefficient of dissolved $CO_2$ in porous medium [$m^2$/s], $D_0^g$ is the diffusivity of $CO_2$ gas in a free atmosphere [$m^2$/s], $D_0^d$ is the diffusivity of dissolved $CO_2$ in pore water [$m^2$/s], $V$ is the pore volume, $r_c$ is the pore radius in which the equilibrated interface of liquid and vapor is created, and $N_k$ is the Knudsen number, which is the ratio of the mean free path length of a molecule of $CO_2$ gas to the pore diameter. Knudsen effect on the gaseous $CO_2$ transport is not negligible in low RH condition, since the porous medium for gas transport becomes finer as relative humidity decreases. As shown in Equation 1.19, the diffusion coefficient $D_{dCO2}$ is obtained by integrating the diffusivity of saturated pores over the entire porosity distribution, whereas $D_{gCO2}$ is obtained by summing up the diffusivity of gaseous $CO_2$ through unsaturated pores. In order to generalize the expression for an arbitrary moisture history, substitution of porosity saturation $S$ for the integrals in Equation 1.19 gives

$$D_{dCO_2} = \frac{\phi S^n}{\Omega} D_0^d, \quad D_{gCO_2} = \frac{\phi \cdot D_0^g}{\Omega} \frac{(1-S)^n}{1 + l_m/2(r_m - t_m)} \quad (1.20)$$

where $n$ is a parameter representing the connectivity of the pore structure, and might vary with the geometrical characteristics of the pores (Chapter 4). In the model, through sensitivity analysis, $n$ is assumed to be 6.0, which is the most appropriate value for expressing the reduction of $CO_2$ diffusivity with the decrease of relative humidity (Figure 1.11). In Equation 1.19, the integral of the Knudsen number is simplified so that it can be easily put into practical computational use. The notation $r_m$ is the average radius of unsaturated pores, and $t_m$ is the thickness of adsorbed water layer in the pore whose radius is $r_m$.

$Q_{CO2}$ in Equation 1.18 and Equation 1.21 is a sink term that represents the rate of $CO_2$ consumption due to carbonation [kg/$m^3$.s]. The rate of $CO_2$ consumption can be expressed by the following differential equation,

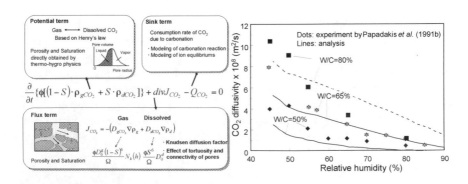

*Figure 1.11* Relationship between $CO_2$ diffusivity and relative humidity.

assuming that reaction is of the first order with respect to $Ca^{2+}$ and $CO_3^{2-}$ concentrations as

$$Ca^{2+} + CO_3^{2-} \rightarrow CaCO_3,$$

$$Q_{CO_2} = \frac{\partial (C_{CaCO_3})}{\partial t} = k[Ca^{2+}][CO_3^{2-}] \qquad (1.21)$$

$$H_2O \leftrightarrow H^+ + OH^-$$

$$H_2CO_3 \leftrightarrow H^+ + HCO_3^- \leftrightarrow 2H^+ + CO_3^{2-}$$

$$Ca(OH)_2 \leftrightarrow Ca^{2+} + 2OH^- \qquad (1.22)$$

$$CaCO_3 \leftrightarrow Ca^{2+} + CO_3^{2-}$$

where $C_{CaCO3}$ is the concentration of calcium carbonate, and $k$ is a reaction rate coefficient. Here, a unique coefficient is applied ($k = 2.08$ [l/mol.sec]), although the reaction rate coefficient involves temperature dependency.

In order to calculate the rate of reaction with Equation 1.21, it is necessary to obtain the concentration of calcium ion and carbonic acid in the pore water at an arbitrary stage. We consider the ion equilibriums, dissociation of water and carbonic acid, and dissolution and dissociation of calcium hydroxide and calcium carbonate as above. Here, the presence of chlorides is not considered, although chloride ions are likely to affect the equilibrium conditions. The formulation including chlorides remains a subject for future study.

As shown in Equation 1.22, carbonation is an acid–base reaction, in which cations and anions act as a Brönsted acid and base, respectively. Furthermore, the solubility of precipitations is dependent on the pH of the pore solutions. Therefore, for calculating the ionic concentration in the pore solutions, the authors formulated an equation with respect to protons [$H^+$], according to the basic principles on ion equilibrium, i.e., laws of mass action, mass conservation and proton balance in the system (Chapter 4). In use of the equation, the concentration of protons in pore solutions can be calculated at an arbitrary stage, once the concentration of calcium hydroxide and that of carbonic acid before dissociation are given.

It has been reported that micro-pore structure is changed due to carbonation. In this modeling, the authors use the empirical formula as (Saeki *et al.* 1991, Chapter 4)

$$\phi' = \phi \left( R_{Ca(OH)_2} \right) \quad \left( 0.6 < R_{Ca(OH)_2} < 1.0 \right)$$

$$\phi' = 0.5 \cdot \phi \qquad \left( R_{Ca(OH)_2} \leq 0.6 \right) \qquad (1.23)$$

where $\phi'$ is the porosity after carbonation, and $R_{Ca(OH)2}$ is the ratio of the amount of consumed $Ca(OH)_2$ for the total amount of $Ca(OH)_2$.

### 1.2.5 Oxygen transport and micro-cell-based corrosion model, $10^{-9}$~ $10^{-6}$ m scale

In this section, a general scheme of a micro-cell corrosion model is introduced based on thermodynamic electro-chemistry (Chapter 6). Corrosion is assumed to occur uniformly over the surface areas of reinforcing bars in a referential finite volume, whereas formation of pits due to localized attack of chlorides and corrosion with macro cell remains for future study. Figure 1.12 shows the flow of the corrosion computation. When we consider the micro-cell-based corrosion, it can be assumed that the anode area is equal to that of the cathode and they are not separated from each other. Then, electrical conductivity of concrete, which governs the macroscopic transfer of ions in pore water, is not explicitly treated.

First of all, electric potential of corrosion cell is obtained from the ambient temperature, *pH* in pore solution and partial pressure of oxide, which are calculated by other subroutines in the system. The potential of half-cell can be expressed with the Nernst equation as (West 1986)

$$\text{Fe(s)} \rightarrow \text{Fe}^{2+}(\text{aq}) + e(\text{Pt})$$

$$E_{\text{Fe}} = E_{\text{Fe}}^{\ominus} + (RT/z_{\text{Fe}}F)\ln h_{\text{Fe}^{2+}}$$

$$O_2(\text{g}) + 2H_2O(\text{l}) + 4e(\text{Pt}) = 4OH^-(\text{aq}) \qquad (1.24)$$

$$E_{O_2} = E_{O_2}^{\ominus} + (RT/z_{O_2}F)\ln(P_{O_2}/P^{\ominus})$$

$$- 0.06\,pH$$

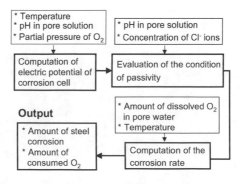

where $E^{\ominus}_{\text{Fe}}$ is the standard cell potential of Fe, anode (V, SHE), $E_{O2}$ is the standard cell potential of $O_2$, cathode (V, SHE), $E^{\ominus}_{\text{Fe}}$ is the standard cell potential of Fe at 25°C (= -0.44V, SHE), $E^{\ominus}_{O2}$ is the standard cell potential of $O_2$ at 25°C (= 0.40V, SHE), $z_{\text{Fe}}$ is the number of charge of Fe ions (= 2), $z_{O2}$ is the number of charge of $O_2$ (= 2), and $P^{\ominus}$ is atmospheric pressure. By assuming an ideal condition, we can overlook the solution of other ions in

*Figure 1.12* Overall scheme of corrosion computation.

pore water on the half-cell potentials. Further study is intended in the multi-scale scheme.

Next, based on thermodynamic conditions, the state of passive layers is evaluated by the Pourbaix diagram, where steel corrodes, areas where protective oxides form, and an area of immunity to corrosion depending on $pH$ and the potential of the steel. From the electric potential and the formation of passive layers, the electric current that involves chemical reaction can be calculated so that the conservation law of electric charge should be satisfied in a local area (Figure 1.13a). The relationship of electric current and voltage for the anode and cathode can be expressed by the following Nernst equation as

$$\eta^a = \left(2.303RT/0.5 \cdot z_{Fe}F\right)\log\left(i_a/i_0\right), \quad \eta^c = -\left(2.303RT/0.5 \cdot z_{O_2}F\right)\log\left(i_c/i_0\right)$$
$$(1.25)$$

where $\eta^a$ is the overvoltage at anode [V], $\eta^c$ is the overvoltage at cathode [V], $F$ is Faraday's constant, $i_a$ is the electric current density at anode [A/m²], and $i_c$ is the electric current density at cathode [A/m²]. Corrosion current $i_{corr}$ can be obtained as the point of intersection of two lines.

The existence of a passive layer reduces the corrosion progress. In this model, this phenomenon is described by changing the Tafel gradient (Figure 1.13b). When chlorides exist in the system, the passive region hitherto ccupied by $Fe_3O_4$ will disappear. In addition, the protective region of $Fe_2O_3$ will be divided into two regions, the lower of which corresponds to truly passive behavior and the upper to an unstable situation in which localized breakdown to pitting attack becomes possible (West 1986). Although the chloride-induced corrosion may involve such a localized attack, an average treatment was introduced in this work, as shown in Figure 1.13b. It has been reported from past researches that, as chlorides increase around reinforcing bars, current density of corrosion becomes larger due to the breakdown of passive films (Broomfield 1997). Here, we assume that the anodic Tafel gradient would become smaller with higher concentration of chlorides. It has been also reported by Broomfield (1997) that there is a chloride threshold for starting corrosion. When the chloride concentration exceeds a concentration

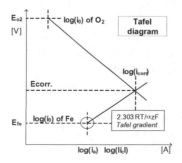

*Figure 1.13a* Relationship of electric current and voltage for anode and cathode.

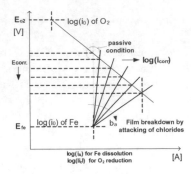

*Figure 1.13b*  Rate of corrosion accelerated by chloride migration.

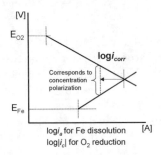

*Figure 1.13c*  Rate of corrosion under $O_2$ diffusion control.

of 0.2% total chloride by weight of cement, corrosion is observed. In this research, the chloride threshold is specified in terms of the amount of free chloride existing in the pore water, and a value of 0.04 [wt% of cement] was given. This threshold value of free chloride was obtained from the chloride equilibrium model in Figure 1.10 by giving a typical concrete mix proportion with ordinary Portland cement used. It is expected that when many chlorides already exist, the rate of corrosion does not increase any more even though new chlorides migrate from the environment. In the model, when the concentration of free chlorides exceeds 0.4 [wt% of cement], the rate of corrosion is assumed to be equivalent to the corrosion rate of the steel without passive films.

When the amount of oxygen supplied to the reaction is not enough, the corrosion rate would be controlled by the diffusion process of oxygen (Figure 1.13c). By coupling with an oxygen transport model, this phenomenon can be logically simulated. Current density $i_{corr}$, which is obtained as the intersection of anodic and cathodic polarization curve, corresponds to the corrosion under sufficient availability of oxygen (Figure 1.13a). When oxygen supply shortens, the rate of corrosion will be limited by the slow diffusion of oxygen. A limited value of current density $i_L$ can be expressed as

$$i_L/z_{Fe}F = O_2^{sup} \tag{1.26}$$

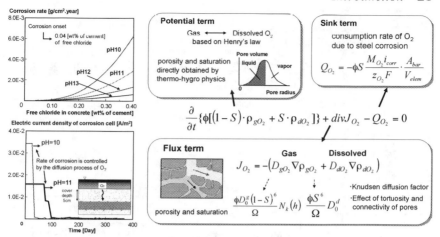

*Figure 1.14*  Oxygen diffusion model linked with micro-corrosion of steel.

where $O_2^{sup}$ [mol/m².s] is the amount of oxygen supplied to the surface of metal, which is obtained by the equilibrium and transport model for oxygen discussed below. In this research, the rate of corrosion under diffusion control of oxygen $i_{corr}$ [A/m²] is assumed to be as (Figure 1.13c)

$$i_{corr} = i_L \tag{1.27}$$

The potential difference between the anode and cathode corresponds to the concentration polarization. Figure 1.14 summarizes the formulation of the key governing equation, which is almost the same as that of carbon dioxide (Chapter 6). Finally, using Faraday's law, the electric current of corrosion is converted to the rate of steel corrosion. These models derive from thermodynamic electrochemistry. Further development and improvement are still needed thorough various verifications of corrosion phenomena in real concrete structures.

### 1.2.6 *Calcium ion transport and leaching, $10^{-9} \sim 10^{-6}$ m scale*

Calcium is one of the main chemicals and in pore solution, $Ca^{2+}$ is equilibrated with $Ca(OH)_2$ and C–S–H solids and other ionic substances as stated in Equation 1.22. Calcium leaching may change pore structures mainly due to lost $Ca(OH)_2$ and long-term performance of cementitious solids including cemented foundation materials would be influenced especially when exposed to pure water and flowing stream. Similar to chloride conservation in Equation 1.17, we have the mass conservation requirement of the total calcium as (Chapter 5)

$$\frac{\partial}{\partial t}(\phi \cdot S \cdot C_{ion}) + \frac{\partial C_{solid}}{\partial t} - div J_{ion} = 0 , \quad J_{ion} = -\left(\frac{\phi \cdot S}{\tau} \cdot \delta \cdot D_{ion}\right) \cdot \nabla C_{ion} + \phi \cdot S \cdot \mathbf{u} \cdot C_{ion} \tag{1.28}$$

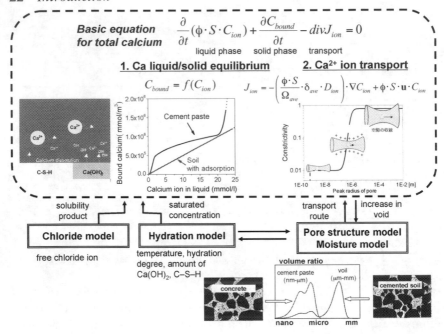

*Figure 1.15* Calcium leaching model coupled with time-dependent thermodynamic states.

where $\phi$ is porosity [m³/m³], $S$ is degree of saturation, $C_{ion}$ is the molar concentration of calcium ions in the liquid phase [mmol/m³], $C_{solid}$ is the amount of calcium in the solid phase [mmol/m³], $J_{ion}$ is the flux of calcium ions [mmol/m². s], $\tau$ is tortuosity, $\delta$ is constrictivity, $D_{ion}$ is the diffusion coefficient of a calcium ion [m²/s], $\nabla^T = [\partial/\partial x\ \partial/\partial y\ \partial/\partial z]$ is the nabla operator, and $u^T = [\ u^x\ u^y\ u^z\ ]$ is the velocity vector of a calcium ion transported by solution flow [m/s].

The isothermal liquid/solid equilibrium of the total calcium is given in the multi-scale system (Figure 1.15) in order to consider the calcium leaching from both Ca(OH)₂ and C–S–H solids. With this global equilibrium, we can systematically take into account the effect of fly ash and other pozzolans on the leaching. The microstructural reform is also simulated and the pore-structure model is extended to the soil foundation for assessment of long-term performance of cemented soil and foundation (Chapter 5, Figure 1.15).

### 1.2.7 Mechanics of cement hydrate and concrete, $10^{-6} \sim 10^{-2}$ m scale

Degree of hydration and moisture in micro-pores greatly influences the solid mechanics of cementitious composites. Constitutive modeling has been proposed with respect to stresses/strains and the micro-pore related solid properties of young concrete composites have been considered in terms of varying elasticity and creep coefficients under specified ambient conditions.

This macroscopic expression of micro-pore development makes practical application possible and has brought great advantage in the past.

Herein, a high-level engineering judgment is required on issues such as ambient related parameters and three-dimensional shape of structural geometry, since the local moisture conditions and micro-pore development are not explicitly known in computation. On this issue, the multi-scale scheme can explicitly deal with the full coupling of thermodynamic equilibrium of moisture and constitutive modeling of solid with micro-pores. The authors present a modeling of cement concrete composites whose micro/meso level solid mechanics is coupled with the nano/micro-pore structures and thermodynamics of micro-climate (Chapter 7).

As the volume of cement paste matrix, unlike aggregates, is highly associated with moisture migration, volumetric interaction of aggregate elastic particles and paste matrix is one of key issues when we construct overall space-averaged constitutive modeling of concrete composite. On the other hand, shear deformational mechanics is thought to be greatly governed by a matrix with less contribution of suspended aggregates. So first, the authors apply mode separation similar to the elasto-plastic and fracturing model of concrete (Maekawa *et al.* 2003b) and second, we compose them into unified constitutive modeling for structural analysis.

Aggregate particles can be modeled as a suspended elastic body by surrounding cement paste, as shown in Figure 1.16. Local stresses developing in both aggregates and cement paste are non-uniform. Here, let us define the referential REV whose scale is 1.5 cm including several kind of gravel and much sand. The volumetric virtual work principle yields equilibrium of stress components and compatibility of strain fields on this REV as

$$\sigma_o = \rho_{ag}\,\sigma_{ag} + \rho_{cp}\,\sigma_{cp}\,, \quad \bar{\varepsilon}_o = \rho_{ag}\,\bar{\varepsilon}_{ag} + \rho_{cp}\,\bar{\varepsilon}_{cp} \tag{1.29}$$

where $\sigma_o$, $\sigma_{ag}$ and $\sigma_{cp}$ are the mean volumetric stresses on concrete, aggregate and cement paste, respectively, and $\varepsilon_o$, $\varepsilon_{ag}$ and $\varepsilon_{cp}$ are the mean volumetric strains, $\rho_{ag}$, $\rho_{cp}$ are the volume fractions of aggregate and cement paste, respectively. As the aggregate phase is assumed elastic, we have

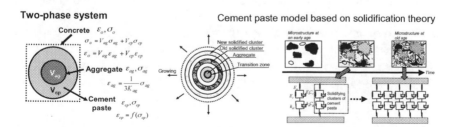

*Figure 1.16* Two-phase modeling of concrete composite – elastic suspension and non-linear matrix – and aging effect by solidified elasto-plastic assembly with different stress/strain histories.

$$\overline{\varepsilon}_{ag} = \frac{1}{3K_{ag}}\overline{\sigma}_{ag}, \quad \overline{\varepsilon}_{cp} = f(\overline{\sigma}_{cp}) \tag{1.30}$$

Here, let us consider the local equilibrium and compatibility between two elements. If the cement paste matrix is a perfect liquid losing resistance to the shear deformation, we have $\sigma_{ag} = \sigma_{cp}$, where the shear stiffness of cement paste becomes zero. This system corresponds to Maxwell chain idealization. If the shear stiffness, $G_{cp}$, is infinitely large on the contrary, it brings no change of geometrical shape and results in $\varepsilon_{ag} = \varepsilon_{cp}$. This system corresponds to the so-called Kelvin chain. As the reality is in between two extremes, the Lagragian method of linear summation is applied as

$$\left(\frac{\overline{\sigma}_{ag} - \overline{\sigma}_{cp}}{G_{cp}}\right) + (\overline{\varepsilon}_{ag} - \overline{\varepsilon}_{cp}) = 0 \tag{1.31}$$

Under the deviatoric shear mode of deformation, shear stress is hardly transferred through the contact link network of aggregates except for pre-packed concrete, and rotational resistance of individual particles is no longer expected when cement paste matrix deforms in shear. Thus, the authors assume that the space-averaged deviatoric stress and strain of cement paste in concrete coincides with those of overall concrete composite. Then, we have

$$S_{ij} = f(e_{ij}) \tag{1.32}$$

where $S_{ij}$ and $e_{ij}$ are the deviatoric stress and strain tensors for both cement paste and concrete composite. The function, $f$, for cement paste is formulated (Chapter 7).

The structural growth of cement paste is rather complex. Upon contact with water, powder particles start to dissolve and reaction products start to form. Due to gradual solidification of hydration products, properties of paste matrix vary with time. Solidification theory (Bazant and Prasannan 1989) proved a potential way of taking this aging effect into consideration.

As shown in Figure 1.16, growth of cement paste is idealized by forming finite fictitious clusters. When a new cluster is formed onto the already formed assembly of old clusters, strain of the cluster results (Chapter 7). The aging process is represented by solidification of the new cluster afterwards. In order to set a criterion for cluster formation, a volume fractional function is introduced. This function is taken as the hydration ratio at a certain time $t$, $\psi(t)$. It is defined as a ratio of hydrated volume of cement powder, $V(t)$, to total volume of cement that is available for hydration, $V_{cp}$, as

$$\psi(t) = V(t)/V_{cp} \tag{1.33}$$

The structure of cement paste at any time is represented using the number of clusters already solidified at that time. According to hydration degree, numbers of clusters, $N$, at a certain time can be determined and when the hydration increment reaches a certain value a new cluster is developed and attached to the older ones. As these clusters share bearing stresses carried by the cement paste, we introduce an infinitesimal stress in each cluster, $S_{cp}$. When a new cluster is formed, these infinitesimal stresses are redistributed among the clusters. It also means that the stress condition in a certain cluster is a function of both current time and the location of this cluster or the time when this cluster formed.

Let $S_{cp} = S_{cp}(t, t')$ denote the average stress in a general cluster, where $t$ is the current time and $t'$ is the time when this cluster is solidified. Total volumetric stress in cement paste at a certain time is the summation of average stresses in all clusters activated at that time and the strain in cement paste is equal to the one induced in each cluster. The solidification concept of cement paste cluster regarding growth of microstructure yields

$$\overline{\sigma}_{cp}(t) = \int_{t'=0}^{t} S_{cp}(t',t)d\psi(t'), \quad S_{ij}(t) = \int_{t'=0}^{t} S_{ij}(t',t)d\psi(t') \tag{1.34}$$

where $S_{cp}(t', t)$ is the mean volumetric stress acting on a certain cluster, $t'$ is the time when this cluster formed, $t$ is the current time, $\psi$ is the hydration degree, $S_{ij}(t', t)$ is the deviatoric stress tensor acting on the cluster concerned. When new cluster is formed, specific cluster stress is null. Then, we have $S_{cp}(t, t) = 0$ and $S_{ij}(t, t) = 0$.

The solidifying mechanical unit as shown in Figure 1.16 is thought to be associated with C–S–H gel grains. Thus, the time-dependent deformation rooted in moisture in gel and interlayer pores should be taken into account in each individual solidifying component. The assembly of solidifying components corresponds to cement paste. Here, the moisture existing in capillary pores between gel grains has to be taken into account in the model, too. The authors assume fictitious time-dependent deformational components embedded in both interlayer/gel pores and capillary ones, as shown in Figure 1.17 (Chapter 7). Moisture transport actually takes place into the capillary and gel pores. Under severe drying, interlayer water is set in motion. Thus, moisture transport related to each of these pore categories can be correlated to a certain aspect of the creep behavior (Glucklich 1959; Neville 1959).

The rate of flow of capillary water is comparatively high and easily reversible. Thus, moisture transport within capillary pores can be assumed as a cause of short-term creep at earlier ages. This creep rate drops steeply with time and is closely related to the hydration process. Moisture transport within gel pores is thought to be slow and prolonged in nature. Kinematics of gel water can be reversible up to a certain limit. It can be assumed as a cause of long-term creep and is responsible for the main part of the unrecoverable creep. Moisture migration within interlayer pores can be assumed to drive creep under comparatively severe conditions. This creep is highly irreversible and path-dependent accompanying so-called disjoining pressure.

In order to reflect three aspects of creep behaviors, a simple rheological model for each cluster component is assumed, as shown in Figure 1.17. The

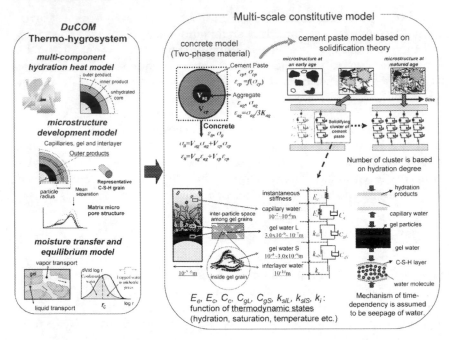

*Figure 1.17* Modeling of solidifying cluster of cement hydrate.

same model is adopted for both volumetric and deviatoric components. Here, the volumetric component is discussed in detail. The total strain of each solidifying cluster is decomposed into instantaneous elastic strain $\varepsilon_e$, visco-elastic strain $\varepsilon_c$, visco-plastic strain $\varepsilon_g$ and instantaneous plastic strain $\varepsilon_l$.

$$\varepsilon'_{cp} = \varepsilon_e + \varepsilon_c + \varepsilon_g + \varepsilon_l \tag{1.35}$$

where $\varepsilon_{cp}$' is the volumetric strain in a general layer.

It should be noted here that $\varepsilon_{cp}$' is the strain induced in an individual cluster after its solidification. As the clusters are assumed to join together, these strains should be common to corresponding volumetric strain in cement paste. However, clusters solidify at zero stress state. Thus, as given by Equation 1.36, this strain at a certain time, $t$, is defined as the difference between the strain of cement paste at time, $t$, and the strain of cement paste at the time when the cluster solidified, $t'$.

$$\varepsilon'_{cp}(t) = \varepsilon_{cp}(t) - \varepsilon_{cp}(t') \tag{1.36}$$

Hereafter, the effect of micro-pore pressure on deformation of the cement paste solid is introduced. The pressure drop of condensed water in micro-pores is equilibrated with capillary surface tension acting between liquid and gas phases, and it is assumed as one of causes of shrinkage (Chapter 7). Since surface tension developing in micro-pores is isotropic, only the volumetric mode of deformation is targeted. Similar to Biot's theorem of two-phase continuum, volumetric stress of cement paste is idealized as being carried by both a skeleton solid and pore pressure (Chapter 7) as

$$\overline{\sigma}_{cp} = \overline{\sigma'_{cp}} + \beta \cdot \sigma_s \tag{1.37}$$

where $\sigma'_{cp}$ is volumetric stress by cement paste skeleton and $\sigma_s$ is pore-water pressure drop. Factor $\beta$ indicates effectiveness to represent the volume fraction where moisture can act. If the whole space is occupied by the water continuum, it shows unity as Biot's two-phase theory reveals.

Pore water pressure can be evaluated by Kelvin's formula and factor $\beta$ is estimated in use of micro-pore water information as

$$\sigma_s = -\frac{2.\gamma}{r_s} = -\frac{\rho.R.T}{M} \ln h , \quad \beta = \frac{\phi_{cap} \cdot S_{cap} + \phi_{gel} \cdot S_{gel}}{\phi_{cap} + \phi_{gel}} \tag{1.38}$$

where $\gamma$ is the surface tension of condensed water, $r_s$ is the pore radius at which the interface is created, $R$ is the universal gas constant, $T$ is the absolute temperature of the vapor–liquid system, $M$ is the molecular mass of the water, and $h$ is the relative humidity, i.e., the ratio of the vapor pressure to the saturated vapor pressure.

This coupling is a key to integrating thermo-hydro dynamics and deformational solid mechanics of concrete composite (Shimomura and Maekawa 1997), and the effect of moisture on deformation is inherently taken into account. Moisture loss may occur due to cement hydration and/or moisture migration brought about by drying climates. The volumetric change caused by both can be systematically included in constitutive multi-scale modeling. Thus, there is no need to assume specific drying/autogenous shrinkage and basic/drying creep. For example, let us study the combination of autogenous and drying shrinkage. Drying accelerates dessication inside micro-pores together with cement hydration. At the same time, moisture loss close to surfaces may retard hydration at an early age accompanying coarser solids and easy loss of moisture. This can be illustrated in Figure 1.18. As the micro-climate is automatically computed in this holistic approach, relative humidity in pores can be shown, too.

### 1.2.8 Mechanics of concrete composite and cracks, $10^{-3} \sim 10^{-0}$ m scale

For simulating structural behaviors expressed by displacement, deformation, stresses and macro-defects of materials in view of continuum plasticity, fracturing and cracking, well-established continuum mechanics is available in multi-scale modeling. Compatibility condition, equilibrium and constitutive modeling of material mechanics are the basis, and spatial averaging of overall defects in a control volume of meso-scale finite elements is incorporated into the constitutive model of quasi-continuum. Here, the size of the referential volume is in the order of $10^{-3} \sim 10^{-1}$ m. The authors adopted a three-

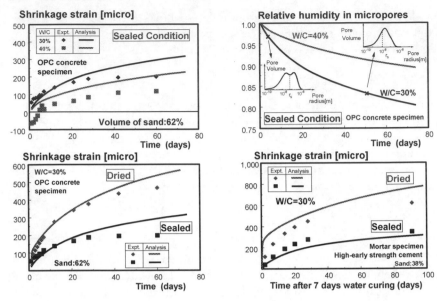

*Figure 1.18* Volume change under coupled autogenous and drying shrinkage and internal states.

dimensional finite element computer code of non-linear structural dynamics (Maekawa *et al.* 2003b, see Figure 1.19).

This frame of structural mechanics interlinks with thermo-hydro physics in terms of the mechanical performances of materials through the constitutive modeling in both space and time (Figure 1.2 and Figure 1.20). The instantaneous stiffness, short-term strengths of concrete in tension and compression, and the free volumetric contraction rooted in coupled water loss and self-desiccation caused by varying pore sizes are considered integral to the creep constitutive modeling of liner convolution. The volumetric change invoked by the hydration in progress and water loss is physically tied with surface tension force developing inside the micro-capillary pores. The micropore size distribution and moisture balance of thermodynamic equilibrium are given from the code *DuCOM* (see Figure 1.2) at each time step as discussed in Section 1.2.1 and Section 1.2.2.

Cracking is the most important damage index associated with mass transport inside the targeted structures. Cracks are assumed normal to the maximum principal stress direction in three-dimensional extent when the tensile principal stress exceeds the tensile strength of concrete. After crack initiation, tension softening on progressive crack planes is taken into account in the form of fracture mechanics. In the reinforced concrete zone, in which bond stress transfer is expected to be effective, the tension stiffness model is brought together. Since the external load level, with which the environmental action is coupled in design, is rather lower than ultimate limit states, compression-induced damage accompanying dispersed micro-cracking is

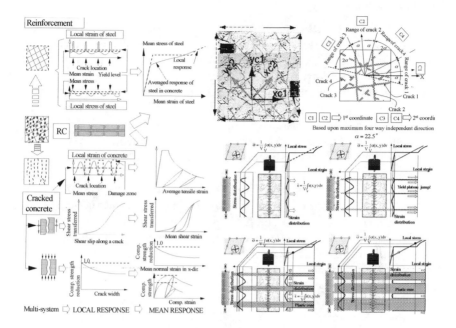

*Figure 1.19* Macroscopic modeling of reinforced concrete with cracking: REV size is about $10^{-3} \sim 10^{-0}$ m.

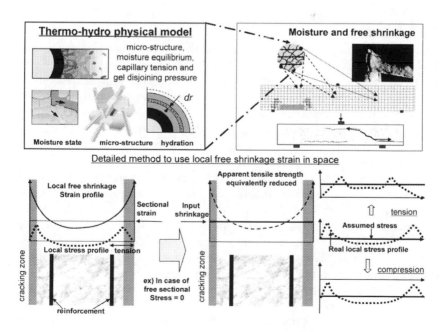

*Figure 1.20* Micro-scale chemo-physics linked with macroscopic mechanics with cracking.

disregarded. Figure 1.19 shows the overall frame of constitutive modeling of macro-scale REV, including cracks. This macroscopic modeling, which can be directly used for behavioral simulation of entire structures, has a mutual link with nanometer-scale structures, as shown in Figure 1.1 and Figure 1.20.

In the case where mechanical actions on structures are dominant, multi-directional cracks intersect orthogonally, because the principal stress hardly rotates in nature. In the case where both mechanical and ambient weather actions are applied to concrete structures, however, principal stress axes can drastically rotate and non-orthogonal cracking easily takes place in the REV domain. As highly non-linear interaction between multi-directional cracking is witnessed, the rotating crack model or single-directional smeared crack approach is hardly used for structural damage analysis under coupled loads and ambient conditions. Thus, in this book, the multi-directional fixed-crack modeling of the reinforced concrete (RC) domain is applied. This modeling was originally developed for reversed cyclic finite element analysis for three-dimensional RC under arbitral ground motions. In fact, multi-directional forces may cause cracks that are non-orthogonal. Here, the active crack method is utilized for simplifying numerical processes with reasonable accuracy for practice, and it is currently used for seismic performance assessment of underground RC and energy facilities.

## 1.3 Numerical simulations

### 1.3.1 Re-hydration and varying permeation

As has been found in laboratories that tested permeation of water through lower water-to-cement ratios, concrete/mortar is time-dependent. Re-hydration and self-curing were reported for concrete with unhydrated cement. Prematurely cured concrete performance can be recovered by re-wetting. Thus, it will be agreed that tested permeability of cementitious composites is not the characteristic constant of material performance. The authors conducted long-term simulation of coupled moisture migration, re-hydration and reforming of micro-pores, as shown in Figure 1.21.

It is clearly observed that micro-pore spaces of larger water-to-cement ratio concrete are entirely occupied by water within a shorter period, except the extreme end exposed to dry air. In contrast, micro-pores of low water-to-cement ratio concrete are hardly filled with moisture and it takes many years for stability in computation. As severe self-desiccation occurs inside the core of concrete, moisture intake is made from the drying boundary. At the other boundary attached to condensed water, re-started hydration makes pores denser and condensed water migration is strongly blocked and vapor diffusivity chiefly conveys moisture inside the analysis domain. Only with short-term tested permeation is it difficult to evaluate the concrete performance against moisture leakage for low water-to-cement ratio concrete. In this analysis, the calcium leaching was not coupled, for simplicity.

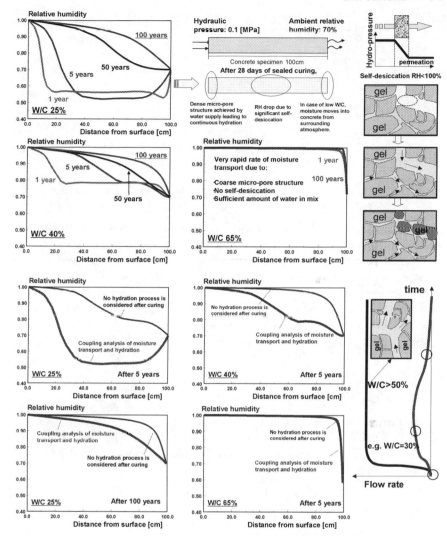

*Figure 1.21* Coupled moisture migration and transient evolution of micro-pore structures for low W/C.

### 1.3.2 *Chloride transport into concrete under cyclic drying–wetting*

Using the proposed method, transport of chloride ion under alternate drying–wetting conditions is presented. For verification, experimental data by Maruya *et al.* (1998a) is used. The size of mortar specimens is $5 \times 5 \times 10$ [cm] and the water-to-powder ratio is 50%. After 28 days of sealed curing, the specimens were exposed to cyclic alternate drying (7 days) and wetting (7 days) cycles. The drying condition was 60% RH, whereas the wetting was exposed to a chloride solution of 0.51 [mol/l] at 20°C. In FEM analysis, mix

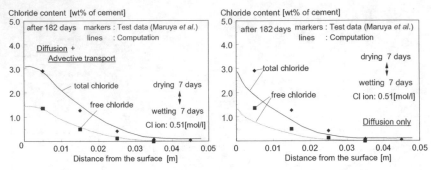

*Figure 1.22* Chloride content profiles in concrete exposed to cyclic drying–wetting–drying.

proportions and chemical compositions of cements are given. The curing and exposure conditions are also defined as boundary conditions for the target structures. All of these input values correspond to the experimental conditions. Figure 1.22 shows the distribution of free and bound chlorides from the surface of exposure. Two cases are compared: one considering only diffusive movement and the other including the advective transport due to the bulk movement of pore water as well as the diffusion process. The distribution of bound and free chlorides can be reasonably simulated with advective transport due to the rapid suction of pore water under the wetting phase.

### 1.3.3  Carbonation phenomena in concrete

Computations are performed to predict the progress of carbonation for different $CO_2$ concentrations and water-to-cement ratios. The amount of $Ca(OH)_2$ existing in cementitious materials can be obtained by the multi-component hydration model as (Chapter 2)

$$2C_3S + 6H \rightarrow C_3S_2H_3 + 3Ca(OH)_2$$
$$2C_2S + 4H \rightarrow C_3S_2H_3 + Ca(OH)_2 \qquad (1.39)$$
$$C_4AF + 2Ca(OH)_2 + 10H \rightarrow C_3AH_6$$

For verification, the experimental data done by Uomoto and Takada (1993) were used. Figure 1.23 shows the comparison of analytical results and empirical formula that was regressed with the square root equation of time. Similar to the previous case, all of the input values in the analysis corresponded to the experimental conditions. Analytical results show the relationship between the depth of concrete in which pH in pore water becomes less than 10.0 and exposed time. The simulations can roughly predict the progress of carbonation for different $CO_2$ concentrations and water-to-powder ratios.

Figure 1.24 shows the distribution of pH in pore water, $CO_2$, calcium hydroxide, and calcium carbonate inside concrete, exposed to the $CO_2$ concentration of 3%. Two different water-to-cement ratios, $W/C = 25\%$ and

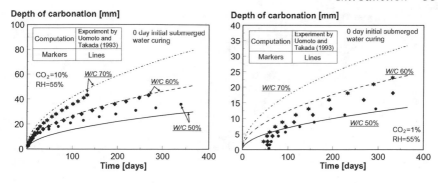

*Figure 1.23* Carbonation phenomena for different $CO_2$ concentrations and water-to-cement ratios.

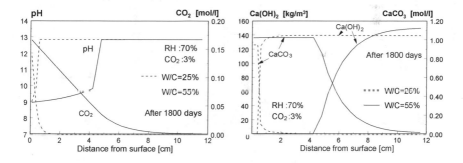

*Figure 1.24* Distribution of pH, calcium hydroxide, and calcium carbonate under the action of carbonic acid.

50%, were analyzed. It can be shown that higher resistance for the carbonic acid action is achieved in the case of a low W/C.

### 1.3.4 Coupled carbonation and chloride-induced corrosion

Attention is directed to steel corrosion in concrete due to the simultaneous attack of chloride ions and carbon dioxide. Provided that concrete is totally submerged, the corrosion rate is generally small due to the much lower supply of oxygen. Thus, drying is also coupled. Interaction of chloride and $CO_2$ sub-phases is modeled such that bound versus free chloride equilibrium is influenced by pH of pore solution. For parametric study, the authors assume pH sensitivity as shown in Figure 1.25 from some literature, and densification of micro-pores caused by carbonation is simply expressed by the reduction factor of porosity. This is to express the replacement of $Ca(OH)_2$ and calcite.

Figure 1.26 solely shows simulation. The carbonated zone close to the surface shows evidence of smaller total chloride due to released bound chloride by carbonation and loss of moisture by drying, but dissolved chloride ion is

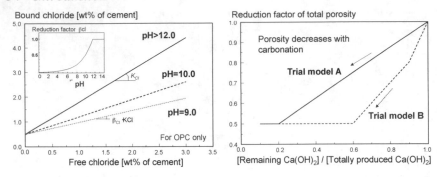

*Figure 1.25* Release of bound chloride and re-densification of pores by carbonation.

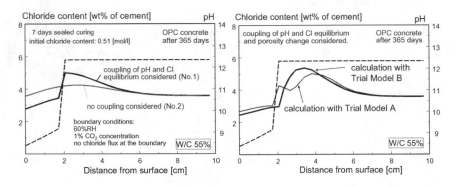

*Figure 1.26* Computation of total chloride contents under coupled drying, carbonation and chloride penetration.

concentrated in the pore solution. Around the carbonation front, a couple of peaks of total chloride are formed. Here, the free chloride is released from formerly bound chloride. Since the carbonation brings densification of pores, chloride ion penetration is also held back and twin peaks can be formed just close to the carbonation front. In the future, detailed discussion of this matter would be beneficial.

For life-span assessment, concrete members with different water-to-powder ratios, W/C = 40, 50, 60%, with only one face exposed to the environment, were considered. The stage where concrete cracking occurs was defined as a limit state with respect to steel corrosion. The progressive period until the initiation of longitudinal cracking was estimated by the equation proposed by Yokozeki *et al.* (1997), which is a function of cover depth. Figure 1.27 shows relationships between cover depth and structural age until cracking due to corrosion obtained by the proposed thermo-hydro system (Chapter 9). It can be seen that the concrete closer to the exposure surface will show an early sign of corrosion-induced cracking, and low W/C concrete has the higher resistance against corrosion.

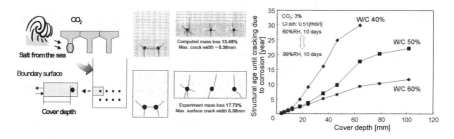

*Figure 1.27* Time till first signs of cracking by corrosion for concrete exposed to $CO_2$ gas and salty water.

### 1.3.5 Moisture distribution in cracked concrete

In order to show the possibility of a unification of structure and durability design, a simple simulation was conducted by using the proposed parallel computational system. It has been reported that there should be a close relationship between moisture conductivity and the damage level of cracked concrete; that is, moisture conductivity should be dependent on the crack width or the continuity of each crack. The proposed system, in which information is shared between the thermo-hydro and structural mechanics processes, is able to describe this behavior quantitatively by considering the inter-relationship between moisture conductivity and cracking properties. For representing the acceleration of drying-out due to cracking, the following model proposed by Shimomura (1998) was used in this analysis.

$$J_w = \begin{cases} J_V + J_L & \textit{before cracking} \\ J_V + J_L + J_V^{cr} + J_L^{cr} & \textit{after cracking} \end{cases} \tag{1.40}$$

where $J_w$ is the total mass flux of water in concrete, $J_V$ and $J_L$ are the mass flux of vapor and liquid in non-damaged concrete respectively, and $J_V^{cr}$ and $J_L^{cr}$ are the mass flux of vapor and liquid water through cracks. In this simulation, only $J_V^{cr}$ is taken into account for the first approximation, since diffusion of vapor would be predominant when concrete is exposed to drying conditions. From the experimental study done by Nishi *et al.* (1999), it has been confirmed that the flux $J_V^{cr}$ can be expressed as

$$J_V^{cr} = -\bar{\varepsilon} \rho_V D_a \nabla h \tag{1.41}$$

where $\bar{\varepsilon}$ is the average strain of cracked concrete, which can be computed by COM3, $\rho_V$ is the density of vapor, $D_a$ is the vapor diffusivity in free atmosphere, and $h$ is relative humidity.

The target structure in this analysis was a concrete slab, which has a 30% water-to-powder ratio using medium-heat cement. Mesh layout and the restraint conditions used in this analysis are shown in Figure 1.28. The volume of aggregate was 70%. After three days of sealed curing, the specimen was exposed to 50% RH. Figure 1.28 shows the cracked elements, the

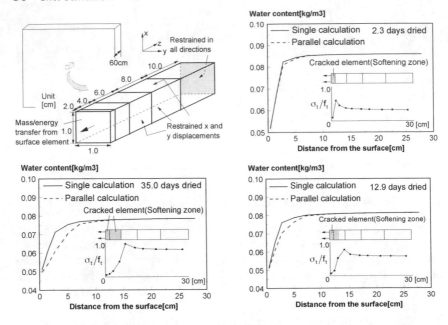

*Figure 1.28* Moisture and internal stress distribution in concrete exposed to drying condition.

distribution of moisture, and normalized tensile stress at each point from the boundary surface exposed to drying conditions. Moisture distribution calculated without stress analysis is also shown in Figure 1.28. As demonstrated in the results, the crack occurs from the element near the surface, and the crack progresses internally with the progress of drying. It is also demonstrated that the amount of moisture loss becomes large due to cracking.

### 1.3.6  Coupling of shear and axial forces invoked by volume change

When thermally induced cracks penetrate through a whole section, a diagonal shear crack may be arrested by pre-existing damage induced by heat and/or drying, as shown in Figure 1.29. The stiffness of the member is certainly reduced by the crack penetrating the entire section, but the shear capacity is, on the contrary, elevated since the shear crack propagation is retarded or has ceased. Non-orthogonal crack-to-crack interaction can be well simulated and the localization pattern of force-induced cracks is greatly affected by the damage level of pre-cracks (Maekawa *et al.* 2003b). This is attributed to shear transfer performance of concrete crack planes. Here, the contact density model for concrete rough cracks is applied. The shear transfer which causes principal stress rotation is one of the crucial mechanics for coupled force–environmental actions.

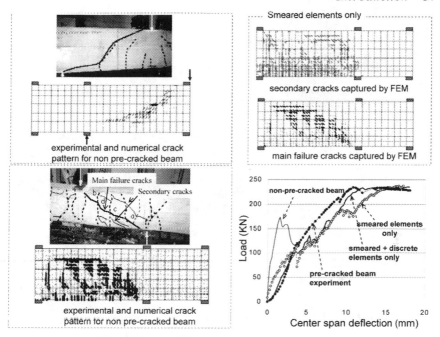

*Figure 1.29* Interaction of non-orthogonal cracking and safety performance simulation.

### 1.3.7 Corrosion-induced cracking and structural capacity

Figure 1.30a shows the chloride ion penetration into the beam subjected to flexure. The bending cracks are distributed around the lower extreme fiber and the averaged tensile strain is transferred to a *DuCOM* sub-system. As the chloride ion diffusivity is formulated based upon both pore-structural characters (micro-information) and crack strain (macro-information) dependent on applied loads, the deep penetration is seen at the center span of the beam.

The cracking pattern of an RC beam subjected to shear and flexure is illustrated in Figure 1.30b. As the shear capacity is less than the flexural one, a localized shear crack band can be seen after failure. First, partial corrosion is numerically reproduced by concentrated chloride with longitudinal cracking and followed by load (Chapter 9). Tension stiffness defined in finite elements with corroded steel is reduced to plain concrete softening since the bond is thought to be deteriorated.

If the corrosion-induced cracking is located around the center span, no interaction of shear and flexural cracks is seen and shear capacity of the beam is computed unchanged. When it is placed around the shear span close to the support, diagonal shear cracking joins this pre-cracking and early penetration of the diagonal crack is computed. Non-orthogonal cracking is the key of this analysis.

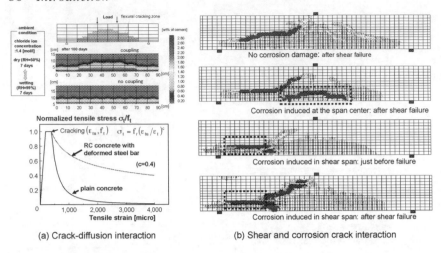

(a) Crack-diffusion interaction   (b) Shear and corrosion crack interaction

*Figure 1.30* Corrosion-induced cracking and remaining structural performance assessment.

### 1.3.8  Time-dependent deflection of beams under drying

Two beams as shown in Figure 1.31 are discussed concerning the load–drying combination. Under pure drying with no external forces, no deflection takes place in nature as no curvature is developed. Under the sealed condition without moisture loss, creep deflection is caused by concrete creep in compression and tension. When two actions are simultaneously combined, time dependency of deflection is significantly stepped up, as shown in Figure 1.31. This

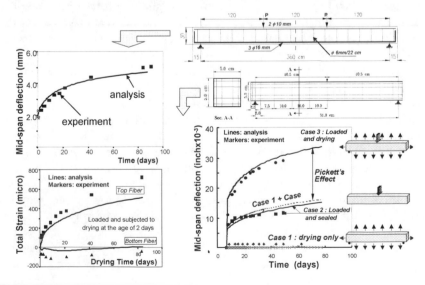

*Figure 1.31* Coupled sustained load and drying onto beams subjected to flexure.

apparent non-linearity is known as the Pickett effect (1956) and practically, the creep function is specified differently from basic creep. In multi-scale analysis, there is no need to change material functions under different ambient conditions (Maekawa and Ishida 2002).

The drying effect is explicitly considered as a moist boundary condition, and the pore-water pressure is computed in *DuCOM*. The self-equilibrated stress and deformation concurrently solved space by space and the creep properties of skeleton stress and strain are evaluated with micro-damage, if any.

## 1.4 Practical applications, most macroscopic $10^0$~$10^{+2}$ m scale

Recent applications of the multi-scale simulation to some real concrete structures are demonstrated. Mainly, it is being used for the performance assessment of existing structures with or without damages, such as corrosion of reinforcement, cracking and alkali aggregate reaction (ASR). Here, the reproduction of the defects in structural concrete is essential. Generally, health inspection of the structural concrete and steel is first conducted. Next, damages and/or deflections are computationally reproduced in the overall structural model so that the internal states may match the reality. This state is regarded as the initial conditions for the remaining life-cycle simulation in future. In general, the crack location and its width are of great importance.

For durability simulation of newly constructed structures, expected ambient conditions and mechanical loads must be defined, as the design values and computational results are used to examine whether the design requirement and expected life are satisfied or not. Examples found in this section can serve as the verification of the largest scale modeling with the monitoring earned in future.

### 1.4.1 Seismic performance assessment

Figure 1.32 and Figure 1.33 show the application examples for the seismic performance assessment of existing RC–steel composite piers to support a long-span bridge in service and an underground aqueduct. Multi-chemo-mechanistic non-linear finite element analyses bring about cost benefits and behavioral simulation of structures of geometrical complexity with soil interaction. In these cases, there is no standard procedure of design to simply decide shapes of constituent structural members, dimensions and boundary conditions. Then, the performance assessment for safety becomes crucial for proposed dimensioning and details.

The large-scale pier in Figure 1.32 was found probably to fail in a sudden shear mode of failure before yield of main reinforcement. In fact, the nominal shear strength is apparently reduced according to increased sizes. This size effect was not known nor taken into account when designed. The computational approach can automatically take this phenomenon into ac-

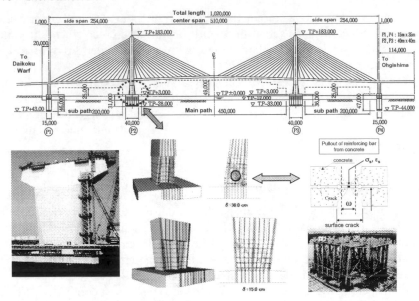

*Figure 1.32*  Multi-scale failure analysis of the steel–RC composite pier of long cable stayed bridge.

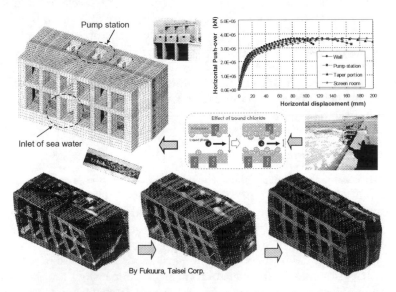

*Figure 1.33*  Push-over analysis of underground aqueduct with and without corrosion of steel in concrete.

count without any special care being necessary. The shear failure of wall members of underground ducts and tunnels results in real challenges for engineers, because only interior planes are accessible for strengthening. Then, the computational performance assessment is expected to invent some acceptable solutions. Otherwise a huge fund for recovery of underground infrastructures has to be prepared.

As the existing aqueduct is normally flooded in the sea-water, detailed inspection can be hardly performed. Thus, the coupling analysis of chloride ion migration, steel corrosion and inelastic push-over was conducted by *DuCOM–COM3*, as shown in Figure 1.33. As the structural redundancy is high, the seismic ductility is hardly affected by the local corrosion of reinforcement. For risk assessment, intentional corrosion of great magnitude was reproduced in the simulation, and the push-over analysis was conducted in view of maintenance.

Figure 1.34 is the recent example of application to the engineering assessment for a 100-year-old railway bridge in the Tokyo metropolis (Sogano *et al.* 2001). Due to uneven settlement of the foundation caused by varying underground water levels and rapid urbanization in 1960s, some initial damage inside the assembled brick masonry was found in the form of cracking in old bricks. Some decades ago, arch ribs was strengthened by additional RC arches on the inside layer. The historical process of structures was assumed and the corresponding mechanistic actions were reproduced in the computational environment. Afterwards, the seismic ground motion was computationally applied to the numerically aged structural concrete and the computed response was used for safety and serviceability assessment in practice. The seismic remaining performance was numerically investigated with initial defects, which were detected by surveying beforehand, and the sustainable life with a light retrofit was judged. Long-term durability as-

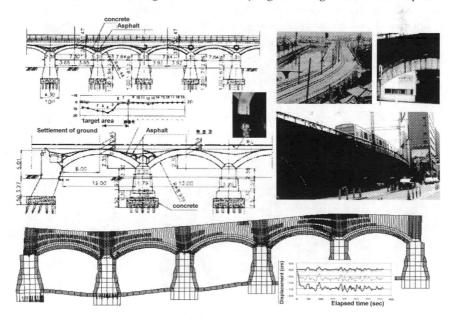

*Figure 1.34* Remaining seismic performance of existing old infrastructures.

sessment and associated damage have much to do with seismic engineering issues. Now, the holistic approach is indispensable for maintenance of urban infrastructures.

In this case, the ground settlement could fortunately be presumed and the associated computed cracking was found to match the reality. If ground settlement was not suspected as a main cause of damage, then it would be inversely detected so that the cracking in the main arch could be computationally created. Here, corrosion risk can be avoided since no reinforcement exists.

### 1.4.2  *Serviceability and risk assessment of damaged structures*

Figure 1.35 shows the analysis for remaining structural safety performance of ASR-damaged RC bridge piers found in the west of Japan (Japan Society of Civil Engineers 2005). As many reinforcing bars are surprisingly ruptured at the inside corners of bent portions, anchorage performance of web reinforcement is thought to have deteriorated (Chapter 9). It is clear that the capacity-predictive formulae in design codes cannot be applied, because these design tools and knowledge have been developed based on the assumption that structural details related to hook, splices, bent radius, concrete cover, etc. are satisfied. At the same time, the effect of dispersed cracking and self-equilibrated pre-stressing caused by ASR expansion has to be considered. Inescapably, the full non-linear structural simulation is just the tool to calculate the remaining capacity in consideration of anchorage deterioration of stirrups and cross-sectional loss of main reinforcement.

In fact, the computed capacity of ASR-damaged members was predicted to gradually increase in accordance with the magnitude of ASR expansion. This is attributed to the pre-stressing effect and self-equilibrated compressive axial force which elevates the shear capacity. If an excessive expansion beyond

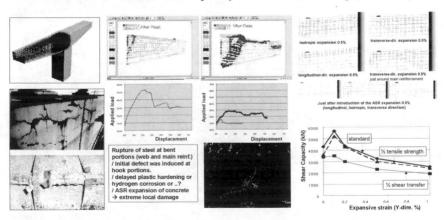

*Figure 1.35* Structural capacity of RC beams with ASR-induced cracking and rupture of reinforcement.

the yield of reinforcement is computationally assumed, as shown in Figure 1.35, the capacity starts to decline together with the structural stiffness. Thus, the way of strengthening and/or repairing must be different according to the induced level of expansion. In this case, the effect of steel rupture at the extreme ends of web reinforcement is comparatively small, because the bond deterioration of reinforcement is fortunately small by volume compared to the size of the damaged structural members. But the rupture of welding portions of main reinforcement is so serious and the failure mode is thought to shift to the shear failure.

Figure 1.36 shows the pre-stressed concrete bridge in which many shrinkage cracks were unexpectedly induced to the main viaducts due to excessive shrinkage of concrete rooted in an intrinsic volumetric change of coarse aggregates (Chapter 7) and unnecessarily heavy reinforcement in terms of serviceability limit states (Japan Society of Civil Engineers 2005). These cracks penetrate through whole sections of damaged members. The actual compliance of the real viaduct in each span was reported to be much greater than the design value. This is attributed to the large loss of structural concrete stiffness, and probably to the initial shapes just after the form striping.

Here, the substantial safety of the bridge system at future earthquakes and the fatigue life were questioned. The JSCE Concrete Committee (2005) investigated the detailed damage and corresponding remaining fatigue life by using the coupled chemo-mechanical simulation (Chapter 9 and Chapter 10). As shown in Figure 1.36, the thermo-hygro analysis was conducted by *DuCOM* to the three-dimensional finite element geometry of the viaducts exposed to the gravity action as well as the recorded temperature and ambient humidity at the construction site. The micro-pore structural character of coarse aggregates (Section 7.4) was taken into account. The cracking and accompanying excessive displacement were well simulated with time dependency (Chapter 8) in comparison with the reality. For verification of the analysis method, the design live load (1,500 kN) was applied on the deck and the incremental deflection was measured as shown in Figure 1.36. The coupled *DuCOM–COM3* (Maekawa and Ishida 2002) shows the reasonable simulation of displacements in each span center. Afterwards, the high cyclic fatigue loads were computationally applied to the whole bridge with damage. It was agreed following these analyses and site investigation that the safety performance would not be sacrificed but the stability of deflection was questionable, so additional PC tendon cables were internally attached to the viaducts.

### 1.4.3 Durability design and remaining life

Multi-scale life-span simulation can be applied to the durability design as well as the risk evaluation of existing structures. Here, the initial set-up of boundary conditions and internal variables are rather easier. Figure 1.37 shows the concrete runway decks made of combined precast PC and in-situ

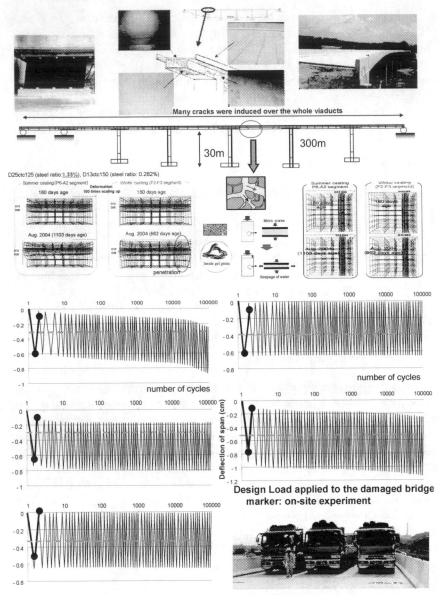

*Figure 1.36* Bridge viaducts heavily cracked by shrinkage of concrete and the structural behavior simulation for planning of repair and strengthening.

RC. The boundary conditions are two million cyclic passages of aircraft of about 5,000 kN by weight and the flying chloride on the bay area of Tokyo (Maekawa *et al.* 2008). In this case, chloride penetration is not the critical limit state, but rather the fatigue rupture of the structural concrete is the dominant issue.

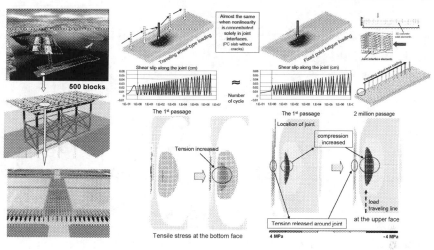

*Figure 1.37* RC–PC composite systems under coupled high-cycle fatigue and environmental actions.

At the first stage, greater cyclic flexure and out-of-plane shear force are applied along the longitudinal support beams. Due to high-cycle repetition of forces, additional pullout of reinforcement from the joint will take place and the central displacement of each panel gradually increases, as shown in Figure 1.37. As the structure is indeterminate in nature, moment shift may occur from the RC joint portion under higher stresses to the center of panels. Then, the amplitude arising along the joint (edge of the analysis domain) is gradually decayed and the risk of fatigue failure can be reduced with longer life. Although the applied bending moment amplitude increases, the fatigue failure risk does not, because the pre-stressing has been already introduced. This structural redundancy was confirmed in view of durability. Corrosion of steel along the RC joints would produce the reduced pullout stiffness which may in turn bring about reduced amplitude. Then, the degraded strength of reinforcement by corrosion can be automatically mildly compensated.

Chemo-physics and mechanical modeling of concrete with greatly different scales of geometry was presented, and synthesized on a unified computational platform from which quantitative assessment of structural concrete performances may derive (Chapter 10). Experimental verifications show its possibility as a holistic approach, while individual modeling of physical chemistry events needs to be enhanced in the future with continuous effort. A great deal of knowledge comes from past development, however, we face difficulty in extracting quantitatively consequential figures from it. The authors expect that the systematic framework on the knowledge-based technology will be extended efficiently and that it can steadily be taken over by engineers in charge.

# 2 Hydration of cement in concrete

The hydration of cement is accompanied by heat generation, development of micro-pore structures and associated mechanical strength in the hardening cement paste, mortar or concrete. Cement hydration is a phenomenon responsible for creating the hardened cement matrix of concrete and the sole cause of the temperature rise during hardening. It is an essential component for evaluating concrete performance in a rational design system that incorporates thermal cracking control, as well as other effects such as autogenous/drying shrinkage and the development of strength and microstructure. For these reasons, a model of the hydration of the cement in concrete must be developed. This chapter focuses on the heat-generation process for the cement in concrete. A strong correlation exists between heat generation and the degree of hydration in each constituent clinker mineral and the other binders. To control the risk of thermal cracking in the early stages as well as evaluate the long-term durability of concrete structures, we must be able to quantitatively predict the hydration process over time. The hydration model presented here adopts a *multi-component system* and applies Arrhenius's law of chemical reaction. When combined with hydration, microstructure development, and moisture transport phenomena, the model faithfully simulates the hardening process for concrete.

As clinker minerals and powder admixtures have complex interactions that have been modeled only in simple ways (Maekawa *et al.* 1999), hydration models have had limited practical application. Recently, chemical analyses of the unreacted remaining chemical compounds have become much more accurate, which provides more reliable data on the degree of hydration over time. The authors propose a multi-component hydration heat model that is more appropriate for various types of cement and pozzolans and a wider range of mix proportions. The proposed model was systematically verified with the appropriate parameter settings and compared with reliable results of experiments in adiabatic temperature rise. The proposed hydration model may become a reliable core technology for evaluating strength development and microstructure formation (Chapter 3).

## 2.1  Multi-component hydration model of cement minerals and pozzolans

### 2.1.1  Hydration exothermic process of cement and its overall quantification

Portland cement consists of pulverized cement clinker and gypsum. Cement clinker consists of four major minerals: alite (mainly $C_3S$), belite (mainly $C_2S$), aluminate phase (mainly $C_3A$) and ferrite phase (mainly $C_4AF$), as well as some impurities. Alite and belite, which are calcium silicates, comprise around 85% of cement clinker, while aluminate and ferrite, which together are called the interstitial phase, occupy the spaces that are created around the calcium silicates when cement clinker is burned. The percentage of each mineral in the cement clinker varies according to the type of Portland cement. The mineral composition of each type of Portland cement tends to be fairly characteristic. For example, $C_3S$ and $C_3A$ are often found in greater proportions in cements that feature early-strength development, such as ordinary Portland cement or early-strength Portland cement. Other types of cement, such as low-heat cement, which is characterized by low-temperature hydration, contain relatively high proportions of $C_2S$ and $C_4AF$.

Portland cement reacts with water to precipitate hydration products, which form a hard and porous structure. The process of cement hydration is often monitored using a conduction calorimeter that can measure the heat liberation rate of cement at a specified constant temperature. The degree of hydration of each clinker mineral over time and at a constant temperature can be detected using X-ray diffraction. The hydration process for cement is not simple, since the rate of hydration and the types of hydrates vary greatly among clinker minerals. Alite and belite produce both calcium silicate hydrates (C–S–H) and calcium hydroxide ($Ca(OH)_2$), while the aluminate and ferrite phases produce ettringite with gypsum. Calcium aluminate hydrates are produced when the gypsum is consumed and the ettringite is converted into a monosulphate. All of these reactions occur simultaneously in the hardening cement paste and generate a certain amount of heat. Each type of clinker mineral also generates a certain amount of heat at complete (pure) hydration. $C_3S$ generates 504 J/g, $C_2S$ generates 260 J/g, $C_3A$ generates 870 J/g, and $C_4AF$ generates 420 J/g.

Physically, cement hydration can be considered a multi-stage process. At the very beginning of the reaction, a protective layer of C–S–H and a hard coating of ettringite are produced on the surface of the cement particles and generate the first peak in the exothermic process of hydration. Few collisions with water occur at this stage. As the reaction progresses, the precipitation of hydrates on the surfaces of the cement particles accelerates and the rate of heat generation shows a second peak in the exothermic process of hydration. Water from the outside first penetrates the cover of the hydrated layer and then reaches the unhydrated particles. At this point, the hydration rate is governed by the diffusivity of the water and ions such as calcium, silica and hydroxyl. As the process nears termination, the surface area of the unhydrated cement particles becomes small. Thus, water and the active molecules of cement have a lower probability of contact because of the thick layer of hydrate products. At this point, the reaction stops.

### 2.1.2 Multi-mineral component concept and general formula

Clinker compounds and gypsum, which is added after the clinker is baked, comprise the major minerals of cement. The proportion of each mineral component varies according to the type of cement. Thus, for a hydration heat model to be applicable to any given type of cement, it must properly describe the exothermic process of the cement under hydration in accordance with its mineral composition. Furthermore, heat generation is usually restrained not only by reducing the amount of Portland cement or changing its type, but also by replacing a certain part of the cement by various types of pozzolans. Thus, the model must be able to handle applications involving pozzolans. The approach employed in this chapter entails dividing the simultaneously occurring reactions into the appropriate units and describing them in as rational a manner as possible. In other words, the hydration reactions are resolved into mineral units and the hydration exothermic process of each component is described according to the interactions among the minerals. The "multi-mineral" reactions also allow for the introduction of other blended powder components.

The hydration heat model based on this multi-mineral concept represents the various types of cement by their clinker mineral compositions. The hydration of each mineral component is modeled separately. A multi-component model developed by Kishi and Maekawa (1995, 1994) calculates the cement's rate of hydration heat generation from the sum of the individual hydration heat rates for the components according to their proportions. The minerals in the cements covered by the model are alite ($C_3S$), belite ($C_2S$), an aluminate phase ($C_3A$), a ferrite phase ($C_4AF$), and gypsum ($C\bar{S}2H$). The exothermic reaction for each of these minerals is described separately. Gypsum is treated as a dehydrate (gypsum dehydrate). For blended cement, blast furnace slag, fly ash, and silica fume are considered single reaction units and are incorporated into the model as individual components. Blast furnace slag, fly ash and silica fume are not classified in the same manner as Portland cement because the reacting part is only a glass phase at normal temperatures and so can be regarded as a homogeneous material. Limestone powder is also treated as a powder material in this model. Although it is regarded as an inert material, the proposed model accounts for its micro-filler effect on other reacting components. The heat rate for the cement as a whole, $H_C$, including the blending powders, is the sum of the heat rates for all reactions.

$$H_C = \sum p_i H_i$$

$$= p_{C_3A}(H_{C_3AET} + H_{C_3A}) + p_{C_4AF}(H_{C_4AFET} + H_{C_4AF}) \qquad (2.1)$$

$$+ p_{C_3S}H_{C_3S} + p_{C_2S}H_{C_2S} + p_{SG}H_{SG} + p_{FA}H_{FA} + p_{SF}H_{SF}$$

where $H_i$ is the heat-generation rate of mineral $i$ per unit weight, $p_i$ is the weight composition ratio, and $H_{C_3AET}$ and $H_{C_4AFET}$ are both heat generation rates during the formation of ettringite. The ettringite formation model is included in the proposed system because ettringite is first formed from $C_3A$ and $C_4AF$ prior to hydration when gypsum is present. Hydration heat is generated by $C_3A$ and $C_4AF$ (and expressed as $H_{C_3A}$ and $H_{C_4AF}$) after the ettringite formation reaction stops due to the disappearance of unreacted gypsum.

The multi-component hydration heat model was originally developed by Kishi *et al.* (1993a, 1993b) to analyze the temperature of massive concrete structures without having to conduct adiabatic temperature rise tests. Thermal stress analysis requires a cement hydration heat model that can be applied to any given condition. Before this model was developed, Uchida and Sakakibara (1987) attempted to predict the hydration exothermic process using a cement hydration model. The hydration reaction model for cement is an excellent approach because it predicts the general heat generation rate of hydration for any temperature history. The hydration heat model must be applicable to the hydration of cement with an arbitrary temperature history. Uchida and Sakakibara (1987) confirmed in studies with cement paste that Arrhenius's law can be applied to cement hydration, which is the reaction of a composite containing multiple mineral compounds, by regulating the thermal activity in terms of the accumulated heat of the cement as a whole. Suzuki *et al.* (1990a, 1990b) successfully developed a quantification technique for the exothermic hydration process of cement in concrete that is dependent on the temperature, and provided a general approach for deriving a hydration-heat model capable of assessing any temperature hysteresis. The feature of this model is that the exothermic hydration properties of cement in concrete are specified by two material functions: thermal activity and intrinsic heat rate. The thermal activity is predetermined by the activation energy, which is an energy barrier to the chemical reaction. The necessary energy to stimulate the reaction is generally supplied by the molecular kinematic energy. In the chemical reaction rate, the reference heat rate represents particle collisions. The accumulated heat is a main parameter for the rate of heat generation as well as the temperature and represents the past hydration process. Arrhenius's law of chemical reaction expresses the temperature dependence of the cement hydration. The activation energy and intrinsic reference heat rate characterize the process of heat generation for the cement in concrete.

Referring to the model proposed by Suzuki *et al.* (1990a, 1990b), the hydration process for each mineral compound is expressed by two material functions: the reference heat rate, which gives the heat rate at a specified constant temperature, and the thermal activity, which describes the temperature dependence of the reaction. Since each mineral-based reaction in the cement is described separately, the model incorporates coefficients that express the interdependence of the reactions. To accommodate the interactions among the reactions of the mineral compounds in the common reaction environment, the effects of the shared environmental temperature, the amounts of mixing water and inert powder/stagnated binder as micro-filler, and the changes in the heat-generation rates for the alite and belite, depending on their proportions, are also taken into account. To make the proposed model applicable to cement blended with blast furnace slag, fly ash, and/or silica fume, the focus should be on the amount of calcium hydroxide $(Ca(OH)_2)$ in the system, because their reactions are strongly dependent on the supply of $Ca(OH)_2$ as an activator of the reaction. Furthermore, the supply of $Ca(OH)_2$ from cement hydration and its consumption by the blast furnace slag, fly ash and silica fume are simultaneously simulated. The delaying effects of the chemical admixture and the fly ash on the cement and slag reaction are also taken into account. Kishi and Maekawa (1996) proposed that these factors are mutual interactions that take place between the exothermic hydration process of the Portland cement and the admixtures, and that they can be rationally treated using the multi-component system adopted here. Consequently, the hydration heat rate of each component can be expressed as follows.

$$H_i = \gamma_i \cdot \beta_i \cdot \lambda_i \cdot \mu_i \cdot H_{i,T_0}(Q_i) \exp\left\{-\frac{E_i}{R}\left(\frac{1}{T} - \frac{1}{T_0}\right)\right\} \tag{2.2a}$$

$$Q_i \equiv \int H_i dt \tag{2.2b}$$

where $E_i$ is the activation energy of component $i$, $R$ is the gas constant, $H_{i,T0}$ is the reference heat-generation rate of component $i$ at constant temperature $T_0$ (and is also a function of the accumulated heat $Q_i$), $\gamma_i$ is a coefficient expressing the delaying effect of the chemical admixture and fly ash in the initial exothermic hydration process, $\beta_i$ is a coefficient expressing the reduction in the heat-generation rate due to the reduced availability of free water (precipitation space), $\lambda_i$ is a coefficient expressing the change in the heat-generation rate of the powder admixtures, such as blast furnace slag, fly ash and silica fume, due to a lack of calcium hydroxide in the liquid phase, and $\mu_i$ is a coefficient expressing changes in the heat-generation rate in terms of the interdependence between alite and belite in Portland cement. Furthermore, the reference heat-generation rate $H_{i,T0}$ is directly modified according to the stipulated rules as stated later, thereby taking into account the micro-filler effect and the effect of the fineness of the powders. Coefficients, $\gamma_i$, $\beta_i$, $\lambda_i$, and $\mu_i$ are assumed to have changing ratios when no effects are due to other factors. The minimum coefficient is used as a reducing ratio when the value of several coefficients is less than unity. $(-E_i/R)$ is defined as thermal activity.

   Taking into account the temperature dependence of the individual mineral component reactions, the exothermic behavior of the cement as a whole during hydration is quantified for any given temperature history.

### 2.1.3 Reference heat-generation rate of components

Figure 2.1 shows the reference heat generation rate $H_{i,T_o}$ at the reference temperature $T_o$, which are material functions for each individual mineral reaction. These rates for the clinker minerals are generally set as values for the mineral composition of ordinary Portland cement. The reference temperature $T_o$ is set at 293K (20°C). The exothermic hydration reaction process in each mineral is defined with reference to the cumulative heat generated. Thus, Stage 1 is defined as the period until 1% of the total heat output is reached. Stage 2 is up to 25% for $C_3S$ and up to 30% for $C_2S$. Stage 3 is defined as percentages above these values. In the exothermic process for $C_3A$ and $C_4AF$, no border is assumed between Stage 2 and Stage 3. Stage 2 in the exothermic process is thought to be a reaction control process. The heat output for $C_3S$ and $C_2S$ during Stage 2 is set so as to fit the initial characteristic temperature increment corresponding to the second exothermic peak in an adiabatic temperature rise test. Stage 3 is a diffusion control process in which the heat-generation rate is considerably slower than that in Stage 2. To be more specific, the reaction rates are arranged in the order of $C_3A > C_3S > C_4AF > C_2S$. The accumulated heat at which the reference heat rate of a particular mineral is zero corresponds to the final heat output achieved by 100% hydration ($Q_{i,\infty}$). This value was determined from the theoretical heat generated by each mineral according to Arai (1984).

   In general, the separate heat-generation rates for the blending admixture are difficult to derive in experiments because the admixture cannot continue to react without an activator. Besides, the heat rate measured when a reagent is added to blended cement is known to differ from that of the blending admixture, since the reagent type affects the reactivity of the admixtures, as reported by Uchikawa (1986). Thus, the reference heat-generation rates for blast furnace slag, fly ash and silica fume were set by comparing the analytic

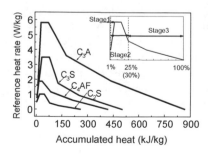
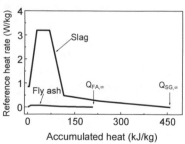

*Figure 2.1*  Reference heat rate for each reaction.

data and the experiment data, as shown later. The total heat generated per unit weight that corresponds to the completed reaction ($Q_{i,\infty}$) in the model can also be derived from qualitative observations of reactions other than those in analytic trials. In the case of blast furnace slag, it is well known from adiabatic tests that replacing up to 50% of the slag has little effect on the ultimate temperature rise of ordinary Portland cement. Rather, the temperature rises gradually in a manner similar to the rise in moderate-heat Portland cement, as reported by Ito and Fujiki (1987). Therefore, the total heat generated by the slag was assumed to be approximately the same as that of ordinary cement and slightly greater than that of moderate-heat cement. In this chapter, 462 kJ/kg is the assumed total heat generated per unit weight of blast furnace slag. On the other hand, little quantitative information exists about the fly ash reaction and the this reaction does not terminate in an identical manner within the adiabatic temperature rise testing period. The total heat generated by the fly ash is temporarily set to a value of 210 kJ/kg in the hydration rate model. The total heat generated by silica fume was set to 565k J/kg as a result of comparing the adiabatic temperature rises for plain concrete and concrete containing silica fume at different replacement ratios.

The reference heat-generation rates for blast furnace slag (SG), fly ash (FA) and silica fume (SF) are valid when sufficient amounts of both water and calcium hydroxide are supplied. The reaction rate for FA, which is the reference heat rate divided by the total heat generated, is much smaller than that for the blast furnace slag and silica fume reaction in the model because the reaction between the glass phase in the fly ash and the $Ca^{2+}$ in the solution is relatively slow and initiated one or three days after mixing at a normal temperature, as reported by Uchikawa (1986), and fly ash has a much smaller total surface area of particles per unit weight than silica fume.

The proposed model also should be applicable to the powder fineness. In general, the reaction rate is faster with finer particles. For example, for early strength development, early hardening Portland cement with enhanced powder fineness and a different mineral composition is used rather than ordinary cement. In the case of blast furnace slag, wide variations in fineness are also produced. Thus, the reference heat-generation rate shown in Figure 2.1 corresponds to a certain fineness related to experiments in parameter identification. Powder fineness may possibly affect the probability of contact between the particles and the surrounding water, with the effect being proportional to the total surface area of reacting binders. Another, more important, effect is that the boundary between Stage 2 and Stage 3, namely the boundary between the reaction control stage and the diffusion control stage, may change as shown in Figure 2.2. Goto and Uomoto (1993) clearly pointed out this fact in both their experiments and analyses. The borderline between the reaction control stage and the ion diffusion control stage is dependent on the thickness of the hydration products surrounding the remaining core of binders. Therefore, the finer particles produce a higher boundary value, and the wider reaction-controlling Stage 2 at the same

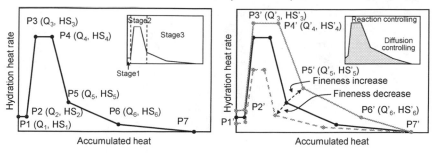

*Figure 2.2* Modification of reference heat rate according to fineness of particles.

thickness of hydration products can be regarded as the ignition of diffusion control Stage 3, and vice versa. This powder fineness effect is systematically considered in this model and expressed as follows.

$$Q'_3 = (x - 4.6) \times 10/24 + 4.6, \ Q'_4 = (x - 4.6) \times 15/24 + 4.6, \ Q'_5 = x, \ Q'_6 = (460 - x) \times 1/3 + x$$

$$x = 460 \times \{1 - \frac{(r - h)^3}{r^3}\}$$

$$HS'_i = HS_i \times (1.2 \times \frac{s_i}{s_{i0}} - 0.2)$$

$$(2.3)$$

where $Q'_i$ is the modified accumulated heat that is a coordinate on the x-axis of point $i$, $r$ is the mean diameter of the cement particles, $h$ is the thickness of the reacted layer around the remaining core in the original particle, $HS'_i$ is modified hydration heat rate that is a coordinate on the y-axis of point $i$, $s_i$ is the Blaine value of component $i$, and $s_{i0}$ is the reference Blaine value. The modifications of the hydration heat rate are shown in Figure 2.2. The reference Blaine values for Portland cement, blast furnace slag, fly ash, silica fume and limestone powder are specified values for the powders used in the experiments and were set by referring to reports by Suzuki *et al.* (1990a, 1990b) and others. They are set to 3,380, 4,330, 3,280, 200,000 and 7,000 cm²/g, respectively, and were derived from the material specifications for the tests.

### 2.1.4 Temperature dependence of mineral reaction

The relation between the overall thermal activity and the accumulated heat for cement is shown in Figure 2.3a (Suzuki *et al.* 1990a, 1990b). Highly reactive minerals mainly undergo rapid reactions in the initial stage, while less reactive minerals undergo reactions later in the process. Thus, the heat-generation characteristics of cement should reflect the characteristics of the minerals that play a major role in each reaction stage. As the temperature dependence of each clinker mineral or admixture is thought to be different, the overall thermal activity of cement will change according to the accumulated heat. This chapter describes the apparently complex thermal activity of

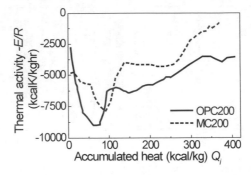

*Figure 2.3a*  Thermal activity of cement from adiabatic temperature rise test.

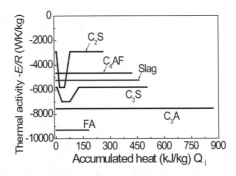

*Figure 2.3b*  Thermal activity for the reaction of each compound.

cement (Suzuki *et al.* 1990a, 1990b) as an integrated result of the intrinsic thermal activity of each mineral expressed by rather simple formulae.

The thermal activity of each mineral included in this model is shown in Figure 2.3b. Consequently, the authors used Suzuki's values as references for the thermal activity of the cement as a whole and determined the thermal activity of each mineral reaction. The overall thermal activity of cement tends to fall from around -6,500 to -2,500 K when the accumulated heat exceeds 105 kJ/kg. This point is thought to be the diffusion control stage (Section 2.1.3). The variation in thermal activity with increasing hydration reflects the major minerals taking part in the reactions at each stage. Taking $C_3A >$ $C_3S > C_4AF > C_2S$ to be the order of mineral activity and assuming a constant thermal activity for all minerals, the overall activity from -6,500 to -2,500 K was mapped. For the sake of simplicity, the thermal activity of each clinker mineral is considered constant throughout the hydration process by assuming that each unit reaction must have its own thermal activity.

The thermal activity of blast furnace slag, fly ash and silica fume were set to 4,500, 12,000 and 12,000 K, respectively, by comparing the results of experiments and analysis of adiabatic temperature rises. The heat generated by fly ash is difficult to ignore as it undergoes an adiabatic temperature rise even if its reaction is not active at a normal temperature. Thus, the fly ash reaction may be greatly accelerated at high curing temperatures. That is why a relatively large thermal activity is assumed. The same is true of silica fume.

### 2.1.5 Hydration and interactions among cement clinker minerals

The mineral reactions affect each other. This interaction can be considered through the common state variables, i.e., the temperature and the free water remaining in the concrete. Czernin (1969) concludes that the amount of water required for the cement to fully hydrate is about 40% of the cement's weight and that some cement remains unhydrated if the water-to-cement ratio is less than 40%. In the hydration reaction model proposed by Tomosawa (1974), the residual concentration of water is used as a parameter to indicate the retarded rate of reaction. This implies that the reaction rate decreases when the amount of remaining water is insufficient, especially in a mix with a low water-to-cement ratio. In other words, the remaining water may be where the additional hydrates precipitate. This concept extends to the pore structure formed under an elevated temperature that is described in Chapter 3. Another consideration is the chemo-physical process interaction between the alite and belite in the cement. This comprises two factors: the physical micro-filler effect of the belite-to-alite reaction and the effect of the calcium ion concentration in solution. The micro filler effect on the hydration rate was originally modeled and implemented for concrete containing limestone.

The ultimate adiabatic temperature rise is reported to be somewhat affected by the initial temperature at casting, with lower casting temperatures producing higher temperature rises (Suzuki *et al.* 1989b). One possible reason for the initial temperature dependence is the microstructure of the hydrate, which exhibits different diffusivity against mass transport, as reported by Moriwake *et al.* (1993). Another possibility is that when the initial temperature is higher, most of the heat generated at the first peak of the exothermic process is lost before it can be measured. Regardless, it is also necessary to accurately model the rapid generation of heat just after mixing. The main source of the rapidly generated heat just after mixing is ettringite formation, which corresponds to the first peak in the exothermic process of cement hydration. In Portland cement containing gypsum, the so-called interstitial materials $C_3A$ and $C_4AF$ react energetically with gypsum to produce ettringite ($C_3A \cdot 3C\bar{S} \cdot H_{32}$). This reaction is accompanied by substantial and rapid heat generation. Thus, the formation of ettringite must be treated separately in the proposed scheme.

#### 2.1.5.1 Ettringite formation and monosulfate conversion

In the early stage of the reaction, ettringite forms very rapidly, but the reaction slows as the remaining $C_3A$ and $C_4AF$ become covered with the ettringite reaction product. Here, it is assumed that $C_3A$ and $C_4AF$ do not hydrate if there is unreacted gypsum in the liquid phase, so ettringite formation continues. On the other hand, ettringite formation halts when there is no more $SO_4^{2-}$ in the liquid phase. This is because the gypsum has been consumed. This disappearance of $SO_4^{2-}$ from the liquid phase undermines the stability of the ettringite and the ettringite layers covering the unreacted parts crumble. The ettringite reacts with unreacted $C_3A$ or $C_4AF$ and easily converts into monosulfate ($C_3A \cdot C\bar{S} \cdot H_{12}$), as reported by Arai (1984). The ettringite will continue to convert into monosulfate until all of it has been converted, at which point hydrates from the unreacted $C_3A$ and $C_4AF$ start to precipitate. Consequently, the heat generated by ettringite formation and the timing of both the ettringite–monosulfate conversion and the hydration of unreacted interstitial materials will depend on the amount of $C_3A$ and $C_4AF$ present and the amount of gypsum added.

The reference heat-generation rates for ettringite-forming reactions involving $C_3A$, $C_4AF$ and gypsum are shown in Figure 2.4, where $Q_{C3AET}$ is the ettringite-forming heat rate for $C_3A$, and $Q_{C4AFET}$ is the ettringite-forming heat rate for $C_4AF$. The thermal activity values for these reactions were assumed to be the same as those for the hydration reactions of $C_3A$ and $C_4AF$ (Figure 2.3b). The modeled ettringite-forming reaction is initiated concurrently with the start of the calculation, and the model expresses the first exothermic peak in the exothermic cement-hydration process as a whole, which occurs when the water is added. The end of ettringite formation is determined by calculating the amount of gypsum remaining. Gypsum consumption can be calculated from the degree of reaction ($Q_{iET}/Q_{iET,\infty}$) and the rate of combination of $C_3A$ and $C_4AF$, and the amount of unreacted gypsum can be determined by deducting this gypsum consumption from the total gypsum content of the cement. The bonding ratios of $C_3A$, $C_4AF$ and gypsum dehydrate during ettringite formation are calculated in this model using the following equation to describe the ettringite formation reaction according to Arai (1984).

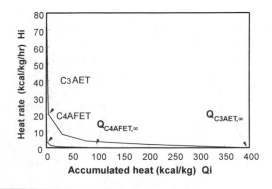

*Figure 2.4* Reference heat rate set for ettringite formation reaction.

$$C_3A + 3C\bar{S}H_2 + 26H \rightarrow C_3A \cdot 3C\bar{S} \cdot H_{32}$$
$$C_4AF + 3C\bar{S}H_2 + 27H \rightarrow C_3(AF) \cdot 3C\bar{S} \cdot H_{32} + CH$$

where $C \equiv CaO$, $A \equiv Al_2O_3$, $F \equiv Fe_2O_3$, $H \equiv H_2O$, $CH \equiv Ca(OH)_2$ and $\bar{S} \equiv SO_3$.

$$(2.4)$$

The casting time from mixing to the beginning of measurement is assumed to be 0.015 days (around twenty minutes) and the heat generated during this period is excluded from the temperature rise analysis. This is because some of the heat generated after mixing is already included in the initial temperature, which is the starting point of the computation. Therefore, in all analyses in this chapter, the heat generated between the start of computation and 0.015 days is not counted in the temperature rise. This means that the temperature rise starts 0.015 days after hydration begins. This model of the ettringite formation reaction continues to operate until all of the gypsum is consumed. Hydration heat starts to be generated in the interstitial phases, expressed by $H_{C3A}$, $H_{C4AF}$, immediately after ettringite formation terminates because no more unreacted gypsum is present.

The proposed model provides for conversion to monosulfate after the formation of ettringite and the hydration of unreacted $C_3A$ and $C_4AF$ using the heat-generation curves for hydration shown in Figure 2.1. In other words, after the collapse of the ettringite layers due to the absence of $SO_4^{2-}$ in the liquid, continuous elution from the unreacted $C_3A$ and $C_4AF$ takes place, and the eluted elements are then used in reactions that convert ettringite into monosulfate for as long as ettringite remains. Upon completion of these reactions, the pattern of reactions taking place among the eluted elements changes to hydration. The timing of the change from monosulfate reactions to hydration reactions is consequently determined by either the amount of $SO_4^{2-}$ in the ettringite or the amount of gypsum that was added. For the sake of convenience, in modeling the hydration heat generated by $C_3A$ and $C_4AF$, as with the ettringite formation model, gypsum consumption and unconverted gypsum are calculated, and conversion to monosulfate is assumed to stop when no unconverted gypsum remains. The following equation represents the conversion from ettringite to monosulfate and calculates the unreacted amounts of $C_3A$ and $C_4AF$.

$$2C_3A + C_3A \cdot 3C\bar{S} \cdot H_{32} + 4H \rightarrow 3\left[C_3A \cdot C\bar{S} \cdot H_{12}\right]$$
$$2C_4AF + C_3(AF) \cdot 3C\bar{S} \cdot H_{32} + 6H \rightarrow 3\left[C_3(AF) \cdot C\bar{S} \cdot H_{12}\right] + 2CH$$

$$(2.5)$$

The reaction heat of these compounds can be calculated from the formation enthalpy of each, providing that an equation for the identified chemical is available. The heat generated in the conversion to monosulfate is determined by Equation 2.5, which shows that 3 mols of monosulfate are formed from 2 mols of unreacted $C_3A$. The standard formation enthalpy for each of these compounds, given by Osbaeck (1992), shows a figure of about 882 joules of heat generated per gram of unreacted $C_3A$, an amount almost equal to the hydration heat of $C_3A$. In the proposed model, 1 mol of $C_3A$ already converted to ettringite is treated as unreacted $C_3A$ when calculating the hydration heat, and the heat generated in the conversion to monosulfate is expressed by multiplying the reference heat rate associated with hydration by two-thirds.

### 2.1.5.2 *Free water as a space for the precipitation of gels*

As the minerals react with the mixing water, the reaction and hydration processes share the free water. Free water is required not only for continued hydration but also to serve as a space in which hydrates can precipitate. As stated previously, the amount of water necessary for the complete hydration of ordinary Portland cement is about 40% of the total cement weight, as reported by Arai (1984). The supply of free water may run short toward the end of the reaction in high-performance concrete when the water-to-cement ratio is around 30% or less.

Given the need for a model that is generally applicable, it is necessary to express the decline in the hydration rate when the supply of free water is inadequate. The reference heat rate set in the proposed model is based on the assumption that hydration proceeds when the supply of free water is ample. However, the actual hydration rate is lower than the reference heat rate when the water-to-cement ratio is low, since it is impossible to assume that sufficient free water exists around the powder particles. In a system with a low water-to-cement ratio, it is highly likely that the reaction will stagnate due to a lack of free water, leaving some material unhydrated once the reaction terminates. It has been demonstrated that mix proportions with lower water-to-binder ratios exhibit a greater drop in the adiabatic temperature increment per unit powder weight, which is probably due to the lack of free water. In the proposed model, the reduction in the heat rate is assumed due to a shortage of free water resulting from the reduced probability of contact between the reacting surface of the particles and the free water. The proposed hydration heat model uses the following equation to express the decline in the heat-generation rate

$$\beta_i = 1 - \exp\left\{-r\cdot\left\{\left(\frac{\omega_{free}}{100\cdot\eta_i}\right)\middle/ s_i^{\frac{1}{2}}\right\}^s\right\} \tag{2.6}$$

where $r$ and $s$ are material constants common to all minerals. A comparison of the results of the experiments and analyses shows that $r = 5.0$ and $s = 2.4$. The coefficient $\beta_i$ represents the reduction in the heat rate and is simply formulated in terms of both the amount of free water and the thickness of internal hydrate layer. The function adopted varies from 0 to 1 and shows a very small reduction when sufficient free water exists and the hydrate layer is thin. On the other hand, it sharply reduces the heat rate when the amount of free water is reduced and thick clusters of hydrates cover the unhydrated particles. The coefficient $\beta_i$ is also a function of the normalized Blaine value, $s_i$, which represents the change in the reference heat-generation rate due to the fineness of the powder. This is because the surface area of the unit weight particles that can come in contact with free water also varies according to the fineness of the powder and the changes in the reference heat rate. $\omega_{free}$ is the free water ratio, $\eta_i$ is the thickness of the internal reaction layer of component $i$. They are defined by the following equations

$$\omega_{free} = \frac{W_{total} - \sum W_i}{C} \tag{2.7}$$

$$\eta_i = 1 - \left(1 - \frac{Q_i}{Q_{i,\infty}}\right)^{\frac{1}{3}} \tag{2.8}$$

where $W_{total}$ is the unit water content, $W_i$ is the water consumed and fixed by the reaction of constituents, $C$ is the unit cement content, $Q_i$ is the accumulated heat of component $i$, and $Q_{i,\infty}$ is the final heat generated (see Figure 2.1). The water used for hydration and the elution of ions is dispersed throughout the hydration formation layer, so the diffusion resistance increases as the formed layer grows. The resistance to diffusion can be represented by the thickness of the internal formation layer $\eta_i$, which has a micro-texture structure.

The amount of water bound by each reaction can be determined by multiplying the bound water ratio obtained from the reaction equation of each mineral by the degree of reaction calculated by the model ($Q_i / Q_{i,\infty}$).

$$\begin{aligned}
C_3A + 6H &\rightarrow C_3AH_6 \\
C_4AF + 2CH + 10H &\rightarrow C_3AH_6 - C_3FH_6 \\
2C_3S + 6H &\rightarrow C_3S_2H_3 + 3CH \\
2C_2S + 4H &\rightarrow C_3S_2H_3 + CH
\end{aligned} \tag{2.9}$$

where $S \equiv S_iO_2$. It must be noted at this point that chemically bound water is not the only water consumed in the various reactions. Water is also physically trapped by the formed texture. Although this constrained water is generally assumed to be dependent on the area and condition of the hydrate surface, it is typically about 15% of the total cement weight (Czernin 1969). In this model, the constrained water ratio was assumed for all minerals undergoing reaction. Consequently, water consumption $W_i$ in Equation 2.7 is the calculated bound water volume plus 15%.

The reactions of the clinker minerals are assumed to terminate when hydration reaches 100% or when the hydration rate drops due to a shortage of free water when only Portland cement is used in the model. In the case of blended cement, however, the reaction rates of blast furnace slag and fly ash drop because of an insufficient supply of $Ca(OH)_2$. Nevertheless, the amount of free water is sufficient. If the shortage of $Ca(OH)_2$ reduces the admixture reaction rates, the free water remaining in the system can act as the reacting object or a space in which hydrates can precipitate. Thus, the free water fulfills its function in the hydration of Portland cement. When the reaction of the admixtures stagnates due to a shortage of $Ca(OH)_2$, the effect on Portland cement of a reduction in the amount of free water is expressed by modifying Equation 2.7 to arrive at the following equations

$$\omega_{free} = \frac{W_{total} - \sum W_i}{C \cdot \left(p_{PC} + m_{SG} \cdot p_{SG} + m_{FA} \cdot p_{FA}\right)} \qquad (2.10)$$

$$m_i = \lambda / \beta_i \qquad (2.11)$$

The water consumed during ettringite formation is not a concern because the reaction ends in the initial period and the water that is bound in the ettringite changes to monosulfate-bound water during the conversion. The mixing water is also assumed to be consumed only by hydration and physical constraint as a result of hydration. Therefore, the simple hydration heat model assumes a virtually sealed condition. When the cement dries, the free water should be solved by combining a hydration heat model with a model of water transport in which the amount of free water can be given as a state variable for the hydration heat model.

### 2.1.5.3 Micro-filler effect by inert powders: limestone powder and stagnated binders

The added inert powder affects the hydration process and the strength of the cement. Yamazaki (1962) showed through a well-arranged series of experiments that mineral fines having no pozzolanic reactivity strengthen concrete by increasing the degree of hydration of the cement in the mixture. Inert powder materials can provide additional precipitation sites for the hydrates, which is the adjacent part of the surface area in contact with the cement particle, by entering the flocculated cement particles. Recent attention has focused on the use of limestone powder, with related research conducted on the hydration heat process for cement. Cement hydration is reported to accelerate according to the amount and fineness of the limestone powder, especially at the second peak of the hydration process (alite reaction). As the quantitative effects of the inert powder on the rate-controlling factors were not apparent, the hydration heat processes accelerated by limestone powder were compared to the accumulated heat, as shown in Figure 2.5a (Uematsu and Kishi 1997). The legends in Figure 2.5a indicate the weight fractions of moderate-heat Portland cement and limestone powder in the cement paste mixture. (The Blaine value of the limestone powder is shown in parentheses.)

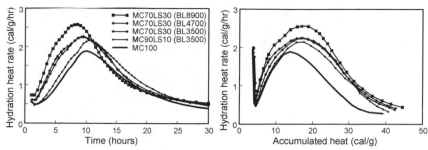

*Figure 2.5a* Accelerated hydration heat of cement in terms of time and accumulated heat.

This allows discussion of the relation between the acceleration effect and the division of the hydration heat process with reference to the same degree of hydration. The acceleration is remarkable in the region of and just after the second peak, where the rate of hydration is mainly governed by the diffusion mechanism of the hydrates that have formed around the unreacted cement particles. Furthermore, the diffusion resistance at the hydrate layer should be mitigated since the surface of the limestone powder shares some part and consequently decreases the cement particle loads.

An appropriate indicator can express the effect of an inert powder on cement hydration. Before such an indicator can be adopted, however, it is necessary to know how the surfaces of inert powder particles contribute to the precipitation of cement hydrates. Using a scanning electron microscope (SEM), Goto and Uomoto (1993) directly observed that hydrates are similarly formed over the entire surface of particles of both alite and limestone powder in the mixture. This may mean that the outer layer of hydrates can precipitate from the eluted ion phase at any location, even locations away from the cement particles, and that all surface areas of the particles can serve as precipitation sites from the beginning of hydration. Furthermore, in the hydration heat of alite mixed with limestone powder, they found that a linear relationship exists between the heat rate at the second peak and the total surface area of the powder particles. These results show that the total surface area of the powder particles can be an indicator for the expression of the acceleration effect on the hydration heat rate.

Figure 2.5a shows that the hydration heat rates around the second peaks, which are represented as Stage 2 in Figure 2.1, are fairly similar in terms of the accumulated heat, although they again stagnate simultaneously in the time domain. This indicates that, at same degree of hydration, the physical effect of the limestone powder affects not only the hydration heat rate but also the regions of their respective stages, which are the reaction and diffusion-controlling stages. In the exothermic process around the second peak, heat generation, which must accompany the precipitation of hydrates, is thought to be attributable to two sources of eluted ions. One is the group of ions that were eluted during the dormant period and stored at over-saturation.

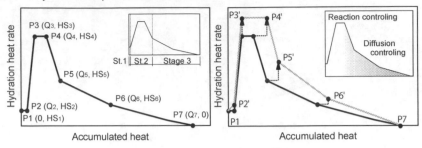

*Figure 2.5b* Modification of the reference heat rate according to the micro-filler effect of the limestone powder.

The other is the new elution of ions accompanied by simultaneous precipitation. Hence, if the diffusion resistance at the layer of hydrates is mitigated, more ions can elute during Stage 2, which expands the region of the stage. These regions, expanded by the limestone powder, should be expressed by assuming that the accelerated hydration heat rate greatly affects the hydrate precipitation rate.

Figure 2.5b (Kishi and Dorjpalamyn 1999) shows a schematic representation of the model of the physical effect of limestone powder on cement hydration. The reference heat rate is graphically expressed as a simple multi-linear function to express the accelerated heat process.

For P2 and P3: $Q'_{ij} = Q_{ij}, \quad HS'_{ij} = HS_{ij} \cdot (1 + k_{H1}r_s)$

For P4 and P5: $Q'_{ij} = Q_{i\max} - ((Q_{i\max} - Q_{ij})/(1 + k_Q r_s)), \quad HS'_{ij} = HS_{ij} \cdot (1 + k_{H2}r_s)$

For P6: $Q'_{ij} = Q_{i\max} - ((Q_{i\max} - Q_{ij})/(1 + k_Q r_s)), \quad HS'_{ij} = HS_{ij} \cdot (1 + k_{H3}r_s)$

$$r_s = (p_{LS} \cdot B_{LS})/(p_{PC} \cdot B_{PC})$$

(2.12)

where $j$ is point number in the reference heat rate function, $Q_{ij}$ and $HS_{ij}$ are the accumulated heat and the heat rate of point $j$ of component $i$, $Q_{i\max}$ is the maximum heat generation of component $i$, and $p_{LS}$, $p_{PC}$ and $B_{LS}$, $B_{PC}$ are the weight fractions and unit weight surface areas of the limestone powder and Portland cement. The parameter denoted by $r_s$ is the ratio of the surface areas of the limestone powder and cement, which is adopted as the indicator to express the acceleration effect. To represent the degree of contribution by limestone powder, coefficients $k_{H1}$, $k_{H2}$, $k_{H3}$, $k_Q$ are multiplied by $r_s$. Through analyses described later, these coefficients are set to 0.3, 0.3, 0.1, 0.05, respectively. Tentatively, this model is applied equally to all four components of the clinker minerals.

## 2.1.5.4 *Interrelation of alite and belite reactions*

The adiabatic temperature rise of ordinary Portland cement or moderate-heat Portland cement indicates that the rapid rise in temperature continues even when the hydration process proceeds to Stage 3 because of the temperature dependence of the reaction. On the other hand, the adiabatic temperature rise of low-heat Portland cement, which is increasingly being used in massive concrete structures, exhibits a slow rise that corresponds to the initial period of Stage 2 (Hanehara and Tobiuchi 1991). This difference is more apparent in cements containing more $C_2S$ and less $C_3S$ than in other types of cement.

The proposed hydration heat model was tested by performing a thermal analysis of low-heat Portland cement using various reference values on the assumption that the reference heat generation of each mineral remains constant regardless of the mineral composition. The test revealed that the reference heat rates poorly predict the adiabatic temperature rise of low-heat Portland cement, unlike ordinary cements. The authors concluded that the heat-generation rate of each mineral changes with its proportion, a finding based upon recent experiments in the field of cement chemistry. The degrees of hydration of $C_3S$ and $C_2S$ in ordinary Portland cement (OPC) and low-heat Portland cement (LPC) were precisely identified, as shown in Figure 2.6. This figure reveals two remarkable features about $C_3S$ and $C_2S$ hydrations. First, the $C_2S$ reaction is faster in OPC than in LPC. A similar tendency can be seen for $C_3S$ at the beginning and final stages, but not the middle stage, of its hydration process. Second, the $C_3S$ reaction in LPC uniquely accelerates in the middle stage of its hydration process, which is separate from the tendency described above for the first feature. These features and their mechanisms can be understood as follows.

The reactivity of calcium silicates such as $C_3S$ and $C_2S$ are dependent on the concentration of calcium ion in solution, thus their reactions are faster in OPC than in LPC. Furthermore, stagnated $C_2S$ particles, due to a shortage of Ca ions in the LPC, seem to produce a micro-filler effect in the $C_3S$ reaction. Figure 2.7 shows an SEM image of hydration products made from a $C_3S$ reaction and precipitated on the surface of an unreacted $C_2S$ particle six

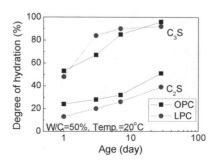

*Figure 2.6* Degree of hydration of $C_3S$ and $C_2S$.

*Figure 2.7* Hydration products generated from C₃S on the surface of a C₂S particle.

hours after mixing. This image indicates that a stagnated C₂S particle can act as a precipitation site for hydration products from C₃S, which is known as the micro-filler effect. Thus, these mechanisms should be systematically incorporated in the proposed model. According to the mineral composition of Portland cement, the heat generation rates for C₃S and C₂S are modified from the original settings by the modification factors shown in Figure 2.8. The ratio between C₃S and C₂S is adopted as the indicator to express the mineral composition of Portland cement. This treatment roughly reflects the calcium ion concentration in the solution.

To express the micro-filler effect on the C₃S reaction by stagnated C₂S particles, the heat rate for C₃S is modified to raise the value and widen the range in Stage 2 if mineral composition becomes relatively rich in C₂S, as shown in Figure 2.9. The ratio of C₂S/(C₃S+C₂S) expresses the intensity of the micro-filler effect, and the heat rate of C₃S is systematically modified according to the mineral composition of the Portland cement.

To verify the overall scheme, five types of Portland cement were tested: OPC, early-hardening cement (HPC), moderate-heat cement (MPC), LPC, and belite-rich cement (BRC). The data for the adiabatic temperature rise was then compared with that of analyses from both the old (original) model and the proposed modified model. The mineral compositions of the cements are

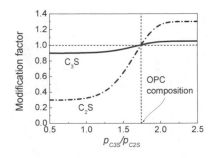

*Figure 2.8* Modification factors for the heat rate of C₃S and C₂S due to the minerals.

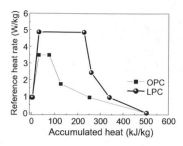

*Figure 2.9* Heat rate of $C_3S$ in OPC and LPC, modified according to mineral composition.

*Table 2.1* Mineral composition of cements

| Type | Composition (mass %) | | | | | Blaine (cm²/g) |
| | $C_3S$ | $C_2S$ | $C_3A$ | $C_4AF$ | $C\bar{S}2H$ | |
|---|---|---|---|---|---|---|
| OPC | 47.2 | 27.0 | 10.4 | 9.4 | 3.90 | 3,380 |
| HPC | 64.2 | 11.8 | 9.2 | 8.9 | 6.45 | 4,210 |
| MPC | 44.4 | 33.7 | 3.7 | 12.5 | 3.90 | 3,040 |
| LPC | 25.0 | 55.0 | 3.0 | 9.0 | 5.59 | 3,390 |
| BRC | 5.2 | 74.7 | 1.7 | 12.2 | 4.09 | 2,970 |

shown in Table 2.1. The mix proportions are shown in Table 2.2. The old model did not explicitly apply the two principal interdependencies for $C_3S$ and $C_2S$ but macroscopically adjusted the temperature history. The effect of the chemical agent is included in these simulations (Section 2.1.6).

The comparisons are shown in Figures 2.10–2.13. The comparisons demonstrate that the proposed model has wider applicability for various types of cements and faithfully reproduces the adiabatic temperature rises.

## 2.1.6 *Hydration and interactions between Portland cement and admixtures*

When analyzing blended cement, the mutual interactions among the Portland cement, blast furnace slag, fly ash and/or silica fume should be taken into account. Powder admixtures such as blast furnace slag, fly ash and silica fume are unable to continue their reactions without an activator, $Ca(OH)_2$, which is produced by $C_3S$ and $C_2S$ hydration in cement. In other words, the reactions of the powder admixtures depend on the hydration of the cement and correlate with $Ca(OH)_2$. Note that the reaction process for multiple blended cement cannot be faithfully simulated unless the blast furnace slag, fly ash and silica fume are appropriately and individually modeled. That is because of the competition for $Ca(OH)_2$ by the blast furnace slag, fly ash and silica

Table 2.2 Mix proportions of concretes

| Type | W/C (%) | Unit weight (kg/m³) | | | | Ad. (%) | Type | W/C (%) | Unit weight (kg/m³) | | | | Ad. (%) |
| --- | --- | --- | --- | --- | --- | --- | --- | --- | --- | --- | --- | --- | --- |
| | | W | C | S | G | | | | W | C | S | G | |
| OPC400 | 39.2 | 157 | 400 | 658 | 1,058 | 0.25* | MPC400 | 39.2 | 157 | 400 | 663 | 1,129 | 0.25* |
| OPC300 | 49.3 | 148 | 300 | 765 | 1,092 | 0.25* | MPC300 | 49.3 | 148 | 300 | 770 | 1,129 | 0.25* |
| OPC200 | 78.5 | 157 | 200 | 862 | 1,071 | 0.25* | MPC200 | 78.5 | 157 | 200 | 865 | 1,089 | 0.25* |
| HPC400 | 38.0 | 152 | 400 | 730 | 1,129 | 0.25 | LPC340 | 55.8 | 145 | 260 | 844 | 1,082 | 0.25 |
| HPC300 | 47.7 | 143 | 300 | 813 | 1,129 | 0.25 | LPC300 | 48.3 | 145 | 300 | 811 | 1,082 | 0.25 |
| HPC250 | 57.2 | 143 | 250 | 869 | 1,089 | 0.25 | LPC260 | 42.6 | 145 | 340 | 779 | 1,082 | 0.25 |
| | | | | | | | BRC | 55.1 | 168 | 305 | 750 | 1,089 | 0.25 |

* Delay-type additive

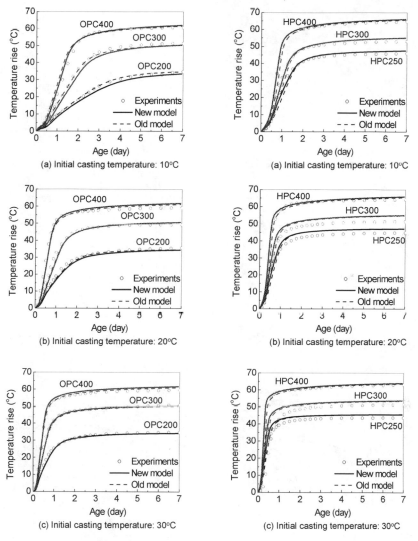

*Figure 2.10* Adiabatic temperature rise of OPC from analyses by new and old models.

*Figure 2.11* Adiabatic temperature rise of HPC from analyses by new and old models.

fume, which strongly governs the related reactions in multiple blended cements. Therefore, not only the interdependence related to $Ca(OH)_2$ but also the individual modeling of the powder admixture should be evaluated to ensure wide applicability. Especially, for concrete containing fly ash, its physical filler effect for other actively reacting minerals was taken into account to enhance rationality and accuracy. Then a balance of reactions between the blast furnace slag and the fly ash can be successfully rationalized in terms of the consumption of $Ca(OH)_2$. For high-strength concrete, a component for

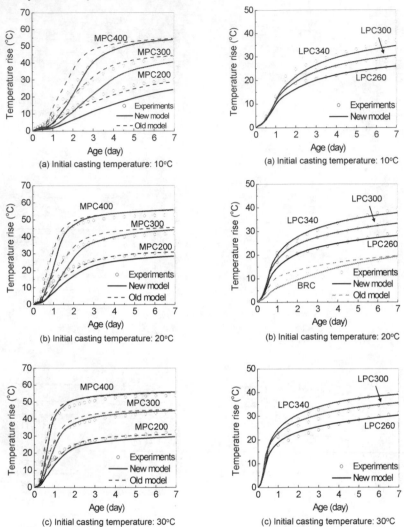

*Figure 2.12* Adiabatic temperature rise of MPC from analyses by new and old models.

*Figure 2.13* Adiabatic temperature rise of LPC from analyses by new and old models.

silica fume is also implemented. The delaying effects of the chemical admixture and the fly ash on other the binders should also be included.

### 2.1.6.1 *Calcium hydroxide produced by Portland cement hydration for admixtures reactions*

In blended cements containing more than two powder admixtures, such as blast furnace slag, fly ash and silica fume, the admixtures simultaneously consume the $Ca(OH)_2$. The analysis should calculate the consumption of $Ca(OH)_2$ step-by-step as it is produced by Portland cement hydration. The reduced reaction rate of the admixtures can be expressed by calculating the quantitative ratio of the amount of $Ca(OH)_2$ remaining in the system and the amount necessary for active reaction of slag and fly ash at that time.

$$\lambda = 1 - \exp\left\{-2.0 \cdot \left(\frac{F_{CH}}{R_{SGCH} + R_{FACH} + R_{SFCH}}\right)^{1.5}\right\} \tag{2.13}$$

where $F_{CH}$ is the amount of $Ca(OH)_2$ that is produced by the hydration of $C_3S$ and $C_2S$ and not yet consumed by $C_4AF$ reaction, and $R_{SGCH}$ and $R_{FACH}$ are the amounts of $Ca(OH)_2$ necessary for the slag and fly ash reactions when a sufficient amount of $Ca(OH)_2$ is available. $F_{CH}$ can be calculated from the reaction degree of each component and the ratios of production and consumption of $Ca(OH)_2$ that are given by the chemical equations adopted in this chapter. $R_{SGCH}$ and $R_{FACH}$ are calculated from the ratio of consumption of $Ca(OH)_2$ by the blast furnace slag and fly ash and their reaction rates when a sufficient amount of $Ca(OH)_2$ is available.

It is not clear how much $Ca(OH)_2$ is consumed by the reacted slag, fly ash and silica fume. In fact, the Ca-to-Si ratio of products also changes according to the amount of $Ca(OH)_2$ that is available and the character of the admixtures (Uchikawa 1986). It is simply assumed that the consumption ratio for $Ca(OH)_2$ is constant throughout the process and is irrelevant to the character of the admixture. It is generally assumed that sufficient $Ca(OH)_2$ can be supplied to the slag reactions up to 60–70% replacement because the adiabatic temperature rise barely changes within this replacement range. On the other hand, the reaction of the fly ash does not proceed without active $OH^-$ ions from the outside, since a high number of silicate ions condense in the glass (Uchikawa 1986). Therefore, $Ca(OH)_2$ consumption per unit weight of fly ash reaction seems to be much higher than in the case of slag and must be even more in the case of silica fume. Thus, the consumption ratios of slag, fly ash and silica fume reactions are assumed to be 22%, 100% and 200% of the reacted mass in the analysis, respectively, as described in the discussions that follow. In ternary blended cement, it is assumed that the $Ca(OH)_2$ in the liquid phase is almost completely consumed by the slag reaction, since the slag reaction is faster than the fly ash reaction.

### 2.1.6.2 Reduction of calcium ion concentration in the solution due to the presence of pozzolans

The concentration of calcium ions in the solution, which is a controlling factor for the hydration rates of the clinker minerals, is restrained when pozzolans are present, and react in the blend mixture as discussed in the previous section. Although this has no substantial effect on the hydration process of ordinary Portland cement, it is especially significant when the base cement is low-heat Portland cement because of the slow dissolution of calcium ions from the clinker minerals. Thus, the concentration of calcium ions is low from the beginning, even in plain concrete. So, changes in the hydration heat rates of alite and belite due to the effect of reduced calcium ion concentrations caused by the presence of pozzolans can be expressed as follows.

Alite $$r_{res,C3S} = 1.1 - (1.1 - 3\delta)\exp\{-0.025(\frac{p_{C3S}}{p_{C2S}})^{7.0}\} \qquad (2.14)$$

Belite $$r_{res,C2S} = 1.3 - (1.3 - \delta)\exp\{-0.025(\frac{p_{C3S}}{p_{C2S}})^{7.0}\} \qquad (2.15)$$

$$\delta = \frac{1}{3 + 0.03p_{SG} + 0.05p_{FA} + 0.1p_{SF}} \qquad (2.16)$$

The reduction factors for C₃S and C₂S are shown in Figure 2.14 and Figure 2.15, respectively. Through the simulation, the effect of fly ash on this

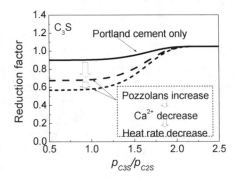

*Figure 2.14* Modification of reduction factor for alite.

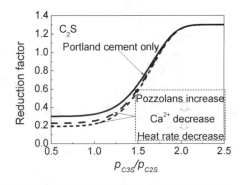

*Figure 2.15* Modification of reduction factor for belite.

phenomenon is set to twice that for blast furnace slag. This is because fly ash consumes a large amount of $Ca(OH)_2$ but has a relatively low reaction rate, while blast furnace slag reacts faster but has a lower $Ca(OH)_2$ consumption ratio. Furthermore, the reduction factor applied to alite is set to one-third that of belite since alite supposedly has a higher potential for self-activated hydration.

### 2.1.6.3 Exothermic process of blast furnace slag

A very important feature of concrete containing slag is that it forms a stiff solid and generates a certain amount of heat even at very high replacement ratios, such as 95%. This is because gypsum can stimulate the blast furnace slag reaction, thus guaranteeing a minimum slag reaction. It was verified that about 10% of the blast furnace slag reaction is due to the presence of gypsum, even when no $Ca(OH)_2$ is present in the solution (see Table 2.3 and Figure 2.16). Thus, in the proposed model, up to 10% of the blast furnace slag reaction occurs freely.

Another important phenomenon is that when higher levels of blast furnace slag are used, a shortage of $Ca(OH)_2$ causes the blast furnace slag reaction to stagnate after just 10% of the reaction has occurred. Because they can serve as nucleation sites for the hydrates, the stagnated reacted slag particles

*Table 2.3* Mix proportions of blast furnace slag with reagents

| Case | W/(SG + CH + Gy) | W | SG | CH | Gy |
|------|------------------|-----|------|-----|-----|
| SG + Gy02 | 0.4 | 40 | 98 | --- | 2 |
| SG + Gy10 | 0.4 | 40 | 90 | --- | 10 |
| SG + Gy20 | 0.4 | 40 | 80 | --- | 20 |
| SG + CH | 0.4 | 40 | 80 | 20 | --- |
| SG + CH + Gy | 0.4 | 40 | 78.4 | 20 | 1.6 |

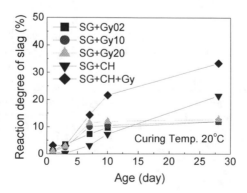

*Figure 2.16* Degree of reaction of slag with reagents.

can have a micro-filler effect on the other reacting clinker minerals. But the number of available nucleation sites always varies with the slag's reaction process, which in turn depends upon the amount of $Ca(OH)_2$ in the solution. To estimate the micro-filler effect, the ratio of the total surface area of the slag to that of the cement is set to the same ratio used for limestone powder, with the efficiency dependent on the degree of stagnation determined by the shortage of $Ca(OH)_2$.

### 2.1.6.4 Exothermic process of fly ash

Fly ash has a much slower reaction rate than blast furnace slag. During model development, it was recognized that the competitive consumption of $Ca(OH)_2$ by blast furnace slag and fly ash was very important when simulating the adiabatic temperature rise of the ternary blended mixture. This is because the simulated slag reaction would improperly stagnate if the consumption of $Ca(OH)_2$ by fly ash, which does not generate higher heat but consumes more $Ca(OH)_2$ than slag does, was overestimated. Thus, the reaction processes for both slag and fly ash must be properly modeled.

In order to trace the very slow fly ash reaction process, the varying $Ca(OH)_2$ in low-heat Portland cement paste containing pozzolans with an adiabatic temperature rise under sealed conditions was analyzed using thermogravimetry-differential thermal analysis (TG-DTA). The mix proportions of the sample paste are shown in Table 2.4. Figure 2.17 shows the $Ca(OH)_2$ time history. The amount of $Ca(OH)_2$ was normalized by the low-heat Portland cement content. This experiment produced the following results. Comparison with the control mixture clearly shows blast furnace slag and fly ash consuming the $Ca(OH)_2$. The slag reaction quickly begins just after mixing and nears termination after three days. The fly ash reaction, however, begins three days after mixing, indicating that it is very slow even under an adiabatic temperature rise. This clearly shows that the blast furnace slag reaction cannot be properly estimated if the fly ash reaction is improperly overestimated, since they competitively consume $Ca(OH)_2$ in the ternary blend mixture. To simulate the very slow fly ash reaction, the minute reference heat rate was set to an extremely small value, which assumes that the reference heat rate of fly ash is proportional to the reacting surface area

*Table 2.4* Mix proportions of LPC base blended paste

| Types | W/B (%) | Mix proportion (mass%) | | |
|---|---|---|---|---|
| | | LPC | SG | FA |
| LPC | 50 | 100 | --- | --- |
| LPC + SG | 35 | 70 | 30 | --- |
| LPC + FA | 35 | 70 | --- | 30 |
| LPC + SG + FA | 27 | 54 | 23 | 23 |

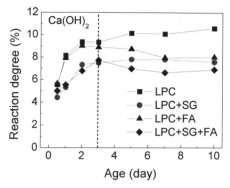

*Figure 2.17* Time-dependent reaction degree of Ca(OH)₂.

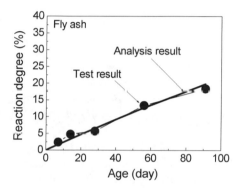

*Figure 2.18* Reaction degree at room temperature.

of the remaining cores. Figure 2.18 shows satisfactory application of the fly ash reaction at room temperature.

Another important feature of the fly ash reaction is temperature dependence. Figure 2.19 shows the different magnitudes of fly ash reaction at 20°C and 40°C. The temperature dependence of the fly ash reaction is clearly very high. The fly ash reaction is drastically accelerated by an increase in the temperature. This is attributed to the accelerated elution of $SiO_4^{2-}$ ion from the fly ash's glass phase. Thus, the thermal activity is set to a high -14,000 W°K/kg.

When the reaction of pozzolans stagnate due to a shortage of Ca(OH)₂ they, like limestone powder, cause a micro-filler effect in the Portland cement. To simulate this acceleration effect due to the inert micro-filler of pozzolans, the following factor for heat generation rate is used.

$$rs_{ad,i} = (1 - \lambda_i)(1 - \alpha_i)\frac{B_i \cdot p_i}{B_{PC} \cdot p_{PC}}$$   (2.17)

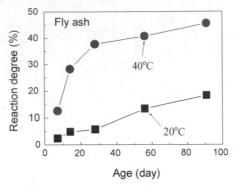

*Figure 2.19* Reaction degree at high temperature.

where $B_i$ is the Blaine value of the surface area of component i, $p_i$ is the weight fraction of component i, $\lambda_i$ is the reduction factor due to the shortage of $Ca(OH)_2$, which is expressed by Equation 2.13, and $\alpha_i$ is the degree of reaction of component i. Figure 2.20 shows the tendency of the acceleration due to the micro-filler effect caused by the inert pozzolans. When the pozzolans stagnate due to a shortage of $Ca(OH)_2$, the micro-filler effect of the stagnated particles becomes high. When they are actively reacting with a sufficient supply of $Ca(OH)_2$, no acceleration effect is produced.

When more than two pozzolans are blended into the mixture, the micro-filler effect of all of the stagnated particles should be taken into account. In this case, the modification factor is extended as follows.

$$rs_{ad,Total} = \frac{(1-\lambda_m)(1-\alpha_m)\cdot B_m \cdot p_m + \ldots + (1-\lambda_n)(1-\alpha_n)\cdot B_n \cdot p_n}{B_{PC} \cdot p_{PC}} \quad (2.18)$$

The proposed model simulates the adiabatic temperature rises of concrete containing fly ash at three different replacement ratios, as shown in Figure 2.29b. Satisfactory application of the model is apparent.

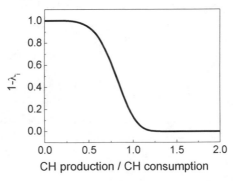

*Figure 2.20* Factor expressing the filler effect produced by pozzolans.

## 2.1.6.5 Exothermic process of silica fume

The heat generated per unit weight of fully reacted silica fume in concrete has yet to be determined. A series of adiabatic temperature rise tests were conducted on concrete containing silica fume at three different ratios (10%, 20% and 30%) using a plain mixture as the control (Figure 2.21). The plain mixture contained the same unit content of Portland cement as the test concretes so that the heat generated by the silica fume could be detected by comparing adiabatic temperature rises. The mix proportions are shown in Table 2.5. Consequently, the heat generated by the silica fume was 539 J/g for the 10% silica fume mixture, 280 J/g for the 20% silica fume mixture, and 180 J/g for the 30% silica fume mixture. Heat generation was not proportional to the amount of silica fume. Silica fume composed of more than 95% silica requires a huge amount of $Ca(OH)_2$ to fully react and thus the amount of $Ca(OH)_2$ was insufficient for complete reaction, even for the 10% silica fume mixture. Thus, the degree of reaction of silica fume should be carefully analyzed at the final stage of the adiabatic temperature rise. Kosuge (1993) quantified the degree of silica fume reaction in cement paste at

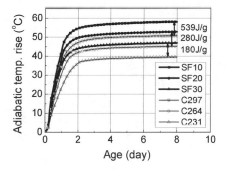

*Figure 2.21* Adiabatic temperature rise of concrete containing silica fume with their control mixtures.

*Table 2.5* Mix proportions of concrete containing silica fume with their control mixtures

| Mark | W/B (%) | Unit weight(kg/m³) | | | | |
|------|---------|-----|-----|-----|-----|-----|
|      |         | W | C | SF | S | G |
| SF10 | 50   | 165 | 297 | 33  | 792 | 999   |
| SF20 | 50   | 165 | 264 | 66  | 780 | 999   |
| SF30 | 50   | 165 | 231 | 99  | 769 | 999   |
| C297 | 55.5 | 165 | 297 | --- | 817 | 1,014 |
| C264 | 62.5 | 165 | 264 | --- | 844 | 1,014 |
| C231 | 71.4 | 165 | 231 | --- | 872 | 1,014 |

room temperature, as shown in Figure 2.22. Though the degree of reaction of silica fume during adiabatic temperature rise was not examined in these tests, the degree of reaction of silica fume at room temperature was extrapolated over a long period. Consequently, the degree of reaction was assumed to be about 95% in the case of the 10% silica fume ratio. Therefore, the amount of heat generated per unit weight of silica fume, which is the maximum value of the accumulated heat in the reference heat rate, would be 565 J/g.

For identifying parameters for silica fume, fly ash is a good reference. The mean particle size and the Blaine value for silica fume are around 0.2 μm and 200,000 cm²/g, respectively. Those values differ by almost two orders from those of fly ash. Considering these conditions, the initial value of the reference heat rate of silica fume was set to be almost two orders higher than that of fly ash. Furthermore, the reference heat rate for silica fume was assumed to be also proportional to the reacting surface area of the remaining cores, so it was set as shown in Figure 2.23. The silica fume reaction also drastically accelerates as the temperature increases, which may be due to the accelerated elution of $SiO_4^{2-}$ ion. This is similar to the scenario for fly ash, so

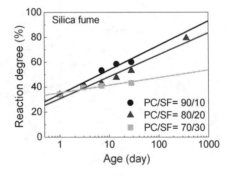

*Figure 2.22* Degree of reaction of silica fume in cement paste over time at room temperature.

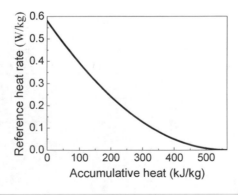

*Figure 2.23* Reference heat rate of silica fume.

the thermal activity value for fly ash, -14,000 W°K/kg, is given to the silica fume reaction.

The reaction of silica fume is dependent on the supply of $Ca(OH)_2$ from the Portland cement. This dependence must be simulated in the model. Kosuge (1993) investigated silica fume reactions for several different blending ratios when $Ca(OH)_2$ is added, as shown in Figure 2.24. The reaction of silica fume clearly completes after a long time in the case of a 2:1 blending ratio of $Ca(OH)_2$ and silica fume, but stagnates at around 85% in the case of a 1:1 blending ratio. Based on this tendency, the ratio of $Ca(OH)_2$ consumption by the silica fume reaction is assumed to be 200% of the weight of the reacted silica fume.

The proposed model was used to simulate the adiabatic temperature rise of three concretes containing different ratios of silica fume, as shown in Figure 2.25. The model's applicability at the initial temperature rise phase was not satisfactory for concretes with higher silica fume ratios, although the

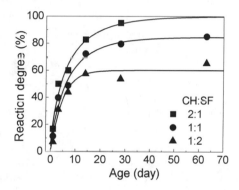

*Figure 2.24*  Degree of reaction of silica fume mixed with $Ca(OH)_2$ over time at room temperature.

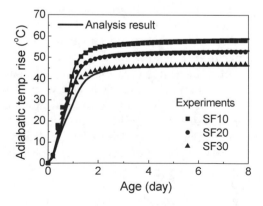

*Figure 2.25*  Adiabatic temperature rises of silica fume concrete simulated without micro-filler effect.

final temperature rises were well reproduced for all cases. A comparison of the results for the three ratios suggests that the fine silica fume particles that stagnated due to a shortage of $Ca(OH)_2$ in the concrete with the higher silica fume ratio must have a micro-filler effect on the reacting cement components. This consideration was also implemented in the proposed model. The results of a second simulation run are shown in Figure 2.29c.

### 2.1.6.6 *Delaying effect of chemical admixtures and fly ash on the hydration exothermic process*

Adding a super-plasticizer delays the initial setting and the active generation of heat, especially in the case of highly fluid concretes such as three-component cements or high-belite cement (Aoki *et al.* 1994). For high-performance concrete, these effects must be taken into consideration because the super-plasticizer is an essential agent. Uomoto and Ohshita (1994) offered an explanation for the delay mechanism. When functional groups of the added super-plasticizer react with the $Ca^{2+}$ formed by cement hydration, a kind of calcium salt is produced. This consumption of $Ca^{2+}$ delays the formation of the crystal cores of $Ca(OH)_2$, which are the triggers for active hydration heat generation, or at least slows the formation of the crystals. Some of the super-plasticizer gets incorporated into the hydrates of the interstitial materials as a result of rapid hydration of the $C_3A$ or other interstitial materials, and it is consumed regardless of the delay. Uchikawa *et al.* (1993) reported that the delaying effect of an organic admixture varies depending on how it is added.

The addition of fly ash to Portland cement is also known to delay hydration (Uchikawa 1986). Santhikumar (1993) measured the hydration heat rate of binary blended cement containing various amounts of fly ash and observed that the exothermic peak is retarded according to the amount of fly ash in the cement. Santhikumar (1993) reported that fly ash also delays the slag reaction. If fly ash is present, the $Ca^{2+}$ is adsorbed by the fly ash particles instead of eluting $Al^{3+}$ just after it comes in contact with water. This is why fly ash delays the initial reaction of the cement (Uchikawa 1986). Thus, the delay mechanism is assumed to be similar to that of an organic admixture.

During Stages 1 and 2, the heat-generation rate is reduced by the delaying effect due to the organic admixture and fly ash. Ozawa *et al.* (1992) reported that the delaying effect from an organic admixture not only extends the dormant period corresponding to Stage 1, but also reduces the second peak of the heat generation rate that corresponds to Stage 2. Figure 2.26 shows the relation between the delaying effect and the amount of organic admixture. After adding the organic admixture, the dormant period elongates and the second peak of the heat-generation rate decreases. On the other hand, in Stage 2, the $Ca^{2+}$ concentration is high since the cement reaction is active. The delaying effect should be weaker during this stage. In addition, both the organic admixture and the fly ash will delay hydration of the blast furnace slag.

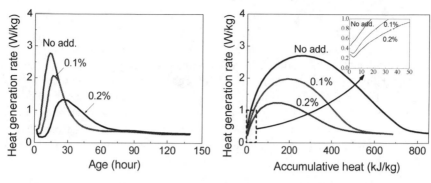

*Figure 2.26* Change in heat rate after mixing with a delaying type water reducer.

In the past, the delaying effect on the blast furnace slag was usually treated in the same manner as the delaying effect on Portland cement. However, since slag generates less $Ca^{2+}$ than Portland cement, in this model, the delaying effect on slag is considered to be more significant.

Uomoto and Ohshita (1994) and Uchikawa *et al.* (1993) reported that the delaying effect depends on the type of chemical admixture used. Given these observations, the multi-component model must take into account the delaying effect imposed by various chemical admixtures. In this chapter, the term "admixture delaying effect" is used to describe this delaying effect, which depends on the admixture's $Ca^{2+}$ consumption. The delaying capacity coefficient $\chi_{sp}$, which represents the delaying effect per unit weight of admixture, is used to describe the characteristics of the admixture. By multiplying coefficient $\chi_{sp}$ by the admixture dosage as a ratio of addition to cement ($C \times \%$), $\vartheta_{total} = p_{sp} \cdot \chi_{sp}$, the total delaying effect of the admixture can be calculated. However, part of the admixture is restrained by $C_3A$, $C_4AF$, slag or fly ash and loses its delaying effect as soon as reaction starts. Therefore, the value obtained by subtracting the invalid value corresponding to the amount of admixture restrained from the total delaying effect is defined as the effective delaying factor of the admixture in the exothermic process, $\vartheta_{SPef}$. It is assumed here that the aforementioned chemical admixture corresponds to the part adsorbed on the surfaces of the particles immediately after mixing, since the delaying effects measured at the time water is added and measured later differ markedly (Aoki *et al.* 1994). Thus, by assuming that part of the organic admixture is adsorbed on the surfaces of the $C_3A$, $C_4AF$, blast furnace slag and fly ash having aluminate composites, the amount of organic admixture consumed can be calculated as follows.

$$\vartheta_{Waste} = \frac{1}{200} \cdot (16 p_{C_3A} \cdot s_{C_3A} + 4 p_{C_4AF} \cdot s_{C_4AF} + p_{SG} \cdot s_{SG} + 5 p_{FA} \cdot s_{FA}) \quad (2.19)$$

where $\vartheta_{Waste}$ is the composite of the chemical admixtures that do not exert a delaying effect due to adsorption on the surfaces of the powder particles. Thus, the effective delaying capability $\vartheta_{SPef}$ can be expressed as follows.

$$\vartheta_{SPef} = p_{SP} \cdot \chi_{SP} - \vartheta_{Waste} \tag{2.20}$$

where $p_{SP}$ is a dosage of organic admixture expressed as an additive ratio to the binder $(C \times \%)$, and $\chi_{SP}$ is a delaying capacity coefficient representing the delaying effect per unit weight of organic admixture. Thus, $p_{SP}\chi_{SP}$ is the total delaying effect produced by the added organic admixture. When the adsorption capacity of the components exceeds the total capacity of the delaying effect produced by the organic admixture, $\vartheta_{SPef}$ is assumed to be zero. Assuming that the delaying effect of the fly ash is proportional to its replacement ratio, the following equation is produced.

$$\vartheta_{FAef} = 0.014 p_{FA} \cdot s_{FA} \tag{2.21}$$

In contrast to the delaying effects of the chemical admixtures and the fly ash, the cement minerals provide $Ca^{2+}$ ions. If the cement provides a copious amount of $Ca^{2+}$, the delaying effect wanes rapidly. On the other hand, if $Ca^{2+}$ is produced slowly, the delaying effect may be long-lived. The elution of $Ca^{2+}$ varies from mineral to mineral. Given the amount of $Ca^{2+}$ and its reactivity, it follows that $C_2S$ supplies a smaller amount than $C_3S$. It may therefore be deduced that low-heat Portland cement, with its relatively poorer elution of $Ca^{2+}$, should be more prone to delaying effects than ordinary Portland cement, even when the same amount of super-plasticizer is added.

This delaying effect of chemical admixtures and fly ash is expressed by incorporating the delaying effect and counteracting it with the rate of $Ca^{2+}$ supply from the minerals. In this case, $C_3S$, $C_2S$, slag and limestone (LS) are treated as minerals supplying $Ca^{2+}$, and corresponding supplying capacities are considered, so the delaying effect of chemical admixtures and fly ash on the heat generation can be expressed by

$$\gamma_i = \exp\left\{-\frac{k_p \cdot \vartheta_{SPef}}{10 p_{C_3S} \cdot s_{C_3S} + 5 p_{C_2S} \cdot s_{C_2S} + 2.5 p_{SG} \cdot s_{SG} + 75 p_{LS} \cdot s_{LS}}\right\} \tag{2.22}$$

$$\gamma_{SG} = \exp\left\{-\frac{k_{SG}(\vartheta_{SPef} + \vartheta_{FAef})}{10 p_{C_3S} \cdot s_{C_3S} + 5 p_{C_2S} \cdot s_{C_2S} + 2.5 p_{SG} \cdot s_{SG} + 75 p_{LS} \cdot s_{LS}}\right\} \tag{2.23}$$

where $\gamma_i$ is the coefficient of heat generation reduction for the mineral compounds of Portland cement, $\gamma_{SG}$ is the coefficient of heat generation reduction for the blast furnace slag, and $p_i$ is the weight proportion of each component.

In Stage 1, the delaying effects are regarded as an extension of the dormant period. In Stage 2, however, the delaying effects are regarded as a reduction of the heat generation rate peak. Both reaction delays are expressed by reducing the heat rate of the relevant components in Stages 1 and 2, which respectively correspond to the dormant period and the peak reduction by means of $\gamma$ in this model. The values for $k_p$ and $k_{SG}$ are shown in Table 2.6.

*Table 2.6* Delaying effect parameters $k_p$ and $k_{SG}$

| Stage - | Stage 1 | Stage 2 |
|---|---|---|
| $k_p$ | 1,000 | 400 |
| $k_{SG}$ | 1,500 | 600 |

The values for $\chi_{sp}$ were determined by testing more than fourteen common water reducers and super-plasticizers in semi-adiabatic temperature rise experiments. Water reducers such as the AE series can be divided into two groups, normal and delaying, with delaying capacity coefficient $\chi_{sp}$ values of 0 and 5, respectively. In the case of super-plasticizer, it was found that $\chi_{sp}$ is linearly related to the initial setting time of the tested mortar specimens. When the water-to-binder ratio is around 30% and the sand-to-cement ratio about 2, the $\chi_{sp}$ can be expressed as follows.

$$\chi_{sp} = 0.18T_{ini,set} - 0.039 \tag{2.24}$$

where $T_{ini,set}$ is the initial setting time of the mortar or concrete. Determining the values of $\chi_{sp}$ corresponding to a larger range of mix proportions is a task for future study.

### 2.1.6.7 Verification

Experiments were conducted on blended cement concretes to verify the performance of the improved multi-component hydration heat model. The mix proportions are listed in Table 2.7. The results of the experiments and analyses calculated by the old and new models are shown in Figures 2.27–2.30.

## 2.2 Strength development modeling of cement hydration products

It was believed that a close relation exists between the strength and the hydrated structure of a cement-hardened matrix, and that strength development can be expressed by the porosity. However, engineering applications cannot be ensured across the whole range of porosity. This section describes the construction of a compressive strength-development model that considers the apparent density, volume and type of hydration products in capillary pores. Furthermore, the proposed model also reflects the formation of hydrated structures under high temperatures and the filler effect of silica fume. Wide applicability is the goal in this section.

### 2.2.1 Changes in the volume of solids according to hydration

A main parameter adopted to describe strength development is the ratio of the volume of the occupying solid to the initial capillary space, which

*Table 2.7* Mix proportion of concretes

| Types | W/B (%) | Unit weight(kg/m³) | | | | | | | | Ad. (C × %) |
|---|---|---|---|---|---|---|---|---|---|---|
| | | W | C | SG | FA | SF | LS | S | G | |
| SG400 | 39.3 | 157 | 240 | 160 | 0 | 0 | 0 | 645 | 1,129 | 0.25 |
| SG300 | 49.3 | 148 | 180 | 120 | 0 | 0 | 0 | 757 | 1,129 | 0.25 |
| SG200 | 78.5 | 157 | 120 | 80 | 0 | 0 | 0 | 854 | 1,089 | 0.25 |
| FA400 | 39.3 | 157 | 320 | 0 | 80 | 0 | 0 | 639 | 1,129 | 0.25 |
| FA300 | 49.3 | 148 | 240 | 0 | 60 | 0 | 0 | 749 | 1,129 | 0.25 |
| FA200 | 78.5 | 157 | 160 | 0 | 40 | 0 | 0 | 852 | 1,089 | 0.25 |
| SG40 | 50.0 | 150 | 180 | 120 | 0 | 0 | 0 | 749 | 1,131 | 0.75* |
| SG60 | 49.3 | 148 | 120 | 180 | 0 | 0 | 0 | 752 | 1,131 | 0.75* |
| SG70 | 49.0 | 147 | 90 | 210 | 0 | 0 | 0 | 755 | 1,131 | 0.75* |
| SG80 | 48.3 | 145 | 60 | 240 | 0 | 0 | 0 | 755 | 1,131 | 0.75* |
| FA15 | 49.0 | 147 | 255 | 0 | 45 | 0 | 0 | 823 | 1,041 | 0.75 |
| FA30 | 49.0 | 147 | 210 | 0 | 90 | 0 | 0 | 806 | 1,041 | 0.75 |
| FA45 | 49.0 | 147 | 165 | 0 | 135 | 0 | 0 | 789 | 1,041 | 0.75 |
| LSG75 | 47.3 | 142 | 90 | 270 | 0 | 0 | 0 | 832 | 1,080 | 0.75 |
| LSF | 47.3 | 142 | 105 | 135 | 60 | 0 | 0 | 765 | 1,105 | 0.75 |
| LSFL | 47.3 | 142 | 90 | 135 | 60 | 0 | 15 | 763 | 1,105 | 0.75 |

* Delay-type additive.

increases with hydration. The ratio of changes in the volume of cementitious mineral compounds to the hydration products can be calculated based on the following chemical equations, specific weights and degree of hydration.

$$2C_3S + 6H \rightarrow C_3S_2H_3 + 3CH \tag{2.25}$$

$$2C_2S + 4H \rightarrow C_3S_2H_3 + CH \tag{2.26}$$

$$2C_3A + C_3A \cdot 3C\bar{S} \cdot H_{32} + 4H \rightarrow 3\left[C_3A \cdot C\bar{S} \cdot H_{12}\right] \tag{2.27}$$

$$2C_4AF + C_3(AF) \cdot 3C\bar{S} \cdot H_{32} + 6H \rightarrow 3\left[C_3(AF) \cdot C\bar{S} \cdot H_{12}\right] + 2CH \tag{2.28}$$

$$C_3A + 6H \rightarrow C_3AH_6 \tag{2.29}$$

$$C_4AF + 2CH + 10H \rightarrow C_3AH_6 - C_3FH_6 \tag{2.30}$$

where C is CaO, S is SiO$_2$, H is H$_2$O, CH is Ca(OH)$_2$, A is Al$_2$O$_3$, C$\bar{S}$ is CaSO$_4$ and F is Fe$_2$O$_3$.

Based on the chemical equations, the ratio of changes in the volume of hydration products to the original mineral compounds can be determined. Table 2.8a shows an example of how to determine these ratios for alite (C$_3$S).

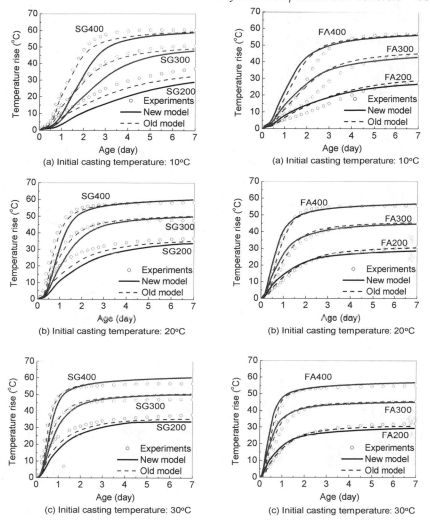

*Figure 2.27* Adiabatic temperature rises of blast furnace slag-blended concrete with analyses by new and old models.

*Figure 2.28* Adiabatic temperature rises of fly ash-blended concrete with analyses by new and old models.

The volume change ratio for solids is calculated by dividing the volume of the hydration product by the volume of the original chemical compound. The volume of each solid phase is calculated from its specific weight (molecular weight) in the chemical equation by its density. In the chemical reactions of aluminate ($C_3A$) and ferrite ($C_4AF$), part of the hydration products will continuously react under further hydration. Thus, when ettringite is converted into monosulphate, as shown in Equations 2.27 and 2.28, the solid volume change ratio is calculated from the volume of the original

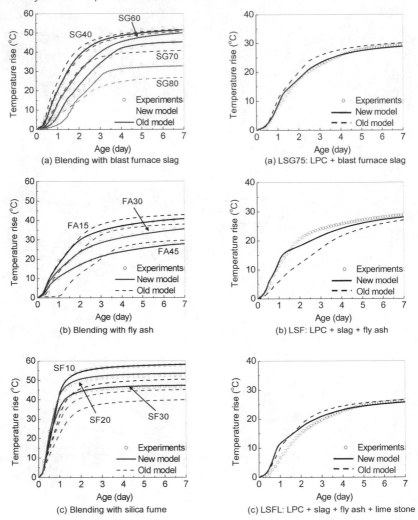

*Figure 2.29* Adiabatic temperature rises of concretes respectively blended with blast furnace slag, fly ash and silica fume.

*Figure 2.30* Adiabatic temperature rises of multi-blended cement concretes.

chemical compounds by subtracting the volume of ettringite from that of monosulphate. Similarly, in the case of $C_4AF$, the solid volume change ratio of $C_4AF$ is calculated by subtracting the volume of calcium hydroxide (CH) consumed during hydration from the volume of the hydration product. The solid volume change ratios for all chemical compounds are shown in Table 2.8b.

The solid volume change is taken into account for a unit volume of cement

*Table 2.8a* Determining the of solid-phase volume change ratio for $C_3S$

| $C_3S$ | $2C_3S$ | + | $6H$ | → | $C_3S_2H_3$ | + | $3CH$ |
|---|---|---|---|---|---|---|---|
| Specific weight (g) | 456 | | 108 | | 342 | | 222 |
| Density (g/cm³) | 3.15 | | 1.00 | | 2.71 | | 2.24 |
| Volume (cm³) | 145 | | 108 | | 126 | | 99 |
| Volume change ratio | | | $Pr_{C3S} = (126 + 99)/145 = 1.55$ | | | | |

*Table 2.8b* Solid volume change ratios for all reactions

| $i$ | $C_3S$ | $C_2S$ | $C_3Am*$ | $C_4AFm*$ | $C_3A$ | $C_4AF$ |
|---|---|---|---|---|---|---|
| $Pr_i$ | 1.55 | 1.51 | 1.20 | 1.53 | 1.74 | 1.75 |

* Solid volume change ratio for conversion of monosulphate.

paste. The initial volume of cement particles in a unit volume of paste is expressed as follows.

$$V_c = \frac{1}{W/C \cdot \rho_C + 1} \tag{2.31}$$

where $W/C$ is the water-to-cement ratio and $\rho_C$ is the density of the cement (g/cm³).

The fraction of the reacted volume of each chemical compound is based on the degree of hydration calculated from the ratio of accumulated heat generated to the ultimate heat generated using the following equation

$$R_i = V_c \cdot p_i \cdot \frac{Q_i}{Q_{i,\infty}} = \frac{1}{W/C \cdot \rho_C + 1} \cdot p_i \cdot \frac{Q_i}{Q_{i,\infty}} \tag{2.32}$$

where $p_i$ is the weight fraction of mineral compound $i$ in the cement, $Q_i$ is the accumulated heat generated by mineral compound $i$, which is provided by the multi-component hydration heat model (J/g), and $Q_{i\infty}$ is ultimate heat generated by mineral compound $i$.

The absolute volume of each hydration product is then calculated as the product of the reacted volume fraction of the chemical compound and the solid volume change ratio. Note that the bulk volume of the hydration product is larger than the absolute volume as it contains fine internal (gel) pores in which no more hydration product can precipitate. In the multi-component hydration heat model, the weight of physically captured water in the gel is assumed nominally equivalent to 15% of each hydration product. Moreover, the volume of the physically captured water is thought to correspond to the fine internal pores in the bulk volume of the hydration products. The specific porosity of the hydration products at room temperature is assumed to be 28% (Chapter 3), which is almost the same as the volume of the physically combined water assumed in the multi-component hydration heat model. The

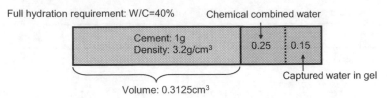

Full hydration requirement: W/C=40%          Chemical combined water

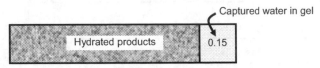

Average volume increase ratio: 1.4 ⟹ Volume after hydration: 0.3125 × 1.4=0.4375cm³

Gel porosity : 0.15/(0.4375+0.15)=0.26 ≒ 0.28

*Figure 2.31* Relation of specific porosity and captured water in gel.

relation of the specific porosity and the captured water in the gel is shown in Figure 2.31. Thus, in the proposed strength-development model, the bulk volume of the total hydration products, which gradually occupies the initial capillary pores and sustains the development of strength, is calculated as follows.

$$V_{hyd.total} = \sum (\mathrm{Pr}_i \cdot R_i + 0.15\rho_i \cdot R_i) \tag{2.33}$$

where $\mathrm{Pr}_i$ is one of the solid volume change ratios listed in Table 2.8b, and $\rho_i$ is the density of each mineral compound. By computing the increase in the bulk hydration products and the decrease in the reacted mineral compounds, the change in the volume fractions of the constituents can be traced. As an example, the volume fraction change of the constituents in an ordinary Portland cement paste with a 55% W/C mix ratio is shown in Figure 2.32. The evolution of space occupation can be analytically estimated by combin-

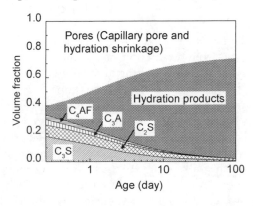

*Figure 2.32* Volume change of mineral compounds and hydration products.

ing the previously described concepts with the multi-component hydration heat model. Any significant amount of water bleeding should be excluded from the initial capillary space to ensure accurate computation.

### 2.2.2 Indicator for strength development

#### 2.2.2.1 Fundamental formulation

Ryshkewitch (1953) proposed a well-known strength-development model described solely by porosity.

$$f_C = f_0 \exp(-BP) \qquad (2.34)$$

where $f_c$ is the strength of cement paste (N/mm²), $f_0$ is the strength at no porosity, $P$ is the porosity (ml/ml), and $B$ is the material constant.

This empirical model overestimates strength development during the initial period of hydration when the porosity is high, which limits its applicability. Ryshkewitch's model is mostly applicable when the porosity is less than 0.35. Another strength development model, proposed by Schiller (1958), improves the applicability to the initial strength (high porosity) by taking into account not only the present porosity but also the initial porosity.

$$f_C = C \ln(P_{cr} / P) \qquad (2.35)$$

where $f_c$ is the strength of cement paste (N/mm²), $P$ is the porosity (ml/ml), $P_{cr}$ is the porosity when the strength is null, and $C$ is the material constant. Schiller set $P_{cr}$ to 0.31 and $C$ to 81.5 by comparing the results of experiments.

Remarkably, Schiller (1958) described strength development as the ratio between the initial porosity at no strength and the decreasing present porosity. This concept is followed in this section. On the other hand, the solid phase of hydration products, not the porosity, is a more rational focus of attention since strength development is sustained by the solid phase, not the porosity. By applying the solid phase of the hydration products, it becomes possible for the model to handle the characteristics of the solids such as their type, size and density. Thus, the volume ratio $D_{hyd.out}$ of the outer hydration products that form outside the original cement particle (Figure 1.4) to the initial capillary space expressed by Equation 2.36 is adopted as an indicator to describe strength development.

$$D_{hyd.out} = \frac{V_{hyd.out}}{V_{cap.ini.}} = \frac{V_{hyd.total} - V_{hyd.in}}{V_{cap.ini.}} \qquad (2.36)$$

where $D_{hyd.out}$ is the ratio of space occupied by the outer bulk hydrates to the initial capillary space, $V_{hyd.out}$ is the volume of hydration products that

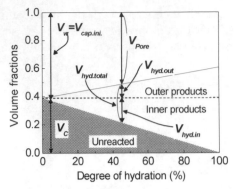

*Figure 2.33* Change in volume of components and its indicator for strength development.

form outside the original cement particle, $V_{hyd.total}$ is the total volume of hydration products, $V_{hyd.in}$ is the volume of hydration products that form inside the original cement particle, which is equivalent to the reacted volume fraction of the mineral compounds, and $V_{cap.ini}$ is the volume of the initial capillary space. Here, the volume of the outer products can be calculated by subtracting volume of the reacted compounds from the volume of the bulk hydration products since it forms outside the original cement particle, as shown in Figure 2.33.

The volume of the initial capillary space $V_{cap.ini}$ is equivalent to the porosity at no strength in Schiller's model (1958). Though Schiller determined this value by fitting an estimate to the results of experiments, in the proposed model, the volume of the initial capillary space $V_{cap.ini}$ is given by the mix's water-to-cement ratio

$$V_{cap.ini.} = \frac{W/C \cdot \rho_C}{W/C \cdot \rho_C + 1} \tag{2.37}$$

In Schiller's model, strength development is expressed by a logarithmic function. In the proposed model, the solid volume ratio, not the porosity, is adopted and an exponential function that implements the concept of the ultimate strength is applied. This results in the following equation

$$f_c = f_\infty \{1 - exp \ (-\alpha D_{hyd.out}{}^\beta) \ \} \tag{2.38}$$

where $f_\infty$ is the ultimate strength (N/mm²) and $\alpha$ and $\beta$ are material constants. These values are temporarily set to 190, 3.0, and 4.0 as a result of sensitivity analysis.

## 2.2.2.2 *Effect of water-to-binder ratio*

This section examines two types of Portland cement (OPC and LPC) and two water–cement ratios (30% and 55%) by weight. The air contents were maintained as close as possible to 1.5% for $W/C = 30\%$ and to 4.5% for $W/C = 55\%$. The specimens were completely sealed to avoid evaporation and were cured at 20°C. Comparisons of the results of the experiments and analyses are shown in Figure 2.34 and Figure 2.35.

Even though the overall tendencies of strength development are almost fully explained by the analyses, at the initial stage (earlier than 20 days), they do not agree well for either the OPC or the LPC. In the current tentative formula expressed by Equation 2.38, strength development is solely dependent on the ratio of space occupied by bulk hydrates to the initial free space denoted by $D_{hyd.out}$. If the values of $D_{hyd.out}$ are the same, then the compressive strengths are also the same. However, the strength of the OPC for $W/C = 30\%$ reached 59.4 N/mm$^2$ on the third day and 37.7 N/mm$^2$ for $W/C = 55\%$ on the 28th day. In both cases, the $D_{hyd.out}$ value was the same: 0.52. Therefore, the

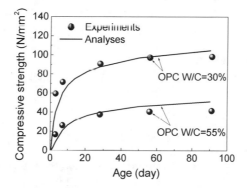

*Figure 2.34* Strength development and tentative estimate for OPC.

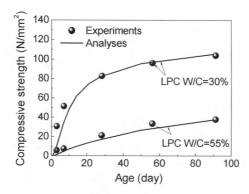

*Figure 2.35* Strength development and tentative estimate for LPC.

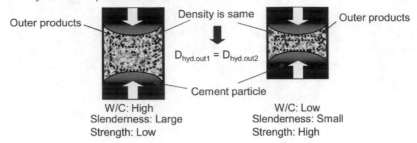

*Figure 2.36* Concept of the effect of particle distance on strength development.

tentative formula is still not accurate enough to compute the compressive strength, and some kind of influence should be considered.

The variety in the size and scale of the hydration products forming between the cement particles may influence the strength. If the gap between the original cement particles is taken as a column and its slenderness ratio is considered, the shorter the column is, the stronger it will be even when the densities of the bulk hydration products in the gap are similar. The lower W/C ratio results in higher compressive strength, even though $D_{\text{hyd.out}}$ is the same, since the gap between the particles in the mixture with the lower W/C ratio is smaller, as shown in Figure 2.36.

The effect of the gap between the cement particles is expressed by powering one-third of the initial capillary space $V_{\text{cap.ini}}$, which converts the three-dimensional volume into a one-dimensional volume.

$$\theta = (V_{\text{cap.ini.}})^{\frac{1}{3}} = (\frac{W/C \cdot \rho_C}{W/C \cdot \rho_C + 1})^{\frac{1}{3}} \tag{2.39}$$

The formula for strength development is modified by implementing the effect of the gap between the cement particles denoted by θ.

$$f_c = f_\infty'[1 - exp \ \{-\alpha'(\frac{D_{\text{hyd.out}}}{\theta})^{\beta'}\} \ ] \tag{2.40}$$

where α' and β' are set to 1.055 and 2.9, respectively, as a result of sensitivity analysis. This allows strength development to reach the ultimate value at a maximum value of $D_{\text{hyd.out}}/\theta$, where it is assumed that the minimum possible W/C ratio is actually 13%.

### 2.2.2.3 Effect of mineral composition of cement

Figure 2.37 and Figure 2.38 are comparisons of experiments and analyses in which the effect of the W/C ratio is implemented as described earlier and the ultimate strength $f_\infty$ is slightly modified to 200 N/mm². Strength development in LPC was found to be nearly completely accurate from the initial stage to the end, while that in OPC was somewhat overestimated, implying that

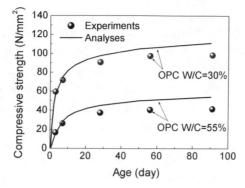

*Figure 2.37* Effect of particle distance on strength development in OPC.

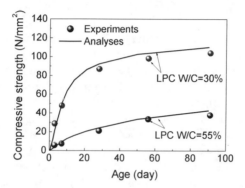

*Figure 2.38* Effect of particle distance on strength development in LPC.

the ultimate strength might differ according to the type of binder used. The proposed model incorporates an appropriate method for achieving this since it is based on solid phase formation. For simplicity, the effects of the characteristics of hydration products on strength development are represented by the types of original minerals. At this moment, two major clinker minerals, $C_3S$ and $C_2S$, are taken into account as contributors to strength development. After these two minerals have hydrated, silicate calcium hydrates ($C_3S_2H_3$) and calcium hydroxide ($Ca(OH)_2$) are produced, and $C_3S_2H_3$ contributes more to strength development than $Ca(OH)_2$. Thus, for the same reacted amounts, $C_2S$ contributes more to strength development than $C_3S$, since the volume fraction of $C_3S_2H_3$ to $Ca(OH)_2$ in hydrates is higher in $C_2S$ than in $C_3S$. Thus, the modified ultimate strength $f_\infty'$ is expressed by the volume fraction between $C_3S$ and $C_2S$ as follows.

$$f_\infty' = A\frac{p_{C3S}}{p_{C3S} + p_{C2S}} + B\frac{p_{C2S}}{p_{C3S} + p_{C2S}}$$

(2.41)

where $p_{C_3S}$ and $p_{C_2S}$ are the volume fractions of C₃S and C₂S, and *A* and *B* express their contributions to the ultimate strength. In the current tentative model, *A* and *B* were set to 215 and 250 N/mm², respectively.

### 2.2.3 Effect of high-temperature curing on strength development

Strength development at a high temperature is known to differ from strength development at a constant room temperature. When curing is done at a higher temperature, the initial strength development becomes drastically faster due to the accelerated hydration of the binder. Subsequent strength development, however, stagnates if the temperature falls to room temperature. Although it is a complex phenomenon, the formation of the microstructure can be seen in SEM images, as shown in Figure 2.39 (LPC, W/C: 55%, degree of hydration: 80%). Many voids remain in the hardened cement paste cured at a high temperature, while only a few voids are present when the cement paste is cured at room temperature. When cement is hydrated at a high temperature, the microstructure in the vicinity of the original cement particles becomes denser and, consequently, the spaces lying at some distance from the particles become larger. The dense layer of hydration products that have formed around the remaining cement particles blocks further hydration, which is accompanied by mutual diffusions of ions and water. This diffusion resistance becomes higher as the curing temperature falls because the ions and molecules have less thermal activity and, therefore, move about less. Furthermore, the non-uniform hydration formation and the remaining voids negatively affect strength. This hydration stagnation effect is due to the higher diffusion resistance of the denser layer of hydrates. This can be expressed in the hydration heat model by decreasing the opportunities for the unreacted cement and water to contact each other.

a) Standard curing      b) High temperature curing

*Figure 2.39* Difference in structure formation between standard curing and high-temperature curing.

### 2.2.3.1 Effect of increased hydrate density due to lower internal porosity

Changes in the bulk density of hydrates and the related ultimate strength are expressed in the proposed model based on the numerical expression of temperature dependence on the internal porosity of hydrates (Chapter 3, Nakarai *et al.* 2007), as shown in Figure 2.40a. If the internal porosity of the hydrates changes as the curing high temperature changes, the bulk volume of the hydration products should correspondingly change. In the proposed strength-development model, the volume of the internal fine pores, represented by the volume of the physically captured water, is used to compute the bulk hydrates.

$$V_{hyd.total} = \sum ( Pr_i \cdot R_i + \omega_{res} \rho_i \cdot R_i ) \qquad (2.42)$$

where $\omega_{res}$ is the ratio of the physically captured water to the weight of the reacted cement and is temperature dependent. $\omega_{res}$ is set to 0.15 when the temperature is lower than 40°C, 0.12 when the temperature is higher than 60°C, and is linearly interpolated for temperatures between these two values. Thus, when the bulk volume of the hydrates becomes compact due to high-temperature curing, the ability to occupy free space also decreases.

In addition, the strength of the bulk hydrates increases as the bulk hydrates become denser. Referring to Ryshkewitch's concept, the effect on the ultimate strength increment due to the denser hydrates is expressed by the modification factor $\psi$.

$$\psi = \frac{\exp( -\varphi_{in} )}{\exp( -0.28 )} \qquad (2.43)$$

where $\phi_{in}$ is as shown in Figure 2.40b (Section 3.4.2).

In high-temperature curing, the volume of the internal porosity in the hydrates decreases. This can be accounted for by decreasing the physically captured water ratio, which releases part of the water for use in further hydration. Figure 2.41 compares this method to calculation of the adiabatic temperature rise in a low W/C ratio mixture. If the physically combined water ratio is kept constant at 15%, the ultimate temperature rise will be a little

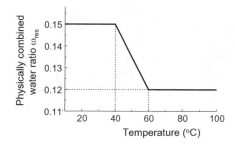

*Figure 2.40a* Temperature-dependent model of physically combined water ratio.

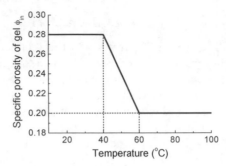

*Figure 2.40b*  Temperature-dependent model of specific porosity.

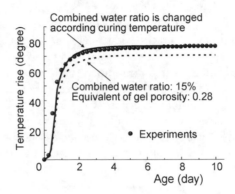

*Figure 2.41*  Estimate of temperature rise in temperature-dependent gel porosity model.

low. In contrast, the results of the experiment can be faithfully represented by the internal porosity of the hydrates (the physically captured water ratio), which decreases as the temperature rises, since the released water can be used for further hydration (Section 3.4.3).

### 2.2.3.2 *Effect of non-uniform formation of hydrates and remaining voids*

If cement hydration proceeds at a high temperature, the spaces in the hydrates will not be homogeneously distributed. First, even though a high temperature can accelerate the dissolution of ions from a cement particle, the hydrates might not diffuse far from the particle and so will precipitate in the vicinity of the particle. Second, the bulk density of the hydrates cured at a high temperature can be increased, which makes the inhomogeneous distribution of spaces in between hydrates more distinct. This mechanism is illustrated in Figure 2.42.

This effect on the inhomogeneous distribution exists to some slight degree even during room-temperature curing since its integrity could be achieved at the extreme close-packed situation. It is also obvious from Figure 2.39b that

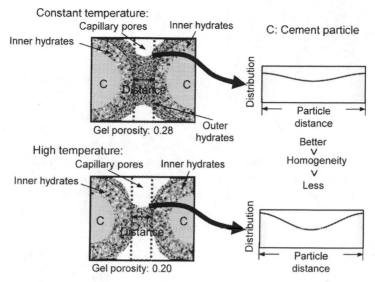

*Figure 2.42* Schematic representation of spatial distribution of hydrates between particles.

this effect on strength reduction is severe during high-temperature curing. Although the intensity of this effect ideally should be numerically formulated, for the sake of simplicity, coefficient $k$ is taken into account in Equation 2.40 to estimate strength development as Equation 2.44, where the value of $k$ is set to 0.9 for curing under a constant temperature and to 0.85 for curing under a semi-adiabatic temperature rise higher than 20°C.

$$f_c = \psi \cdot f_\infty{}'[1 - exp \; \{-\alpha'(k\frac{D_{hyd.out}}{\theta})^{\beta'}\} \; ] \tag{2.44}$$

### 2.2.3.3 Effect of denser hydrate cluster formation on hydration stagnation

Under high-temperature curing, the bulk density of the hydrates that formed around the remaining cement core becomes greater, which increases the diffusion resistance through this layer during further hydration. If a high temperature is maintained, the high diffusion resistance due to the dense microstructure will not matter because the high temperature also increases the motion of the ions and water molecules. If temperature then falls to room temperature, the high diffusion resistance due to the dense microstructure will drastically stagnate during further hydration. In the multi-component hydration heat model, the stagnation of hydration due to the shortage of free water and the increment of hydrates cluster is expressed as follows.

$$\beta_i = 1 - exp \left\{ -r\left(\frac{\omega_{free}}{100 \cdot \eta_i}\right)^s \right\} \tag{2.45}$$

$$\eta_i = 1 - (1 - \frac{Q_i}{Q_{i,\infty}})^{\frac{1}{3}} \tag{2.46}$$

where $\omega_{free}$ is the remaining free water ratio, $\eta_i$ is the normalized thickness of the inner hydrates layer of component $i$, $r$ and $s$ are material constants set to 5.0 and 2.4, respectively.

Based on this modeling process, $\eta_i$ must be increased to express how the microstructure becomes denser during high-temperature curing and the effect of the depressed motions of ions and molecules at room temperature, which can be expressed by decreasing $r$. The material constants adopted in Equation 2.45 were set by examining the applicability of the multi-component hydration heat model to the data from experiments on adiabatic temperature rise. Thus, those material constants are set for high adiabatic temperatures. In principle, to simulate room-temperature curing, $\eta_i$ must be moderated and $r$ must be decreased. For simplicity, the original set of material constants are also used for the case of room-temperature curing, since any changes in those two constants according to relative conditions can be ignored. On the other hand, the effect on hydration stagnation when the temperature falls to normal must also be appropriately expressed. At this moment, for simplicity, this effect is expressed by nominally decreasing constant $r$ from 5.0 to 0.8. This tentative treatment should be systematically improved in the future.

### 2.2.4 Extension for high-strength concrete containing silica fume

For extremely high-strength concrete over 100 N/mm², a low water-to-binder ratio of less than 25% and the use of silica fume are indispensable. Silica fume's micro-filler effect modifies the freshness property and enhances the strength of the concrete. To expand its applicability, this section discusses specific issues related to silica fume. First, the solid-volume change ratio for the silica fume reaction must be set. The chemical equation for the silica fume reaction has never been clearly determined, so this value was tentatively set to 1.2 as a result of sensitivity analysis. Although the analysis does not explicitly monitor the reduced amount of $Ca(OH)_2$ for the silica fume reaction, its effect is implicitly reflected in the solid-volume change ratio for the silica fume reaction (a relatively lower 1.2).

Second, the effect of the silica fume should be reflected in the ultimate strength. The silica fume reaction replaces part of the $Ca(OH)_2$ with C–S–H, which is stronger than $Ca(OH)_2$. In addition, the $CaO/SiO_2$ ratio of C–S–H produced by the silica fume reaction is reported to be around 1.3, which is significantly lower than that of the C–S–H from $C_3S$ and $C_2S$, which is around 1.7. Thus, silica fume apparently contributes more to strength development than other mineral compounds do. This can be expressed as follows.

$$f_{\infty}' = A \cdot p_C \frac{p_{C3S}}{p_{C3S} + p_{C2S}} + B \cdot p_C \frac{p_{C2S}}{p_{C3S} + p_{C2S}} + C \cdot p_{SF} \qquad (2.47)$$

where $p_C$ and $p_{SF}$ are the weight fractions of the Portland cement and the silica fume, $p_C + p_{SF} = 1$, and $A$, $B$, and $C$ are the material constants that express the contributing factors of the components of strength development, which are respectively set to 215, 250 and 300.

The third issue concerns the micro-filler effect of silica fume. The mean diameter of a silica fume particle is around 0.1μm. Thus, silica fume particles can fill the gaps among the cement particles. Consequently, the absolute solid volume in paste can be increased and the water-to-binder ratio lowered while maintaining satisfactory workability. Since silica fume particles can be cores for hydrate formation, the fine silica fume particles in the gaps among the cement particles can enhance strength development by serving as precipitation sites. Due to this effect, the gaps among the cement particles are efficiently packed with hydrates, thereby enhancing the homogeneity of the hardened structure. Thus, coefficient $k$, which represents the homogeneity and the effect of defects on the strength, is set to 0.95 for concrete containing silica fume (0.90 for conventional concrete).

Figure 2.43 shows an example of high-temperature curing that follows appropriate pre-curing at room temperature. In practice, this process allows concrete containing silica fume to develop maximum strength, since very fine tobamorite, which is produced by silica fume cured at a high temperature, can almost completely fill the gaps among the hydrates. The homogeneity of the spatial distribution of the hydrates has the integrity at the previously stated situation and then coefficient $k$ is set to unity. The development of strength in concrete containing silica fume and subjected to a curing process for factory products, as shown in Figure 2.43, can be analytically determined using the proposed model. Figure 2.44 shows that the calculated strength development can exceed 200 N/mm² in that situation. Thus, the proposed model can be applied to extreme situations.

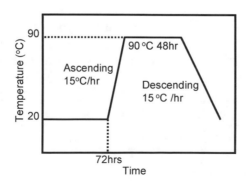

*Figure 2.43* Typical temperature history applied to pre-cast high-strength concrete.

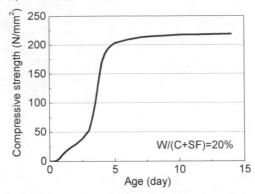

*Figure 2.44* Calculated compressive strength development of pre-cast high-strength concrete containing silica fume.

### 2.2.5 Verifying the strength development model

#### 2.2.5.1 Strength development at constant room temperature

The strength development predicted by the model was compared with data from experiments under constant room temperature. The analyzed concretes included OPC, MPC and LPC. The mineral compositions of these cements were nominally calculated using Bogue's equation. The results are shown in Table 2.9. Two water-to-cement ratios were adopted: 55% for the conventional concrete and 30% for the high-strength concrete. The mix proportions are shown in Table 2.10. The target air contents are 4.5 % for W/C = 55% and 1.5% for W/C = 30%. The coarse aggregate was hard sandstone, the strength of which exceeded the strength of the target hardened cement paste matrix and did not affect the concrete's strength development.

The analytical results for OPC, MPC and LPC are shown with the test results in Figure 2.45, Figure 2.46 and Figure 2.47. The specimens used for the strength test were fully sealed and cured at 20°C, and the same conditions were applied to the analyses. The analytical results show that the model successfully reproduces the test results for various types of Portland cement, and has a wide water-to-cement ratio range. The proposed model has satisfactory applicability to Portland cement under standard curing conditions, although

*Table 2.9* Mineral composition of Portland cements

| Type of cement | Mineral composition (mass %) | | | | | Blaine value (cm²/g) |
|---|---|---|---|---|---|---|
| | $C_3S$ | $C_2S$ | $C_3A$ | $C_4AF$ | $C\bar{S}2H$ | |
| OPC | 53.0 | 22.0 | 12.0 | 9.0 | 3.90 | 3,310 |
| MPC | 41.2 | 36.7 | 3.7 | 12.5 | 4.08 | 3,280 |
| LPC | 24.0 | 56.0 | 3.0 | 12.0 | 5.16 | 3,290 |

Table 2.10 Mix proportions of concrete

| Specimen | W/C (%) | Unit weight(kg/m³) W | C | S | G | Ad. (Cx%) |
|----------|---------|---|---|---|---|------------|
| OPC30 | 30.0 | 165 | 550 | 895 | 808 | 1.55* |
| OPC55 | 55.0 | 165 | 300 | 840 | 994 | 0.25 |
| MPC30 | 30.0 | 165 | 550 | 890 | 808 | 1.40* |
| MPC55 | 55.0 | 165 | 300 | 843 | 997 | 0.25 |
| LPC30 | 30.0 | 165 | 550 | 893 | 808 | 1.20* |
| LPC55 | 55.0 | 145 | 264 | 881 | 1,042 | 0.25 |

\*  super-plasticizer

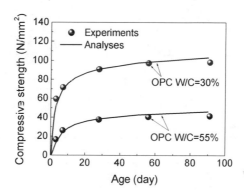

*Figure 2.45* Applicability of proposed model for strength development of OPC.

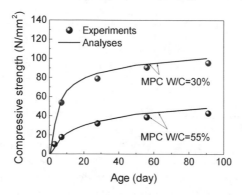

*Figure 2.46* Applicability of proposed model for strength development of MPC.

the proposed model does not take into account the effect of the air content on strength development. The effect of air content should be implemented in the future, especially for cases in which the air content is very different from that in the mixture used for verification in this chapter.

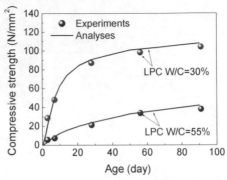

*Figure 2.47* Applicability of proposed model for strength development of LPC.

*Table 2.11* Mineral composition of LPC containing silica fume

| Cement | Chemical composition (mass %) | | | | | Blaine value (cm²/g) |
|--------|------|------|------|------|------|------|
| | $C_3S$ | $C_2S$ | $C_3A$ | $C_4AF$ | $C\bar{S}2H$ | |
| LPC | 23.0 | 57.0 | 2.0 | 10.0 | 4.95 | 3,210 |

*Table 2.12* Mix proportions of LPC containing silica fume

| Type | W/C (%) | Unit weight(kg/m³) | | | | | Ad. (C×%) |
|------|------|------|------|------|------|------|------|
| | | W | C | SF | S | G | |
| LS24 | 24.0 | 140 | 526 | 58 | 885 | 837 | 1.4 |
| LS20 | 20.0 | 140 | 630 | 70 | 788 | 837 | 1.4 |
| LS16 | 16.0 | 140 | 787 | 88 | 640 | 837 | 2.0 |

The applicability of the proposed model to high-strength concrete containing silica fume and cured under the same temperature conditions was verified. The silica fume had a Blaine fineness of 226,000 cm²/g and a $SiO_2$ content of 93.4%. The base cement was LPC with the mineral composition shown in Table 2.11. The mix proportions of the concrete containing silica fume are shown in Table 2.12. Three water-to-binder ratios with a 9:1 blending ratio of LPC and silica fume were adopted. The results of the analyses and tests, shown in Figure 2.48, agree well, thereby confirming the applicability of the proposed model for the adopted treatments and parameter sets for concrete containing silica fume, including the solid-volume change ratio, the contributing factors to ultimate strength, and the micro-filler effect on the homogeneity of the hydrates.

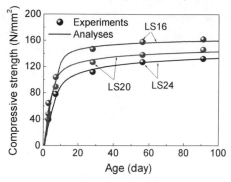

*Figure 2.48* Applicability of the proposed model to concrete containing silica fume.

### 2.2.5.2 Strength development at high temperature

The applicability of the proposed model is verified for cases having an adiabatic temperature rise history or cases such as factory products that have undergone initial high-temperature curing. For cases having an adiabatic temperature rise history, the mix proportions are listed in Table 2.10, which targets two water-to-cement ratios and three types of Portland cement. The specimens were cured in a temperature-controlled chamber that applied the same adiabatic temperature rise history for each mixture, as shown in Figure 2.49. In the case of initial high-temperature curing, which is usually applied to factory products, the same three types of Portland cement were used but the water-to-cement ratio was only 30%. These parameters are listed in Table 2.10. Immediately after casting, eighteen specimens were put in a temporary handcrafted container made of styrofoam so that the initial temperature elevations were caused by their own hydration heat generation. The temperature history is shown in Figure 2.53. The specimens for both cases – adiabatic temperature rise history and initial high-temperature curing – were cured under fully sealed conditions.

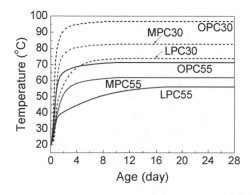

*Figure 2.49* Adiabatic temperature rises applied for the series of verifications.

Comparisons of the results of the analyses and tests on the OPC, MPC and LPC concretes cured under an adiabatic temperature rise are shown in Figure 2.50, Figure 2.51 and Figure 2.52, while those cured at an initially

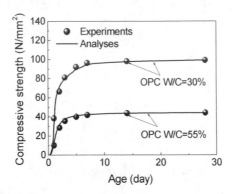

*Figure 2.50* Applicability of the proposed model for OPC under adiabatic temperature curing.

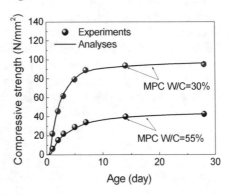

*Figure 2.51* Applicability of the proposed model for MPC under adiabatic temperature curing.

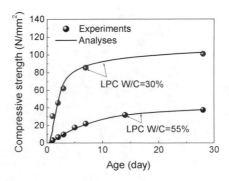

*Figure 2.52* Applicability of the proposed model for LPC under adiabatic temperature curing.

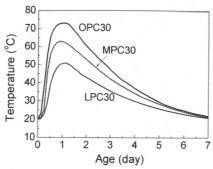

*Figure 2.53* Transient high-temperature histories applied for the series of verifications.

high temperature are shown in Figure 2.54, Figure 2.55 and Figure 2.56. To deal with the effect of high-temperature curing on strength development, the proposed model takes into account several considerations, such as the reduc-

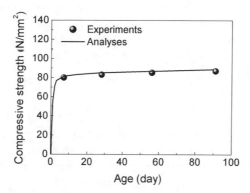

*Figure 2.54* Applicability of the proposed model for OPC30 under initial high-temperature curing.

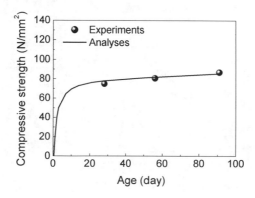

*Figure 2.55* Applicability of the proposed model for MPC30 under initial high-temperature curing.

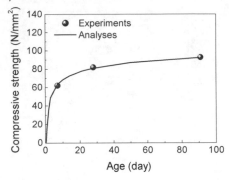

*Figure 2.56* Applicability of the proposed model for LPC30 under initial high-temperature curing.

tion in the internal porosity of the hydrates, the accompanying increase in the ultimate strength of the bulk hydrates, the reduction in the homogeneity of the special distribution of the hydrates, and the stagnation of hydration due to increased diffusion resistance by the denser clusters of hydrates around the remaining cement. These figures confirm the adequate applicability of the proposed model for cases involving high-temperature curing.

### 2.2.5.3 Other verifications

Finally, the overall applicability of the proposed model to various types of Portland cement, high-strength cements containing silica fume, and arbitrary water-to-binder ratios and curing temperature histories is verified. Figure 2.57 shows all of the data for the comparisons. As shown in Figure 2.57, the standard deviation for all data comparisons is 2.95 N/mm², which confirms

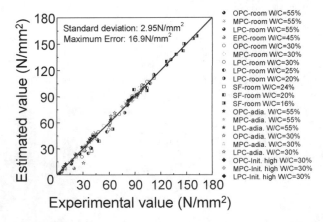

*Figure 2.57* Verification of the applicability of the proposed model for arbitrary cement types and time ranges.

that the proposed model has fair applicability to a wide range of materials, mix proportions and temperatures.

However, there are still shortcomings that require additional study. Although all of the mechanisms considered in this chapter on strength development concern phenomena in hardening cement paste, the strength of the cement paste correlates directly to the strength of the concrete in the proposed model. Other factors, such as the strength of the aggregate, the effect of the interfacial transition zone (ITZ) around the aggregate, the air content, etc., which affect the actual concrete strength will be taken into account in the future. Moreover, the tensile strength, not the compressive strength, may be much more important from a structural viewpoint, even though from a practical viewpoint the compressive strength is very important for quality control and inspection.

# 3 Micro-pore structure formation and moisture transport

As the development of microstructure, hydration and moisture transport are intrinsically coupled (Chapter 1), an integrated approach is required for macro-scale computations of temperature and related stress models that make use of the sophisticated micro-scale material models. Therefore, the goal of this chapter is to develop analytical models of microstructure development based upon the hydration model of cement (Chapter 2) and fundamental solid models pertaining to each physical process. At the same time, an important consideration is that the formulations should be simple enough to be incorporated in a real-time computational scheme of heat and moisture transport in a full-sized structural concrete scenario from a practical view point. The development of a microstructure formation model is also necessary since mass transport characteristics are strongly dependent on it. This chapter presents the multi-scale modeling of micro-pore structures, moisture equilibrium and its transport as mutually interrelated thermodynamic events. The development of the pore structure at early ages is obtained using a pore structure development model based on the average degree of hydration. The predicted computational pore structures of concrete would be used as a basis for moisture transport computation. In this way, by applying a dynamic coupling of pore-structure development to both the moisture transport and hydration models, the development of strength along with moisture content and temperature can be traced with the increase in the degree of hydration for any arbitrary initial and boundary conditions.

## 3.1 Basic modeling of micro-pore structure development

The hydration products and various phases are represented as computer-based digital pixels of different types and properties. The rate of their deposition and transport in the pore solution phase is governed either by a simple stoichiometric balance of various phases or is based upon some variation of the homogenous nucleation models proposed by Dhir *et al.* (1996). These digital models generally provide a useful qualitative insight into the nature of hydration products and paste microstructure, but due to the digital nature of the simulations, these methods are limited in their resolution and require large computer resources.

Though continuum models, such as those proposed by Breugel (1991) and by Ulm and Coussy (1996), are sophisticated and useful for research, some are not incorporated in computational schemes that take into account the effect of temperature and moisture availability in a real-time basis over the entire structure domain. Without the dynamic incorporation of such models in the integrated scheme of coupled heat and mass transfer solutions, it is hard to predict macroscopic structural behaviors under arbitrary conditions of casting and curing. In this section, the authors formulate the micro-pore structure as an assembly of randomly connected multi-scale micro-pores which are expressed by statistical functions, and apply the requirement of thermodynamics to the moisture held in each pore.

Early-age development phenomena of concrete structures have been extensively studied in terms of the development of macroscopic characteristics, such as permeability, strength, stiffness, etc., with time. Studies dealing with the micro-aspects of pore structure development in cementitious materials have been primarily confined to experimental realms. Moreover, little effort has been made to convert the findings of these experimental studies into a computational system that could be used as a simulation tool for predicting concrete performance, especially related to durability of concrete structures. The primary aim of this book is to create bridges between various theories that deal with the prediction of moisture content, degree of hydration, strength, and so on.

### 3.1.1 Stereological description of the initial state of mix

The particle sizes of commercially available cement are spread over at least two orders of magnitude. Thus, manufactured pastes contain particles that range from colloidal sizes to particles that are about 150 μm in size. Just after mixing with water, many of the smaller particles dissolve completely, leaving room for the larger particles to grow as the hydration proceeds. In structure development, it appears that the larger particles play an important role in deciding the morphological features of the paste. It might be possible to represent the entire particle distribution of powder materials by a representative, volumetrically averaged particle size. As a starting point for the basic model, the authors assume that the powder material can be idealized as consisting of uniformly sized spherical particles. An important implication of this assumption is that the characteristics of the entire particle size distribution are lumped into a specific particle size. This may appear to be a gross approximation. However, the initial colloidal dispersion characteristics of the paste are more important than the particle size distribution itself.

For W/B (water-to-binder) ratio pastes encountered in practice, the distance between powder particles is about the same order of magnitude as their average size. The theoretical arguments point to the fact that initially, a completely homogenous and dispersed system should exist. However, there are observations that point to the agglomeration and flocculation of particles, which might result from the adherence of smaller powder particles to the larger ones. In such a case, the average size of the flocculated particles is to be considered as the representative size of the particle. In any case, the same-size assumption is simple enough to deal with any initial configuration of the dispersed system, although a system considering a particle size distribution would be more theoretically complete. When the initial average size of the powder particles is known, it is assumed that the system comprising these colloidal dispersions of powder particle in the mix is homogenous. In a uniformly dispersed system of spherical particles of radius $r_o$, the average volumetric concentration $G$ and the mean spacing between outer surfaces of two particles $s$ is linked by the following relationship

$$G = \frac{G_o}{\left(1 + s/2r_o\right)^3} \tag{3.1}$$

where $G_o$ is the maximum volumetric concentration or the limiting filling capacity that can be achieved. The theoretical maximum of $G_o$ is $\pi/3\sqrt{2}$. However, in our model, $G_o$ is computed from the following empirical relationships, considering a large spread in the size of particles as

$$G_o = 0.79(BF/350)^{0.1} \quad G_o \leq 0.91 \tag{3.2}$$

where $BF$ represents the Blaine fineness index. For usual cements, $r_o$ is taken as 10 μm and the average $BF$ index is taken as 340 m²/kg. If the W/B ratio of the paste is $\omega_o$ then the average spacing $s$ between powder particles can be obtained as (using Equation 3.1)

$$s = 2r_o\left[\left\{G_o\left(1 + \rho_p\omega_o\right)\right\}^{1/3} - 1\right] \tag{3.3}$$

where $\rho_p$ denotes the average specific gravity of the powder materials and can be obtained as a weighted average for blended powders. In a dispersed system of particles, each particle has on average a free cubical volume, $l^3$, available for expansion, as shown in Figure 3.1. At a certain stage of hydration, expanding outer clusters of each particle will start touching. This will be followed by further expansion and merging as the new hydration products start filling the empty spaces available in a cubical volume (cell). To make the analytical treatment of such a process simple, initial volume of the cubic cell is transformed into a representative spherical cell of radius $r_{eq}$, which can be simply obtained as

$$r_{eq} = \left(\frac{3}{4\pi}\right)^{1/3} l = \chi l \tag{3.4}$$

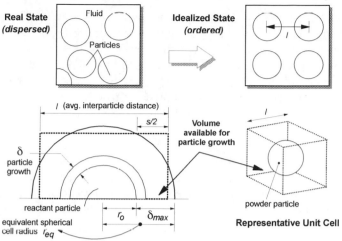

*Figure 3.1*  A representative unit cell of powder particle.

where $l$ is the length of the edge of the cubic cell and $\chi$ is the stereological factor. In such a representation, we can find that the touching of adjacent particles would occur when the thickness of expanding cluster $\delta$ equals the spacing $s$ (Figure 3.1). Furthermore, a complete filling of the cubic cell, or the equivalent representative spherical cell, would occur when the thickness of the expanding cluster equals $\delta_{max}$ given by

$$\delta_{max} = r_{eq} - r_o = r_o\left[2\chi\left\{G_o\left(1+\rho_p\omega_o\right)\right\}^{1/3} - 1\right]$$  (3.5)

The geometrical parameters, $r_{eq}$ and $\delta_{max}$, represent the information pertaining to the stereological description of the initial state of monosized particle dispersion.

### 3.1.2 *Macroscopic volumetric balance of hydration products*

Fresh cement paste is plastic in nature, consisting of a network of powder particles suspended in water. However, once the cement is set, very small changes in the overall volume take place. In this analysis, we ignore small changes in the gross volume that might take place due to bleeding or contraction while the paste is still plastic. Also, at any stage of hydration, the matrix contains the hydrates of various compounds, such as gel, unreacted powder particles, calcium hydroxide crystals, traces of other mineral compounds and large void spaces partially filled with water. As the hydration proceeds, new products are formed and deposited in the large water-filled spaces known as capillary pores. Moreover, the gel products also contain interstitial spaces, called gel pores, which are at least one order of magnitude smaller than the large voids of capillary pores. Due to the chemical reactions of hydration, water available in the capillary voids combines with the still unreacted powder minerals, leading to the formation of new hydration products.

As a physical basis for pore structure development, let us subdivide the overall pore space into three broad categories of different scales. These are the *interlayer*, *gel* and *capillary* porosity capacities. Capillary porosity exists in the large inter-particle spaces of powder particles, whereas gel porosity exists in the interstitial spaces of gel products or more specifically calcium silicate hydrate (C–S–H) grains. Interlayer porosity comprises the volume of water residing between the layer structures of C–S–H (Figure 3.2). For hydration reactions, only the water available in capillary pores is relevant since it represents the space where new hydration products can be formed. The moisture existing in interlayer porosity is a part of the physical structure of C–S–H and strongly bound to the surfaces. It is believed to be removable only under severe drying conditions, such as oven drying. The degree of hydration seems to be a useful parameter that can assist with the compution of the volume and mass of hydration products from a macroscopic point of view.

Bearing in mind these points, weight, $W_s$, and volume, $V_s$, of gel solids per unit volume of the cement paste can be computed, provided average degree of hydration, $\alpha$, and the amount of chemically combined water, $\beta$, per unit weight of hydrated powder material, are known. The amount of chemically combined water is dependent on the chemical characteristics of the powders. Here, the degree of hydration denoted by $\alpha$ is obtained as a ratio of the amount of heat liberated to the maximum possible heat of hydration for the mix (Chapter 2). The amount of chemically combined water denoted by $\beta$ is obtained based upon the stoichiometric balance of chemical reactions.

Computation of these parameters is discussed in the multi-component hydration heat model of cement (Chapter 2). As reported by Powers (1968b) and Feldman and Sereda (1968), we assume a layer structure for C–S–H mass with an interlayer spacing of a single water molecule. Moreover, it is assumed that the gel products have a characteristic porosity of $\phi_{ch}$ which is constant

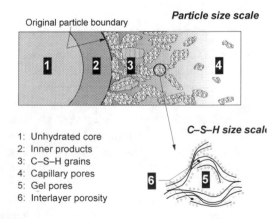

1: Unhydrated core
2: Inner products
3: C–S–H grains
4: Capillary pores
5: Gel pores
6: Interlayer porosity

*Figure 3.2* Schematic representation of porosity components of mortar.

at all the stages of hydration (Section 2.2.1). A value of 0.28 for $\phi_{ch}$ is usually reported for ordinary cement pastes (Neville 1991). The characteristic porosity contains the interstitial gel porosity as well as the interlayer porosity. Then, at any arbitrary stage of hydration, volume $V_s$ of the gel products in a unit volume of the paste can be obtained as

$$V_s = \frac{\alpha W_p}{1 - \phi_{ch}} \left( \frac{1}{\rho_p} + \frac{\beta}{\rho_w} \right) \tag{3.6}$$

where $W_p$ is the weight of powder materials per unit paste volume and $\rho_w$ is the density of chemically combined water ($\sim 1.25$ g/cm$^3$) (Neville 1991). The interlayer porosity ($\phi_l$) and gel porosity ($\phi_g$) existing in the gel products are computed from the following expressions

$$\phi_l = \left( t_w s_l \rho_g V_s \right)/2, \quad \phi_g = V_s \phi_{ch} - \phi_l \tag{3.7}$$

where $\rho_g$ is dry density of gel products $= \rho_p \rho_w (1+\beta)(1-\phi_{ch})/(\rho_w + \beta \rho_p)$, $t_w$ is the interlayer thickness (2.8Å), and $s_l$ is the specific surface area of the interlayer.

At this stage, the dependence of any specific surface area of interlayer on the mineralogical characteristics of the powder materials is not clear. While studies have reported values of the order of 500–600 m$^2$/g for Portland cements, only unreliable data exist for blended powder materials. In this chapter, we tentatively propose $s_l$ [m$^2$/gm] as

$$s_l = 510 f_{pc} + 1500 f_{sg} + 3100 f_{fa} \tag{3.8}$$

where $f_{pc}$, $f_{sg}$ and $f_{fa}$ denote the weight fractions of the Portland cement, blast furnace slag and fly ash in the mix, respectively. Lastly, from the overall volume balance of the paste, capillary porosity ($\phi_c$) or the volumetric ratio of larger voids can be obtained as

$$\phi_c = 1 - V_s - (1-\alpha)\left( W_p / \rho_p \right) \tag{3.9}$$

Equations 3.7 and 3.9 macroscopically describe the microstructure of the hydrating paste in terms of the porosity of different phases, once the average degree of hydration and the amount of chemically combined water due to hydration are known.

### 3.1.3 Expanding cluster model

After initial contact with water, powder particles start to dissolve and the reaction products are precipitated on the outer surfaces of particles and in the free-pore solution phase. Figure 3.2 shows a schematic representation of various phases at any arbitrary stage of hydration. Precipitation of the pore solution phase on the outer surfaces of particles leads to the formation of outer products, whereas so-called inner products are formed inside the original particle boundary where the hydrate characteristics are assumed to be more or less uniform. The hydration products have been reported to appear like needles, fibers, tubes, rods, foilitic, tree-shaped, prismatic crystals, tabulated, acicular, scalular, and so on. Moreover, the development of the cementitious microstructure involves the formation of various hydration products such as ettringite, C–S–H and CH at different degrees of hydration and time (Figure 3.3). The physical shape and size of these products might depend on various parameters such as cement composition, temperature, degree of hydration or the presence of additives. Here, it is indispensable to select a couple of primary factors in the development of physical chemistry characteristics of the microstructure in a multi-scale modeling that could also be put into practical computational use.

Thus, a simplified picture of the development of porosity and associated microstructure is presented. The microstructural properties of inner products are assumed to be constant throughout the process of pore structure formation. Moreover, it is assumed that hydrates of similar characteristics are formed at all the stages of hydration. While this might not be strictly accurate in the very early stages of hydration, there is evidence pointing to be the fact that the hydration products formed in the later stages are almost identical (Copeland 1956).

The hydration products deposited outside the original particle boundary are assumed to have similar characteristics to the inner products. It is assumed that the conventional capillary porosity exists between the grains of hydrates of the outer product as well as in the spaces between growing clusters. The grains of C–S–H hydrates in outer products as well as the inner products account for all of the finer gel and interlayer porosity. The characteristic porosity of the C–S–H mass, $\phi_{ch}$, and the granular characteristics of the gel, are assumed to be constant throughout the hydration process. Here, a value of 0.28 is assumed for $\phi_{ch}$. It has to be noted that this porosity in-

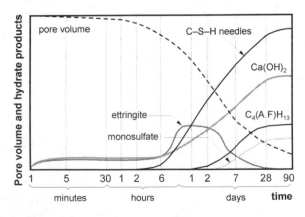

*Figure 3.3* Formation of several hydration products with time.

cludes the interlayer as well as the interstitial gel porosity. For this model of a growing cluster, the porosity distribution in the cluster around a partially hydrated particle would appear as shown in Figure 3.4. Moreover, the density at any point in this cluster, $\rho_\delta$, would depend on the mass of the gel and the bulk porosity at that point as

$$\rho_\delta = \frac{(1+\beta)(1-\phi_\delta)}{(1/\rho_p + \beta/\rho_u)} \tag{3.10}$$

where $\phi_\delta$ is the bulk porosity at any point in the outer product and $\rho_u$ is the specific gravity of the un-evaporable or chemically combined water. The bulk porosity distribution in the outer products is modeled by a power function as (Figure 3.4)

$$\phi_\delta = (\phi_{ou} - \phi_{in})\left(\frac{\delta}{\delta_m}\right)^n + \phi_{in} \tag{3.11}$$

where $\delta$ denotes the distance of any point in the cluster from the outer boundary of the particle, $\delta_m$ is the cluster thickness at the instant of hydration and $\psi_{in}$ is the porosity of inner products. This would be the characteristic porosity of the gel, $\phi_{ch}$, and $\phi_{ou}$ is the porosity at the outermost boundary of the expanding cluster and equals unity before the expanding cluster reaches the outermost end of the representative spherical cell of the particle. The parameter $n$ is the power-function parameter to take into account a generic pattern of deposition of products around the particle and might be used to take temperature effects into account. However, in the computations, we have adopted $n$ as unity.

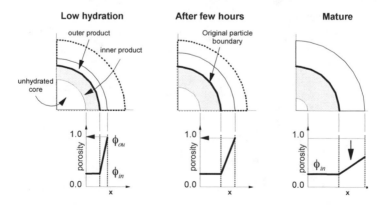

*Figure 3.4* Bulk porosity function of outer product and cluster expansion model.

### 3.1.4 Surface area of capillary and gel components

The macroscopic balance of the hydration products provides us with an estimate of the macroscopic quantities of various phases at different stages of hydration. However, several phenomena linked to the durability of concrete, such as transport processes, also require additional microstructure information, e.g., the specific surface areas or porosity distributions of each phase. It is not possible to obtain these parameters from the macroscopic discussions alone.

The specific surface area of gel can be estimated from adsorption experiments. Furthermore, if it is assumed that the hydration products formed at different stages of hydration are more or less identical, the porosity distribution parameter of gel products would be constant at all the stages of hydration. An important implication of this fact is that gel microstructural property can be estimated without any knowledge of the progress of hydration. From a view point of transport processes, however, the larger capillary pores are of great importance. Unlike gel porosity, no direct estimation of the capillary porosity distribution is possible, since capillary pores undergo dynamic changes, such as subdivisions, branching and filling-up with new hydrates, as the hydration proceeds. The specific surface area of capillary porosity must be computed at each stage of the hydration. This is accomplished by considering the hydration degree based on the expanding cluster model described in the previous section.

Let us consider Figure 3.5, which shows a reactant particle at an arbitrary stage of hydration. In this case, the weight of the hydrates existing in the inner and outer products should be equal to the total amount of hydrates formed up to that stage of hydration. The total amount of hydrates, $W_T$, as well as the amount of hydrates residing as inner products, $W_{in}$, can be obtained as

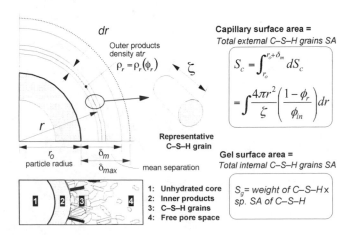

*Figure 3.5* A reactant particle at an arbitrary stage of hydration.

$$W_T = \frac{4}{3}\pi r_o^3 \alpha \rho_p (1+\beta) W_{in} = \frac{4}{3}\pi r_o^3 \alpha \left[ \frac{(1+\beta)(1-\phi_{in})}{(1/\rho_p + \beta/\rho_u)} \right] \tag{3.12a}$$

The amount of hydrate material existing in the outer products, $W_{ou}$, can be obtained by integrating the mass present at any arbitrary location in the cluster over the entire cluster thickness as (Figure 3.5)

$$W_{ou} = \int_{r_o}^{r_o+\delta_m} 4\pi r^2 \rho_r dr = \frac{4\pi(1+\beta)}{1/\rho_p + \beta/\rho_u} \int_0^{\delta_m} [1-\phi_\delta](\delta + r_o)^2 d\delta \tag{3.12b}$$

By applying the conservation of mass condition to Equation 3.12a and Equation 3.12b, such that $W_T = W_{in} + W_{ou}$ and simplifying the resulting expressions, we have the mass compatibility condition as

$$A\delta_m^3 + B\delta_m^2 + C\delta_m + D = 0$$
$$A = \{n(1-\phi_{in}) + 3(1-\phi_{ou})\}/\{3(n+3)\}$$
$$B = \{n(1-\phi_{in}) + 2(1-\phi_{ou})\}r_o/(n+2) \tag{3.13}$$
$$C = \{n(1-\phi_{in}) + (1-\phi_{ou})\}r_o^2/(n+1)^2$$
$$D = -(\alpha r_o^3/3)[\phi_{in} + \beta\rho_p/\rho_u]$$

In the cluster model described above, two different conditions of interest would arise. The first case would involve an unhindered free expansion of the cluster. The second condition would involve the expanding front of the cluster reaching the limit of the representative spherical cell, i.e., when $\delta_m = r_{eq}$. These two cases are discussed as follows.

### 3.1.4.1 Case 1: Free expansion of the cluster

The outer product cluster thickness, $\delta_m$, can be simply computed from Equation 3.13 by substituting the outermost cluster porosity, $\phi_{ou}$, as unity and solving the resulting cubic equation.

### 3.1.4.2 Case 2: Cluster thickness equals free space available for expansion

The bulk porosity at the outermost boundary of the product, $\phi_{ou}$, would gradually reduce so as to accommodate new hydration products. The porosity at the outermost end of the cluster, $\phi_{ou}$, can be obtained by substituting $r_{eq}-r_o$ as the cluster thickness, $\delta_{max}$, in Equation 3.13. Such an exercise leads to the following expression for $\phi_{ou}$.

$$\phi_{ou} = 1 - \frac{X+Y}{Z}\delta_{max} = kr_o$$

$$X = -n(1-\phi_{in})\left[\frac{k^3}{3(n+3)} + \frac{k^2}{n+2} + \frac{k}{n+1}\right] \tag{3.14}$$

$$Y = \frac{\alpha}{3}\left[\phi_{in} + \beta\frac{\rho_p}{\rho_u}\right]Z = \frac{k^3}{n+3} + \frac{2k^2}{n+2} + \frac{k}{n+1}$$

An implicit assumption, once again, is that the inner mass develops inside the original particle boundary with constant properties throughout the hydration process, whereas representative hydrate crystals of constant and uniform properties are deposited in the outer pore solution phase. With the maturity of the hydrating matrix, microstructural properties tend to become uniform in the outer as well as inner products. Consequently, difference of microstructural properties in the outer and inner products diminish with hydration. Once the variation of bulk porosity in outer products is known, we can proceed to determine the capillary surface area contained in the outer products. Let us consider a region of thickness $dr$ located at a distance $r$ from the particle center (Figure 3.5). If $dV_g$ and $\rho_g$ represent the real volume and dry density of hydrates in this region, and $\phi_r$ is the bulk porosity of the expanding cluster at $r$, then by equating the mass of hydrates contained in this region, we have

$$\rho_g dV_g = 4\pi r^2 dr\frac{(1+\beta)(1-\phi_r)}{1/\rho_p + \beta/\rho_u} \tag{3.15a}$$

We assume that the volume-to-surface area ratio of the solid grains of hydrates contained in this region is $\zeta$. This ratio may in fact be proportional to the average size (radius) of hydration products. Since the C–S–Hs constitute the major proportion of the gel, it can be assumed that the overall average hydrate size would correspond to the C–S–H grain size. The average size of the particles of cementitious materials has been reported to be of the order of 0.01 μm. While the C–S–H size may range from 0.002 μm to 0.1 μm over the complete history of hydration, the later stages of hydration are dominated by finer grain sizes. Powers and Brownyard (1947) reported the average gel particle size of 0.01 μm to 0.015 μm. Considering these discussions, we assume that $\zeta$, which represents the ratio of volume to the external surface area of a typical grain, is constant during the course of hydration. Note that an assumption on the actual size of these grains is not made. In this book, the following empirical expression is used to obtain $\zeta$ (nm) for blended powders

$$\zeta = 19.0f_{pc} + 1.5f_{sg} + 1.0f_{fa} \tag{3.15b}$$

Using the volume-to-surface area concept, the capillary surface area $dS_c$ contained in the region $dr$ is obtained by applying the chain rule of differentiation to $dV_g$ in Equation 3.15a and further simplifying as

$$dS_c = \frac{dS_c}{dV_g} dV_g = \frac{4\pi r^2}{\zeta} \left( \frac{1-\phi_r}{1-\phi_{in}} \right) dr \qquad (3.15c)$$

The specific surface area of capillary pores, $S_c$, per unit volume of the matrix can be obtained by integrating the differential Equation 3.15c over the entire cluster thickness and then normalizing it to the volume of the representative spherical cell. Similarly, the surface area of gel, $S_g$, is simply obtained as a product of the mass of gel products to the specific surface area of gel mass. Then, it yields,

$$S_c = \frac{3\delta_m}{\zeta r_{eq}^3 (1-\phi_{in})} \left( A\delta_m^2 + B\delta_m + C \right) S_g = W_s s_g \qquad (3.16)$$

where $s_g$ is the specific surface area of the hydrates and $A$, $B$, $C$ are similar to Equation 3.13. The specific surface area of hydrates is assumed to be 40 $m^2/g$ in this chapter (Neville 1991).

### 3.1.5 Computational model of the total paste microstructure

The size of pores present in the cement paste is distributed over several orders of magnitude, and the distribution of moisture in such a porous system is different for each different range of pore sizes considered. As the hydration proceeds, large spaces initially taken up by water are replaced by the increasing network of hardening mass which is colloidal in nature (Neville 1991). This hardening mass suspended in the liquid constitutes what is called a cement gel. Hydration progress leads to more and more liquid space being taken up by the gel and eventually a discontinuity occurring in the network of large voids filled with water. The pore structure thus formed in the hydrated mass is highly complex due to its complicated shape, interconnectivity and spatial distribution. These pores have traditionally been classified according to their sizes. For example, the complete pore classification for hardened cement paste proposed by Comité Euro-International du Béton (1992) is shown in Figure 3.6, which applies the International Union of Pure and Applied Chemistry (IUPAC) classification. The capillary pores can be considered as the remnants of water-filled space existing between the original cement grains, while micro-pores are part of the C–S–H component. The usual measurement of these pores using a mercury intrusion or helium pycnometry technique assumes a uniform pore shape, whereas the pores form a network of highly irregular geometry.

From a viewpoint of moisture transport, the authors subdivide the overall cementitious micro-pore structures into three basic components: capillary

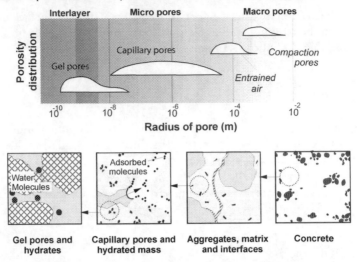

*Figure 3.6* A schematic representation of porosity classification in concrete.

pores, that are actually empty spaces left between partially hydrated cement grains; gel pores, that are formed as an internal geometrical structure of the C–S–H grains; and the physically fixed interlayer moisture that is a structural part of the C–S–H gel. The moisture transport takes place only into the gel and capillary components of the cementitious microstructure, since the moisture physically bound as interlayer water is unable to move under the usual pressure gradients. However, a local rearrangement of moisture from the interlayer structure to gel pores might take place, depending on the severity of drying or wetting conditions. The interlayer porosity might not influence the transport behavior of fluids that do not interact with the layer structure of the C–S–H gel. Then, for analytical purposes, total porosity, $\phi$, of hardened cement paste can be mathematically expressed as a summation of the capillary, gel and interlayer porosity. If the distribution functions of porosity for these porosity systems are known, we can obtain the total porosity distribution of the paste as a function of the pore radius.

As the exact characteristics of the interlayer component of total porosity are under investigation, interlayer porosity is simply lumped with the porosity distributions of gel and capillary porosity, called *external* or *capillary-gel porosity*, to obtain the total porosity distribution of cement paste as

$$\phi(r) = \phi_{lr} + \phi_{cg}(r) \tag{3.17}$$

where $r$ is pore radius, $\phi_{lr}$ is interlayer porosity and $\phi_{cg}(r)$ is combined porosity distribution of gel-capillary pores. In reality, $\phi_{cg}(r)$ is a complex and continuous distribution function. It can be imagined to be similar to the distributions as obtained by mercury intrusion porosimetry (MIP) tests on hardened cement paste. In the multi-scale modeling, however, we have categorized this system broadly into capillary and gel pores. With such a classification, the non-interlayer porosity distribution $\phi_{cg}(r)$ of cement paste can be obtained as

$$\phi_{cg}(r) = \frac{1}{\left(\phi_{cp} + \phi_{gl}\right)}\left[\phi_{cp}(r) + \phi_{gl}(r)\right] \tag{3.18}$$

where $\phi_{cp}$ is the total capillary porosity and $\phi_{gl}$ is the total gel porosity. Each of the capillary and gel porosity distributions, $\phi_{cp}(r)$ and $\phi_{gl}(r)$, is represented by a simplistic Raleigh–Ritz (R–R) distribution function as

$$V = 1 - \exp(-Br), \quad dV = Br\exp(-Br)d\ln r \tag{3.19}$$

where $V$ represents the fractional pore volume of the distribution up to pore radius, $r$, and $B$ is the sole porosity distribution parameter, which in fact represents the peak of porosity distribution on a logarithmic scale. A simple porosity distribution has been chosen to facilitate numerical computations and parametric evaluations. A bimodal analytical porosity distribution of the cement paste can therefore be obtained as

$$\phi(r) = \phi_{lr} + \phi_{gl}\left(1 - \exp\left(-B_{gl}r\right)\right) + \phi_{cp}\left(1 - \exp\left(-B_{cp}r\right)\right) \tag{3.20}$$

where the distribution parameters, $B_{cp}$ and $B_{gl}$, correspond to the capillary and gel porosity components, respectively. Interlayer porosity, $\phi_{lr}$, is believed to exist in much smaller pores than gel porosity does. Apart from the cement paste pores, there may be a distinct class of pores present in the aggregates. However, it is generally thought that these pores do not directly take part in the transport process of liquid and gases. Moreover, the aggregate cement paste interface shows distinct zones of high porosity and coarse microstructure. Characteristics of such pores will again be discussed in Section 3.5 and Chapter 7. These pores may play an important part with regard to the transfer of liquid between cement paste and aggregates. Our immediate target is to obtain the pore structure parameters that will enable us to compute the dynamic porosity distributions with time, based upon the degree of hydration.

The use of a simple computational pore structure enables easy evaluation of surface area parameters. Let us consider a unimodal R–R distribution function given by Equation 3.19. If we assume a cylindrical pore shape in such a distribution, then the pore distribution parameter $B$ can be obtained from the following relationship, if $S$, the surface area per unit volume of the matrix, is known (Figure 3.7), i.e.

$$S = 2\phi\int r^{-1}dV = 2\phi\int_{r_{min}}^{\infty} B\exp(-Br)d\ln r \tag{3.21}$$

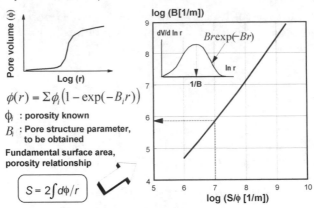

*Figure 3.7* Relationships of pore structure parameter $B$ and $S/\phi$.

where $r_{min}$ is the minimum pore radius. Unfortunately, the above expression cannot be evaluated analytically as a closed-form solution. Therefore, the authors use an explicit relationship which has been obtained by fitting the accurate numerical evaluations of the above integral for a large number of data-sets which relate $B$ as a function of $S/\phi$.

From earlier discussions, the overall cementitious microstructure can be represented as a bimodal porosity distribution by combining the inner and outer product contributions to the total porosity function $\phi(r)$ as

$$\phi(r) = \phi_l + \phi_g\{1 - \exp(-B_g r)\} + \phi_c\{1 - \exp(-B_c r)\} \qquad r : pore\ radius$$

(3.22)

The distribution parameters $B_c$ and $B_g$ can easily be obtained from Equation 3.16, Equation 3.21 and Figure 3.7, once the surface areas and relevant porosity of the gel and capillary components are computed. The overall scheme of pore structure development is outlined in Figure 3.8. This numerical model of cementitious microstructure is dynamic in nature, so the effects of hydration can be dynamically traced with time.

Figure 3.9 shows a comparison among the computed micro-pore structures of mortar of water-to-cement ratios of 25%, 45% and 65%. The pore structure has been computed at a degree of hydration of 0.15 and the maximum achievable for each mortar in sealed curing conditions (60 days). It can be seen that an increase in the gel pore volume fractions occurs gradually, and the ratio of finer porosity to coarser porosity increases for low W/B mortars. It appears that generally, the peak of porosity distribution lies between 20 nm to 120 nm for most of the mature concrete. The microstructure development model described in this chapter will enable us to consider the coupling of moisture transport and the hydration phenomenon in a rational way, where the hydration-dependent effects are automatically considered in the overall computational scheme.

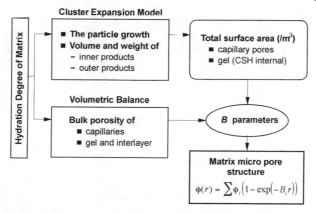

*Figure 3.8* Schematic outline of the pore structure development computation.

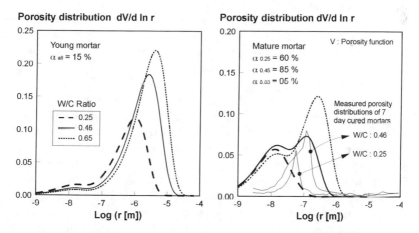

*Figure 3.9* Comparison of computed microstructures for different mortars.

## 3.2 Basic modeling of moisture transport and thermodynamic equilibrium

In this section, a moisture transport model is formulated to consider the multi-phase dynamics of liquid and gas in multi-scale pores, and the molecular diffusive movement is combined with the bulk motion of moisture. The macroscopic moisture transport characteristics of flow, such as conductivity and isotherms, are obtained directly from the microstructure of porous media. This is achieved by considering the thermodynamic equilibrium conditions which are typical for extremely slow flow and the random geometrical nature of the pores spread over a large order of magnitude in cementitious materials. Special attention is directed to wetting and drying cycles for practical use of the model for concrete performance under ordinary conditions.

In particular, attention is drawn to phenomena usually not well investigated. These include the hysteresis phenomena observed in the moisture isotherms of porous media and the anomaly in observations of the intrinsic permeability of hardened cement paste. The hysteresis is accounted for by analytical models that consider the effect of entrapment of water in pores. The same logic is used to predict the moisture content of concrete under an arbitrary wetting or drying history.

### 3.2.1 Coupled liquid and vapor transport formulation

The driving forces for species transport in isothermal porous media are viscous forces, gravity, hydrodynamic dispersion and pressure forces. Viscous forces may arise due to the motion of the species and resulting shear stress acting along the fluid surface. Hydrodynamic dispersion involves mechanical and molecular diffusion, which may occur due to gradients of concentration and temperature. Pressure forces also arise in an unsaturated system due to capillary forces. The capillary forces depend upon the interfacial tension between different phases, on the angle of contact of the interface to the solids and on the radius of curvature of the interface. In general, it is the difference in energy or potential that causes the mass transport, that is to say, movement always occurs from higher to lower energy states. In this section, the viscous and capillary driving forces are the focus of discussion.

The process of moisture transport in a porous medium can be divided into five major steps. For the sake of simplicity, let us idealize the porous medium as a single pore with a neck at each end (Figure 3.10). The very first step (a) involves the adsorption of water molecules to the solid pore surface. Until a complete monolayer of water molecules is formed, vapor flux cannot be transmitted. Under quasi-equilibrium conditions, the thickness of the adsorbed layer depends upon the vapor pressure in the pore. Subsequent to this step, an unimpeded flux of vapor takes place (b) where the vapor behaves like an ideal gas. As the vapor pressure gradually increases due to the vapor

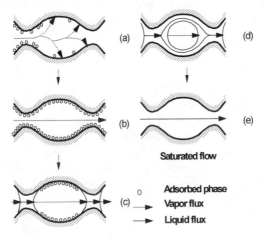

*Figure 3.10* Progress of moisture sorption in an idealized pore.

flux in the pore, condensation occurs (c) at the neck portions of the pore. In this situation, the system is impervious to inert gases and pervious to vapor through a process of distillation, in which necks act as short-circuited paths for vapor movement. This step can also be termed *liquid-assisted vapor movement*. In stage (d), with substantial thickness of the surface adsorbed water formed, flow in thin liquid films may occur. Eventually, there is a transition to the final stage (e) where vapor pressure increases to a point such that the condensation occurs within the pores. Subsequent to this step, saturated flow begins in the porous system. The actual porous medium contains pore sizes spread over several orders of magnitude, therefore, at any given humidity, different steps of moisture transport would be in progress in a small representative elementary volume (REV). Flow in the saturated pores would generally be in the laminar regime and can be explained by Hagen–Posuille flow behavior. Integrating the moisture transport over all the saturated and unsaturated pores present in the REV of concern, the total moisture transport behavior in the porous medium can be obtained.

### 3.2.1.1 Mass and momentum conservation

In this section, general equations of mass and momentum conservation for the moisture transport in an isotropic porous medium are developed under an assumption of an isothermal and non-deformable porous medium. A more general case for moisture transport in a developing microstructure will be discussed later and modeling of moisture equilibrium and transport at arbitrary temperature will be present in the next section. All the equations discussed in subsequent sections are a representation of macroscopically averaged properties, such as velocity, saturation or pressure. In other words, porous media are viewed as a continuum and balance equations are developed over a representative elementary volume which represents an average of the properties existing at microscopic level. All of the discussion is limited to the one-dimensional case only. However, the extension of these equations to three dimensions is straightforward.

A general notation is used in the formulation for easy identification of various terms. A single subscript denotes a particular component or a phase of concern, e.g., $l$ represents a liquid water or liquid phase, $g$ represents the gaseous phase, $a$ represents the dry air component of the gas phase, and $v$ represents the vapor (steam) component of the air phase. Italic script is used to denote the major properties of interest in porous media. The three major properties identified here are given in Table 3.1.

#### 3.2.1.1.1 MASS CONSERVATION EQUATIONS

In the REV as shown in Figure 3.11, only liquid and gas phases are present. They are assumed to fill up the network of pores. Thus, the volumetric compatibility equation can be stated as

*Table 3.1* General notations

| Symbol | Property | Units [dimension] |
|--------|----------|-------------------|
| $S_i$ | Saturation of the *i-th* phase | $m^3/m^3$ [$L^3/L^3$] |
| $P_i$ | Pressure of the *i-th* phase | Pa [$L^{-1}MT^{-2}$] |
| $u_i$ | Velocity of the *i-th* phase or component | m/sec [$LT^{-1}$] |

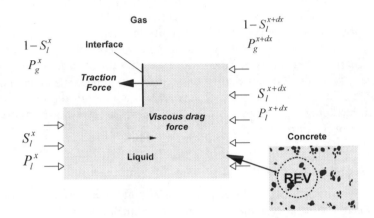

*Figure 3.11* Macroscopic equilibrium of multi-phases in REV.

$$S_l + S_g = 1 \tag{3.23}$$

where $S_g$ represents the fraction of pores filled up by the gaseous phase and $S_l$ represents the volumetric saturation level of pores due to liquid water. Using this relationship, mass conservation for liquid and gaseous phase components can be obtained in a differential form (liquid) as expressed below.

$$\frac{\partial}{\partial t}\left(\rho_l \phi S_l\right) + \frac{\partial}{\partial x}\left(\rho_l \phi S_l u_l\right) + v = 0 \tag{3.24}$$

where $\phi$ is the bulk porosity of concrete, $S_l$ is the volumetric saturation of pores due to liquid water expressed as a fraction, $\rho_l$ is the density of liquid water, $u_l$ is the averaged velocity of pore water in the REV, $v$ is the rate of phase transition (mass transformation) of moisture from liquid to the vapor phase per unit of porous volume. A moisture loss term due to hydration and changes in bulk porosity will be discussed later.

The gas phase actually consists of vapor and air components. Then, mass conservation of these two components can be added to give total mass conservation of the gaseous phase. Here, it is assumed that none of these components, including $H_2O$, $O_2$, $N_2$ or other possible oxides, is reacting with the solid porous matrix. Then, we have

$$\frac{\partial}{\partial t}\left(\rho_g \phi(1-S_l)\right) + \frac{\partial}{\partial x}\left(\rho_g \phi(1-S_l)u_g\right) - v = 0 \tag{3.25}$$

where $\rho_g$ is the bulk density of the gas phase, and $u_g$ is the averaged gaseous phase velocity in the REV. The gas phase velocity can be determined from the molar concentrations of vapor and dry air components present in the gas phase. Next, the momentum conservation for gaseous and liquid phases, which describes the equation of motion of multi-phases in porous media, is discussed.

3.2.1.1.2 MOMENTUM CONSERVATION OF PHASES

Primary mechanisms considered here for liquid and vapor transport are pressure forces that may arise in an unsaturated system due to capillary tension or bulk hydraulic heads for a completely saturated system. Also, viscous forces will act on the phase of interest by virtue of the motion of species and resulting shear stresses. If $P_l$ and $P_g$ denote the bulk liquid and gas pressure, we can obtain the momentum conservation of the liquid and gas phase as described below (see Figure 3.11).

$$\frac{\partial}{\partial t}\left(\rho_l S_l u_l\right) + \left(\rho_l S_l \frac{\partial u_l}{\partial x}\right)u_l + u_l \frac{\partial}{\partial x}\left(\rho_l S_l u_l\right) + v u_l = -\frac{\partial}{\partial x}\left(P_l S_l\right) + P_g \frac{\partial S_l}{\partial x} - u_l K_{Dl} S_l + T_{gl}$$

$$\tag{3.26}$$

where $K_{Dl}$ represents the average viscous drag force acting over a unit volume of the liquid phase per unit of liquid velocity. Here, a linear relationship of viscous drag with velocity is assumed, since the rate of moisture progress is very slow and the flow can be assumed to be well within the laminar regime. Also, $T_{gl}$ denotes the averaged traction force acting over the liquid phase arising due to surface tension forces. This force is assumed to be the primary driving force for the liquid moisture movement in an unsaturated porous medium. Equation 3.26 can be further simplified by using the mass balance Equation 3.24 as

$$\rho_l S_l \frac{Du_l}{Dt} = -\frac{\partial}{\partial x}\left(P_l S_l\right) + P_g \frac{\partial S_l}{\partial x} - u_l K_{Dl} S_l + T_{gl} \tag{3.27}$$

where $D/Dt$ is the substantial derivative term and equals $\partial/\partial t + u_l \partial/\partial x$. The viscous drag coefficient $K_{Dl}$ would generally depend on the porous medium micro-structural properties and the viscosity characteristics of the liquid. It must be noted here that this parameter may depend on the interaction of fluid with the surrounding porous medium. The resulting implications of this phenomenon on the fluid conductivity as well as overall transport will be taken up in a later section. As in the liquid phase, momentum conservation for the gas phase can be given by

$$\rho_g(1-S_l)\frac{Du_g}{Dt}=-\frac{\partial}{\partial x}\left(P_g(1-S_l)\right)-u_gK_{Dg}(1-S_l) \tag{3.28}$$

where $K_{Dg}$ is the average viscous drag coefficient acting over the unit volume of the gas phase and $u_g$ is the average convective bulk gas phase velocity. Bulk pressure gradients are assumed to be the only driving force for the above derivation. The average velocity of vapor and air components of the gas phase can be obtained as

$$u_v=u_g-\frac{D_v}{\rho_v}\frac{\partial\rho_v}{\partial x}u_a=\frac{1}{\rho_a}\left(\rho_gu_g-\rho_vu_v\right) \tag{3.29}$$

where $u_v$, $u_a$ are bulk vapor and air velocity respectively, $\rho_v$, $\rho_a$ are the bulk density of vapor and air components, and $D_v$ is the bulk molecular diffusivity of the vapor component in porous media.

### 3.2.1.2 Reduction to classical formulation

The system of equations presented in previous sections contains many variables as well as numerous unknown constitutive laws involving various terms. Without compromising significantly on accuracy, we make some approximations concerning the nature of mass transport. The first approximation is that the total gas pressure in porous media is assumed to be constant and equal to the atmospheric pressure. Since concrete is a porous medium consisting of a very fine porous network, the change in bulk liquid pressures is usually a few orders of magnitude greater than the change in gas pressure. Then, the constant gas pressure assumption appears to be adequate for practical cases of interest. Thus, we assume the gas phase in motion due to molecular diffusion and insignificant convection. Then, the flux of moisture in the vapor phase can be obtained by

$$q_v=-D_v\frac{\partial\rho_v}{\partial x} \tag{3.30}$$

Second, we assume that the total mass of vapor present in the REV can be disregarded compared to the liquid mass. In other words, to obtain the total mass conservation of moisture (liquid and vapor), the conservation equations for liquid and vapor phases (Equations 3.24 and 3.25) are added up such that the phase transition term is cancelled out and the mass of the vapor component is disregarded. Then, we have the total mass conservation for moisture as

$$\frac{\partial\theta_w}{\partial t}+divq_w=0 \tag{3.31}$$

where $\theta_w \approx \phi\rho S$ is the total mass of water present in a unit volume of concrete expressed in kg/m$^3$ (as the mass of moisture in vapor form is disregarded). Here, $q_w$ is the total flux expressed in weight (kg) per unit area per second.

Third, the velocity of the liquid phase is assumed to be extremely small, with no acceleration as it moves inside the porous medium. Then, the substantial differential terms of Equation 3.27 by $D/Dt$ are disregarded. In other words, convective terms of the liquid phase are ignored. With these assumptions, Equation 3.27 reduces to

$$-\frac{\partial}{\partial x}(P_l S_l) + P_g \frac{\partial S_l}{\partial x} - u_l K_{Dl} S_l + T_{gl} = 0 \tag{3.32}$$

Furthermore, the equilibrium condition of the non-accelerating liquid–gas interface enables us to obtain an expression for the traction force $T_{gl}$ for a unit volume of porous body as

$$\left(P_g - P_l\right)\frac{\partial S}{\partial x} = -T_{gl} \tag{3.33}$$

Thus, there is no driving force either when saturation gradient is zero or when capillary pressure difference equals zero. Substituting the above expression for traction force into the approximate momentum balance of Equation 3.32, we have the following expression for the averaged flux, $q_l$, of moisture in the liquid phase.

$$q_l = \rho\phi S u_l = -\frac{\rho\phi S}{K_{Dl}}\frac{\partial P_l}{\partial x} = -K_l \frac{\partial P_l}{\partial x} \tag{3.34}$$

where $K_l$ is the saturation content-dependent permeability coefficient and a parameter representing the ease with which a liquid can flow through the complex porous network. Using Equation 3.30 and Equation 3.34, we have total moisture conservation for concrete in one dimension as

$$\frac{\partial \theta_w}{\partial t} = \frac{\partial}{\partial x}\left[D_v \frac{\partial \rho_v}{\partial x} + K_l \frac{\partial P_l}{\partial x}\right] = \frac{\partial}{\partial x}\left[D(\theta_w)\frac{\partial \theta_w}{\partial x}\right] \tag{3.35}$$

where $D(\theta_w)$ is the bulk diffusivity of concrete. This parameter can be obtained by considering Kelvin's thermodynamic equilibrium condition, which gives a relationship between the relative liquid pressure $P_l$ and the relative humidity of the atmosphere with which it is in equilibrium as

$$\Delta P_c = P_l = \frac{\rho RT}{M}\ln h \tag{3.36}$$

where $\Delta P_c$ is the pressure difference due to capillarity, $\rho$ is the density of the liquid, $M$ is the molecular mass of the liquid, $R$ is the universal gas constant,

$T$ is the absolute temperature in K, and $h$ is the relative humidity, i.e., the ratio of vapor pressure to the fully saturated vapor pressure. The above expression is valid only under the assumptions made when defining the capillary pressure difference. The change in liquid pressure given by Equation 3.36 is in fact equivalent to the change in free energy of the homogenous bulk phase, i.e., water. Using Equation 3.36, we can obtain the conventional diffusivity parameter $D(\theta_w)$ as

$$D\left(\theta_w\right) = \left(\rho_s D_v + K_l \frac{\partial P_l}{\partial h}\right)\frac{\partial h}{\partial \theta_w} \tag{3.37}$$

where $\rho_s$ is the saturated vapor density such that $h = \rho_v/\rho_s$. The diffusivity parameter, $D(\theta_w)$, combines the contributions of moisture transfer in the liquid as well as the vapor phase. Also, the underlying parameter contains all the details regarding the geometrical characteristics of pore structure and the moisture retention characteristics defined by the moisture isotherms. In the conservation Equation 3.35, secondary effects, such as gravity terms, are ignored since the capillary pore pressure gradients are usually a few orders of magnitude higher than the gravity terms.

For analyzing the generic saturated–unsaturated flow in a porous medium, a pore pressure-based formulation is usually desirable, since a diffusivity based model is only valid in the unsaturated flow regime. A pore pressure-based formulation is also more robust as the transport coefficients are continuous with respect to pore pressure. In the transformation of Equation 3.35 to a pore pressure-based formulation, special attention must be paid to the hydration and microstructure development phenomena. During the early-age hardening of concrete, pore water content is dependent not only on the pore water pressures but also on the micro-structural characteristics of the cementitious matrix. Based on our definition, the rate of change of unit water content $d\theta_w/dt$ in concrete can be expressed as a summation of the rate of change of bulk porosity $d\phi/dt$, density of water $d\rho/dt$ as well as the degree of saturation $dS/dt$.

Furthermore, change of the microstructure will result in a varying degree of saturation, $S$, even if the pore pressures are kept constant. If the cementitious microstructure at a particular stage of hydration could be represented by the microstructure-related parameter $B_m$ (Section 3.1), then the degree of saturation could be expressed as a function of the pore pressure $P$ as well as the microstructure, denoted by $B_m$. The generic form of unit water content of concrete would therefore be expressed as $\theta_w = \rho\phi S(P, B_m)$. If a sink term of moisture content changes in a control volume, e.g., $Q_p$ which represents the rate of loss of moisture in a porous medium due to chemical fixation during hydration, changes in bulk porosity and microstructure or other such processes are also included. Then Equation 3.35 can be extended using the above-mentioned discussions and alternatively expressed in liquid pore pressure terms as

$$\alpha_P \frac{\partial P}{\partial t} - div(D_P \nabla P) + Q_P = 0 \tag{3.38a}$$

$$\alpha_P = \phi\left(S\frac{\partial \rho}{\partial P} + \rho\frac{\partial S}{\partial P}\right)$$

$$D_P = K_l + D_v\frac{\partial \rho_v}{\partial P} = K_l + K_v \tag{3.38b}$$

$$Q_P = Q_{pd} + Q_{hyd} + Q_{other}$$

$$Q_{other} \approx 0$$

where $\alpha_P$ is the specific fluid mass capacity parameter that approximately represents the amount of water that the concrete can absorb or release for a unit change in the liquid pore pressure potential. It can be noted that the extended definition of unit water content of concrete has been used in deriving Equations 3.38. The terms denoting the change in the shape of the microstructure during hydration contribute significantly in Equation 3.38b and must be included for a microstructure-based mass transport theory in porous media. Here $D_P$ represents the macroscopic moisture conductivity of concrete owing to the pore pressure potential gradients. Temperature dependencies in moisture capacity and moisture conductivity terms will be discussed later (Section 3.3). A summarized representation of the approach adopted in this book to obtain various coefficients of Equation 3.38 is shown in Figure 3.12. Various constitutive material models of transport behavior in concrete are explained later.

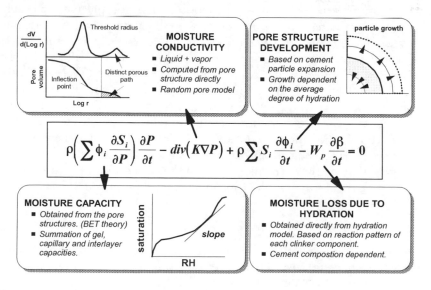

*Figure 3.12* Schematic representation of moisture transport modeling in concrete.

It should be noted that the sink term $Q_P$ represents the rate of internal moisture loss due to hydration and related effects. It is obtained by considering the changes in the bulk porosity distribution as well as the amount of water bound to the hardened cement paste structure by dynamically coupling a moisture transport model to a multi-component heat of hydration model and a microstructure formation model, discussed in Chapter 2 and Section 3.1.

During early-age hydration, a gradual reduction in the bulk porosity of concrete takes place, accompanied by a shift of porosity distribution to finer pore radii. Also, a substantial amount of free moisture is consumed and firmly fixed as chemically combined water of hydration products. The first term of $Q_P$ in Equation 3.38, $Q_{pd}$, represents the contributions due to the change in the bulk porosity distribution of the hardened cement-paste matrix. It includes changes in the bulk porosity as well as distributions of the interlayer and gel-capillary components. This reduction in the change of individual porosity results from the continual formation of new hydration products around cement grains, depending on the degree of hydration of each mineral clinker component of cement. The updated porosity at any stage of hydration can be computed from the pore structure formation model. Therefore, the first term is explicitly obtained as

$$Q_{pd} = \rho \left( S \frac{\partial \phi}{\partial t} + \phi \frac{\partial S}{\partial B_m} \frac{dB_m}{dt} \right) \tag{3.39}$$

A rigorous treatment of the effect of the dynamic nature of porosity distribution on the moisture transport formulation can be found in Section 3.2.5. Primary coupling of the hydration and moisture transport phenomena occurs due to the second term of $Q_P$ in Equation 3.38b, i.e. $Q_{hyd}$, which represents the rate of fixation of pore water as a chemically combined part of the C–S–H due to hydration and requires an accurate computation, especially during the early stages. The rate of moisture loss due to self-desiccation, $Q_{hyd}$ ($= d(W_p \alpha \beta)/dt$), is dependent on the degree of hydration, $\alpha$, which is in turn dependent on the available free water (taken as capillary condensed water in this model, Section 2.1.5). The incremental amount of combined water due to hydration is obtained from Equation 2.7 (Chapter 2) using a multi-component heat of hydration model of cement. This interdependency of moisture transport and hydration along with a dynamic change in the cementitious microstructure makes the early-age hydration problem dynamically coupled.

### 3.2.2 Isotherms of moisture retention

Due to the various deterioration processes associated with the presence of moisture in concrete, an accurate prediction of moisture content in concrete under arbitrary environmental conditions is essential for developing a rational and quantitative performance evaluation system for concrete struc-

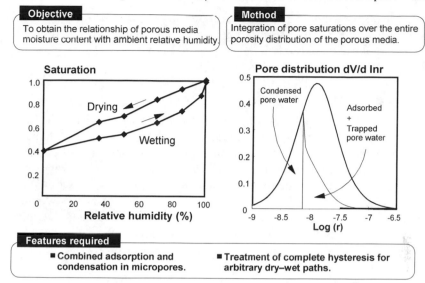

*Figure 3.13* Definition of the problem of concrete water content prediction.

tures. Moisture in the cementitious microstructure can be present in both the liquid and vapor phases. Usually, the volumetric content and thermodynamic behavior of these two phases of moisture dispersed in capillary and gel pores are determined by Kelvin's equation, which expresses the thermodynamic equilibrium between liquid water and the vapor phase. The problem of predicting the water content of concrete under arbitrary environmental conditions is addressed at this stage by considering the saturation states of different classes of micro-pores of concrete and integrating their saturation. A schematic description of the isotherm computation problem and the methodology required is illustrated in Figure 3.13.

### 3.2.2.1 Thermodynamic equilibrium of phases

Under normal conditions, some of the pores of the cementitious microstructure will be completely filled with water whereas others might be empty or contain a thin adsorbed layer of water molecules on the internal solid surfaces. Under static and isothermal conditions, the liquid moisture dispersed in the cementitious microstructure will generally be in equilibrium with the vapor phase surrounding the completely filled pores. For such a system, a small change in the vapor pressure or relative humidity of the gas phase present in microstructure will cause a corresponding change in the liquid pressure phase, so as to satisfy the equilibrated energy state principle of thermodynamic equilibrium. If we ignore the surface energy effects of the adsorbed water, the change in liquid pressure can be treated as an equivalent of the change in the free energy of the homogenous bulk phase, i.e., pore water. By arbitrarily assigning the saturated state of porous media to a zero capillary pressure or the reference state, we can obtain the pressure difference due to capillarity across a liquid–vapor interface as

$$\Delta P_c = P_l = \frac{\rho RT}{M} \ln h \tag{3.40}$$

where $\Delta P_c$ is the pressure difference due to capillarity, and $P_l$ is the liquid pressure (relative to the vapor pressure that is usually small and constant, and hence usually set as zero for computational ease). Equation 3.40 predicts a large drop in liquid pressure associated with a small decrease in vapor pressure, especially at low relative humidity (*RH*). From a microscopic point of view, the pressure difference between the liquid–vapor phases would be balanced by the curved vapor–liquid interfaces formed in numerous pores of the microstructure. The stress continuity across such liquid–vapor phase interfaces, which shows a large difference in bulk pressures, can be obtained by considering the interface equilibrium. We will primarily confine our discussion of interface equilibrium to the system of hydrated solid matrix and moisture only. The continuity of stress field across such an interface is modified by the effect of surface tension. For a generic curved interface, the pressure on one side differs from the pressure on the other by (Figure 3.14)

$$p_2 - p_1 = P_g - P_l = \gamma \left( \frac{1}{R_a} + \frac{1}{R_b} \right) \tag{3.41}$$

where $\gamma$ is the surface tension of the liquid, and $R_a$ and $R_b$ are radii of curvature of the interface in a pair of perpendicular directions. If the surface tension is uniform, the shear stress is continuous across an interface. However, the presence of impurities or surface active agents or temperature gradients can set up the gradients of surface tension. These effects can be disregarded for an ideal isothermal system. Equation 3.41 may also be termed the equilibrium form of the surface momentum equation at the micro-scale. If the vapor phase pressure is set as constant and equal to zero, we obtain the famous Kelvin's equation by combining Equation 3.40 and Equation 3.41 as

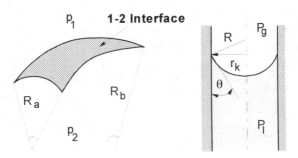

*Figure 3.14* Interface equilibrium in an idealized pore.

$$P_l = -\gamma\left(\frac{1}{R_a} + \frac{1}{R_b}\right) = \frac{\rho RT}{M}\ln h \qquad (3.42)$$

where $P_l$ is the relative liquid pressure. The curvatures of the interface would depend on the assumptions that we make regarding the geometrical description of how a pore is filled. Clearly, Equation 3.42 defines the potential or energy state of water in terms of relative liquid vapor pressure.

Next, we assume that the complex micro-pores are cylindrical in shape. For an interface formed in such a pore, radii $R_a$ and $R_b$ are equal to the radius of the pore $r_s$. Furthermore, under ideal conditions, the angle of contact of the liquid water to the solid surface is assumed to be zero. Then, from Equation 3.42, we have

$$r_s = -\frac{2\gamma M}{\rho RT}\frac{1}{\ln h} \qquad (3.43)$$

If the porosity distribution of the microstructure is known, Equation 3.43 enables us to obtain the amount of water present in the microstructure at a given ambient relative humidity (RH). This is because, to satisfy the equilibrium conditions, all the pores of radii smaller than $r_s$ would be completely filled whereas others would be empty. By integrating the pore volume that lies below pore radius $r_s$, we obtain an expression of water content in a porosity distribution $V$ given by Equation 3.19 as (Figure 3.15)

$$S = \int_0^{r_s} dV = 1 - \exp(-Br_s) \qquad (3.44)$$

It might be possible to obtain the water content of the concrete if the same principle as above is applied to a representative microstructure of concrete. Such a model completely disregards a significant amount of moisture in concrete present as the adsorbed phase. Also, this model predicts an explicit

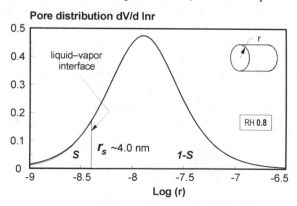

*Figure 3.15* Moisture distribution in a porosity distribution.

one-to-one relationship of water content and relative humidity. That is, the water content as obtained from this model would not be dependent on the history of drying and wetting. However, it is experimentally known that the water content in concrete is different under drying and wetting stages, even if exposed to the same relative humidity. This phenomenon is called *hysteresis* and can be explained by various mechanisms, such as the "ink-bottle effect", differences in the energy of adsorption and desorption of water molecules to micro-pore surfaces, and so on. A rational and quantitative method of expressing this irreversible hysteresis phenomenon for practical evaluations of moisture content in a porous body does not exist. Therefore, in the subsequent sections, we attempt to build a generic, path-dependent absorption–desorption model of concrete, which considers the adsorption phenomenon as well as the path dependence of water content under arbitrary drying–wetting conditions.

### 3.2.2.2 *Ideal adsorption models and absorption isotherms*

The moisture in a cementitious microstructure is generally dispersed into interlayer, gel and capillary porosity (Section 3.1.1). The total moisture content of the hardened cement paste can be obtained by summing the amount of water present in each of these components. While moisture content in the capillary and gel components of porosity can probably be explained by physical thermodynamic models of condensation and adsorption, the same cannot be said about the interlayer component of porosity. This is because the interlayer water probably exists in extremely small spaces between C–S–H sheets, which are of the order of only a few water molecule diameters thick (Neville 1991). The exact behavior of such water in terms of its entry and exit into this layer structure cannot be explained by a physical theory. For this reason, the contribution of the interlayer component of isotherms is obtained by direct empirical isotherms and is explained later.

The molecules at the surface of a solid porous matrix have higher energy due to the unattached molecules. In presence of molecules of a different substance, physical binding of these molecules takes place due to Van der Waal's forces. This process of physical bond formation is called *adsorption*. Due to this interaction, a layer of finite thickness of molecules of different substance or adsorbate is attached over the adsorbent surfaces. For hardened cement paste moisture systems, various models have been proposed that compute the thickness of this adsorbed water film. Bradley's equation, for which the constants for silicate materials have been obtained by Badmann *et al.* (1981), can be stated as

$$t_a = (3.85 - 1.89\ln(-\ln h)) \times 10^{-10} \tag{3.45}$$

where $t_a$ is the statistical thickness of the adsorbed water layer in meters, and $h$ is the actual pore relative humidity. Another theory that deals with the general case of adsorption of water molecules over a plane surface was developed by Brunauer *et al.* (1938) and is popularly known as BET theory. The main drawback of this theory relating to porous media is that the pores can be assumed to be a planar surface of infinite radius. A modification in the original theory to take into account of the shape effect of pores was proposed by Hillerborg (1985). With this model, adsorption and condensation phenomena, classically separated as different phenomena, become essentially a single continuous mechanism where no distinction is made between the adsorbed and condensed water. For an ideal case, disregarding osmotic effects, the thickness denoted by $t_a$ of adsorbed layer is given by

$$t_a = \frac{0.525 \times 10^{-8} h}{(1 - h/h_m)(1 - h/h_m + 15h)} \tag{3.46}$$

where $h_m$ is the humidity required to fully saturate the pore. If we assume a cylindrical shape of pores, then from Kelvin's equation and Figure 3.16 we have

$$h_m = \exp\left(\frac{-2\gamma M}{\rho R T r_1}\right) \tag{3.47}$$

where $r_1 = r - t_a$ is the actual interface radius and is smaller than the actual pore radius. As the relative humidity inside the pore increases, the thickness of the adsorbed layer increases. At the same time, humidity, $h_m$, required to completely fill the pore decreases. Therefore, at some stage the situation becomes unstable and the pore becomes fully saturated. The degree of saturation, $S_r$, of a cylindrical pore in an unsaturated state can be obtained as

$$S_r = 1 - \left(\frac{r - t_a}{r}\right)^2 \tag{3.48}$$

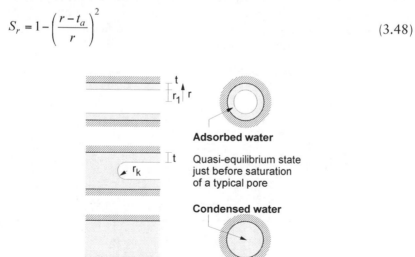

*Figure 3.16* Adsorption phenomenon in a single pore.

The degree of saturation of individual pores can be integrated over a given porosity distribution to obtain the total saturation of the porous medium. This integration can be significantly simplified if we consider the total saturation as the sum of contributions from fully filled and partially filled pores. Therefore, the degree of saturation of a model porosity distribution, V, given by Equation 3.19 is obtained as

$$S = \int_0^{r_c} dV + \int_{r_c}^{\infty} S_r dV = S_c + \int_{r_c}^{\infty} Br S_r \exp(-Br) d\ln r \tag{3.49}$$

where $r_c$ denotes the pore radius with which the equilibrated interface of liquid and vapor is created. In other words, pores of radii smaller than $r_c$ are completely saturated, whereas larger pores are only partially saturated. It should be noted that the pore radius, $r_c$, is larger than the pore radius, $r_s$, as determined by Equation 3.43. An adsorbed film of thickness, $t_a$, exists in other unsaturated pores. Figure 3.17 shows moisture distribution in a pore structure considering adsorption and condensation. Comparing this figure with Figure 3.15, it can be noted that no consideration of adsorption would lead to an underestimation of moisture content, since a significant amount of moisture exists in the pores that are not completely filled but contain an adsorbed layer of water. This difference would generally be larger at higher humidity. Figure 3.18 shows the absorption saturation characteristics of individual pores as well as the moisture profiles in a model pore distribution at different RH.

Let us now consider some computational simplifications. The Hillerborg model of adsorption, Hillerborg (1985), given by Equations 3.46–3.48, is implicit in $t_a$. Therefore, some iterative scheme, such as the Newton–Raphson method, must be used to generate the isotherms for each individual pore. Moreover, for a given relative humidity, $h$, evaluation of $r_c$ is not a straightforward task but instead requires the evaluations of isotherms for each pore.

Clearly, this is a daunting task and poses a practical limitation to the use

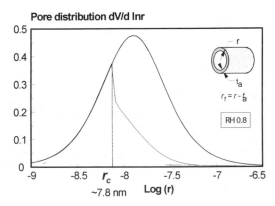

*Figure 3.17* Moisture profile in a model pore distribution considering unified adsorption and condensation phenomenon.

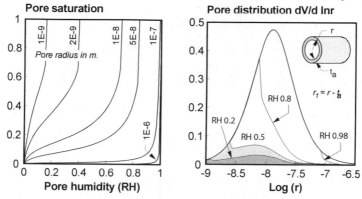

*Figure 3.18* Moisture profiles in a model PD at various humidities.

of this model in a dynamic computational scheme where a change in the microstructure itself with time makes necessary frequent evaluations of the total isotherms. For this reason, the authors propose a simplification to the scheme of obtaining total saturation, where $r_c$ is obtained from an explicit relationship given as

$$r_c = C \cdot r_s = -\frac{2C \cdot \gamma \cdot M}{\rho \cdot RT} \frac{1}{\ln h}, \quad C = 2.15 \tag{3.50}$$

The constant $C$ has been obtained after numerous comparisons of the analytical predictions of equilibrated interface radii, $r_c$, as obtained from modified BET theory and $r_s$ as given by Kelvin's equation. The relationship holds good over the range of pore radii encountered in cementitious microstructures and does not appears to be very sensitive to temperature. Also, the contribution from unsaturated pores is obtained not by evaluating the complete integral of Equation 3.49 but from a modified expression as given below

$$S_{ads} = \int_{r_c}^{\infty} S_r dV \approx t_m \int_{r_c}^{r_{0.99}} 2r^{-1} dV = t_m A_s \tag{3.51}$$

where $t_m$ is the thickness of the adsorbed film of water given by Equation 3.46 in a pore of radius $r_m$, $r_m$ is the geometrical mean of $r_c$ and $r_{0.99}$, $A_s$ is the surface area of pores lying between those of radius $r_c$ and $r_{0.99}$, and $r_{0.99}$ corresponds to the pore radius below which 99% porosity of the porosity distribution exists. Saturation as given by Equation 3.51 requires only one iterative evaluation of $t_a$ and gives reasonable estimates of moisture contents compared to the exact but highly computation-intensive Equation 3.49.

### 3.2.2.3 Computational model of hysteretic behavior of isotherms

Usually, the absorption and desorption curves of typical moisture isotherms are observed to follow different paths, not only in concrete but also in other porous media. The irreversibility of isothermal paths of water content in concrete under cyclic drying–wetting conditions is shown in Figure 3.19. All desorption curves lie above the adsorption curves and hysteresis loops can be observed. In the past, the thermodynamic equilibrium condition given by Equation 3.43 was usually used to describe the state of moisture in concrete. However, such approaches are not adequate, since they fail to address the issue of hysteretic behavior under generic drying–wetting conditions.

The hysteresis behavior can be partially addressed if we consider a moisture isotherm model which takes into account the effect of entrapped liquid water in the microstructure during drying. So far, we have considered both the condensed and adsorbed phases of liquid water. For the sake of simplicity, the adsorbed component of moisture will be ignored in the following discussions as it has no effect on the hysteresis model explained later. Moreover, the moisture content contributions of the adsorbed water can be simply added to the contributions from fully saturated pores to give the total saturation. Further, to illustrate the hysteresis model, discussions are limited to a notional porous medium that has a representative porosity distribution, as given in Equation 3.19.

#### 3.2.2.3.1 ENTRAPMENT OF LIQUID WATER IN THE MICRO-STRUCTURE (INK-BOTTLE EFFECT)

As already mentioned, the very first drying curve of the moisture isotherm is always found to be higher than the corresponding wetting curve. To describe this behavior, we focus primarily on the entrapment of liquid water in larger pores, which might be one of the main reasons of hysteresis. Owing to the complex geometrical characteristics of a random microstructure, many of the larger pores have interconnections only through pores that are much smaller

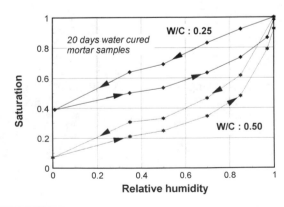

*Figure 3.19* Irreversibility of S-h paths in typical moisture isotherms.

in terms of pore radii. For such pore systems, there exists a possibility of liquid becoming entrapped in the larger pores for arbitrary drying–wetting histories. The concept of entrapment of pore water in an idealized pore is shown in Figure 3.20, which shows that during drying we can expect some additional trapped water in the pores whose radii $r$ are larger than $r_c$. Such pores have external openings only through the pores of smaller radii, and therefore cannot lose moisture as long as the connecting pores remain saturated. The additional water trapped in such pores gives rise to the hysteresis effect observed in isotherms. By adding the volume fraction of trapped water, $S_{ink}$, in such pores, to the volume fraction of water, $S_c$, as obtained in Section 3.2.2.2, we can obtain the total saturation of the porous medium under an arbitrary history.

To consider the volume of entrapped water in such pores, we define a probability parameter, $f_r$, to take into account the hysteresis (Figure 3.21). The parameter $f_r$ denotes the probability of water entrapment in a pore of

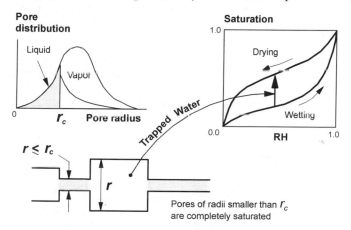

*Figure 3.20* Entrapment of water in pores with restricted openings.

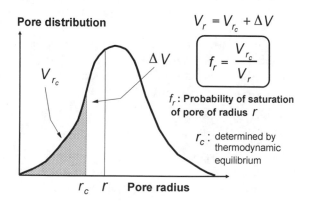

*Figure 3.21* The definition of probability of entrapment parameter $f_r$.

radius $r$ larger than the pores of radii $r_c$. In other words, $f_r$ is the probability that a pore of radius $r$ would be connected only to the pores whose radius is smaller than $r_c$. Obviously, this probability will be dependent on the chance of intersection of the larger pore to the smaller, completely filled pores. In its simplest form, we take this probability to be proportional to the ratio of the volume of completely saturated pores and the volume of pores with a radius less than $r$. Based on this assumption, a mathematical definition of $f_r$ which statistically represents the structure of pore connection can be obtained as

$$f_r = \frac{V_{r_c}}{V_r} \tag{3.52}$$

where $V_{rc}$ is the volume of the pores with radius less than $r_c$, which is in fact the volume of fully saturated pores, and $V_r$ is the volume of the pores with radius less than $r$ in the porosity distribution. The moisture isotherms of porous media can be completely described for any arbitrary drying–wetting histories, if we consider four main patterns of the drying–wetting isothermal paths. These are:

1   virgin wetting paths, deriving initially from a completely dry state (primary wetting loop);
2   virgin drying paths, deriving initially from a completely wet state (primary drying loop);
3   scanning curves with a completely dry state as their starting point;
4   scanning curves with a completely wet state as their starting point.

3.2.2.3.2 PRIMARY DRYING–WETTING LOOPS OF THE MOISTURE ISOTHERM: WETTING STAGE

First, let us deal with the virgin wetting curve where absorption starts from a completely dry state of the porous medium. This condition is similar to the one discussed in Section 3.2.2.1. Under equilibrium conditions, the pores of radii smaller than $r_c$ would be completely saturated, whereas larger pores would only contain moisture in the adsorbed phase, which is disregarded here for the sake of brevity. Therefore, total saturation, $S_{total}$, of porous media can be obtained by integrating the individual micro-pore saturation over the entire porosity distribution function as (Figure 3.22)

$$S_{total} = S_c = \int_0^{r_c} \Omega dr = \int_0^{r_c} dV = 1 - \exp(-Br_c) \tag{3.53}$$

where $\Omega = dV/dr$ is a representative porosity distribution function. In other words, total saturation in this case is obtained by simply summing the condensed pore volumes and does not include any contribution due to hysteresis.

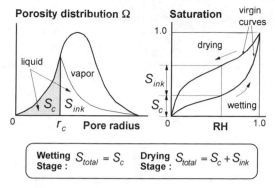

*Figure 3.22* Hysteresis along virgin drying and wetting paths.

3.2.2.3.3 DRYING STAGE

The degree of saturation, $S_{ink}$, due to the additional water present in the trapped pores should be computed so that total saturation of porous media can be obtained. This is done by summing the most probable degree of saturation of all pores having a radius greater than $r_c$, over the porosity distribution ranging from pore radius $r_c$ to infinity. Then, we have

$$S_{ink} = \sum_{r=r_c}^{\infty} f_r \Omega \Delta r = \int_{r_c}^{\infty} f_r \, dV = \int_{r_c}^{\infty} \left( \frac{S_c}{V} \right) dV = -S_c \ln(S_c) \tag{3.54}$$

Therefore, total saturation under virgin drying conditions can be obtained by simply adding $S_{ink}$ to $S_c$, which is the volume of completely saturated pores below radius $r_c$, as (Figure 3.22)

$$S_{total} = S_c + S_{ink} = S_c [1 - \ln S_c] \tag{3.55}$$

It will be noticed that the additional term $-S_c \ln S_c$ is always positive, which confirms the observation that drying curves are always higher than the absorption curves. It is important to note that this term, which represents the additional quantity of trapped pore water, does not depend on our assumptions regarding the mathematical description of the porosity distributions. In other words, Equation 3.55 will give us an estimate of the trapped water, as long as we know $S_c$ or the virgin wetting history of the porous medium in question.

3.2.2.3.4 SCANNING CURVES OF THE MOISTURE ISOTHERMS

In the previous section, a computational model was presented for a quantitative representation of virgin wetting–drying behavior of moisture in concrete. However, actual concrete structures are often exposed to variable environmental conditions, which are not virgin loops. Therefore, we need to extend the application of this concept from virgin wetting–drying paths to arbitrary environmental conditions, such as complicated cyclic wetting–drying conditions. As with the virgin wetting and drying stages, we consider the following two cases for discussion, which cover all possible scenarios of drying–wetting paths.

### 3.2.2.3.5 FROM A VIRGIN WETTING PATH TO DRYING AND SUBSEQUENT LOOPS

During the monotonous wetting phase, the absorption path of the moisture isotherm would follow a similar curve to the virgin wetting curve (state a to state c in Figure 3.23). However, when the relative humidity decreases, the desorption curve will never return to the virgin curve, but instead traces an inner scanning drying curve (state c to state d in Figure 3.23). For scanning drying curves, the ink-bottle model similar to the former drying case is applied and the saturation owing to ink-bottle effect, $S_{ink}$, can be obtained as

$$S_{ink} = \int_{r_c}^{r_{max}} \left( \frac{S_c}{V} \right) dV = S_c \left( \ln S_{r_{max}} - \ln S_c \right) \tag{3.56}$$

where $r_{max}$ is the pore radius of the largest pores that experienced a complete saturation in the wetting history of the porous medium, and $S_{rmax}$ is the highest saturation experienced by the porous medium in its wetting history. In fact, $r_{max}$ is the pore radius that corresponds to the saturation state, $S_{rmax}$, of the porous medium (state c in Figure 3.23). In Equation 3.56, the summation to obtain the trapped water is carried out up to $r_{max}$, since pores whose radius is greater than $r_{max}$ would never have experienced a complete saturation in their wetting history and therefore would not contain any trapped pore water. Thus, total saturation as a sum of the usual condensed and entrapped water is

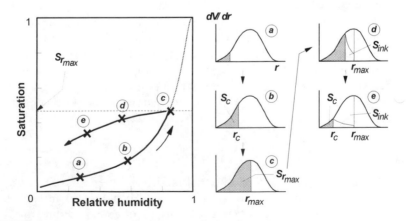

*Figure 3.23* Scanning paths of isotherms along wetting paths.

$$S_{total}=S_c+S_{ink}=S_c\left(1+\ln S_{r_{max}}-\ln S_c\right) \tag{3.57}$$

In the inner loops, absorption and desorption processes are assumed to be reversible, so the inner scanning curves follow a similar path. In other words, subsequent hysteresis in the scanning curves is disregarded. We have assumed the reversibility of the inner loops primarily to obtain a closed form analytical solution of the hysteresis model. Of course, based on the ink-bottle concept discussed earlier, it is possible to trace exact hysteresis behavior in the inner loops. Here, this would require keeping track of all the turning points in the drying–wetting history and would therefore limit the practical applicability of the model. In the history of the porous medium, if wetting proceeds such that $r_c$ exceeds $r_{max}$, the adsorption path will return to the virgin wetting loop.

3.2.2.3.6 FROM VIRGIN DRYING PATH TO WETTING AND SUBSEQUENT LOOPS

As in the previous case, during the first drying, saturation will decrease along the virgin drying loop (state a to state c in Figure 3.24). When the ambient relative humidity increases, the scanning absorption loop will be formed and a gradual moisture re-entry from the filling of smaller pores will take place (state c to state d in Figure 3.24). Then, we have

$$S_{total}=S_c+S_{ink}=S_c+\int_{r_c}^{\infty}\left(\frac{S_{r_{min}}}{V}\right)dV=S_c-S_{r_{min}}\ln S_c \tag{3.58}$$

where $r_{min}$ is the pore radius of the smallest pores that experienced an emptying-out in the drying history, and $S_{rmin}$ is the lowest saturation of the porous medium in its wetting–drying history (state c in Figure 3.24). The integration in this case is carried out from pore radius, $r_c$, since all the pores below $r_c$ would be completely saturated (Figure 3.15). Moreover, the prob-

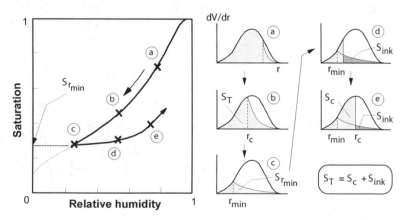

*Figure 3.24* Scanning paths along virgin drying curves.

ability parameter, $f_r$, in this equation takes $S_{rmin}$ as the volume of completely filled pores, since the probability of entrapment of liquid in the pores of radii above $r_c$ would correspond to a state of the porous medium (state c in Figure 3.24) when it experienced its lowest saturation. The scanning curves of absorption and desorption paths are assumed to be similar owing to the reasons explained earlier. As drying proceeds such that $r_c$ becomes smaller than $r_{min}$, the desorption path will again return to the virgin drying loop.

The results obtained are fundamental in nature and do not depend on the assumptions of the porosity distribution. The probability of the water entrapment model for $f_r$ is sufficient to build a complete description of the hysteresis. Consideration of porosity distribution functions simply assists visualization of the phenomenon. The final results in the previous derivations do not include the contributions from the adsorbed phase of pore water, which we have kept out of the discussion for the sake of clarity. Therefore, actual saturation of the porous media $S$ is given by

$$S = S_{total} + S_{ads} \tag{3.59}$$

where $S_{total}$ depends on the history of drying and wetting and can be obtained from Equations 3.53–3.58. The adsorbed component of saturation, $S_{ads}$, can be obtained from Equation 3.51.

### 3.2.2.4 Interlayer moisture and its contribution

While the discussion in the previous section can explain the hysteresis behavior of isotherms at high relative humidity, it cannot fully explain the behavior actually observed for cementitious materials at low relative humidity. Various reasons have been advanced to account for this difference at low relative humidity. One reason is thought to be found in the gel structure of hydrated Portland cement (Figure 3.25). The hysteresis is accounted for by the distinction of water present in the C–S–H structure as interlayer hydrate water. It has been reported that very large internal surface areas as predicted by BET theory are not possible for pore distributions as obtained by mercury intrusions or other similar means. Moreover, the values of internal surface areas of hardened cement paste as computed from water adsorption experiments are significantly higher than those computed from nitrogen adsorption experiments. This difference was accounted for by Feldmann and Sereda (1968) in the interlayer water concept. Interlayer water is a component of the gel structure and resides in the layer structures of C–S–H, accounting for a very high surface area as computed from the BET equation. The removal process of water in such layers is different from its re-entry during sorption and thus accounts for the hysteresis.

As shown in Figure 3.25, gradual removal of interlayer water from the edges occurs between the range of about 30% to 10% RH. Further removal from which probably occurs in the range of 10% to 2% RH, is accompanied

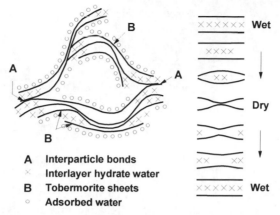

| | |
|---|---|
| **A** | **Interparticle bonds** |
| ×  | **Interlayer hydrate water** |
| **B** | **Tobermorite sheets** |
| ○ | **Adsorbed water** |

*Figure 3.25* A schematic representation of C–S–H gel structure.

by a large change in the length of the specimen. The important point to note is that a significant amount of interlayer water resides in the C–S–H structure even at very low humidity during drying and severe drying conditions are required to remove this water. During the subsequent wetting process, re-entry of interlayer water takes place gradually, with most of it occurring at higher humidity.

Summarizing, during the drying phase most of the interlayer water can be removed only at very low RH whereas the re-entry is gradual over the entire range of RH. The exact thermodynamic behavior associated with these processes is not clearly understood. Therefore, the authors propose an empirical set of equations to describe the generic drying–wetting behavior of interlayer component on the same lines as above by considering four scenarios of drying–wetting histories. These relationships correlate the interlayer saturation to the ambient humidity for different cases of drying–wetting paths and are described as (Figure 3.26),

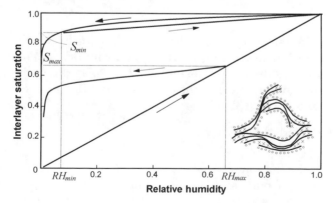

*Figure 3.26* The computational isotherm model of an interlayer.

$$S_{lr} = h \qquad\qquad\qquad \text{Virgin wetting loop}$$

$$S_{lr} = h^{0.05} \qquad\qquad\qquad \text{Virgin drying loop}$$

$$S_{lr} = 1 + (h-1)\left(\frac{S_{min}-1}{h_{min}-1}\right) \quad \text{Scanning curve, initially fully wet} \qquad (3.60)$$

$$S_{lr} = S_{max} h^{0.05} \qquad\qquad\qquad \text{Scanning curve, initially fully dry}$$

The constants in Equation 3.60 have been chosen after comparison with test results. The model is very simple in that it does not include the effect of temperature (Section 3.3.3.2). The completely dry state in above model corresponds to the state of hardened cement paste which is dried to constant weight at 105°C. Clearly, heating to higher temperatures would lead to a higher loss of the structural component of C–S–H water. In this book, we have based the definition of porosity of hardened cement paste as that corresponding to a maximum dried state of 105°C. Moreover, due to the inclusion of the interlayer phase as a part of the total hardened cement paste porosity, the porosity discussed throughout this book is that of suction moisture only.

### 3.2.2.5 The total moisture isotherm of hardened matrix

Various models discussed in the previous sections can be combined together to predict the actual moisture isotherms of a cementitious matrix. In our computational model, the micro-porosities of the capillary and gel pores are represented by simple mathematical distributions (Equation 3.19). These can be combined together to give the non-interlayer porosity distribution, $V_{cg}$, of hardened cement paste as

$$V_{cg} = \frac{1}{\left(\phi_{cp}+\phi_{gl}\right)}\left\{\phi_{cp}\left(1-\exp\left(-B_{cp}r\right)\right)+\phi_{gl}\left(1-\exp\left(-B_{gl}r\right)\right)\right\} \qquad (3.61)$$

The degree of saturation, denoted by $S_{cg}$, of the porosity distribution given by Equation 3.61 at any arbitrary drying–wetting history would be given by the isotherm model discussed (Equation 3.59). Adding this to the interlayer component of moisture gives us the total moisture content of hardened cement paste denoted by $\theta$ as

$$\theta = \rho_l\left(\phi_{lr}S_{lr}+\left(\phi_{cp}+\phi_{gl}\right)S_{cg}\right) \qquad (3.62)$$

The pore distribution parameters $B_{gl}$ and $B_{cp}$ as well as the gel, capillary and interlayer porosity required in Equation 3.62 are obtained from a combined pore-structure development, hydration and moisture transport model as described in Chapter 1. However, experimental data on porosity and an approximation of $B_{gl}$ and $B_{cp}$ based on MIP results can also be used to predict the moisture content of hardened cement paste for any arbitrary drying–wetting path.

### 3.2.3 Permeability of concrete

Darcy's law of saturated flow through porous media states that the flux through a porous medium is proportional to the pressure gradient applied to it, i.e.

$$Q = -\frac{k}{\eta}\frac{dP}{dx} \tag{3.63}$$

where $k$ is the intrinsic permeability of the porous medium, a material property, and $\eta$ is the viscosity of the fluid. For extremely small rates of flows, typical for porous media, fluid flow characteristics are taken care of by the viscosity term. The parameter $k$ is actually a transport coefficient dependent on the characteristics of porous media only and hence a constant. Since transport coefficients and pore structures of porous media are interrelated, there are many approaches that build upon this correlation. However, the abundant diversity in the geometrical as well as the physical chemistry characteristics of porous structures has limited the universal applicability of these methods. The aim of this section is to give a brief overview of the existing microstructure-based permeability models and to propose a simple microstructure-based mathematical model of the permeability of concrete, derived in part from work in soil science.

Many of the models put forward to interrelate permeability and microstructure include certain correction factors that apply only to a similar group of materials. Even for a specific material, such as concrete, different researchers have suggested different microstructure properties and empirical correction factors as the parameters in their models to predict permeability. The empirical formulations have traditionally taken porosity and critical pore diameter as being among basic parameters. The theoretical approach varies from assuming simple Hagen–Posuille flow in the capillaries, Reinhardt and Gaber (1990), to numerical models based on computer simulations. In this section, we confine our discussion to only those models that consider porous media microstructure as the basis of development. Mehta and Manmohan (1980) suggested a rather empirical formula for the prediction of permeability as

$$K_l = \exp(3.8V_1 + 0.2V_2 + 0.56 \times 10^{-6} TD + 8.09 MTP - 2.53) \tag{3.64}$$

where $K_l$ is the permeability coefficient, $V_1$ and $V_2$ are pore volumes in the >1320Å and 290–1320Å range, respectively, $TD$ is the threshold diameter, and $MTP$ is the total pore volume divided by the degree of hydration. It was suggested by Nyame and Illston (1981) that the maximum continuous pore radius has a close relationship with permeability. However, median or average pore radius of the porosity distribution has been suggested as the controlling factor by Katz and Thompson (1987). Garboczi (1990) reviewed classical and current pore structure theories and suggested KT theory (Katz and Thompson 1987) as a permeability model with universal appeal. This model as such has no adjustable parameters and all the quantities can be measured experimentally.

$$k = \frac{1}{226} \frac{d_c^2}{F} \tag{3.65}$$

where $k$ is intrinsic permeability, $d_c$ is the critical diameter obtained from the threshold pressure in a mercury injection experiment, and $F$ is the formation factor, which is the ratio of brine conductivity in the pore space to porous media conductivity (measured experimentally). It appears that the KT model gives reasonable estimates of intrinsic permeability.

However, criticism of this theory has been directed towards the difficulty in measuring accurately the inflection point or threshold pressure in the mercury injection experiment, which, due to the crack-like pore characteristics of concrete, may not be present at all in some cases. Moreover, the theory is not so successful in predicting the permeability values of mortar with water-to-cement ratio (W/C) ratios less than 0.4 (Halamickova *et al.* 1995). A different approach is chosen here for permeability modeling, because the KT theory itself does not indicate anything about relative permeability or permeability of partially saturated porous media. It implies that the theory cannot easily be used in a computational scheme that combines the partially saturated transport, microstructure formation and hydration in concrete (Chaube *et al.* 1993). The model described in the next section does not give a very different result compared to that obtained by KT theory, but it is more conducive to a coupled computational scheme of hydration, mass transport and pore-structure formation.

### 3.2.3.1 *Intrinsic permeability model of a porous media*

With regard to the transport of material within the complex system of randomly dispersed pores, two characteristics are of paramount importance. These are: (1) total open porosity, which constitutes a continuously connected cluster of pores and (2) pore size distribution, which influences the rate of the transport. Any of the pore structure theories (PSTs) should consider these parameters in some way or other. For intrinsic permeability, we consider both of the above-mentioned characteristics, since the probability of permeation depends not only on total open porosity but also on the possibility of interconnecting the pores that contribute to the total porosity.

Let us consider a section of porous medium of small but finite thickness,

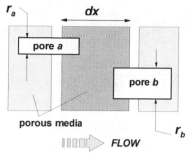

*Figure 3.27* Flow across a finite thickness of porous media.

$dx$, as shown in Figure 3.27. Our goal is to deduce the effective intrinsic permeability parameter of the porous medium based on permeation probability. Let $\Omega_A$ represent the average aerial distribution function of pores exposed on any arbitrary face cut perpendicular to the flow, such that the total aerial porosity $\phi_A$ is equal to the term $\int \Omega_A \, dr$. Aerial and volumetric porosity are related to each other as

$$\phi = n\phi_A \tag{3.66}$$

where $n$ is tortuosity factor $= (\pi/2)^2$ for a uniformly random porous medium, $\phi_A$ is aerial porosity, and $\phi$ is volumetric porosity of the porous medium. The fractional area of pores of radius $r_a$ and pores of radius $r_b$ at either face of the section can be obtained as

$$dA_a = \Omega_A dr_a, \quad dA_b = \Omega_A dr_b \tag{3.67}$$

From a statistical viewpoint, the probability of permeation $dp_{ab}$ through $a$ and $b$ pores of this section is a product of the normalized areas $dA_a$ and $dA_b$ (or aerial porosity) of the pores of radius $r_a$ and $r_b$.

$$dp_{ab} = dA_a dA_b \tag{3.68}$$

It has to be noted that we assume an independent arrangement of pores over the section. For such a case, integrating Equation 3.68 yields the total penetration probability as $\phi^2$. If the arrangement of pores in the porous medium were to be constant, the penetration probability $dp_{ab}$ as given by Equation 3.68 would be unity for a pair of pores with the same pore radius and zero for others, resulting in a total penetration probability of $\phi$, which is the classical assumption usually followed.

Next, we consider a steady-state laminar flow through the porous medium, which is idealized as consisting of a bundle of straight capillary tubes. For such a simple model, total flow through the porous medium $Q$ can be obtained simply by Feldman and Sereda (1968) as

$$Q = -\left(\frac{1}{8\eta}\int_0^\infty r^2 \Omega_A dr\right)\frac{dP}{dx} \tag{3.69}$$

where $\eta$ is fluid viscosity and $dP/dx$ is fluid pressure gradient. In reality, pores are not a continuous bundle of capillaries, and a permeation probability is associated with flow across any section and various other geometrical and scale effects can significantly alter flow behavior. The average flow behavior through a joint of pores of different radii might best be represented by a pore whose radius is the geometric mean of two pore radii. Then, we assume that the flow through the system of pores of radii $r_a$ and $r_b$ can be represented by an equivalent pore of radius $r_{eq}$ given by

$$r_{eq}^2 = r_a r_b \tag{3.70}$$

This assumption appears to give a more balanced weight to the entire distribution and to better estimate the permeability as compared to the conventional methods that give a rather skewed weight to the pore distribution (Reinhardt and Gaber 1990 and Nyame and Illston 1981). Taking these factors into account, we obtain a modified expression for total flow through the porous medium by integrating the flow contributions from all possible pair combinations of pores of the porous medium as

$$Q = -\left(\frac{1}{8\eta}\int_0^\infty\int_0^\infty K r_{eq}^2 dA_a dA_b\right)\frac{dP}{dx} \tag{3.71}$$

where $K$ is an unknown parameter that is dependent on the geometrical and scale characteristics of the porous medium. Due to the difficulty in the analytical treatment of this parameter, its value has been fixed as unity for the rest of this derivation. Simplifying Equation 3.71 and comparing it to Darcy's law gives an expression of intrinsic permeability of the porous as

$$k = \frac{1}{8}\left(\int_0^\infty r dA\right)^2 \approx \frac{\phi^2}{50}\left(\int_0^\infty r dV\right)^2 \tag{3.72}$$

where $dV = ndA/\phi$ is the incremental normalized volumetric porosity. The only parameter in the above expression is the porosity distribution of the porous medium.

A note of caution is in order when applying Equation 3.72 to predict intrinsic permeability of hardened cement paste, using the MIP method. From a theoretical viewpoint, almost all of the conductivity models emphasize the role of larger pores. That is, larger pores are the ones that contribute most to the flow if connected continuously. For this reason, it is possible to get erroneous results if, for example, MIP data are used without applying any correction for larger pores. The MIP experimental method itself is not very accurate for larger pore-diameters, since the test samples are crushed before testing and might contain inadvertent macro-scale defects. For hardened cement paste samples, the authors recommend that the pores above the first inflection point in the MIP curve or 1000 nm, whichever is greater, should not be used in the analysis. The first inflection point is the one on the MIP curve (where cumulative intruded volume $V$ is plotted against the log of the pore radius $r$) which shows a distinct rise in the volume $V$ of intruded mercury.

### 3.2.3.2 Inconsistency in intrinsic permeability observations

Ideally, intrinsic permeability is a basic micro-structural characteristic and should not be dependent on the fluid used to measure it. However, it has been extensively reported that different values of intrinsic permeability of a porous medium are obtained for different fluids, after applying the density and viscosity normalization. This anomaly can be attributed to the micro-structure and pore-fluid interaction. It might occur due to a change in the chemo-physical state of porous media or a gradual change in the viscous properties of the fluid being transported, depending on its thermodynamic state or mass transport history. For example, it has been widely reported by Hearn *et al.* (1994) that even well-hydrated concrete shows a reduction in water permeability over a long period of exposure to moisture.

On the other hand, the authors found that exposure to relatively inert fluids, such as acetone, yields constant permeability or drag coefficients with time. Figure 3.28 shows data on intrinsic permeability of different concrete as reported by various researchers. For this data, permeability measurements for different concrete were taken by gas and water, and converted to intrinsic permeability by applying the density and viscosity normalization. This difference has been attributed to various causes, namely: (1) gas slippage theory (Bamforth 1987), (2) plasticity of water (the visco-plastic nature of

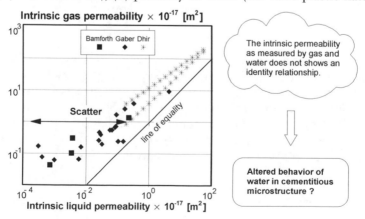

*Figure 3.28* Intrinsic permeability of concrete.

pore liquid), (3) long-range intra-molecular force theory (Luping and Nilsson 1992), (4) yield shear of pore fluid (a shearing force acting on pore walls), (5) swelling of hydrates (Dhir *et al.* 1989), (6) resumption of dormant hydration and self-sealing (Dhir *et al.* 1989), and (7) altered viscosity of pore-water (Volkwein 1993).

In addition, experimental problems in measuring extremely small values of water permeability have made it practically impossible to trace the exact cause of the discrepancies in intrinsic permeability as obtained by gas and water measurements. However, it must be pointed out that the permeability values obtained by using gas, oil or even alcohol as the measuring fluid have traditionally produced more reliable, reproducible and consistent results. It has also been found and reported that usually it takes more than 20 days to obtain steady-state flow conditions in water permeability measurement (Hearn *et al.* 1994). Even so, a reduction in water flux or self-sealing behavior of concrete when exposed to water is also reported by Dhir *et al.* (1989) and Hearn *et al.* (1994). For low W/C ratio mix concrete, the scenario is even grimmer.

For this reason, the authors have tested each of the above-stated causes against a large set of databases of water and gas permeability available in the literature, but none of the theories can explain the discrepancy completely. Furthermore, counter-evidence is also available in the literature, which negates some of the theories, such as the delayed hydration of unhydrated C–S–H in the presence of water, as described in Hearn *et al.* (1994). All of these observations point out that it is the pore water and cementitious microstructure interaction that needs to be properly addressed since, for other fluids, reasonable agreement between observations and predictions by using the rational microstructure-based theories can be obtained. It is believed that the altered nature of pore water, which probably changes the pore water viscosity in a time-dependent manner, is perhaps the most likely cause of large differences in intrinsic permeability as measured by gas and water.

### 3.2.3.3 *Modified water conductivity of concrete*

The water present in cementitious microstructures is in far from ideal conditions. Such a non-ideal situation may result from the dissolution of salts, long-range forces exerted on the water molecules from the surrounding porous medium, the effects of the polarity of water, or a complex combination of any of these factors. From a thermodynamic point of view, however, the theory of viscosity states that the actual viscosity $\eta$ of a fluid under non-ideal conditions at an absolute temperature $T$ is given by Welty *et al.* (1969) as

$$\eta = \eta_i \exp(G_e / RT) \tag{3.73}$$

where $G_e$ is the free energy of activation of flow in excess of that required for ideal flow conditions, $\eta_i$ is the viscosity under ideal conditions, and $R$ is the universal gas constant. The actual permeability of water can be evaluated if the effect of non-ideal viscosity of water is taken into account in Equation 3.72.

Peschel (1968) reported a viscosity that is about one or two orders higher than the ideal viscosity of water when measured under thin quartz plates (Figure 3.29). The exact physical cause of this phenomenon is not known, but the authors have considered a phenomenological thermodynamic approach to explain this mechanism. When concrete is exposed to water, a transient phase of apparent reduction in the water permeability is observed until it reaches a final value, which is smaller than the permeability as expected under ideal conditions. One-dimensional water sorption experiments in concrete also show a deviation from the square-root law of water absorption after a few days (Volkwein 1993). These deviations are not a result of the anomalous diffusion as explained by the percolation theory for porous media where the percolation probability is much lower than critical percolation probability (Stauffer and Aharony 1992). This is because absorption experiments with alcohol by the authors show a near square root behavior, whereas water sorption for the same mortar shows a large deviation from the ideal behavior (Figure 3.30). It must be noted that drying out the specimens has been shown to restore the initial conditions, which indicates that the time-dependent change of the water permeability of concrete is indeed a state-dependent thermodynamic phenomenon.

To account for such time-dependent behavior, we hypothesize that there exists a pore-water and microstructure interaction, which is a time-lag phenomenon, bringing about the change in the state of pore water from ideal to non ideal state, and roughly depends on the history of pore humidity. That is, a change in pore humidity brings about a gradual interaction or altered state of the microstructure and pore-water system, leading to delayed changes in the liquid viscosity and hence observed permeability. This change is not immediate, as would be apparent from a one-dimensional water sorption experiment or water permeability experiments. In a non-ideal state, the

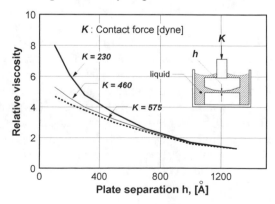

*Figure 3.29* Relationships of water viscosity and spacing of quartz plates.

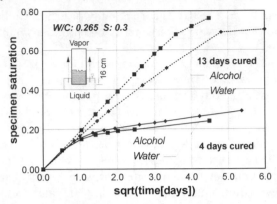

*Figure 3.30* The sorption behavior of alcohol and water in concrete.

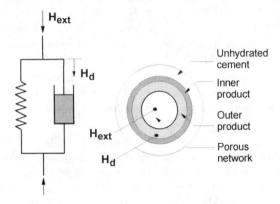

*Figure 3.31* The Kelvin chain representation to compute concrete's pore water viscosity.

additional energy for the activation of flow, $G_e$, may be dependent on the altered state of the system and can be imagined to bring about a fictitious delayed pore humidity, $H_d$. In turn, $H_d$ would be dependent on the actual pore humidity, $H$, history and micro-structure characteristics.

In the computational model, this phenomenon is represented by a simple Kelvin chain model (Figure 3.31), where the dashpot viscosity $\eta_d$ represents the responsiveness of the microstructure to changes in the actual pore humidity, $H$ (stress analogy), that brings about a change in $H_d$ (strain analogy). Our goal through this model is to obtain the extra energy, $G_e$, required for the activation of flow at any point and time such that effective non-ideal viscosity of the pore fluid, $\eta$, and the effective permeability can be computed (Figure 3.32). The computational model is given by

$$G_e = G_{\max} H_d, \quad \ddot{H}_d + \left(\frac{1+\ddot{\eta}}{\eta}\right)H_d = \frac{H}{\eta}$$

$$\eta = a\left(1 + bH_d^c\right), \quad a = 1.59\left(\frac{\phi - \phi_m}{\phi_m}\right) + 0.85, \quad b = 2.5a, \quad c = 2.0$$

(3.74)

where $G_{\max}$ is the maximum additional Gibbs' energy for the activation of flow = 3,850 kcal/mol (assumed constant for all cases), $\phi$ is the total suction porosity, and $\phi_m$ is the porosity accessible by mercury intrusion. The salient point of this model is that the additional energy for activation of flow, $G_e$, is assumed to be proportional to a fictitious humidity parameter, $H_d$. Moreover, the responsiveness of the microstructure to an external change represented by $\eta$ depends on the nature of its porosity distribution. For example, the microstructures that have higher proportions of large pores or capillary porosity are more responsive or react more quickly to changes, and the microstructures with a higher fraction of finer pores react more slowly to become a non-ideal state.

### 3.2.3.4 Preliminary verification of intrinsic permeability models

The intrinsic permeability model was verified by comparing the computed permeability value with those obtained from experiments. The experiments were a part of the research project on the long-term performance of concrete in which several kinds of mortars were tested. The W/B ratio (by volume) ranged from 78% to 190%. The sand volume fractions by compaction ratio

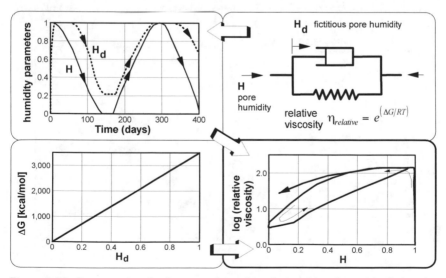

*Figure 3.32* Computational scheme to show the relationship of non-ideal viscosity of water with pore humidity.

($S/S_{\text{lim}}$) ranged from 0.2 to 0.8. Some mortar mixes with high lime volume fractions of up to 17% of the mix volume were also prepared. After sufficient curing, one pair of specimens was exposed to accelerated calcium leaching tests. Gas permeability measurements and pore structure measurements using the MIP method, involving both the exposed and unexposed groups of specimens, were done. Figure 3.33 shows a comparison of the measured and predicted intrinsic permeability values for many specimens. In the analysis, the porosity distribution lying above 1.1 $\mu m$ pore radius was ignored for all specimens, due to experimental errors usually associated with large pore radii data in the MIP method. The figure shows a reasonable agreement between the theory and experiment.

### 3.2.4 Transport coefficients for unsaturated flow

In most of the concrete structures exposed to the atmosphere, the ingress of moisture and various harmful agents involves the transport of both the liquid and gas phases. For a complete system of modeling concrete performance, we need a transport theory that can deal with the unsaturated conditions of real-life concrete structures. In this regard, conductivity of the liquid as well as the vapor phase under arbitrary saturated conditions must be obtained.

### 3.2.4.1 Liquid conductivity in unsaturated flow

The discussion in Section 3.2.3 on the permeability of hardened cement paste dealt with the saturated conditions of the paste. The main emphasis was on understanding, in an analytical way, the role that the complex microstructure of cementitious materials plays in the overall transport of fluids. The same

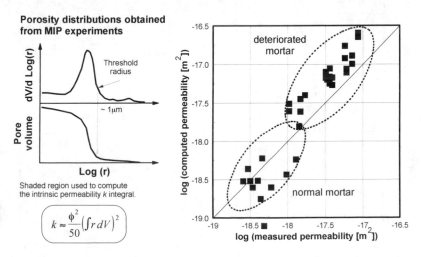

*Figure 3.33* Computed and measured intrinsic permeability of mortars, experiment by Saito *et al.* (1996).

concepts can be extended if we take into account the thermodynamic equilibrium conditions which exist in the porous medium as discussed in Section 3.2.1 for an arbitrary state of moisture content. Under unsaturated conditions, only those pores that are completely saturated would contribute to the flow. Knowledge of the contribution to the total flow from these pores would enable us to compute the relative permeability of the porous medium. For a microstructure-based computational system, this means that the intrinsic permeability model Equation 3.72 can be used if we know the pore radius, $r_c$, of the microstructure in which the equilibrated interface of liquid and vapor is created. Obviously, for such a case, the permeation probability should be obtained only across those pores that are completely filled and completely connected, i.e., those pores lying below $r_c$. For a given pore relative humidity $h$, $r_c$ can be obtained from Equation 3.50. Thus, an expression for unsaturated permeability of water can be obtained as

$$K_l = \frac{\rho\phi^2}{50\eta}\left(\int_0^{r_c} r\,dV\right)^2 \tag{3.75}$$

The viscosity term in the above equation corresponds to the time-dependent viscosity of water in a cementitious microstructure and can be obtained from Equation 3.74. Although some mechanisms have been proposed to account for the surface flow in the adsorbed water layers, we have disregarded the surface flow component completely.

### 3.2.4.2 Vapor diffusivity in unsaturated flows

At lower humidity, moisture flux in cementitious materials is dominated by the flux in the vapor phase. The vapor flow in a pore is governed by mechanisms that depend upon the vapor pressure or relative humidity and radius of the pore, $r$. For very small pores whose radius is of a comparable size to the mean free path length of a water molecule ($l_m$), an overall reduction in the observed flux occurs due to a hindered path and a greater number of collisions. This phenomenon is known as the Knudsen effect (Welty *et al.* 1969), and is dependent on the Knudsen number, $N_k$, which is the ratio of $l_m$ to the effective pore diameter $2r_e$.

$$N_k = \frac{l_m}{2r_e} \tag{3.76}$$

where $r_e$ is the actual pore radius, $r$, minus the thickness of adsorbed layer of water, $t_a$. Using a modified form of Fick's law, the flux of the vapor phase, $q_v^r$, in a single pore of effective radius, $r_e$, can be obtained as

$$q_v^r = \frac{\rho_s D_o}{1+(l_m/2r_e)}\frac{\partial h}{\partial x} \tag{3.77}$$

where $\rho_s$ is the saturated vapor density, and $D_o$ is the vapor diffusivity in a free atmosphere. As the saturation of the porous medium increases, fewer paths are available for vapor to move along. Therefore, only unsaturated pores should be considered while evaluating total vapor flux through the porous medium. Furthermore, the tortuosity effects should be considered to take into account the actual path over which the loss of vapor pressure gradient takes place. At any arbitrary relative humidity, effective vapor flux, $q_v$, can therefore be obtained as

$$q_v = \frac{\rho_s \phi D_o}{\Omega} \int_{r_c}^{\infty} \frac{dV}{1+N_k} \frac{dh}{dx} \tag{3.78}$$

where $\Omega = (\pi/2)^2$ accounts for the tortuosity in three dimensions. The integral in the above expression takes into account the dependence of vapor diffusivity on pore-structure saturation.

### 3.2.4.3 *Computational model of moisture conductivity*

The integrals of Equation 3.75 and Equation 3.78 are too complicated to be put into practical computational use. Usually, real-life numerical schemes involving finite elements, or some other methods, require that conductivity coefficients be computed quite frequently, especially since the problem is non-linear. For practical purposes, we have simplified the conductivity models discussed above sufficiently that they can be implemented easily into regular finite-element routines. For a porosity distribution given by Equation 3.61, the liquid conductivity, $K_l$, can be simply obtained by inserting the computational microstructure model of the cementitious matrix into Equation 3.75 as

$$K_l = \frac{\rho \cdot \phi_{cg}^2}{50\eta} \left(A_{cp} + A_{gl}\right)^2 \quad A_{cp} = \frac{\phi_{cp}}{\phi_{cg}} \frac{\left\{\exp\left(-B_{cp}r_c\right)\left(-B_{cp}r_c - 1\right)+1\right\}}{B_{cp}} \quad Sly. \, for \, A_{gl} \tag{3.79}$$

Similarly the vapor conductivity is obtained as

$$K_v = \frac{\rho_v \phi D_o}{(\pi/2)^2} \frac{1-S}{1+l_m/2(r_m - t_m)} \left(\frac{Mh}{\rho RT}\right) \tag{3.80}$$

where $t_m$ corresponds to the thickness of an adsorbed layer of water in a pore of radius $r_m$, and $r_m$ is the average size of unsaturated pores expressed as a geometric mean. All other symbols have their usual meaning. The above expressions can readily be computed without involving time-consuming numerical integration schemes over a complete porosity distribution. The conductivity curve for a typical pore distribution is shown in Figure 3.34, which also shows the reduction in conductivity under long-time fully saturated

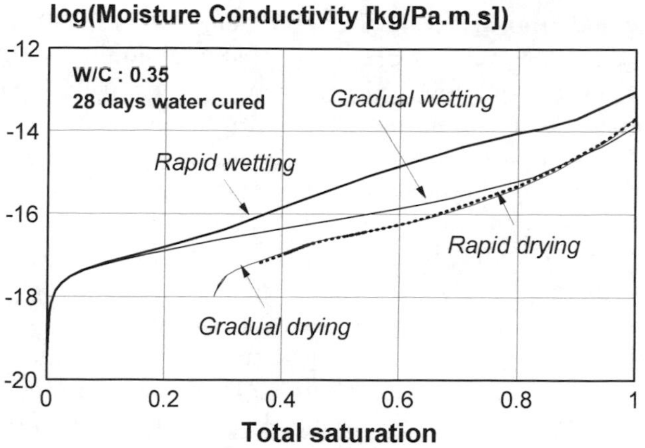

**log(Moisture Conductivity [kg/Pa.m.s])**

*Figure 3.34* Simulated moisture conductivity curves for a typical cementitious porosity distribution.

conditions. Unfortunately, the dynamic nature of concrete permeability dependent upon the duration and conditions of moisture exposure is a relatively unexplored area in the field of concrete science.

### 3.2.5 Consideration of an altering micro-structure in the moisture transport formulation

In Section 3.2.1.2, the influence of moisture loss due to self-desiccation and a change in the microstructure on the moisture transport process was discussed briefly. In this section, we focus on a more rigorous analysis by considering the effect of a dynamically varying microstructure on transport behavior. The changes in the microstructure could be brought about by hydration or any other mechanism, including calcium hydroxide leaching or even carbonation. Our main concern will be the determination of the rate of change of the unit water content of concrete as influenced by a change in the microstructure. The change of microstructure will be considered not only in terms of the change in total porosity but also the changes in the shape of the porosity distribution itself.

Based upon a multi-component division of porosity, the unit water content of concrete can be expressed as a linear summation of the water content of interlayer, gel and capillary porosity components. The change in total water content can therefore be expressed as

$$\frac{\partial \theta_w}{\partial t} = \sum \phi_i S_i \frac{\partial \rho}{\partial t} + \rho \left( \sum \phi_i \frac{\partial S_i}{\partial t} + \sum S_i \frac{\partial \phi_i}{\partial t} \right) \tag{3.81}$$

where $\phi_i$ and $S_i$ denote the porosity and degree of saturation of the $i$-th porosity component (i.e., interlayer, gel and capillary components). In the case of concrete under normal conditions, the change in pore water density can be ignored due to the very small compressibility of liquid water. The second and third terms in Equation 3.81 are the primary contributing terms to the rate of change of total water content. Generally, in a static microstructure, only the change in pore water pressure (or pore humidity) would result in a corresponding change in the degree of saturation as governed by thermodynamic equilibrium conditions (Section 3.2.2).

However, in a dynamic microstructure, the degree of saturation will also be influenced significantly by a change in the shape and size of the pore radii. For example, under conditions of thermodynamic equilibrium and similar pore humidity, a finer porosity distribution will contain more condensed water than a coarser distribution. Therefore, the rate of change of pore distribution parameters that define the porosity distributions of gel and capillary components should be included in Equation 3.81. In this treatment, the interlayer properties are assumed to be constant throughout the aging process of concrete. Therefore, the degree of saturation of the interlayer is essentially a function of the pore humidity or pore water pressure only (Equation 3.60). The gel and capillary porosity components are represented by the pore structure parameters $\phi$ and $B$ (Section 3.2.2.5). The rate of change of porosity of these two components, $d\phi_i/dt$, can be obtained directly from the microstructure development model described in Section 3.1, which is based upon the average degree of hydration of reacting powder materials. The last term in Equation 3.81 can therefore be resolved in a direct stepwise manner by simultaneously applying the microstructure development model and the multi-component cement heat of hydration model to the moisture transport model.

To obtain the rate of variation of the degree of saturation of the gel and capillary components ($dS_i/dt$), we express the saturation as a function of both the pore water pressure, $P$, and a pore structure parameter, $B$. Since, the capillary and gel distributions are similar as regards the physical and analytical treatment, we consider a representative porosity distribution that is similar in analytical description to the gel and capillary components. The steps used for this representative distribution can be similarly applied to gel and capillary components in Equation 3.81. The pore water in a typical cementitious microstructure exists in the condensed and the adsorbed states. Therefore, the rate of variation of the total degree of saturation in the representative porous medium is a summation of the rate of change in condensed and adsorbed water, i.e.

$$\frac{\partial S_T}{\partial t} = \frac{\partial S}{\partial t} + \frac{\partial S_{ads}}{\partial t} \tag{3.82}$$

where $S_T$ is the total degree of saturation of the representative porosity distribution, $S$ is the pore water volume fraction in the condensed state, and $S_{ads}$ is the pore water volume fraction in the adsorbed state. Based on the descriptions of hysteretic nature of condensed moisture and consideration of adsorption, these terms could be expressed as

$$S = S_C + S_{ink}(S_C) \qquad S_{ads} = k(P)U(P,B)$$
$$S_C = S_C(P,B)$$

(3.83)

where $S_C$ is the volume fraction of the primary condensed water, $S_{ink}$ is the volume fraction of pore water existing due to ink-bottle effect (Section 3.2.2.3), $k$ is the adsorbed water film thickness, and $U$ is the exposed internal pore structure surface area per unit pore volume. It is important to note the functional dependencies in Equation 3.83. Applying these definitions to Equation 3.82 results in,

$$\frac{\partial S_T}{\partial t} = \left( A\frac{\partial S_C}{\partial P} + U\frac{dk}{dP} \right)\frac{dP}{dt} + Q$$
$$Q = A\frac{\partial S_C}{\partial B}\frac{dB}{dt} + k\frac{dU}{dt} \quad A = 1 + \frac{dS_{ink}}{dS_C}$$

(3.84)

From the above, it can be seen that in a pore pressure-based formulation (Equation 3.38), the sink terms, $Q_P$, would contain several additional terms resulting from a consideration of the dynamic nature of microstructure. In effect, for a microstructure becoming finer, Equation 3.84 states that a suitable amount of moisture should be deducted to maintain moisture mass compatibility conditions. However, for a pore structure getting coarser, a suitable amount of moisture would be released to maintain thermodynamic compatibility conditions. Of course, for a static microstructure, this formulation would again reduce to the usual moisture transport formulation in a porous medium.

## 3.3 Extended modeling of temperature-dependent moisture transport and equilibrium

In the basic modeling of moisture transport presented in Section 3.2, the authors formulated the flux of both liquid and vapor driven by pore pressure and vapor density. Moisture state in the system can be obtained by combining thermodynamic theory and computed micro-pore structure. This methodology enables us to simulate the moisture profiles under arbitrary drying–wetting history. These models, however, were verified mainly at normal temperature (20°C). This section aims to generalize moisture transport and moisture equilibrium models with respect to temperature. As real concrete structures are used under various temperature conditions, it is important from an engineering viewpoint to expand the scope of applicability. The purpose of this part is to quantify the temperature sensitivity of the state and transport of moisture in nano-meter to micro-meter scale pores, both experimentally and theoretically, and to enhance a system for internal moisture in motion.

### 3.3.1 Mathematical formulations applicable to arbitrary temperature conditions

#### 3.3.1.1 Governing equation for moisture

The law of mass conservation governing the moisture balance in a system is expressed by Equation 3.85. The accuracy and applicability of the analysis system depends on the modeling of each term in the equation

$$\frac{\partial \theta_w}{\partial t} + div\big(\mathbf{J}(\theta_w, T, \nabla \theta_w, \nabla T)\big) + Q = 0 \tag{3.85}$$

where $\theta_w$ is the mass of moisture in a unit volume of concrete [kg/m³], $J$ is moisture flux [kg/m².s], $T$ is temperature [K], and $Q$ is the sink term corresponding to water consumption due to hydration [kg/m³.s].

In Section 3.2, the potential term (the first term in Equation 3.85), which represents moisture capacity of the material, and the flux term (the second term in Equation 3.85) have been verified only for normal temperature conditions (20°C). To widen their applicability, temperature-dependent moisture transport and equilibrium models are introduced as follows.

#### 3.3.1.2 Equilibrium between liquid and vapor phases of water under arbitrary temperatures

The potential term for the moisture in a porous material can be expressed as

$$\frac{\partial \theta_w}{\partial t} = \frac{\partial(\rho_l \phi S)}{\partial t} \tag{3.86}$$

where $\rho_l$ is the density of liquid water [kg/m³], $\phi$ is the porosity, and $S$ is the degree of saturation of porosity. The mass of water vapor is ignored in this term as it is negligible compared to that of liquid water. The density of liquid water $\rho_l$ is temperature dependent and simply expressed as (National Institute of Natural Sciences 2002, Figure 3.35a),

$$\rho_l = 1.54 \times 10^{-8} \cdot T^3 - 1.85 \times 10^{-5} \cdot T^2$$
$$+ 6.65 \times 10^{-3} \cdot T + 2.47 \times 10^{-1} \tag{3.87}$$
$$(273 < T < 373)$$

In order to compute the degree of saturation $S$, it is necessary to describe the relation of liquid water and water vapor in the pores. Liquid water in

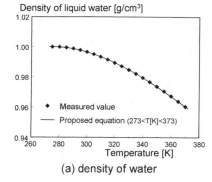

(a) density of water

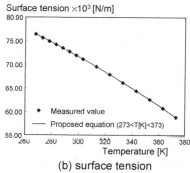

(b) surface tension

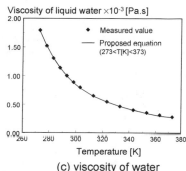

(c) viscosity of water

*Figure 3.35* Relationship between temperature and density of liquid water.

the microstructure is always under negative pressure $P_l$ [Pa] due to its surface tension. By assuming that a pore is cylindrical, a relationship between this pressure and the surface tension $\gamma$ of liquid water can be calculated as (Section 3.2.2.1)

$$P_l = -\frac{2\gamma}{r} \tag{3.88}$$

where $r$ is the pore radius [m], and $\gamma$ is the surface tension of liquid water [N/m]. Since the surface tension of liquid water also varies with temperature, an equation derived by the regression of measured values (National Institute of Natural Sciences 2002) is used in a similar way to that for the density of liquid water (Figure 3.35b).

$$\gamma(T) = 2.66 \times 10^{-4} \cdot T^2 + 3.17 \times 10^{-3} \cdot T$$
$$+ 9.46 \times 10^{1} \quad (273 < T < 373) \tag{3.89}$$

Absolute vapor pressure $p_{vap}$ [Pa] in the pores becomes lower than that in the atmosphere due to the formation of meniscus. In a state of phase equilibrium, Gibbs' free energy in the gas and liquid phases are equal. Then, we have

$$RT \ln \frac{p_{vap}}{p^*} = V_l \cdot P_l \tag{3.90}$$

where $R$ is the gas constant [J/mol.K], $p^*$ is the saturated vapor pressure [Pa] and $V_l$ is the molar volume of liquid water [m³/mol]. The ratio $p^*/p_{vap}$ corresponds to relative humidity. Substituting $V_l = M_w/\rho_l$ by using the molar volume $M_w$ [kg/mol] and density $\rho_l$ [kg/m³] of liquid water and rewriting, the equation finally becomes

$$P_l = \frac{\rho_l RT}{M_w} \ln \frac{p_{vap}}{p^*} \tag{3.91}$$

where the saturated water vapor pressure, $p^*$, varies with temperature. The value of $p^*$ at an arbitrary temperature can be calculated from the Clausius–Clapeyron equation as

$$\frac{d \ln p}{dT} = \frac{\Delta H_{vap}}{RT^2} \tag{3.92}$$

where $\Delta H_{vap}$ is the heat of evaporation [kJ/mol] of liquid water. If the range of temperatures to which the proposed model is applicable is from 273[K] to 373[K], the heat of evaporation of liquid water may be considered to be almost constant regardless of temperature. Therefore, integrating Equation 3.92 assuming that the heat of evaporation gives

$$\begin{aligned}
p_{vap} &= p^* \exp\left(\frac{P_l M_w}{\rho_l RT}\right) \\
&= p_0 \exp\left\{-\left(\frac{\Delta H_{vap}}{R}\right)\left(\frac{1}{T} - \frac{1}{T_0}\right)\right\} \exp\left(\frac{P_l M_w}{\rho_l RT}\right)
\end{aligned} \tag{3.93}$$

where $p_0$ and $T_0$ are reference pressure and reference temperature, respectively.

### 3.3.1.3 Modeling of moisture flux

In order to generalize the modeling of moisture flux with respect to temperature, a flow driven by both the pore pressure gradient and the temperature gradient is to be considered. Here, the moisture flux $J$ [kg/m².s] for both vapor and liquid water can generally be expressed as

$$J = -\left(D_p \nabla P_l + D_T \nabla T\right) \tag{3.94}$$

where $D_p$ is moisture conductivity [kg/Pa.m.s] with respect to the pore pressure gradient, and $D_T$ is moisture conductivity [kg/K.m.s] with respect to the temperature gradient.

The basic modeling of vapor transport in the preceding section applicable to the isothermal condition formulates the flux $q_v$ [kg/m².s] driven by the relative humidity gradient as follows (Section 3.2.4.2).

$$q_v = -\frac{\rho_v^{sat} \phi D_0}{\Omega} \int_{r_c}^{\infty} \frac{dV}{1 + N_k} \nabla h$$

$$N_k = \frac{l_m}{2(r - t_a)}$$

(3.95)

where $\rho_v^{sat}$ is saturated vapor density [kg/m3], $D_0$ is vapor diffusivity [m²/s] in free atmosphere at 293 [K], $\Omega$ is the parameter representing tortuosity of pore (= $(\pi/2)^2$), $r_c$ is the pore radius at which an interface between liquid and vapor is created, $N_k$ is the Knudsen number, $l_m$ is the mean free path of gas molecules [m], $V$ is the normalized pore volume, $h$ is relative humidity, and $t_a$ is the thickness [m] of an adsorbed layer in a pore, calculated by using modified BET theory (Hillerborg 1985).

When dealing with vapor flow under various temperatures, the gradient of relative humidity cannot be defined as transport potential. In other words, relative humidity at different temperatures does not represent the driving force accurately, since the saturated vapor pressure depends on temperature, which leads to different relative humidity even though vapor density of the system is the same. Thus, Equation 3.95 is generalized for arbitrary temperature as

$$q_v = -\frac{\phi D_0(T)}{\Omega} \int_{r_c}^{\infty} \frac{dV}{1 + N_k} \nabla \rho_v = -D_v \nabla \rho_v$$

$$N_k = \frac{l_m}{2(r - t_a)}$$

(3.96)

where $D_v$ is the vapor diffusivity in concrete [m²/s]. The proposed vapor transport model is driven by the gradient of absolute vapor density $\rho_v$ [kg/m³] in the system. The product of the relative humidity $h$ and $\rho_v^{sat}$ in Equation 3.95 corresponds to this $\rho_v$. Under an isothermal condition, Equation 3.96 is equivalent to Equation 3.95. As the vapor diffusivity $D_0$ in free space is dependent on temperature (Welty *et al.* 1969), we have

$$\frac{D_0(T_1)}{D_0(T_2)} = \left(\frac{T_1}{T_2}\right)^{3/2} \left(\frac{\Omega_{D,T2}}{\Omega_{D,T1}}\right)$$

(3.97)

where $\Omega_D$ is the collision integral at temperature $T_1$ or $T_2$, given as a function of the Boltzmann constant (= $1.38 \times 10^{-23}$ [J/K]), temperature, and so on.

Next, let us derive a formula for liquid water. The flux of liquid water $q_l$ [kg/m².s] can be calculated by using the following model

$$q_l = -\frac{\rho_l \phi^2}{50\eta} \left( \int_0^{r_c} r dV \right)^2 \nabla P_l = -K_l \nabla P_l \tag{3.98}$$

where $K_l$ is the liquid conductivity [kg/Pa.m.s]. With regard to temperature changes, the most significant factor affecting liquid water transport is the viscosity $\eta$ [Pa.s] of condensed water. The viscosity $\eta$ is expressed as

$$\eta = \eta_i \exp\left(\frac{G_e}{RT}\right) \tag{3.99}$$

where $\eta_i$ is the viscosity of liquid water under ideal conditions, and $G_e$ is additional Gibbs' energy required for liquid flow under non-ideal conditions. Because the viscosity $\eta_i$ of liquid water, like density and surface tension, is dependent on temperature, we have a regression formula based on measured values (National Institute of Natural Sciences 2002, Figure 3.35c) as

$$\begin{aligned}
\eta_i = &\, 3.38 \times 10^{-8} \cdot T^4 - 4.63 \times 10^{-5} \cdot T^3 \\
&+ 2.37 \times 10^{-2} \cdot T^2 + 5.45 \cdot T + 4.70 \times 10^2 \\
&(273 < T < 373)
\end{aligned} \tag{3.100}$$

Then, we have the moisture fluxes as

$$\begin{aligned}
J &= -\left(D_v \nabla \rho_v + K_l \nabla P_l + K_T \nabla T\right) \\
&= -\left\{ D_v \left( \frac{\partial \rho_v}{\partial P_l} \nabla P_l + \frac{\partial \rho_v}{\partial T} \nabla T \right) + K_l \nabla P_l + K_T \nabla T \right\} \\
&= -\left( D_v \frac{\partial \rho_v}{\partial P_l} + K_l \right) \nabla P_l - \left( D_v \frac{\partial \rho_v}{\partial T} + K_T \right) \nabla T \\
&= -\left( D_p \nabla P_l + D_T \nabla T \right)
\end{aligned} \tag{3.101}$$

The term expressed as $K_T \nabla T$ in Equation 3.101 represents thermal diffusion known as the Soret effect (Welty *et al.* 1969). For simplicity, this thermal diffusion is ignored ($K_T \approx 0$) since contribution of thermal diffusion to the total flux is not well known, and this phenomenon normally plays a minor role in diffusion compared with moisture transfer driven by pore pressure and vapor pressure gradient (Welty *et al.* 1969).

### 3.3.1.4 Calculation of the degree of saturation in a porous system

The relations among pore pressure, saturated vapor pressure and absolute vapor pressure at arbitrary temperatures have been obtained as stated in the preceding section. By combining existing multi-scale models for water (Figure 3.36) and the moisture equilibrium equations (Equations 3.88 and 3.91), the

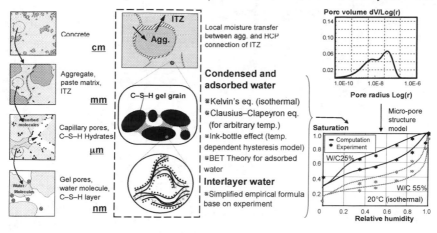

*Figure 3.36* Multi-scale modeling of moisture existing in capillary, gel and interlayer pores.

authors compute moisture isotherms at 20°C and 60°C. When the degree of saturation in gel and capillary pores is calculated, it is assumed that all pores whose radii are smaller than the radius $r_c$ in which a vapor–liquid interface is created are filled with condensed water. The saturation $S_c$ by such condensed water can be determined from $r_c$ derived from the moisture equilibrium equations and the pore distribution function. Then, in order to take into account the contribution of adsorbed water in the unsaturated pores, the degree of saturation $S_{ads}$ is calculated by integrating the adsorbed water layer thickness determined by BET theory with respect to pores. Then, as the sum of quantities of water thus obtained, the degree of saturation in the wetting phase is calculated. In the drying phase, the overall degree of saturation is calculated by adding additional moisture $S_{ink}$ due to the ink-bottle effect. Moisture isotherms determining the state of interlayer water are expressed by the same functional equations regardless of temperature. From these physical models, the total degree of saturation $S_{total}$ is calculated as

$$S_{total} = \frac{\phi_{cp} \cdot S_{cp} + \phi_{gl} \cdot S_{gl} + \phi_{lr} \cdot S_{lr}}{\phi_{cp} + \phi_{gl} + \phi_{lr}} \tag{3.102}$$

where $\phi_{cp}$ is the capillary porosity, $\phi_{gl}$ is the gel porosity, $\phi_{lr}$ is the interlayer porosity, $S_{cp}$ is the degree of saturation of capillary pores, $S_{gl}$ is the degree of saturation of gel pores, and $S_{lr}$ is the degree of saturation of interlayer pores. The pore structure development model (Section 3.1) gives each of these porosities for an arbitrary stage of hydration. Figure 3.37 shows the computed moisture isotherms. Although temperature-dependent surface tension and density are introduced, the computed moisture isotherms at different temperatures do not show any significant differences.

Figure 3.38 shows moisture loss under 20°C and 60°C. Prismatic mortar

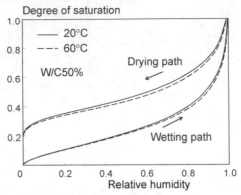

*Figure 3.37* Computed moisture isotherm.

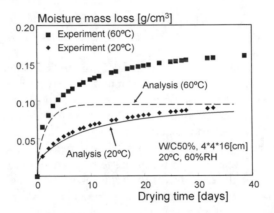

*Figure 3.38* Moisture loss behaviors at 20°C and 60°C.

specimens (4 × 4 × 16 cm) with a water-to-cement ratio of 50% were prepared. After the specimens were cured under sealed conditions for 38 days, they were dried in a controlled chamber at 60% RH. Moisture loss at normal temperature (20°C) is accurately predicted, but the calculated values for 60°C do not capture the actual trends. Although relative humidity in the external environment is 60% RH in both cases, the absolute vapor density gradient $\nabla\rho_v$ at higher temperatures is large because of the temperature dependence of saturated vapor pressure. Consequently, the computed moisture loss behavior in the early stages of drying are high, as in reality, but the continuous moisture loss cannot be seen in the analysis. As shown in Figure 3.37, the moisture isotherms, which determine the water content of the specimens, do not differ significantly at the two temperatures. Then, the amount of moisture loss does not show any significant difference.

These results indicate that the most important issue for future consideration is an appropriate expression of moisture equilibrium based on a microscopic viewpoint. In the moisture isotherm model used in the calculation, the effects of temperature for the two-phase equilibrium of condensed liquid and vapor (Equation 3.88 and Equation 3.93) and adsorbed water on the pore wall surfaces (BET theory, Section 3.2.2.2) have already been given according to thermodynamic theory. The hysteresis model based on the hypothesis of the ink-bottle effect and the interlayer water model have only been verified through the observed behaviors at room temperature (20°C). It was decided, therefore, to extract these temperature effects from the following systematic experiments. The reason for this is that if attention is paid only to the total summation of moisture from a macroscopic point of view, simple phenomena behind apparently complex behavior might be overlooked.

### 3.3.2 Temperature effect on internal moisture state

#### 3.3.2.1 Experimental investigation

Table 3.2 shows the mix proportions of experimental specimens. Cement paste specimens with a water-to-cement ratio of 50% were prepared using ordinary Portland cement. To prevent bleeding, 15% or 40% of limestone powder by volume was mixed into the cement paste. As the filler effect of limestone powder did not apparently influence the experimental results, no distinction is made between the two types of mixes in Table 3.2. Cylindrical molds used to cast cement paste mixes were 10 cm in diameter and 20 cm in height. Molded cement paste was cured under sealed conditions, and was removed after one day of curing. Then, in a climate-controlled room kept at 20°C and 60% RH, the molded cement paste blocks were cut into about 1 cm cubes using a wet-type concrete cutter, and these cubes were water cured for 80 days at 20°C. The aim of this method was to ensure the progress of hydration reaction and to eliminate the influence of the hydration-induced consumption of free water during wetting–drying testing and changes in pore structure on moisture equilibrium. The specimens were wetted or dried at three temperature levels (20°C, 40°C and 60°C). Before the experiment, specimens were stored for one week in water kept at 20°C, 40°C or 60°C.

After water curing, the specimens were wetted or dried under the conditions as shown in Table 3.3 and Table 3.4 in a climate-controlled chamber. During the experiment, temperature in the chamber was kept within ±0.5°C of the specified level and relative humidity was kept within ±2%. The experiments were conducted under different temperatures and relative humidity

*Table 3.2* Mix proportion of cement paste specimen

| W/C (%) | Unit mass [kg/m³] | | |
|---|---|---|---|
| | Water | Cement | Lime powder |
| 50 | 520 | 1,040 | 405 |
| | 367 | 734 | 1,080 |

*Table 3.3* Wetting conditions in the experiment

| Temperature | Relative humidity | Duration of wetting (days) |
|---|---|---|
| 20°C | 40%, 55%, 70%, 85% | 7 |
| 40°C | 40%, 55%, 70%, 85% | 7 |
| 60°C | 30%, 60%, 90% | 7, 14, 28 |

*Table 3.4* Drying conditions in the experiment

| Temperature | Relative humidity | Duration of drying (days) |
|---|---|---|
| 20°C | 30%, 60%, 90% | 7, 28, 60 |
| 40°C | 60% | 7, 14, 28 |
| 60°C | 30%, 60%, 90% | 7, 14, 28, 60 |

conditions for the drying process starting from the complete saturation and the wetting process starting from the complete dispersion of liquid water and interlayer water. Here, an oven-dry condition achieved by 105°C drying is defined as a zero-saturation state in which neither liquid water nor interlayer water exists.

In order to measure liquid water and interlayer water in the specimens separately, an organic solvent-based water extraction method was used (Fujii and Kondo 1974). In the method, internal moisture is leached out by immersing a hardened cementitious material in a hydrophilic organic solvent. The most important requirement is to properly separate liquid water and interlayer water. A study by Fujii and Kondo (1974), however, has pointed out that the amount of water that can be leached out varies depending on the type of solvent used. One possible factor is differences in the size, structure or other attributes of solvent molecules that influence the ability of solvents to leach out water from hardened cementitious materials. The structure of C–S–H gel and the distribution of interlayer water and gel water are usually illustrated as in Figure 3.25 (Feldman and Sereda 1968). In reality, however, there is no clearly discernible boundary, and measured values may vary depending on solubility.

In this study, the authors used ethanol as an organic solvent after the preliminary trials with methanol, ethanol and propanol. Table 3.5 lists the data by ethanol. After 1 cm-thick specimens were cured in water for three months, the specimens were immersed in the organic solvent until the water concentration in the solvent reached a state of equilibrium (immersed for 40 days or, only in the cases where the water-to-cement ratio was 25%, for 70 days). Both the curing in water and the immersion in ethanol were carried out at 20°C. It is likely that after three months of water curing, all pores are saturated with liquid water and interlayer water, and that the amount of water can be extracted by the organic solvent is equal to the amount of liquid water existing in the capillary and gel pores. As shown in Table 3.5, measured

*Table 3.5* Measured and computed values of liquid water existing in capillary and gel pores

| | Amount of liquid water (%) (measured) | Amount of liquid water (%) (computed) |
|---|---|---|
| Cement paste (W/C 65%) | 75.7 | 81.1 |
| Cement paste (W/C 50%) | 70.0 | 73.0 |
| Mortar (W/C 50%) | 72.4 | 73.0 |
| Cement paste (W/C 25%) | 57.7 | 61.5 |

and calculated values of the different water-to-cement ratios designed to vary the capillary, gel and interlayer water composition show several percentage points of differences, but they show fair agreement. Thus, the measured values of physical quantities and the calculated values for condensed water obtained from the model can be regarded as being mostly in agreement. Then, the water that is leached out by ethanol is defined as liquid water (condensed water + adsorbed water) existing in the capillary and gel pores, and the remaining water is defined as interlayer water.

After the specimens were subjected to various wetting and drying episodes, they were immersed in ethanol in sealed containers kept at 20°C for 40 days, and the extracted water regarded as liquid was measured with a trace moisture meter using the Karl Fischer titration method. The degree of saturation $S_{lw}$, which is the ratio of the volume of liquid water to the volume of pore space, was calculated by

$$S_{lw} = \frac{V_{lw}}{V_{lw}^{sat}}$$   (3.103)

where $V_{lw}^{sat}$ is the volume of liquid water per unit oven-dry mass of specimen at saturation (20°C) [ml/g], and $V_{lw}$ is the volume per unit oven-dry mass of liquid water existing in a specimen subjected to a specified period of drying and wetting [ml/g]. Samples were put into three containers for immersion in ethanol, and the measured values were averaged.

### 3.3.2.2 Kinematics of liquid water under different temperatures

Figure 3.39 shows the degrees of saturation of liquid water subjected to specified environmental conditions for seven days. In the wetting process, specimens in an oven-dry condition achieved after 105°C drying were subjected to different relative humidity. Drying was started from full saturation. For comparison, calculated values obtained from the analysis model are also shown in Figure 3.39. As the initial and boundary conditions for the analysis, the mix proportions (water-to-cement ratio, volume fraction of aggregate, air content), properties of the materials used (chemical composition of binder,

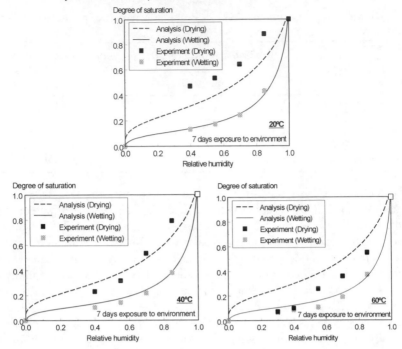

*Figure 3.39* Drying and wetting isotherm under different temperatures.

specific gravity, Blaine fineness index, etc.) and curing conditions (boundary conditions for heat and moisture transfer) used in the test were given.

For the wetting process, the measured saturation and the calculated values obtained from the analysis model show fair agreement, indicating that the moisture in the hardened cementitious material in the wetting phase is in a state of equilibrium described by thermodynamic theory. This means that the state of internal moisture can be expressed as the sum of condensed and adsorbed water that can be determined by vapor–liquid equilibrium and BET theory.

The situation differs considerably, however, with respect to the drying phase. As temperature rises, measured values become smaller than the calculated ones and gradually approach the values in the wetting phase. This tendency is particularly pronounced in the low humidity range. The analysis model tries to explain the moisture hysteresis during drying and wetting by the ink-bottle effect (Section 3.2.2.3). The model takes this approach on the assumption that moisture trapped in ink-bottle-shaped pores brings higher water content in the drying phase than that in the wetting phase. One possible cause of the disappearance of hysteresis as temperature rises or as humidity decreases is the stability of moisture trapped in the ink-bottle-shaped pores.

The requirement for the third law of thermodynamics dictates that if stability is to be maintained for an infinite period of time, ink-bottle-shaped pores in which moisture is trapped must be completely closed spaces. If, however, numerous micro-pores are interconnected in many directions, it is very unlikely that completely closed spaces are formed. It is logical to assume that moisture trapped during the drying process will gradually be dispersed over a long period of time. The rates of transport and dispersion are strongly dependent on temperature. It is reasonable to assume that as temperature rises to 40°C and to 60°C, the moisture trapped in the ink-bottles becomes increasingly subject to transport and diffusion so that the isotherm path in the drying phase gradually becomes closer to that in the wetting one.

Figure 3.40 shows previously reported experimental results relevant to this discussion (Taylor 1997). Here, 1 mm-thick cement paste specimens with a water-to-cement ratio of 50% were prepared by using purely synthetic C₃S, and these specimens were cured for 5.8 years in order to complete the hydration process. Then, the specimens were dried at 25°C from a saturated condition, and the total water content was measured. As shown in Figure 3.40, even in the case of very thin specimens, the water content continues to change for as many as 170 days at normal temperature, indicating that in the drying phase, a long period of time is required before equilibrium is reached.

In order to investigate in detail the time dependence described above, a series of tests was conducted under different temperature conditions for different exposure periods. Figure 3.41 and Figure 3.42 show the results. In the wetting phase (Figure 3.41), similar values were obtained regardless of exposure periods, indicating a temporally stable state of equilibrium. Figure 3.42 shows the degrees of saturation of specimens dried at 20°C, 40°C and 60°C. Let us first pay attention to the influence of temperature. Examination of behavior at the relative humidity of 60% reveals that the rate of dispersion during each period increased as temperature rose. For the 60°C case,

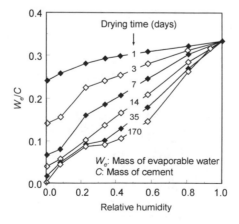

*Figure 3.40* Desorption curves for an initially saturated C₃S paste (Taylor 1997).

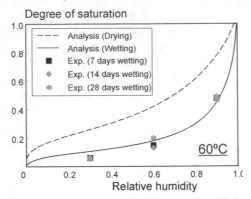

*Figure 3.41* Degree of saturation in the wetting process at 60°C for different exposure periods (liquid water).

the degree of saturation decreased gradually as time passed, and on and after day 28 reached the level of values typically found in the wetting phase, resulting in a state of equilibrium as given by the model. These experimental data suggest that even the moisture trapped by the ink-bottle effect caused by interconnected micro-pores continues to disperse gradually toward the thermodynamic state of equilibrium so as to reach a state of equilibrium similar to that in the wetting phase. From the fact that the degree of satura-tion decreased quickly as temperature increased, it can be inferred that the dispersion of moisture trapped in ink-bottle-shaped pores is strongly related to the chemical potential of internal water molecules.

Next, let us consider the effect of relative humidity. Comparison revealed that under all temperature conditions and for all drying periods, the rate of dispersion increased as relatively humidity decreased. At the temperature of 60°C, for example, a state of equilibrium was reached on the wetting curve in 30% and 60% humidity, while at the relative humidity of 90%, there was a difference between the drying and wetting curves even after 60 days of dry-ing. At the relative humidity of as high as 90%, more pores are in a saturated condition than under other conditions. In a situation like this, there are only limited dispersion paths for the moisture trapped by the ink-bottle effect (i.e., a higher likelihood of the formation of completely closed pore spaces).

### 3.3.2.3 Behavior of interlayer water under different temperature conditions

Figure 3.43 shows the relationship between the amount of extracted inter-layer water and relative humidity. The amount of interlayer water is cal-culated by subtracting the amount of liquid water extracted by the solvent extraction method after seven days of exposure under different conditions from the total amount of water. The degree of saturation of interlayer water, $S_{iw}$, is calculated from Equation 3.104 as

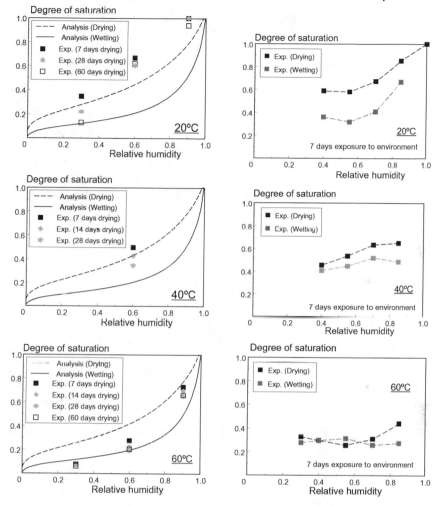

*Figure 3.42* Degree of saturation in the drying process at different temperature and exposure periods (liquid water).

*Figure 3.43* Degree of saturation at different temperatures (interlayer water).

$$S_{iw} = \frac{V_{iw}}{V_{iw}^{sat}} \tag{3.104}$$

where $V_{iw}^{sat}$ is the volume of interlayer water per unit oven-dry mass of specimen at saturation (20°C) [ml/g], and $V_{iw}$ is the volume of interlayer water per unit oven-dry mass existing in a specimen subjected to specified periods of drying and wetting [ml/g]. The amount of interlayer water in saturated condition at 40°C and 60°C was not measured; instead, the degree of saturation at 20°C was taken as a reference value and calculated as shown in Equation

3.104. The values calculated under different temperature are relative values based on the saturated condition at 20°C.

The amount of interlayer water in the specimens subjected to up to 60 days of wetting and drying, as well as seven days of exposure, was measured. The specimens, however, exposed to a high temperature (60°C) and a medium relative humidity (60% RH) show increases in the mass of the specimen, particularly after 28 days. Thermal gravimetric analysis (TGA) revealed the production of calcium carbonate, suggesting the possibility of mass increases due to carbonation. Eliminating the influence of carbonation is important because the amount of interlayer water is obtained by measuring the total mass of the specimen and then subtracting the mass of moisture extracted by the solvent extraction method. Only data that did not indicate the production of calcium carbonate were chosen in advance through TGA.

One issue worthy of note about the experiments is that the apparent behavior of interlayer water does not indicate the drying–wetting hysteresis at 40°C or 60°C. Also noteworthy is that the degree of saturation is more or less constant regardless of relative humidity. For example, the measured degrees of saturation of interlayer water ranged from 30% to 40%. These values indicate the ratio of the amount of interlayer water stably existing at 20°C to the measured amount of interlayer water in a saturated condition at the same temperature. The moisture in a stable state varied with temperature. At 20°C, there are path differences between the drying and wetting phases. Some time is required in order for interlayer water to recover from an oven-dry condition (Asamoto and Ishida 2003; Asamoto *et al.* 2006). Part of interlayer water, therefore, did not reach a state of equilibrium after seven days of wetting, and it can be inferred that the degree of saturation in the wetting phase was undermeasured.

### 3.3.3 Generalized model for moisture equilibrium

#### 3.3.3.1 Condensed liquid water

The discussion described in the preceding section has confirmed that the test results concerning the wetting phase show good agreement with the total amount of liquid water calculated as the sum of the amounts of condensed water and adsorbed water expressed by Kelvin's equation and BET theory. In the drying phase, however, behavior varies considerably depending on the temperature and humidity to which the specimen is exposed. There is a difference in the time elapsed before reaching equilibrium depending on temperature and humidity, but it can be assumed that ultimately the relationship between the degree of saturation in the drying phase and relative humidity follows the equilibrium curve for the wetting phase.

The ink-bottle effect has been already mentioned as a cause of the short-term drying–wetting hysteresis. That is, the requirement for the third law of thermodynamics dictates that water trapped in the "ink-bottles" can exist only in closed spaces. The probability of occurrence of completely closed spaces in an irregular and complex pore structure is low. It can be inferred that the moisture trapped by the ink-bottle effect will gradually disperse into adjacent connected pores at the rates corresponding to ambient temperature and relative humidity. This section proposes a model that expresses the process of the transition to moisture equilibrium in the drying phase.

A relatively short-term monotonic drying process at normal temperature, though depending on the size of the object to be analyzed, has been accurately modeled as follows (Section 3.2.2).

$$S = S_c + S_{ads} + S_{ink} \tag{3.105}$$

where $S$ is the degree of saturation with condensed water and adsorbed water, $S_c$ is the degree of saturation with moisture existing in pores below radius $r_c$, $S_{ads}$ is the degree of saturation due to adsorbed water, and $S_{ink}$ is the degree of saturation with trapped water due to the ink-bottle effect. As is evident from the discussion so far, the model can be enhanced overall by introducing the influence of temperature on the degree of saturation associated with the ink-bottle effect. With the aim of taking into account the time-dependent dispersion of "ink-bottle" water, it was decided to incorporate the effect of temperature in the simplest form while following the basic principles of the existing model

$$S' = S'_c + S_{ads} + k \cdot S_{ink} \quad (0 \leq k \leq 1) \tag{3.106}$$

where $k$ is a parameter that takes the value of 1.0 immediately after the start of the drying process and decreases with the progress of drying (Figure 3.44). When the wetting curve is reached after the passage of sufficient time, $k$ takes the value of zero. As indicated by the test results shown in the preceding chapter, the rate of decrease of this parameter is dependent on temperature and is also influenced by ambient humidity. Since Equation 3.106 is a model that describes local equilibrium in pores, results obtained from specimens of finite dimensions (1 cm cubes) cannot be taken as the properties of infinitesimal volume elements. The only logical way to directly determine the properties of infinitesimal volume elements is to conduct tests on infinitesimal volume specimens. For this reason, sensitivity to temperature and humidity was back-analyzed through finite element analysis consistent with the test conditions. To be more specific, a 1 cm mesh was prepared, temperature and humidity conditions identical to the test conditions were given as conditions for analysis, and parameters were determined so that the averages of the calculated values within the finite dimensions were consistent with the macroscopic test results shown earlier (Figure 3.42). As a result, the following equation was obtained:

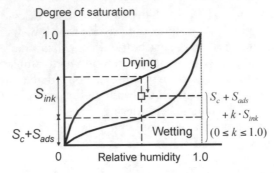

*Figure 3.44* Enhanced moisture isotherm model considering the time-dependent dispersion of "ink-bottle" water.

$$\frac{dk}{dt} = -C \cdot a_{ink}^T \cdot a_{ink}^h \cdot k$$

$$a_{ink}^T = \exp\left(-1.5 \times 10^4 / T\right) \qquad (3.107)$$

$$a_{ink}^h = 0.05 \cdot \left(100.0 - 100.0 \cdot h\right)^{0.81}$$

where $C$ is a constant ($3.0 \times 10^{13}$ [1/sec]), $a_{ink}^T$ is a coefficient expressing sensitivity to temperature, $a_{ink}^h$ is a coefficient expressing sensitivity to relative humidity, $T$ is temperature [K], and $h$ is relative humidity.

### 3.3.3.2 Interlayer water

The analysis system uses an empirical model (Section 3.2.2.4) of the hysteretic behavior of interlayer water. In order to expand the scope of application to cover a wider range of temperature conditions, it is necessary to take into account the temperature dependence of the equilibrium of not only liquid water but also interlayer water. Thus, the authors introduce a function for expressing an envelope curve on the moisture isotherm of interlayer water (Figure 3.45), whereas the conventional model assumed a constant temperature of 20°C.

$$S_{lr}^{env} = \frac{a \cdot h^{n_2} + b \, (h \geq 0.8)}{h^{n_1} + c \quad (h < 0.8)}$$

$$a = \frac{1.0 - h_t^{env}}{1.0 - 0.8^{n_1}}, \quad b = \frac{-0.8^{n_2} + h_t^{env}}{1.0 - 0.8^{n_1}} \qquad (3.108)$$

$$c = h_t^{env} - 0.8^{n_2}, \quad h_t^{env} = -1.5 \times 10^{-2} \cdot T + 5.4$$

$$n_1 = 25.0, \quad n_2 = 0.05$$

In the above equation, the temperature dependence of interlayer water $S_{lr}^{env}$ is expressed with the parameter $h_t^{env}$, which shows the ratio of interlayer

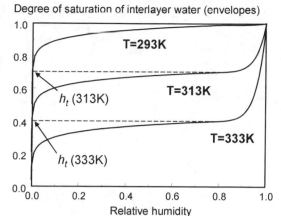

Degree of saturation of interlayer water (envelopes)

*Figure 3.45* Moisture isotherm model of interlayer water under different temperature conditions.

water existing in a stable condition as a criterion for a saturated state at 20°C. For 20°C, parameter values consistent with the existing model shown in Figure 3.26 were given. For 40°C and 60°C, parameters were determined according to the test results reported in Figure 3.43. As Equation 3.108 indicates, the model expresses the tendency of interlayer water to reach a stable range when humidity reaches 80% and then to disappear gradually.

The measured values for interlayer water at 20°C in the drying phase are smaller than the values given by the proposed model. In the test, water extracted by the solvent extraction method was defined as liquid water, and the difference between the total amount of water and the amount of liquid water was taken as the amount of interlayer water. However, the possibility cannot be denied that the values for the components isolated by solvent extraction differ from the definitions used in the analysis. As the reason for this, there is no denying the possibility that the ethanol that has seeped into unsaturated pores after the completion of the normal-temperature drying process extracts part of the interlayer water, particularly the kind of interlayer water that would disperse at 40°C or 60°C. The authors think that there is a need for further discussion on differences in the solvating ability of different solvents and the amount of bound water that can be extracted, as well as discussion on micro-pore structure change after solvent extraction.

### 3.3.4 Analysis and verification

#### 3.3.4.1 Moisture loss

First, let us discuss moisture loss at different temperatures. The experimental results as shown in Figure 3.38 are analyzed by using the newly proposed model. Figure 3.46 shows calculated moisture loss behaviors of mortar

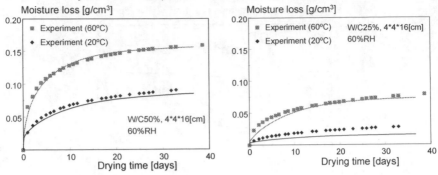

*Figure 3.46* Computed and measured moisture loss under different temperature conditions (W/C = 50% and 25%).

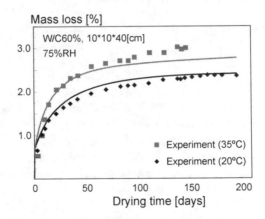

*Figure 3.47* Computed and measured moisture loss behaviors under different temperature conditions (Ayano and Sakata 1997).

specimens with water-to-cement ratios of 50% and 25%. In both cases, the specimens were dried at a relative humidity of 60%. The proposed model captures moisture loss behaviors accurately under different temperature conditions. Figure 3.47 shows moisture loss at 20°C and 35°C (Ayano and Sakata 1997). Prismatic mortar specimens (10 × 10 × 40 cm) with a water-to-cement ratio of 50% were prepared. After 14 days of water curing, specimens were dried at 75% RH. Reasonable agreement can be seen for both temperature conditions.

Figure 3.48 shows moisture loss of concrete drying at 38°C and 40% relative humidity (Cano-Barrita *et al.* 2004). Prior to drying, different curing conditions were given to cylindrical specimens having one exposure surface: 5C (no moist curing), 4C (moist curing for one day), 3C (moist curing for seven days) and 2C (moist curing for 28 days). The proposed model reasonably reproduces moisture loss behaviors for 2C, 3C and 4C cases, whereas it underestimates the amount of mass loss for 5C. In case of no moist curing, moisture evaporation at the surface is much accelerated, which leads to

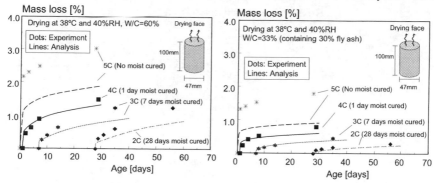

*Figure 3.48* Computed and measured moisture loss behaviors at high temperature under different curing conditions (Cano-Barrita *et al.* 2004).

delay of hydration and increased cracking due to shrinkage. Larger moisture loss measured in the 5C case may be caused by such cracks at the exposure surface, which is not taken into account in the numerical model.

### 3.3.4.2 Correlation of internal relative humidity and the progress of hydration

Next, let us investigate the progress of self-desiccation in different thermal environments. The authors focus on the correlation among the hydration process under sealed condition, temperature and humidity in pores, and the moisture state inside the material. In the experiment (Park and Noguchi 2002), decreases in humidity in pores due to self-desiccation were measured at three temperatures (20°C, 40°C, 60°C). The ranges of the hygrometer used were 0–100°C and 15%–95%, and its accuracy was ±1°C and ±3% RH. Here, attention is addressed to changes in relative humidity of 28% W/C cement paste made of ordinary Portland cement. Generally, hydration at early ages is promoted as curing temperature rises. In fact, the amount of calcium hydroxide generated by hydration reaction increased as the curing temperature rose (Park and Noguchi 2002). The degree of self-desiccation, however, did not have a one-to-one relationship with curing temperature or the degree of hydration. At the highest temperature of 60°C, relative humidity decreased considerably at early stages, but it began to increase at a certain point of time (Figure 3.49, Park and Noguchi 2002).

A possible mechanism by which relative humidity rose in a 60°C environment is the temperature dependence of moisture stability. Figure 3.50 shows the analysis results. Relative humidity first decreased to 90% because of hydration reaction at the early stages, but afterwards, it increased gradually. The analysis reproduces a tendency obviously differing from the tendency of humidity changes at 20°C characterized by a monotonic decrease with the progress of hydration. At higher temperatures, water in the ink-bottle spaces is released and redistributed as condensed water with the passage

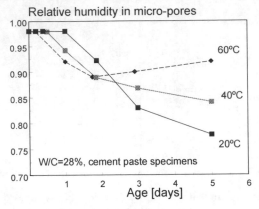

*Figure 3.49* Measured humidity under sealed conditions for different temperatures (Park and Noguchi 2002).

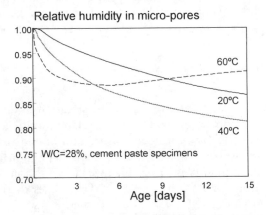

*Figure 3.50* Computed humidity change under sealed conditions for different temperatures.

of time (Figure 3.51). This is why internal relative humidity rises in the analysis. It simulates the trends of relative humidity with fair accuracy, but there are small differences between the analysis and the reality. In order to simulate real phenomena accurately and quantitatively, it is necessary to accurately quantify interdependent factors such as early hydration reaction, water consumption, pore structure development, and water equilibrium and redistribution. This is one of the subjects for future research.

## 3.4 Extended modeling of temperature-dependent micro-pore structure

In the case of massive or low W/C concrete, inner temperature of structures can reach 70°–80°C, which increases the risk of thermal cracking. In order to evaluate this risk and the long-term properties of concrete, the analytical

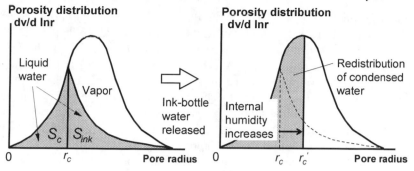

*Figure 3.51* Schematic representation of redistribution of trapped "ink-bottle" water and increase in relative humidity at high temperature.

system is expected to be applied to a wide range of temperatures. Regarding the hydration model (Chapter 2), referential hydration heat rate and activation energy of each mineral compound are defined in a manner that considers temperature dependency. In the microstructure model (Section 3.1), however, temperature dependency is not taken into account despite experiments showing that the curing temperature strongly influences the properties of hydration products and the formed microstructure (Kjellsen *et al.* 1990; Goto and Roy 1981; Famy *et al.* 2002). In addition, Kishi and Maekawa (1997) and Kishi *et al.* (2001) revealed that the hydration process for low W/C concrete under adiabatic temperature conditions differs from that for normal W/C concrete containing adequate water for hydration. The peculiar behaviors of low W/C concrete are tentatively explained by the increase in free water for hydration due to the decrease in absorbed water associated with self-desiccation. Further investigation of microstructures is undertaken in this section because concrete with a low water-to-cement ratio and high performance is quite widely being used in real-life situations.

### 3.4.1 Temperature dependency

#### 3.4.1.1 Redefinitions of pores in the basic modeling of micro-pore structure

In the basic modeling on micro-pore structure (Section 3.1), pores in the cement paste are classified into interlayer, gel and capillary pores. They are determined according to whether or not the pore space is available for precipitating cement hydrate. Interlayer and gel pores are formed with cement hydrates during the hydration process and are unavailable for hydrate precipitation. Capillary pores are free spaces for precipitation of hydrates. These definitions are based on the Powers–Brownyard model (1947), which defines a gel pore as a pore in hydrate products and a capillary pore as the remnant of an initially water-filled space. As proposed by Taylor (1997), "hydration product" replaces "cement gel" in the Powers–Brownyard gel-pore model, as the hydration product includes crystals of calcium hydroxide (CH).

Since the gel pore in the Powers–Brownyard model consists of different kinds of pores (interlayer, micro, and fine meso-pores) according to Taylor (1997), which differs somewhat from the general definition associated with pore size (Daimon *et al.* 1977 and Uchikawa *et al.* 1991), the definitions of pores need to be clarified in detail before discussing the extended model. The gel pore in the basic model (Section 3.1) can be more accurately redefined as two kinds of pores: an intra-gel pore and an inter-hydrate pore (Table 3.6). This redefinition is necessary to discuss the macroscopic characteristics of hydration and the porosity distribution of concrete materials for arbitrary thermal actions.

The intra-gel pores are assumed to be located inside the cement gel, while the inter-hydrate pores are assumed to be located in the spaces between the hydrate products, which include C–S–H gel and CH. In the definition based on pore size (Table 3.6), intra-gel pores correspond to the smaller gel pores in the basic model that have a radius smaller than around 2 nm. The inter-hydrate pores correspond to the larger gel pores in the basic model that have a radius larger than around 2 nm. The inter-hydrate and capillary pores correspond to the capillary ones by Uchikawa *et al.* (1991).

The intra-gel and inter-hydrate pores are basically considered to be dead space for precipitating new hydrate products, similar to the gel pore in the basic model (Section 3.1). In order to evaluate the material performance of concrete under arbitrary temperature histories, this section focuses on the additional functions of the inter-particle pores as free space for the precipitation of hydrate products under high temperatures.

*Table 3.6* Classification of pores

| Powers and Brownyard (1947) | Taylor (1997) | Daimon et al. (1977) | Uchikawa et al. (1991) | Maekawa et al. (1999), Section 3.2 | Modified modeling, Section 3.4 |
|---|---|---|---|---|---|
| Gel | Interlayer | Intra-crystallite (<0.6 nm) | | Interlayer | Interlayer |
| | Micro (<2 nm) | Inter-crystallite (0.6–1.6 nm) | Gel (0.5–1.5 nm) | Gel | Intra-gel |
| | Fine meso (>2 nm) | Inter-gel particle (1.6–100 nm) | Capillary (1.5 nm–15 μm) | | Inter-hydrate |
| Capillary | | | | Capillary | Capillary |

### 3.4.1.2 Preparatory review of the multi-scale model of hygro-chemo physics

This section briefly discusses issues associated with temperature dependencies using the redefinitions of pores proposed in the basic micro-pore structure development model. In the multi-component cement hydration model (Chapter 2), the referential hydration heat rate and activation energy in the equation of reaction kinetics are defined in a manner that considers the temperature dependency. Mutual interactions among the reacting constituents during hydration are quantitatively formulated. The effect of free water on the hydration rate is modeled using the hard shell concept of a hydrated cluster. The degree of heat generation rate decline in terms of both the amount of free water and the thickness of the internal hydrates layer is formulated. The amount of free water can be calculated from the microstructure and the moisture distribution. In the basic model in Section 3.1 (Maekawa *et al.* 1999), only the condensed water in the capillary pores was considered as free water for hydration (Model A).

Kishi *et al.* (2001) tentatively proposed an enhanced model that added a part of the moisture in the inter-hydrate pores to the free water and considered the change of adsorbed water associated with relative humidity (Model B). The purpose of this proposal was to explain the continuous hydration process of low W/C concrete under adiabatic temperature conditions. The large temperature rise was predicted by considering the increase in the amount of free water for hydration. In calculating the amount of free water in Model B, the following water is assumed to be constrained water that cannot be used for hydration:

- the water in inter-hydrate pores that have a radius of *hydRmin* or smaller;
- a part of the condensed water present within 1 nm of the walls of saturated inter-hydrate and capillary pores;
- the absorbed water in unsaturated pores;
- the interlayer water and the water in the intra-gel pores.

Here, *hydRmin* is the minimum radius of the available space in an inter-hydrate pore for hydrate precipitation. It was tentatively assumed to be 1.0 nm in Model B.

The hydration models can be verified by measuring the heat generated in the adiabatic temperature rise test or by performing TGA of the chemically bound water. For normal W/C (above approx. 40%), both models produce the same degree of hydration since enough free water is present for hydration. The applicability of the hydration model for the normal W/C concrete has already been verified through a comparison with the results of several experiments (W/C = 38.0–78.5%, Maekawa *et al.* 1999). On the other hand, for low W/C, the two models produce different results due to the lack of free water. Figure 3.52 shows the test results of an experiment and an analysis of adiabatic temperature rise in the case of concrete T25 (W/C = 25%, Taniguchi *et al.* 2000). The analysis shows that Model B generated more heat than Model A generated. The continuous hydration result from the analysis of Model B is calculated from the increase in the amount of free water. Figure 3.53 shows the results of the experiment and analysis of bound water in cement paste Y30 (W/C = 30%, Maekawa *et al.* 1999). The analysis shows that Model B had higher hydration than Model A.

However, the value is much larger when the experiment is compared with the adiabatic temperature rise test. While the temperature of the specimen in the adiabatic temperature test exceeded 90°C, the specimen in the bound water test was cured at a temperature of 20°C. As the curing temperatures in the two tests were very different, the temperature dependency is thought to be a key issue. The mix proportions of T25 and Y30 are shown in Table 3.7.

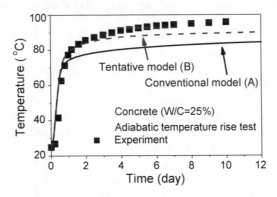

*Figure 3.52* Hydration process under adiabatic temperature condition, experiment by Taniguchi *et al.* (2000).

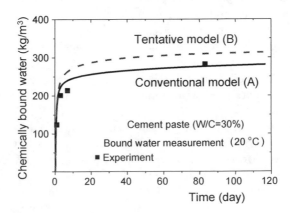

*Figure 3.53* Hydration process under normal temperature conditions (20°C), experiment by Maekawa *et al.* (1999).

*Table 3.7* Mix proportion of concrete, paste and mineral components of cement

| No. | Ref. | W/C (%) | Unit mass (kg/m³) | | | | Mineral component | | | |
|---|---|---|---|---|---|---|---|---|---|---|
| | | | W | C | S | G | C₃S | C₂S | C₃A | C₄AF |
| T25 | Taniguchi *et al.* (2000) | 25 | 150 | 600 | 876 | 730 | 40.4 | 37.7 | 3.9 | 11.8 |
| Y30 | Maekawa *et al.* (1999) | 30 | 485 | 1,620 | 0 | 0 | 49.7 | 23.9 | 8.8 | 3.4 |
| P25 | - | 25 | 440 | 1,762 | 0 | 0 | 54.37 | 19.86 | 9.42 | 8.79 |
| 25N | Ito *et al.* (2002) | 25 | 441 | 1,765 | 0 | 0 | 60.55 | 14.23 | 8.22 | 8.86 |

In the basic moisture model (Section 3.2), the pore pressure, relative humidity, and moisture distribution are mathematically simulated using a moisture transport model that considers both the vapor and liquid phases of mass-transport as defined by thermodynamics. The moisture dispersed in the capillary, intra-gel, inter-hydrate and interlayer pores in the cement hydrate are classified as condensed water, absorbed water, or interlayer water. They are modeled according to their properties. The hysteresis behaviors of isotherms under cyclic drying–wetting conditions take into account the ink-bottle effect in the capillary, intra-gel and inter-hydrate pores and the process of dispersing the interlayer water. Figure 3.54 and Figure 3.55 show the analytical results of the relative humidity in concrete T25 and cement paste Y30, respectively. The relative humidity calculated by Model B is lower than that calculated by Model A. It is also lower than the results of other experiments. For example, Persson (1997) reported that the relative humidity of low W/C concrete (W/C = 25%) was around 75% after 450 days under sealed conditions. Park and Noguchi (2002) showed that the relative humidity increases as the curing temperature rises, i.e. the humidity at two days becomes around 90% at 60°C. Then, it is necessary to investigate the validity of the calculated relative humidity at the end of the hydration process by

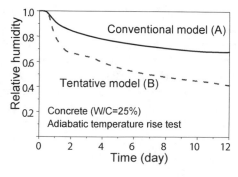

*Figure 3.54* Relative humidity under adiabatic temperature conditions.

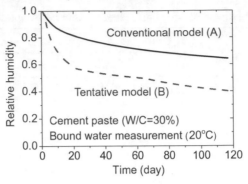

*Figure 3.55* Relative humidity under normal temperature conditions (20°C).

considering the temperature dependency of the microstructure and moisture profile (Kishi *et al.* 2001).

Recent research presents a temperature-dependent hysteresis model (Section 3.3). Here, the water trapped by the ink-bottle effect is redefined as water that can gradually disperse according to the temperature of the pore system. The parameter, which numerically expresses the time-dependent dispersion of ink-bottle water, is dependent on temperature and is influenced by the ambient humidity. The sensitivity to temperature and humidity was back-analyzed through finite element analysis. The rate of trapped water loss increases at higher temperatures and lower humidity. The proposed model was verified by comparing the results with the reality. In the adiabatic temperature test (T25), high curing temperatures can affect the moisture profile. Then, it is necessary to use the proposed temperature-dependent model to investigate the effect of temperature.

Also proposed is a temperature-dependent model for the interlayer water. This model is based upon the results of experiments (Section 3.3). In order to expand the scope of application to cover a wider range of temperature conditions, a function for expressing an envelope curve on the moisture isotherm of interlayer water has been introduced.

The basic micro-structure model (Section 3.1) assumes mono-sized particle dispersion and uses a particle expansion model. It is based on the degree of hydration and the amount of chemically bound water calculated by the hydration model. The effect of particle size is taken into account as the effect of average distance between particles. Assuming linear variations in porosity in the expanding cluster, the pore size distributions are computed over time. In this calculation, the micro-structure model uses parameters that represent the properties of hydrates, such as the density of the bound water and the specific surface area of the gel particles. The temperature dependencies of the properties are not considered in the basic model, although other researchers have reported on them (Kjellsen *et al.* 1990; Goto and Roy 1981; Famy *et al.* 2002). The effect of the temperature dependencies on the formed microstructure will be discussed in this section. The intrinsic porosity will be one of the most important parameters in the micro-pore structure model.

The intrinsic porosity of the hydrates includes the intra-gel, inter-hydrate and the interlayer porosity. The porosity is defined as the volume ratio of the intra-gel, inter-hydrate and interlayer pores to the hydrates, including the pores.

$$\phi_{ch} = \frac{\phi_{lr} + \phi_{ig} + \phi_{ih}}{V_s} \tag{3.109}$$

where $\phi_{ch}$ is the intrinsic porosity, $\phi_{lr}$ is the interlayer porosity [$m^3/m^3$], $\phi_{ig}$ is the intra-gel porosity [$m^3/m^3$], $\phi_{ih}$ is the inter-hydrate porosity [$m^3/m^3$], and $V_s$ is the volume of hydrates including the interlayer, intra-gel, and inter-hydrate pores [$m^3/m^3$]. In the basic model (Section 3.1), the intrinsic porosity is 0.28 (Powers 1964) and, for the sake of simplicity, it is a constant value independent of temperature. This value is equivalent to the minimum porosity that hydrate products cannot fill (Taylor 1997; Powers 1964).

### 3.4.1.3 Sensitivity analysis as preparation for modification

Table 3.8 summarizes the current issues regarding the calculations for low W/C. Neither Model A nor Model B can explain the overall behavior of low W/C concrete under different experiment conditions. In this section, the parameters that need to be improved in terms of their temperature dependencies on the hydration process are investigated by means of sensitivity analysis. In the analytical scheme, the strong interrelationships among the cement hydration, the micro-pore structure development and the moisture transport/equilibrium are taken into account by real-time sharing of the material characteristic variables across each model. Therefore, the influence of a certain parameter in each model on the overall behavior is automatically considered. Through sensitivity analysis, it is found that the following parameters have strong influence on concrete performance under different temperature conditions.

It was reported (Kjellsen *et al.* 1990; Goto and Roy 1981; Famy *et al.* 2002) that a coarser microstructure is formed under higher curing temperatures when W/C is relatively high. This is thought to be due to the formation

*Table 3.8* Evaluation of calculation results in the case of low W/C

| Temperature condition | Index | Model A | Model B |
|---|---|---|---|
| Adiabatic (high temperature) | Heat generation | Underestimate | Good |
| | Relative humidity | Underestimate | Underestimate |
| 20°C (normal temperature) | Bound water | Good | Overestimate |
| | Relatives humidity | Good | Underestimate |

of a dense hydrate layer around the cement particle under high temperatures. This temperature effect on the microstructure can be expressed by the decrease in the intrinsic porosity in the numerical analysis. In other words, a decrease in the intrinsic porosity under high temperatures is assumed in the sensitivity analysis. When the intrinsic porosity decreases, denser hydrate product is formed near the cement particles, and coarser capillary pores remain distant from the cement particles (Chapter 2).

In the sensitivity analysis, constant porosities are assumed as the first approximation throughout the hydration process. Figure 3.56 shows the results of T25 using Model A. The smaller intrinsic porosity produces the higher temperature rise in the final stage. Figure 3.57 shows the computed pore size distribution after 10 days. When the intrinsic porosity is assumed to decrease under high curing temperatures, the microstructure does not become dense despite the increase in the degree of hydration. The decrease in the intrinsic porosity causes the decrease in intra-gel and inter-hydrate porosity and the

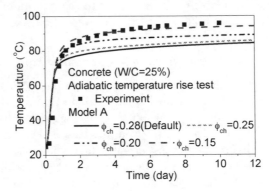

*Figure 3.56* Sensitivity analysis (Intrinsic porosity), experiment by Taniguchi *et al.* (2000).

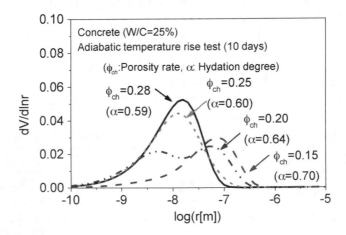

*Figure 3.57* Influence of intrinsic porosity on pore size distribution.

increase in capillary porosity. Since the change in the microstructure affects the moisture distribution, the amount of free water in the capillary pore increases as the intrinsic porosity decreases. As a result, the shortage of free water is resolved and the large temperature rise is simulated. In addition, when the intrinsic porosity decreases, the relative humidity does not decrease very much despite the large increase in the degree of hydration (Figure 3.58). This is due to the increase in the size of pores at the liquid–vapor interface, because of the decrease in intra-gel and inter-hydrate porosity. These results are computed by taking into account the strong interaction of microstructure formation, moisture distribution and cement hydration.

The extended moisture model was proposed as a way to simulate the temperature dependency on water trapped by the ink-bottle effect (Section 3.3). Since the temperature rose to around 95°C in the adiabatic temperature rise test for T25, the temperature dependency of the moisture might be significant. The sensitivity analyses are performed using Model A and Model B with two moisture models. The first is the basic model that disregards the loss of trapped ink-bottle water. The second is the extended model. The analytical results for hydration indicate that the different moisture models had no significant effect on the hydration process, since the amount of free water was almost the same. On the other hand, Figure 3.59 shows that different moisture models have a strong effect on the relative humidity under high curing temperatures. The enhanced moisture model calculates the transfer of trapped ink-bottle water from large to small pores under sealed conditions and high temperatures. As a result, the size of the pores at the liquid–vapor interface increases and the relative humidity also increases.

It was reported that the dispersion of interlayer water is significant under high temperatures (Section 3.3), and an extended model for this phenomenon is proposed (Section 3.3.3.2). Sensitivity analyses are performed to compare the conventional model and the enhanced one in both Model A and Model B (see Figure 3.60). This difference can be attributed to the increase in free water. This in turn is caused by the decrease in interlayer water, which can-

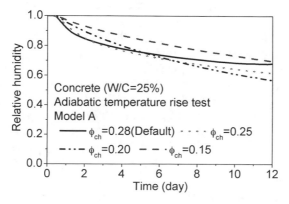

*Figure 3.58* Influence of intrinsic porosity on relative humidity.

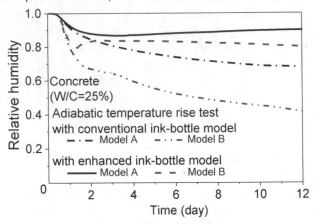

*Figure 3.59* Influence of the ink-bottle effect on relative humidity.

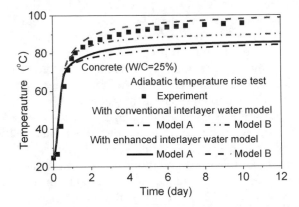

*Figure 3.60* Sensitivity analysis (enhanced interlayer water model), experiment by Taniguchi *et al.* (2000).

not be used for hydration, as well as the increase in water in the intra-gel, inter-hydrate and capillary pores. Since the tentative Model B treats a part of the water in the inter-hydrate pores as the free type, it creates a large heat generation. In addition, the increased water in the intra-gel, inter-hydrate and capillary pores greatly reduces the relative humidity (Figure 3.61).

The minimum radius of the available space in an inter-hydrate pore for hydrate precipitation (*hydRmin*) is an important parameter to calculate the amount of free water for hydration (Section 3.4.1.2). It is assumed to be 1.0 nm in Model B. The sensitivity analysis indicates that an inter-hydrate pore behaves as part of the available space for hydrate precipitation in the case of low W/C concrete (Kishi *et al.* 2001), although the inter-hydrate pore is originally determined to be one of the gel pores in the basic model where hydrate cannot precipitate. This idea is rooted in Powers and Brownyard (1947) and Powers (1964). While the specimens by Kishi *et al.* (2001) were

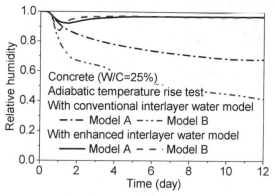

*Figure 3.61* Influence of interlayer water on relative humidity.

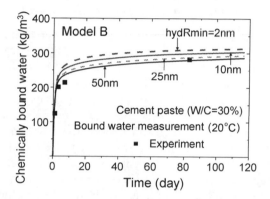

*Figure 3.62* Sensitivity analysis (minimum radius of available space), experiment by Maekawa *et al.* (1999).

tested under adiabatic temperature conditions, the specimens in Powers (1964) were cured under normal temperature. Then, the reason for the different hydration functions of the inter-hydrate pores in the two studies is thought to be temperature dependency. The increase in temperature causes a decrease in the minimum radius of the available space for hydration. In this section, the sensitivity analysis for the minimum radius is conducted for Y30 using Model B.

The results of the sensitivity analysis show that an increase in the minimum radius reduces cement hydration (Figure 3.62). When the minimum radius is 50 nm, the calculated result agrees with experiments. In this calculation, hydrate almost fails to precipitate in the inter-hydrate pores. The decrease in the relative humidity is also reduced by the increase in the minimum radius of the available space as shown in Figure 3.63. This is caused by the reduced hydration.

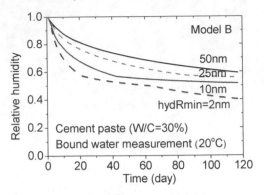

*Figure 3.63* Influence of minimum radius of available space on relative humidity.

### 3.4.2 Extended modeling in terms of temperature dependency

Applying the results of the sensitivity analysis in the preceding section, this section proposes an extend model for temperature dependency on the micro-structure (Nakarai *et al.* 2005, 2007). The proposed model consists of two temperature-dependent sub-models, i.e., the intrinsic porosity of the hydrate and the minimum pore radius of the available space for hydrate precipitation in inter-hydrate pores. In the proposed model, the increase in the curing temperature causes a decrease in the intrinsic porosity based on result A of the sensitivity analysis, and the number of capillary pores remaining increases. Since the amount of free water in the capillary pores increases under sealed conditions, cement hydration of low W/C continues for a longer period. Once the capillary pores are filled with cement hydrates, the precipitation of new hydrates may start in the inter-hydrate pores under high temperatures based on result B of the sensitivity analysis. As a result, continuous hydration under high temperature is computed. The extended moisture models for ink-bottle water and interlayer water are used in the computation.

#### 3.4.2.1 Temperature-dependent model for the intrinsic porosity of the hydrate

The temperature-dependent model for the intrinsic porosity of the hydrate is formulated. When the curing temperature is 30°C or lower (normal temperature), the intrinsic porosity in the proposed model is assumed to be 0.28, which is consistent with the value in the conventional model based upon research by Powers (1964). When the curing temperature increases, the porosity decreases. Here, the decrease in the inter-hydrate porosity is implicitly assumed to be the decrease in the porosity under high temperatures. When the curing temperature is 60°C or higher, the porosity is assumed to be 0.20. The inflection points are at 30°C and 60°C, since the results of the experiment (Ito *et al.* 2002), which will be discussed later, indicate a strong temperature dependency between 40°C and 60°C. The minimum value of porosity at high temperature is determined to be 0.20 by considering the results of sensitivity analysis. Here, the varying intrinsic porosity associated with the cement hydration in progress was investigated on the adiabatic temperature rise and relative humidity.

### 3.4.2.2 Temperature-dependent model for the minimum radius of free space for hydration

The temperature-dependent model is proposed for the minimum pore radius of the available space for hydrate precipitation in inter-hydrate pores. When the curing temperature increases, the minimum radius decreases. When the curing temperature is 30°C or lower, the minimum radius is assumed to be 50 nm. This means that the hydrate product almost fails to precipitate in the inter-hydrate pores at normal curing temperatures. When the curing temperature is 60°C or higher, the minimum radius is assumed to be 2 nm, which is the boundary between intra-gel and inter-hydrate pores described in Section 3.4.1.1. The inflection points are the same as ones in the temperature-dependent model for the intrinsic porosity.

### 3.4.2.3 Calculating the micro-structure

Based on the average hydration degree of powder and the amount of bound water calculated by the hydration model, the pore size distribution over time is calculated using the microstructure model. This section explains the two stages of the procedure in the proposed computations (Figure 3.64).

#### 3.4.2.3.1 CALCULATING THE POROSITY OF THE FIRST STAGE (HYDRATE PRECIPITATION INTO CAPILLARY PORES)

First, the total volume of the hydrate products, including the interlayer and the intra-gel and inter-hydrate pores, is calculated. Since the intrinsic porosity varies with temperature in the proposed model, the calculation must consider the temperature history with the hydration process over time. The total volume in a unit of cement paste is calculated by accumulating the incremental changes in volume over time.

$$V_s^{n+1} = V_s^n + \Delta V_s \tag{3.110}$$

where $V_s^{n+1}$ and $V_s^n$ are the total volume [m$^3$/m$^3$] of hydrate in the $n+1$th and $n$th step, respectively, and $\Delta V_s$ is the increased volume [m$^3$/m$^3$] from the $n$th step to the $n+1$th step. The increased volume at any arbitrary stage of hydration can be obtained as follows.

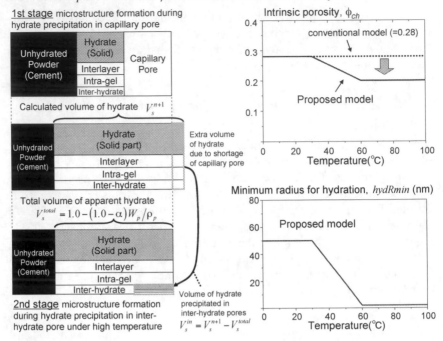

*Figure 3.64* Computations of microstructure formation in the proposed model.

$$\Delta V_s = \frac{1}{1-\phi_{ch}^{n+1}} \left( \frac{\Delta \alpha \cdot M_p}{\rho_p} + \frac{\Delta M_{chem}}{\rho_u} \right) \qquad (3.111)$$

where $\phi_{ch}^{n+1}$ is the intrinsic porosity for the temperature at the $n+1$th step, $\Delta \alpha$ is the increased average degree of hydration from the $n$th step to the $n+1$th step, $M_p$ is the mass of the powder materials per unit of paste volume, $\rho_p$ is the density of the powder materials, $\Delta M_{chem}$ is the increased mass of the chemically bound water from the $n$th step to the $n+1$th step and $\rho_u$ is the density of the chemically bound water (= $1.25 \times 10^3$ kg/m³). The changes of hydration degree and bound water are calculated by the hydration model based on reaction kinetics over time. The capillary porosity at $n+1$th step, $\phi_c^{n+1}$, can be calculated from the overall volume balance of the paste as follows.

$$\phi_c^{n+1} = 1 - V_s^{n+1} - (1-\alpha)\frac{M_p}{\rho_p} \qquad (3.112)$$

3.4.2.3.2 CALCULATING THE POROSITY OF THE SECOND STAGE (HYDRATE PRECIPITATION INTO INTER-HYDRATE PORES)

When the value of the capillary porosity is negative in the above computation, the following additional calculation is necessary to compute the precipitation of the hydrate products into the inter-hydrate pores as shown in Figure 3.64. This computation of the second stage is based upon the temperature-dependent model for the minimum radius of the available space only when the concrete containing the inter-hydrate pores for the precipitation is being cured at a high temperature.

The extra hydrate that cannot precipitate in the capillary pores fills the free space in the inter-hydrate pores under high temperatures. This is based on the extended model that takes into account the minimum radius of the available space. As a result, the total volume of apparent hydrate products, including the interlayer and the intra-gel and inter-hydrate pores ($V_s^{total}$), becomes smaller than the total volume that is calculated by integrating the volume of the hydrate products ($V_s^{n+1}$). The apparent total volume $V_s^{total}$ is equal to the volume of the paste excluding unhydrated powder materials.

$$V_s^{total} = 1 - (1-\alpha)\frac{M_p}{\rho_p} \tag{3.113}$$

The difference between $V_s^{total}$ and $V_s^{n+1}$ is the volume of hydrate that precipitates in the inter-hydrate pores. The extra volume $V_s^{in}$ is calculated as follows.

$$V_s^{in} = V_s^{n+1} - V_s^{total} \tag{3.114}$$

3.4.2.3.3 CALCULATING PORE SIZE DISTRIBUTION

The surface area and the pore size distribution in the microstructure can be obtained from the porosities calculated above. The temperature dependencies of the average size and specific surface area of the hydrate products for calculating the surface area of pore are determined related to the minimum size of free space for hydration. Finally, the pore size distributions of the capillary, intra-gel, and inter-hydrate pores are represented by simplistic Raleigh–Ritz (R–R) distribution functions (Section 3.1). Here, the intra-gel and inter-hydrate pores are treated by a shared distribution function for the sake of simplicity, as mentioned in Section 3.4.1.1.

### 3.4.3 Numerical simulation

#### 3.4.3.1 Hydration process under adiabatic and normal temperature conditions

The proposed model was verified with two types of experiments: the adiabatic temperature rise test (T25) and measurement of the chemically bound water (Y30). Those experiments represent hydration processes under both high and normal temperature conditions. Figure 3.65 and Figure 3.66 show the results of the experiments and analyses of the hydration process.

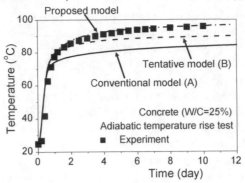

*Figure 3.65* Adiabatic temperature rise calculated by proposed model, experiment by Taniguchi *et al.* (2000).

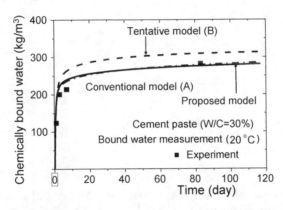

*Figure 3.66* Chemically bound water calculated according to the proposed model under normal temperatures, experiment by Maekawa *et al.* (1999).

Compared to the conventional model, the proposed model can accurately predict the results of both experiments under different temperature conditions. In the adiabatic temperature rise test, continuous hydration is simulated by the proposed model for two main reasons. The first is the decrease in intrinsic porosity, which results in a larger amount of free water for hydration. The second is the increase in the space available for hydration under high temperatures. The decrease in the average intrinsic porosity in the adiabatic temperature test resulted in the formation of denser hydrate product near the cement particles. This result is consistent with the results of experiments using backscattered electron imaging that show lighter products forming under high curing temperatures (Famy *et al.* 2002).

For the bound water measurements, both the proposed and conventional models give reasonable analytical results, since the conventional model has been verified by experiments at normal temperatures (mainly 20°C).

Figure 3.67 and Figure 3.68 show the results of an analysis of the changes in relative humidity inside the specimen. In adiabatic temperature conditions,

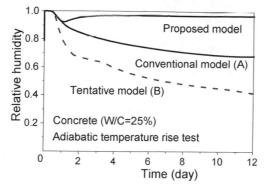

*Figure 3.67* Calculated relative humidity under adiabatic temperature conditions.

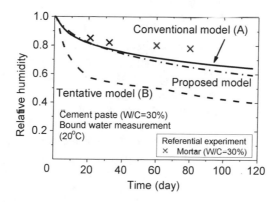

*Figure 3.68* Calculated relative humidity under normal temperatures, experiment by Paillere *et al.* (1990).

the decrease in the relative humidity calculated by the proposed model was very small. The calculated relative humidity at 12 days was around 95%. Although there are no data from experiments that directly verify these results, the following two experiments indirectly validate the calculation. The first experiment investigated the effect of the curing temperature on the relative humidity (Park and Noguchi 2002). The relative humidity was measured at temperatures of 20°C, 40°C and 60°C. The results show that the relative humidity increases as the curing temperature rises. The relative humidity at two days was around 90% at 60°C. Since the maximum temperature in the adiabatic temperature test was around 95°C, the small decrease in the calculated relative humidity under adiabatic conditions seems to be validated. The second experiment was a trial. The authors measured the relative humidity under quasi-adiabatic temperature conditions. The relative humidity was measured at the center point of a cube specimen whose sides were 400 mm. The specimen was covered with Styrofoam having a thickness of 100 mm. The measured relative humidity was 100% during the seven-day experiment, although the value is too high to ensure the accuracy of the measuring device.

In sealed conditions at 20°C, the results of the analysis showed a decrease in the relative humidity similar to that seen in the experiment, although the decrease in the analysis was a little greater than in the experimental result. The mortar test used for the verification was conducted by Paillere *et al.* (1990), which has the same W/C calculated in this book.

### 3.4.3.2 *Hydration process under late-elevated temperature conditions*

This section investigates the effect of late-elevated temperatures on the hydration process of cementitious materials, which were sufficiently cured under normal temperatures in advance. Two different curing temperature histories were adopted in the experiment. In the first part of the experiment (P25a), a constant temperature of 20°C was given. In the second part of the experiment (P25b), ambient temperature was elevated instantaneously from 20°C to 60°C at the age of 15 days, and then temperature was further elevated from 60°C to 80°C at the age of 33 days. The bound water was measured in cylindrical specimens having a radius of 100 mm and a height of 200 mm. The mix proportion is shown in Table 3.7. The specimens were cured under sealed conditions until they were measured. They were then broken, and samples of about 20 mg were collected for TGA. In the TGA, the temperature was increased in steps of 10°C up to 100°C, where it was kept constant for five minutes. Subsequently, the temperature was increased at a rate of 10°C/s up to 1,200°C. The loss of mass in the sample between 100°C and 1,000°C was considered to be caused by the loss of un-evaporable bound water. Figure 3.69 shows the results of the experiment and analysis. The experiment revealed a large increase in the amount of bound water due to the increase in the curing temperature. While the conventional model underestimated the large increase in the bound water under the elevated

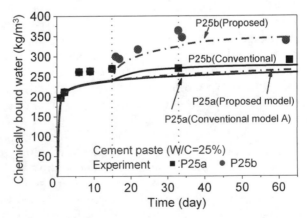

*Figure 3.69* Change of bound water under late-elevated temperature conditions.

temperature conditions, the proposed model could accurately predict the hydration process in the experiment. In the proposed calculation, the increase in the available space for hydration under high temperatures resulted in additional hydration. In this case, the decrease in the intrinsic porosity under elevated temperatures did not contribute to an increase in hydration since almost no capillary pores remained.

### 3.4.3.3 Influence of curing temperature on the micro-structure (normal W/C)

This section describes the influence of high-temperature curing on the formation of the microstructure in cement paste having a normal W/C. Here, normal W/C means a W/C with enough free water for hydration (around 40% or more). It is well known that the microstructure becomes coarser with a high curing temperature. Goto and Roy (1981) reported the results of an experiment in which the pore size distribution was measured using the MIP method. The specimens (W/C = 40%) were cured at either a normal temperature (27°C) or a high temperature (60°C) for four weeks. In the case of high-temperature curing, the fine porosity on a $10^{-9}$ m scale was higher and the coarse porosity on a $10^{-7}$ m scale lower than was seen in the case of normal-temperature curing. Figure 3.70 shows the analytical results. In the conventional model, the microstructure became dense under high-temperature curing since the degree of hydration was high. No large capillary pores remained. On the other hand, in the proposed model, a coarse microstructure was observed in the case of high-temperature curing. The decrease in the intrinsic porosity under high temperatures produces an increase in the number of large capillary pores. The increase in available space for hydration under high-temperature curing did not affect the hydration process since an adequate volume of capillary pores remained.

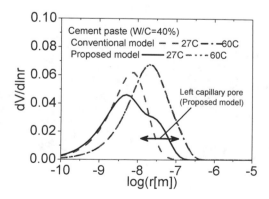

*Figure 3.70* Pore size distribution in case of normal W/C (calculation).

### 3.4.3.4 *Influence of curing temperature on the microstructure (low W/C)*

This section describes the influence of high-temperature curing on the formation of the microstructure in cement paste having a low W/C. Ito *et al.* (2002) measured the pore size distribution of cement paste (W/C = 25%) under different temperature histories. After first curing at 20°C for 24 hours, four temperature conditions (20°C, 40°C, 60°C and 80°C) were applied as environmental conditions. The specimens were sealed to prevent moisture loss. At two and 57 days after the casting of specimens, the specimens were broken, and the pore size distributions were measured using the MIP method. The mix proportion is shown in Table 3.7.

Figure 3.71 (upper) shows the results of experiments conducted by Ito *et al.* (2002). At 20°C and 40°C, the peaks of the pore size distributions were observed between $10^{-8}$ m and $10^{-7}$ m, and the porosity decreased as hydration proceeded. On the other hand, at 60°C and 80°C, the peaks of the pore size distributions were between $10^{-9}$ m and $10^{-8}$ m, which are smaller than they were at 20°C and 40°C.

Figure 3.71 (lower) shows the results calculated by the proposed model. The proposed model can roughly simulate the different tendencies between normal-temperature curing (20°–40°C) and high-temperature curing (60°–80°C). However, the analysis does not precisely express the clear peak seen in the experiment. The reason for the gentle peak in the analysis is that

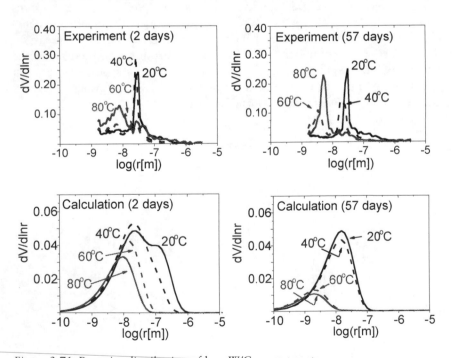

*Figure 3.71* Pore size distribution of low W/C cement paste.

the system used in the analysis represents the average pore size distribution by applying a simplistic Raleigh–Ritz distribution function. Several verifications indicate that this simplification can still be applied to solving practical issues such as thermal stress, shrinkage and mass transfer (Maekawa *et al.* 1999, 2003a).

## 3.5 Concrete: a multi-component composite porous medium

Transport properties of concrete have generally been analyzed by treating concrete as an isotropic, homogenous and uniform porous medium (Hall 1989; Dhir *et al.* 1989). In this approach, the physical characteristics of various components of concrete – like aggregates, cement paste matrix, aggregate cement paste interfaces and bleeding paths – are lumped into a single representative porous medium. This approach can treat analytically most of the cases of practical interest involving concrete made of normal mixes and low-porosity aggregates. However, it fails to treat rationally the moisture transport in special concrete, for example, the type that contains high-porosity lightweight aggregates. A different analytical approach is also desirable for the cases where, due to high aggregate content, a distinct aggregate–cement matrix interface might be present and enhance the rate of mass transport.

It is the aim of this section to identify various components of concrete relevant to moisture transport and then to combine the moisture transport in these components into a unified moisture transport theory of concrete composite. This treatment would not only enable us to consider rationally different components of concrete but also help us to identify the conditions in which a single, uniform porous medium assumption can be applied with confidence to analyze the moisture transport in concrete. The previous section discussed the theoretical aspects of moisture transport in concrete by treating it as a uniform porous medium. This section extends those concepts to multi-components. Depending on the micro-structural characteristics and distributions of porosity in each of these components, the overall moisture transport behavior in concrete might be influenced significantly. The authors intend to incorporate these additional systems of moisture migration into the generic moisture transport formulation of concrete.

### 3.5.1 Multi-components of concrete

The modeling of concrete performance identifies three distinct mediums existing in concrete through which moisture mass transport can take place in liquid and gaseous phases. These three basic mediums of mass transport will hereafter be referred to as the components of concrete. These components are (a) hardened cement paste matrix (*hcp*), (b) aggregate cement paste matrix interface and bleeding paths, and (c) aggregates. These three basic mediums of moisture transport are schematically represented in Figure 3.72.

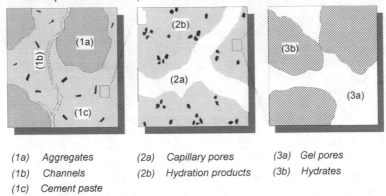

| | | |
|---|---|---|
| (1a)   Aggregates | (2a)   Capillary pores | (3a)   Gel pores |
| (1b)   Channels | (2b)   Hydration products | (3b)   Hydrates |
| (1c)   Cement paste | | |

*Figure 3.72* The multi-components of concrete.

### 3.5.1.1 *Hardened cement paste matrix*

Under usual conditions, this component is responsible for the bulk of the mass transport through concrete and includes most of the concrete bulk porosity. It consists of the various products of hydration and includes the so-called gel and capillary pores. The capillary or larger pores existing primarily between products of hydration serve to transport most of the bulk flow at high relative humidity (due to the presence of condensed liquid), whereas gel pores play an important role at low relative humidity when most of the flow occurs in vapor form. The pores in this component usually follow a very tortuous path and constitute a complex network of random interconnections. The overall tortuosity of flow path may also increase with the addition of aggregates.

### 3.5.1.2 *Aggregate matrix interfacial zones and bleeding paths*

Here, we consider the large pores existing at the aggregate and bulk matrix interfaces as well as some additional settlement or bleeding paths that might be formed due to an insufficient compaction in large water-to-powder ratio mix concrete and/or segregation of water from cement paste mix. The water-to-powder ratio near the aggregate surfaces is usually quite large relative to the bulk water-to-powder ratio, primarily because of the so-called wall effect. This essentially means that cement powder particles near a wall-like surface cannot pack as efficiently as they do in the bulk medium, leading to a lower powder density near the aggregate surfaces (Breugel 1991). This effect coupled with the one-side-restricted growth of powder particles during hydration (Bentz and Garboczi 1991) leads to a zone of coarse microstructure and high porosity near the aggregate surface, the thickness of which depends on the ratio of the size of aggregates to powder particles. For the smaller of the fine aggregates, the thickness might be practically non-existent as these particles are stereologically similar to powder particles, whereas large aggregates would have an interfacial zone of uniform thickness primarily dependent on the powder particle size distribution. The thickness of interfacial zones was reported to be lying in the range of 15 to 50 μm (Breugel 1991; Synder *et al.* 1990). The probability that the interfacial zone would form a completely connected path or percolate through the concrete is dependent on aggregate volume and grading. An analytical treatment of the wall effect and its impact on the zone of higher local porosity near the aggregate surface can be seen in Breugel (1991).

Insufficient compaction and bleeding may also lead to the formation of additional coarser paths. Usually the total volume fraction of the pores contained in this component is very small. However, owing to their large size, they may exert subtle influence on permeation characteristics such as the diffusivity and permeability of concrete. For example, under fully saturated wet face conditions of concrete, a large amount of liquid might be carried through this porous network and transferred to the hardened matrix pores and aggregates, thus increasing the rate of moisture transfer within concrete and thereby aiding the deterioration of concrete.

### 3.5.1.3 Aggregates

The aggregate pores constitute a distinct range of pore size distribution. Since aggregates are usually dispersed inside the paste matrix, they do not constitute a continuous porous path. Their role in the moisture transport can be idealized as a buffer zone which either stores or releases moisture during interaction with the surrounding paste matrix. It is believed that no net movement of moisture occurs through these pores, or that it is very small compared to the amount of moisture carried through channel or cement paste components. Thus, only local moisture transfer is assumed to take place between aggregates and the surrounding paste matrix or channels or both. Obviously, this component needs to be considered for moisture transport only when a significant amount of concrete porosity exists in aggregates, i.e., when high-porosity aggregates like artificially produced lightweight products are used in the concrete mix. The additional components responsible for mass transport in concrete besides *hcp* (hardened cement matrix) described above would generally change the pattern and behavior of moisture transport in concrete (Section 7.4). The difference as compared to a *hcp* system only may vary depending on the distribution characteristics of these additional components.

### 3.5.2 Moisture transport formulation in a composite

Detailed discussion of moisture transport in concrete has occurred in Section 3.2 and Section 3.3, where the concrete composite was treated as a uniform isotropic, homogenous porous medium. In developing the transport model, flow contributions from vapor and liquid phases were also included.

Furthermore, an assumption regarding the thermodynamic equilibrium of different phases was assumed. In other words, it was assumed that various transportable phases exist in an equilibrium state distribution inside the complex porous network of the composite. These assumptions may be valid if the aggregate matrix interfaces or channel structures do not percolate across the composite, or when the aggregates are practically impervious and their porosity does not constitute a significant amount of the total composite bulk porosity. However, while these conditions might exist for most of the usual concrete, to consider a generic treatment of moisture transport, the transport occurring in interfaces and aggregates should also be integrated into the unified moisture transport formulation.

To consider the moisture transport behavior in concrete composite as a multi-component system, three basic components – namely, bulk hardened cement paste matrix, channels or interfaces and aggregates – need to be considered. It is assumed that the thermodynamic equilibrium assumption holds good for each of the above porosity components considered individually. Thus, each of these components can be treated as an effective homogenous porous medium in itself or each of the porosity components can be viewed as a continuum. For this reason, the moisture transport model originally developed for an isotropic and homogenous porous medium can be applied to each of the porosity components individually. Also, for a complete description of the issue, the mutual interactions of these components among themselves need to be considered. The bulk transport behavior of the composite can be obtained by summing up the transport behavior of its components.

### 3.5.2.1 Mass and momentum conservation

The assumptions of thermodynamic equilibrium adopted in the modeling of moisture transport in cementitious materials would be applied as well. Of the three basic components of concrete, the channel (interface) component is similar to the bulk matrix system, with the only difference being higher porosity and relatively larger pore distribution. Therefore, analytically it can be treated in a similar way as the cement paste pore system (uniform porous media).

However, aggregate porosity differs due to the nature of moisture transport modeled through the aggregates. Essentially, the modifications required in the formulation described in the previous section are minimal. It is basically only the mass balance equations that require reconsideration. Furthermore, since no movement of moisture is assumed to take place through aggregates across a representative elementary volume (REV), then the momentum balance of moisture for the aggregate component is not required. Accompanying moisture transport in any individual component of concrete would be a mutual interaction, effectively a moisture exchange. Mass transport equations for these components would therefore be coupled with some interaction terms. These terms would define the inter-component rate of moisture exchange within a REV and could be called *local mass transfer* rate terms. The word *local* is used since the rate of moisture exchange is said to be dependent on the thermodynamic state of moisture, such as relative humidity, in individual components at any given location and time. Therefore, when considering overall moisture balance in each of the components, an account must be given of the respective mass exchange rates due to interaction with other components.

In the development outlined below, all the succeeding differential equations are written over a REV of interest, which is assumed to be very small compared to the specimen size and quite large compared to the microscopic scale. A continuum approach is adopted in the modeling of concrete performance. For example, aggregates are modeled as being uniformly and equally distributed in the REV according to their volume ratio. A similar approach will be applied later on for the development of a local mass transfer model. It is assumed that the mass of vapor in a REV can be disregarded compared to the mass of liquid and that the total gas pressure is constant everywhere. We also assume that the liquid phase does not interact with the surrounding solid cementitious mass, so no chemical fixation or permanent loss of moisture within a REV takes place.

It is believed that the deformation field in a porous medium is unconnected to the moisture transport issue, so moisture transport would be independent of the deformation fields in the composite material. A general subscript and superscript notation is used. A subscript denotes the phase of interest. For example, $l$ indicates the liquid phase, whereas $v$ denotes the vapor phase. Total water present in the REV within a component is represented by subscript $w$. A superscript denotes the component in the composite of interest. Cement paste matrix, channels and aggregate components are abbreviated as: $cp$, $ch$ and $ag$, respectively. Using these conventions, the appropriate moisture balance equations for the $hcp$ matrix, channels and aggregates can be obtained via general laws of mass conservation. The term *moisture* is used to represent the liquid and vapor phases of water collectively.

3.5.2.1.1 MOISTURE IN THE CHANNELS AND CEMENT PASTE MATRIX

The rate of moisture change in channel and cement paste matrix components in a given REV would be a consequence of the flow of moisture in the liquid and vapor forms into the REV due to potential gradients and phase transitions from one phase to other. Moreover, there would be a local exchange of moisture within the components leading to a redistribution of moisture locally. Therefore, a mass balance of liquid moisture in the cement paste matrix by considering it as an independent porous component yields

$$\frac{\partial \theta_l^{cp}}{\partial t} + divq_l^{cp} + v^{cp} + R_{cp.ag} - R_{ch.cp} = 0 \qquad (3.115)$$

where $\theta_l^{cp}$ is the volumetric liquid water content in kg/m³ of total concrete, i.e., $\theta_l^{cp} = \rho_l \phi^{cp} S_l$, $\phi_{cp}$ is the bulk porosity of hardened cement paste (*hcp*) matrix, i.e., the pore volume of cement paste matrix per unit volume of the concrete composite, $\rho_l$ is liquid density and $S_l$ denotes the degree of saturation of the porous network of *hcp* matrix with liquid water, $q_l$ is the flux of liquid water per unit area of the concrete, $v_{cp}$ is the rate of transition from the liquid to the vapor phase. $R_{cp.ag}$ and $R_{ch.cp}$ denote the rate of moisture exchange per unit volume of concrete between *cp* to *ag* and *ch* to *cp* components, respectively. We have a similar expression for channel structures as

$$\frac{\partial \theta_l^{ch}}{\partial t} + div q_l^{ch} + v^{ch} + R_{ch.ag} + R_{ch.cp} = 0 \qquad (3.116)$$

with all the terms having a similar meaning as they did in the case of the cement paste matrix.

It should be noted that the above equation would be included in the overall moisture transport issue of the concrete composite only if the channels formed a percolated path across the composite. The porosity $\phi_{ch}$ in such a case would correspond with the porosity existing in the percolated interfaces per unit volume of concrete. When a percolated or continuously connected path of channels does not exist, concrete can effectively be treated as consisting of aggregate and *hcp* matrix components only and the channel component need not be considered.

The mass balance of moisture existing as the vapor phase in channel and hardened cement paste matrix yields

$$\frac{\partial \theta_v^i}{\partial t} + div q_v^i - v^i = 0 \qquad (3.117)$$

where *i* denotes either of the *cp* or *ch* components, and $\theta_v$ is the mass of moisture present in the vapor phase in component *i* per unit volume of concrete. This quantity is usually very small compared to the liquid water content, $\theta_l^i$. Therefore, the total moisture content (in absolute mass terms) of a component in the composite can be approximately taken as the mass of the liquid phase of moisture only. Term $q_v$ represents the flux of moisture across the composite in component *i* in the vapor form. Considering that the total gas pressure is assumed to be constant, it is the vapor pressure gradient and diffusive flux of water vapor that are responsible for the movement of the vapor phase in the porous microstructure.

### 3.5.2.1.2 MOISTURE IN THE AGGREGATES

The role of aggregates can be unified in the same framework of moisture transport modeling using the concept of local moisture transfer described earlier. It can easily be observed that aggregates usually form a sort of porosity region that is isolated each from the other. Thus, aggregate porosity, unlike hardened cement paste matrix and interfacial zones, does not form a continuously connected path across the composite and any change in its moisture content would result from the interaction with other continuously connected components only. The aggregates can be idealized as reservoir zones that either release or store moisture locally due to their interaction with the surrounding cement paste matrix and interfaces. Thus, we have

$$\frac{\partial \theta_l^{ag}}{\partial t} = R_{cp.ag} + R_{ch.ag} \tag{3.118}$$

where $\theta_l^{ag}$ is the volumetric liquid water content of aggregates per unit volume of concrete, such that $\theta_l^{ag} = \rho_l \, \phi^{ag} \, S_l$, $\phi^{ag}$ is the porosity of aggregate relative to the unit volume of concrete, i.e., it is the volume of aggregate pores in a unit volume of concrete. Mathematically, $\phi^{ag} = G\phi_{abs}$ where $G$ is the volumetric fraction of aggregates in the composite and $\phi_{abs}$ is the absolute porosity of aggregates, while $S_l$ denotes the degree of water saturation of aggregate pores. Terms $R_{cp.ag}$ and $R_{ch.ag}$ denote the rate of moisture exchange with the cement paste matrix and channels, respectively, per unit volume of concrete. The phase transition term $v$ has been discarded, since the mass of vapor in aggregates is considered to be negligible.

### 3.5.2.2 Reduction to simplified classical form

To obtain the total mass transport behavior in concrete, Equations 3.115–3.118 need to be solved simultaneously. The basic target of interest in this formulation is the moisture content of each component. The total moisture content of concrete can be obtained by a simple summation of the moisture content of each component. In the basic mass conservation problem defined above, we have, in all, seven primary variables (including $v$) and only five conservation equations. It is assumed that the remaining terms like flux $q$ and local mass exchange terms could be expressed as a function of the primary variables. The constitutive law defining the phase transition terms $v$ of each component must be developed, so that for a given problem above equations provide a unique solution. While it is possible to derive the moisture flux models and local moisture exchange rates from the micro-structural characteristics of components and thermodynamic equilibrium conditions, a generic rational model that defines the rate of phase transition between liquid and vapor phases remains to be found. Although, a few models defining this relationship have been proposed in the field of soil science (Connell and Bell 1993), they are primarily empirical and at best qualitative in nature.

An attempt has been made to redefine the problem such that the phase transition terms are not required in the overall formulation while at the same time basic features of the model are retained. The model is restated in terms of the *total* moisture contained in individual components. That is, we are now concerned with the conservation of total moisture instead of individual phases like liquid water and vapor separately in each component. This is achieved by summing up the conservation equations of liquid and vapor phases and diregarding the mass of vapor present in a REV but retaining the flux contribution terms. This manipulation leads to a cancelling out of the phase transition terms and we obtain a mass conservation equation of total moisture in a component. The assumption seems to be reasonable enough, since the mass of liquid water in a REV would be many times more than the vapor mass, which can therefore be ignored. Using this redefinition, we have reformed the moisture balance for *cp*, *ch* and *ag* components as (i.e., adding Equations 3.115 and 3.116 to Equation 3.117)

$$\frac{\partial \theta_w^{cp}}{\partial t} = -div\, q_w^{cp} - R_{cp.ag} + R_{ch.cp}$$

$$\frac{\partial \theta_w^{ch}}{\partial t} = -div\, q_w^{ch} - R_{ch.cp} - R_{ch.ag} \tag{3.119}$$

$$\frac{\partial \theta_w^{ag}}{\partial t} = R_{cp.ag} + R_{ch.ag}$$

where $\theta_w^{cp}$, $\theta_w^{ch}$ and $\theta_w^{ag}$ denote the total moisture content of *cp*, *ch* and *ag* components, respectively, expressed in kg/m³ of concrete. $q_w^{cp}$ and $q_w^{ch}$ denote the flux of total moisture in *cp* and *ch* components, including the moisture flux contributions from both the liquid and vapor phases. Once again, it is stressed here that the channel component needs to be considered only when it constitutes a percolated path across the concrete composite. In non-percolated composites, we are essentially left with only two primary unknowns, $\theta_w^{cp}$ and $\theta_w^{ag}$ (Equation 3.119). The classical constitutive law defines the moisture flux $q$ in a porous medium as (Dhir *et al.* 1989; Connell and Bell 1993),

$$q = -D(\theta)\, \nabla \theta \tag{3.120}$$

where $D(\theta)$ is the so-called *diffusivity* parameter of the porous medium and usually expressed as a non-linear function of the moisture content $\theta$ of that medium. Since it is the only parameter that defines the transport characteristics of the porous medium, we can find many studies devoted to analytical description of this relationship. However, the suggested relationships have generally been empirical in nature (Shimomura and Maekawa 1997). Furthermore, in the realm of the empirical models, it appears that no consensus exists on the nature of the $D$-$\theta$ relation. This has led to several different experimental measurements of the diffusivity of concrete under various exposure conditions, which shows that indeed diffusivity is a complex parameter and cannot be simply specified as a function of moisture content. Factors like the age of concrete, exposure history, maturity and temperature among others have a direct bearing on the transport properties of concrete. The empirical models of diffusivity cannot take into account all of these factors.

For these reasons and because of a certain level of arbitrariness involved in choosing the values of the so-called material parameters of these models, the issue of the momentum balance of moisture in the concrete porous network was reworked in the previous sections. It was found that making certain assumptions, it is indeed possible to arrive at a similar form of the moisture flux equation as given by Equation 3.120. The only difference is that through this approach, the diffusivity function of a porous medium can be analytically evaluated in a non-empirical way, once the micro-structural properties of the porous medium, its exposure history and the thermodynamic state of moisture inside the porous network are known. The details of this development of conductivity and diffusivity models has been described in the previous chapter. Diffusivity $D(\theta_w{}^i)$ of component $i$ (*cp* or *ch*) can be evaluated from the general concepts of adsorption of vapor molecules, surface tension, thermodynamic equilibrium and Hagen–Poiseuille flow characteristics through a random network of micro-pores. In a compact form, $D(\theta_w{}^i)$ can be expressed as

$$D(\theta_w^i) = \left( \frac{\rho RT}{Mh} K_l(\theta_l^i) + \rho_v D_a f(\theta_w^i) \right) \frac{\partial h}{\partial \theta_w^i} \tag{3.121}$$

where $R$ is the universal gas constant, $T$ is absolute temperature, $M$ is the molecular mass of water, $h$ is the relative humidity in the REV of component $i$, $K_l$ is the permeability of liquid and depends on the liquid content (degree of saturation) of the component, $\rho_v$ is vapor density, $D_a$ is the vapor diffusivity in free atmosphere. The factor $f(\theta_w^i)$ accounts for the effect of tortuosity and varying moisture content on vapor diffusion. The slope of the relative humidity with moisture content is obtained from the isotherms of individual components. An extensive discussion of these parameters can be found in Section 3.2 and Section 3.3, where the basic constitutive models related to the moisture transport in concrete are analyzed.

### 3.5.2.3 Local moisture transport behavior

Associated with the diffusive and bulk movements of moisture through the components of concrete, there exists an associated phenomenon of rearrangement and redistribution of moisture among the components of concrete composite locally. Thus, the exchange of moisture may take place between aggregates, channels and continuous hardened cement paste, as shown in Figure 3.73. An accurate formulation of the mechanisms for this local moisture transfer needs to be worked out. In this book, however, guided by the general nature of moisture transport into a porous medium, the pressure potential level of moisture in each of the components in a given REV is taken

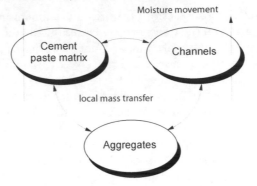

*Figure 3.73* The local moisture exchange concept.

as a key parameter controlling the inter-component moisture transfer. The formulation of local moisture transfer can be obtained by recognizing the difference in these potentials.

In the most simplistic scenario, the rate of inter-component exchange can be taken to be linearly proportional to the difference of liquid pressure potentials of interacting components. Alternatively, this result can be obtained as follows. Consider a system of two conducting components in contact with each other, as shown in Figure 3.74. $\delta_1$ and $\delta_2$ represent the average spatial dimensions of the components considered in a unit volume of the porous medium. If $P_1$ and $P_2$ represent liquid pressure at the centroids of components 1 and 2, respectively, then the moisture exchange rate $R_{12}$ from component 1 to component 2, in a unit volume of the overall solid matrix, can be obtained as

$$R_{12} = q_{12}A_{12} = -\frac{2A_{12}}{\displaystyle\sum_{i=1}^{2}(\delta_i/K_i)}(P_2 - P_1)$$

(3.122)

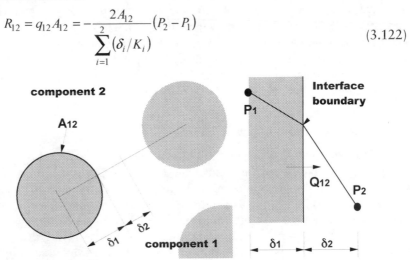

*Figure 3.74* Formulation of the local moisture transfer model.

where $A_{12}$ represents the actual area of contact per unit volume between the two components, $q_{12}$ is the corresponding flux, and $K_i$ represents the unsaturated permeability of the $i$-th component. It has to be noted that a linear variation of liquid pressures $P_1$ and $P_2$ is assumed in the individual components considered above. Moreover, this formulation is based upon quasi steady-state assumptions. Expressing the rate of mass transfer as being linearly proportional to the difference of pressure potentials, the local mass transfer coefficient $\alpha_{12}$ can be obtained as

$$\alpha_{12} = \frac{2A_{12}}{\displaystyle\sum_{i=1}^{2}(\delta_i/K_i)} \tag{3.123}$$

Primarily in this work, three major components of moisture mass transport have been identified. Thus, the local mass transfer coefficients for the three interacting pairs, namely $\alpha_{cp.ag}$, $\alpha_{cp.ch}$ and $\alpha_{ag.ch}$, would be derived next. These derivations should be treated at best as qualitative in nature since a large amount of uncertainty is involved in computing the actual areas of contact between different components. Moreover, the accuracy of length scales δ for different components is unclear. Therefore, a number of assumptions and simplifications are inherently involved in the derivation of these coefficients. Nevertheless, such an exercise would give us an idea of the orders of magnitude of these coefficients and related sensitivities of the various relevant parameters.

### 3.5.2.3.1 AGGREGATE AND FINE POROSITY MATRIX

Aggregates can be modeled as spheres of the same radius, $r$. If the volumetric concentration of aggregates in the total hardened concrete is $G$, $s$ is the average distance between outer surfaces of two spheres and $G_o$ is the limiting filling capacity of the aggregates. Then, we have

$$G = G_o/(1 + s/2r)^3 \tag{3.124}$$

Average separation $s$ represents the length scale of the fine porosity matrix. The area of contact between two components in a unit volume would be $3G/r$. Therefore, the local mass transfer coefficient is obtained as

$$\alpha_{cp.ag} = \frac{3G}{r^2}\left(\frac{1}{2K_{ag}} + \frac{(G_o/G)^{1/3} - 1}{K_{cp}}\right)^{-1} \tag{3.125}$$

Generally, permeability of the finer porosity component is very small compared to the permeability of aggregate pores. Also, the actual effective area of contact may be smaller than the estimate obtained from the above expressions. Thus, a parameter $\beta_1$ is introduced which represents the apparent effectiveness of the moisture transfer across actual contact areas. Using this and the assumption that $K_{cp} \ll K_{ag}$, $\alpha_{cp.ag}$ becomes

$$\alpha_{cp.ag} = \frac{3\beta_1 G K_{cp}}{r^2\left\{(G_o/G)^{1/3} - 1\right\}} \qquad (3.126)$$

### 3.5.2.3.1.1 FINE POROSITY MATRIX AND INTERFACES/CHANNELS

Coarser porosity paths are modeled as circular cylindrical paths with the same radius, $r$. With a similar convention of symbols and assumptions as earlier, the local moisture transfer coefficient $\alpha_{cp.ch}$ can be obtained as

$$\alpha_{cp.ch} = \frac{2\beta_2 G K_{cp}}{r^2\left\{(G_o/G)^{1/2} - 1\right\}} \qquad (3.127)$$

where $\beta_2$ represents the effectiveness of moisture transfer across the areas of contact between fine and coarse porosity paths. The limiting packing factor $G_o$ can be taken as 0.91 in this case. Length scale $r$ for coarse porosity paths was reported to be 10 to 50μm (Breugel 1991; Synder *et al.* 1990). A tentative value of 20μm has been assumed.

### 3.5.2.3.2 INTERFACES/CHANNELS AND AGGREGATES

It is extremely difficult to ascertain the area of contact of these two components. When a complete percolation of the interfacial zone occurs, the contact area would be of similar magnitude to the external surface areas of aggregates. At any intermediate stage, it would be approximately proportional to a fraction of the total surface area of aggregates that are in the percolating interface. Also, the permeability of both the components can be assumed to be roughly similar. Thus we have

$$\alpha_{ag.ch} = \frac{3\beta_3 G}{r\left(r/2K_{ag} + r_{ch}/K_{ch}\right)} \qquad (3.128)$$

where $G$ is the volume fraction of aggregates, and $\beta_3$ is the fraction of effective area of contact between the interfaces and aggregates. The approximation that $K_{ag} \approx K_{ch}$ and $r_{ch} \ll r$ (radius of aggregates), yields,

$$\alpha_{ag.ch} = \left(6\beta_3 G K_{ch}\right)/r^2 \qquad (3.129)$$

From the expressions for local moisture transport coefficients, it can be observed that the local mass transfer rate is quite sensitive to the average spatial dimensions of aggregate and interface components. A linear sensitivity can be expected for other parameters involved in the above model. It will be restated here that the transport coefficient models discussed above are quite primitive in nature, considering the finite limitations of our understanding of the complex micro-structure formed in a hydrated sample of the concrete and the difficulty involved in their exact analytical description. The coefficients discussed above would, however, help us to make informed and educated guesses about the moisture transport in a multi-component system.

### 3.5.3 Simulations of moisture transport

The system of partial differential equations (Equation 3.119) describes the governing equations of moisture transport in a non-deformable, isothermal composite porous medium. Compared to the formulation for an isotropic and homogenous porous medium, the multi-component formulation is complex and involves a higher degree of non-linearity due to the pressure potential based on local mass exchange terms. The inter-component moisture transfer models are based on coefficients derived in the previous section. Through these simulations, the role of each component on the moisture transport process as a whole is clarified. Parameters for which a numerical parametric study is done are: (a) local mass transfer coefficients, to measure their sensitivity (b) the effect of aggregate volume fractions and aggregate porosity. Both the drying and wetting processes have been included. The initial and boundary conditions adopted in the simulations are shown in Table 3.9. Perfect moisture transfer across the surface has been assumed. That is, the surface emmisivity parameter used is a very high number compared to the diffusivity $D$ of the bulk matrix.

As discussed in Section 3.5.2.3, the length scale for individual components as well as the permeability of the cement paste component is very important in the computation of inter-component moisture transfer rate coefficients. At this point, only an approximate guess can be made. Another way to deal with such a difficulty is to make an inverse analysis. That is, from a given set of experimental data, values can be chosen to predict a larger set of experimental data. Once a pore distribution is selected, a more or less accurate guess for permeability can be made. The only parameters that need investigation are the length scales ($\delta_i$) and effectiveness of the area of contact between various

*Table 3.9* Initial and boundary conditions adopted in the simulation

| Process | Initial condition | Boundary condition | Length |
|---------|-------------------|--------------------|--------|
| Wetting | RH = 1% for all components | Face 1: S = 1 for all components<br>Face 2: Atmospheric relative humidity = 90% | L = 5 cm for (a)<br>L = 20 cm for (b) |
| Drying | S = 1 for all components | Atmospheric relative humidity = 50% | L = 5 cm for (a)<br>L = 20 cm for (b) |

components ($\beta_i$). Some approximate estimate for $\delta_i$ is possible, based on the results of past researchers. The parameters that are left are $\beta_i$. A sensitivity analysis for $\beta_1$, $\beta_2$ and $\beta_3$ is done with an experimental comparison and an estimate for these coefficients is made. The length scales required are radius of the aggregates and radius of the coarser porosity zone, tentatively assumed as 2 mm and 20 μm, respectively.

First, let us observe the *virtual concrete* behaviors of water sorption. Analysis of the wetting case for different values of $\beta_1$, $\beta_2$ and $\beta_3$ from 0 to 1 is shown in Figure 3.75a–c. The moisture passes through both the interfacial channels and paste matrix with pores. At the same time, the moisture will be supplied from the channel to the paste matrix provided that the diffusivity of the channel system is predominant. Similarly, analysis of the drying case for different values of $\beta_1$, $\beta_2$ and $\beta_3$ is shown in Figure 3.75d–f. The point is the averaged overall diffusivity as a composite.

For the wetting case, it can be seen that increasing $\beta_1$ for aggregate–cement paste interaction gradually increases the rate of sorption (Figure 3.75a–c). However, total normalized saturation and the rate of sorption decreases with the addition of aggregates, due to the addition of non-participating porosity of the aggregate component. As aggregate versus cement paste interaction is increased, the aggregate porosity starts participating in the moisture

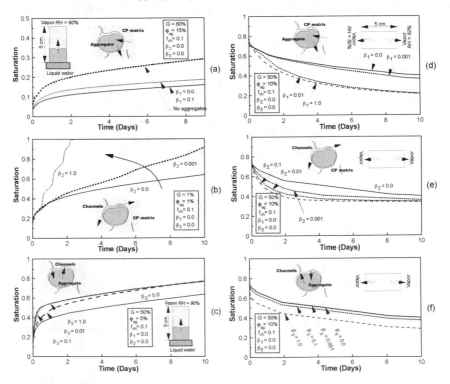

*Figure 3.75* The parametric study for local mass transfer coefficients.

transport by virtue of the local mass transfer. Therefore, the rate and level of normalized sorption increase gradually from the minimum value but, even at maximum interaction, this value is less than that for the case where there are no aggregates. Clearly, this effect will be significantly smaller for usual aggregates. Since usual aggregate porosity is around 2%, the contribution of aggregate porosity to the total concrete porosity would be very small compared to the contribution from a hardened cement mass. Overall sorption is *not* very sensitive to the $\beta_1$ parameter for usual aggregates.

Figure 3.75e–f shows the effect of $\beta_2$ (cement paste–channel interaction) on the rate of sorption. It can be seen that the overall rate of sorption is very sensitive to $\beta_2$. The rate of sorption increases rapidly with increasing values of $\beta_2$. This is due to the fact that ingress of moisture into the concrete through channel systems is very rapid, because of its high diffusivity. A rapid interaction with the cement paste component transfers a large amount of moisture to the paste matrix, increasing the apparent rate of sorption. It has to be noted that higher values of $\beta_2$ approximate closely the usual assumption of equilibrated moisture distribution inside the REV of a porous network. It is difficult to understand why $\beta_2$ should be very small where interfaces are closely dispersed and form a percolated path, as in a typical concrete containing a high aggregate volume fraction. Therefore, we can infer that when interfaces form a continuously connected path, the net effect of rapid moisture suction can also be represented by a single matrix system – one which has a higher apparent diffusivity than the real diffusivity of the cement-paste matrix in concrete.

Coarser-aggregate porosity interaction shows a similar trend to that observed for *cp–ch* interaction. If the aggregate porosity is small, the sensitivity of $\beta_3$ on the total normalized weight gain curve would not be large. It has to be noted that for a $\beta_3$ value greater than 0.01, the results are almost identical for all subsequent higher values of $\beta_3$. That is, even a 100-fold increase in $\beta_3$ does not effect the overall result very much. This shows the importance of percolation. That is, once a continuous connected interaction path of aggregates and channels is established, change in the effective area of contact does not cause significant change in the overall sorption behavior. Thus, the establishment of continuous connected interaction clusters of aggregate and coarser paths (which may occur at some critical aggregate content) is important and it will have considerable impact on the life expectancy of concrete structures. In this regard, computer simulation models that describe the probability of connectivity of channel–aggregate paths with an increasing volume fraction of aggregates are indispensable.

Similarly, for the drying process, it can be seen that all the interaction constants follow an almost identical trend in sensitivity as for the wetting case, although the magnitude of variation is decreased for aggregate–channel and channel–cement paste constants.

# 4 Transport of carbon dioxide and carbonation

It is a well-known fact that steel corrosion reduces the serviceability and safety performance of reinforced concrete in general (Chapter 9 and Chapter 10). Usually, concrete maintains high alkaline conditions, which leads to a passive layer forming on the steel surface. However, several types of environmental actions such as carbonic, sulphuric and nitric acid can cause the reduced alkalinity of pore water in concrete. Under low-pH conditions, the passive layer around the steel surface breaks down, and rust readily appears on the surface. In order to evaluate performance factors of a concrete structure in a corrosive environment, it is therefore essential to quantify the neutralization of concrete, that is to say, the pH reduction due to such environmental acids. In this chapter, the authors mainly focus on the carbonation phenomenon, and aim to predict pH fluctuations and material degradations under carbonic acid attack. This is also the basis of corrosion prediction of steel in concrete (Chapter 6).

## 4.1 Simplified modeling of carbonation

Gaseous carbon dioxide readily dissolves into the pore water of concrete, and it turns into a carbonic acid solution. Carbonic ions dissociate from the carbonic acid solution, releasing protons, which then react with calcium ions to form calcium carbonate. In this reaction, carbonic acid neutralizes alkalis in the pore water, mainly consuming calcium hydroxide, the solubility of which is the highest in cement hydrates. Consequently, as the calcium hydroxide is consumed, the pH value drops. From a thermodynamic point of view, the carbonation reaction is sure to progress under the condition that carbon dioxide exists, since the free energy of calcium carbonate is lower than that of calcium hydroxide (Japan Concrete Institute 1993).

In the conventional approach to the carbonation phenomenon, empirical formulae are generally used for its prediction. These models have originated from the assumption that carbonation would progress in proportion to exposure time as (Japan Concrete Institute 1993; Sakai 1993)

$$X_c = b\sqrt{t} \tag{4.1}$$

where $X_c$ is the depth of carbonation, $b$ is the carbonation rate coefficient, and $t$ is the exposure time. In fact, depth of carbonation $X_c$ can be obtained by solving a linear diffusion equation for $CO_2$ gas. Thus, Equation 4.1 involves the assumption that the depth of carbonation is associated with the amount of carbon dioxide diffusing through the concrete materials. In these methods, the carbonation rate coefficient $b$ is empirically determined as a function of the water-to-cement ratio, strength, carbon dioxide concentration in the external surroundings, temperature, relative humidity, water content, and types of finishing materials (Hamada 1969; Kishitani 1963; Nagataki *et al.* 1987; Uomoto and Takada 1993). This type of methodology, however, cannot easily be extended to construct a universal carbonation model, since it is difficult to consider all of the various factors and their interactions in such a simple empirical equation.

Thus, in order to predict carbonation progress under arbitrary conditions, it is useful to define the equations governing the carbonation phenomenon in concrete based on microscopic mechanisms, and to employ a deductive method involving numerical analysis techniques. From this perspective, in recent years numerical analytical methods, which can provide a solution for an arbitrary space and time, have been applied to the prediction of carbonation (Maeda 1989; Papadakis *et al.* 1991a, 1991b; Saeki *et al.* 1991; Saetta *et al.* 1993, 1995). In these studies, diffusion processes in concrete (moisture, carbon dioxide, and so on) and the carbonation process are modeled, and a linear or a non-linear diffusion equation is solved in the analysis. Depth of carbonation is evaluated by the amount of remaining calcium hydroxide and/or produced calcium carbonate in the concrete. Some researchers also aim to solve ion equilibriums and evaluate the pH value of pore water (Osada *et al.* 1997; Matsumoto *et al.* 1998).

In this chapter, the authors introduce a generalized computational method that can deal with pH fluctuations of pore water and degradation of the microstructure due to carbonation for arbitrary initial and environmental conditions. To simulate carbonation phenomena in concrete, the equilibrium between gaseous and dissolved carbon dioxide, their transports, ionic equilibriums and the carbonation reaction process are formulated based on thermodynamics and chemical equilibrium theory. The material properties of concrete, which are necessary for computing the formulations, are evaluated considering the inter-relationships among hydration, moisture transport and pore-structure development processes based on fundamental physical material models. With this methodology, the proposed system can be applied not only to the carbonation phenomenon, but also to the prediction of the pH profile in pore solutions attacked by other acids, including sulphuric acid and nitric acid.

In this section, the authors introduce a simplified modeling of carbonation, which considers only the carbonation reaction of calcium hydroxide and deals with the carbonation process under an isothermal environment (25°C). Thus, in implementing the model, unique values are given to Henry's constant describing $CO_2$ equilibrium, the reaction rate coefficient of carbonation, equilibrium constants of ionic concentration, and solubility. An enhanced modeling including carbonation reaction of C–S–H gel and the carbonation reaction at arbitrary temperature will be discussed in Section 4.2.

### 4.1.1 Modeling of carbon dioxide transport and equilibrium

#### 4.1.1.1 Mass conservation law for carbon dioxide

When dealing with mass, energy and momentum flows in a control volume, the starting point is to build mass balance equations. In other words, the summation of the rate of mass efflux from the control volume, the rate of mass flow into the control volume, and the rate of accumulation of mass within the control volume should all be zero. In this section, the mass balance conditions for carbon dioxide in a porous medium are formulated. Two phases of carbon dioxide in concrete are considered, i.e., gaseous carbon dioxide and carbon dioxide dissolved in pore water. By solving the mass balance equation under given initial and boundary conditions, the non-steady state conduction of carbon dioxide is quantified. The mass balance equation for a porous medium can be expressed as (Figure 4.1)

$$\frac{\partial}{\partial t}\{\phi[(1-S)\cdot\rho_g + S\cdot\rho_d]\} + div J_{CO_2} - Q_{CO_2} = 0 \tag{4.2}$$

where $\phi$ is porosity, $S$ is the saturation of porosity, $\rho_g$ is the density of gaseous carbon dioxide[kg/m³], $\rho_d$ is the density of dissolved carbon dioxide in pore water [kg/m³], and $J_{CO2}$ is the total flux of dissolved and gaseous carbon dioxide [kg/m².s]. The first term in Equation 4.2 represents the rate of change

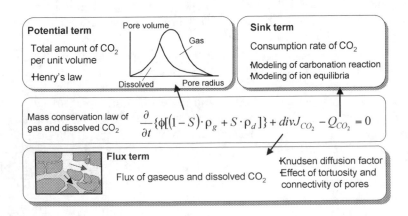

*Figure 4.1* Governing equation and constituting models.

in the total amount of carbon dioxide per unit of time and volume, the second term is the flux of carbon dioxide, and the third term, $Q_{CO_2}$, is a sink term. Equation 4.2 gives the concentrations of gaseous and dissolved carbon dioxide with time and space. In the micro-pore structure development model (Section 3.1 and Section 3.4), the pores in the cement paste are classified into three types: interlayer, gel pores and capillary pores. Here, the target is carbon dioxide existing in hardened cement, so only capillary and gel pores, which can act as transport paths or locations for carbonation reactions, are considered. In other words, the porosity $\phi$ of Equation 4.2 is the sum of the capillary pores and the gel pores.

The conservation law must be satisfied in all material systems and so it applies to the field of concrete materials. In the following sections, each term will be modeled based on the specific characteristics of concrete materials, i.e., the equilibrium and transport of gaseous and dissolved $CO_2$, and the consumption rate of carbon dioxide.

### 4.1.1.2 Equilibrium conditions for gaseous and dissolved carbon dioxide

The local equilibrium between gaseous and dissolved carbon dioxide is represented here by Henry's law, which states the relationship between gas solubility in pore water and the partial gas pressure. In this modeling, we assume that the system will instantaneously reach local equilibrium between the two phases as (Welty *et al.* 1969)

$$P_{CO_2} = H'_{CO_2} \cdot \rho'_d \tag{4.3}$$

where $P_{CO2}$ is the equilibrium partial pressure of carbon dioxide in the gas phase, $\rho'_d$ is the mole fraction of gaseous $CO_2$ [mol of $CO_2$/total mol of solution], and $H'_{CO2}$ is Henry's constant for carbon dioxide (= $1.45 \times 10^8$ [Pa/mol fraction] at 25°C). For one cubic meter of dilute solution, the moles of water in the solution $n_{H2O}$ will be approximately $5.56 \times 10^4$ [mol/m³]. Accordingly, the concentration of dissolved carbon dioxide per cubic meter of solution $\rho_d$ [kg/m³] can be expressed as

$$\rho_d = \frac{P_{CO_2}}{H'_{CO_2}} \cdot n_{H_2O} \cdot M_{CO_2} = \frac{P_{CO_2}}{H_{CO_2}} \tag{4.4}$$

where $M_{CO2}$ is the molecular mass of carbon dioxide (= 0.044[kg/mol]). The complete perfect-gas equation is, then,

$$P_{CO_2} = \frac{\rho_g RT}{M_{CO_2}} \tag{4.5}$$

where $\rho_g$ is the concentration of gaseous carbon dioxide [kg/m³], $R$ is the gas constant [J/mol.K], and $T$ is temperature [K]. From Equations 4.4 and

4.5, the equilibrium relationship between gas and dissolved $CO_2$ can be expressed as

$$\rho_g = \frac{M_{CO_2}}{RT} \cdot H_{CO_2} \cdot \rho_d = K_{CO_2} \cdot \rho_d \tag{4.6}$$

After dissolving into solution, carbon dioxide reacts with calcium ions, and the concentration of dissolved $CO_2$ can fluctuate from the above equilibrium condition. Strictly speaking, therefore, the equilibrium condition cannot be formulated by Henry's law alone; as well, it is necessary to determine the amount of dissolved $CO_2$ based on the rate of chemical reactions, which represents kinetic fluctuations dependent on the distribution of $CO_2$ concentration. However, it seems difficult to take into account such kinetic fluctuations because the rate of $CO_2$ gas dissolution will be faster when the partial pressure of $CO_2$ gas becomes large. For these reasons, in our model we assume that the amount of dissolved $CO_2$ can be approximately described by Henry's law (Saeki *et al.* 1991; Papadakis *et al.* 1991b).

### 4.1.1.3 Modeling of carbon dioxide transport

The transport of $CO_2$ is considered for both dissolved and gaseous carbon dioxide phases. The $CO_2$ gas can move through unsaturated pores, whereas dissolved $CO_2$ is transported within pore liquid water. In the model, it is assumed that all pores have a cylindrical shape.

Diffusion in a porous body may occur via one or more of three mechanisms: molecular (Fick's) diffusion, Knudsen diffusion and surface diffusion. In the model, molecular diffusion and Knudsen diffusion are considered, whereas the contribution of surface diffusion is ignored, since this type of diffusion takes place when molecules which have been adsorbed are transported along the pore wall, and normally it plays a minor role in diffusion within concrete materials under typical environmental conditions (Section 3.2.4.2; Kunii and Hurusaki 1980; Kobayashi and Syutto 1986).

As the relative humidity in a pore decreases, the porous medium for gas transport becomes finer. If the pores through which the gas is traveling are quite small, the molecules will collide with the walls more frequently than they do with each other. This is known as Knudsen diffusion. The conditions transiting to this type of diffusion are expressed by the following equation using Knudsen number $N_k$ as

$$N_k = \frac{l_m}{2r_e} > 1.0 \tag{4.7}$$

where $l_m$ is the mean free path length of a molecule of gas, and $r_e$ is the actual pore radius, i.e., its radius minus the thickness of the adsorbed layer of water obtained by BET theory (Section 3.2.2.2). With consideration of both molecular diffusion and Knudsen diffusion, the one-dimensional gaseous flux through a single pore of radius $r$ can be expressed as

$$J_g^r = -\frac{D_0^g}{1 + (l_m/2r_e)} \frac{\partial \rho_g}{\partial x} \tag{4.8}$$

where $D_0^g$ [m²/s] is the diffusivity of $CO_2$ in a free atmosphere (= $1.34 \times 10^{-5}$, Welty *et al.* 1969). Similarly, the flux of dissolved carbon dioxide can be obtained as

$$J_d^r = -D_0^d \frac{\partial \rho_d}{\partial x} \tag{4.9}$$

where $D_0^d$ [m²/s] is the diffusivity of dissolved $CO_2$ in pore water (= $1.0 \times 10^{-9}$, Welty *et al.* 1969).

The total flux within porous bodies can be obtained by integrating Equation 4.8 and Equation 4.9 over the entire porosity distribution. For example, let us consider the equilibrium condition of moisture during a monotonic wetting phase, in which there is no ink-bottle effect (Section 3.2.2.3). In this case, based on the thermodynamic conditions, a certain group of pores whose radii are smaller than the specific radius $r_c$ at which a liquid–vapor interface forms are completely filled with water, whereas larger pores remain empty or partially saturated, so these pores can be routes for the transfer of $CO_2$ gas. Therefore, by integrating the gaseous and dissolved fluxes of $CO_2$, respectively, the entire flux of carbon dioxide is formulated as

$$J_{CO_2} = -\left( \frac{\phi D_0^d}{\Omega} \int_0^{r_c} dV \frac{\partial \rho_d}{\partial x} + \frac{\phi D_0^g}{\Omega} \int_{r_c}^{\infty} \frac{dV}{1 + N_k} \frac{\partial \rho_g}{\partial x} \right) \tag{4.10}$$

where $V$ is the pore volume, and $\Omega = (\pi/2)^2$ accounts for the average tortuosity of a single pore as a fictitious pipe for mass transfer. The latter parameter considers the tortuosity of a hardened cement paste matrix, which is uniformly and randomly connected in a three-dimensional system (Section 3.2.3). The first term of the right-hand side in Equation 4.10 denotes the diffusive component of dissolved carbon dioxide in the pore liquid, whereas the second term represents the component of gaseous diffusion. In order to generalize the expression for an arbitrary moisture history, the substitution of porosity saturation $S$ for the integrals in Equation 4.10 gives,

$$J_{CO_2} = -\left( D_{dCO_2} \nabla \rho_d + D_{gCO_2} \nabla \rho_g \right) = -\left( D_{dCO_2} + D_{gCO_2} \cdot K_{CO_2} \right) \nabla \rho_d \tag{4.11}$$

$$D_{dCO_2} = \frac{\phi S}{\Omega} D_0^d \ , D_{gCO_2} = \frac{\phi \cdot D_0^g}{\Omega} \frac{(1-S)}{1 + l_m/2(r_m - t_m)} \tag{4.12}$$

where $D_{gCO_2}$ is the diffusion coefficient of gaseous $CO_2$ in a porous medium [$m^2/s$], and $D_{dCO_2}$ is the diffusion coefficient of dissolved $CO_2$ in a porous medium [$m^2/s$]. In Equation 4.12, the integral of the Knudsen number is simplified so that it can easily be put into practical computational use, $r_m$ is the average radius of unsaturated pores, and $t_m$ is the thickness of the adsorbed water layer in the pore whose radius is $r_m$.

It has to be noted that the formulations do not include the complete effects of the connectivity of pores on diffusivity. That is to say, considering one-dimensional transport of gaseous phases, all unsaturated pores, which are routes for gas movement, are assumed to be connected with perfect continuity (Figure 4.2, left). In actual pore structures that have complicated connectivity, however, the movement of gas will be blocked by saturated pores containing liquid water (Figure 4.2, right). Here, as the saturation of pores decreases, the open pore space gains higher connectivity so total diffusivity increases non-linearly. One potential model for this phenomenon would be based on percolation theory (Cusack 1987), however, we adopt the following model for the sake of simplicity.

Let us consider a finite field that consists of segments small enough to have continuity of pores, as shown in Figure 4.2, right. In the cross-section of each unit, the ratio of gas transportation paths will be $(1-S)$. If we assume that the probability of unsaturated pores being connected to each other would be proportional to the ratio of the volume in a cross-section, the overall flux can be expressed as

$$D_{gCO_2} = \frac{\phi \cdot D_0^g}{\Omega} \frac{(1-S)^n}{1 + l_m/2(r_m - t_m)}, D_{dCO_2} = \frac{\phi S^n}{\Omega} D_0^d \qquad (4.13)$$

where $n$ is a parameter representing the connectivity of the pore structure, and might vary with the geometrical characteristics of the pores. However, at this stage, it is difficult to take account of the exact connectivity situation. Through sensitivity analysis, $n$ is tentatively assumed to be 6.0, which is the most appropriate value for expressing the reduction of $CO_2$ diffusivity with the decrease of relative humidity (Figure 4.3).

In order to evaluate the hysteresis behavior of moisture saturation during drying–wetting cycles, the authors have developed a moisture isotherm model that considers the connectivity of pores with two different radii (Section

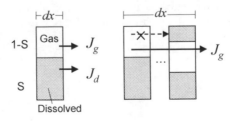

*Figure 4.2* Effect of connectivity of pores on $CO_2$ transport.

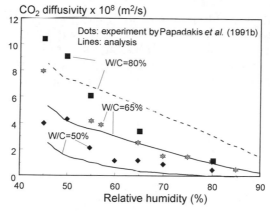

*Figure 4.3* Comparison of experiment and analytical prediction using two different values of connectivity parameter (n = 6).

3.2.2.3). The expression of saturation S in Equation 4.13 involves the blockage effect due to the connection of pores in its simplest form. However, the connectivity factor considered in the moisture isotherm model cannot alone evaluate the non-linear behavior in the $CO_2$ diffusion process with relative humidity change. This might be because of the difference between the characteristics of $CO_2$ gas transport and moisture transport. In the case of moisture transfer, the local equilibrium between liquid and vapor is maintained, and the movement of the vapor phase is dominant, whereas the diffusive movement of $CO_2$ gas is completely blocked by the existence of pore water.

Figure 4.3 compares the $CO_2$ diffusivity calculated by the model and by an empirical formula obtained by diffusion tests (Papadakis 1991a). It can be seen that the non-linear behavior of $CO_2$ diffusivity is reasonably well predicted.

### 4.1.2 Modeling of carbonation reaction

#### 4.1.2.1 Formulation of carbonation reaction rate

The carbonation phenomenon in cementitious materials is simply described by the following equation of ionic reaction.

$$Ca^{2+} + CO_3^{2-} \rightarrow CaCO_3 \tag{4.14}$$

Calcium ions resulting from the dissolution of calcium hydroxide are assumed to react with carbonate ions, whereas the reaction of calcium silicate hydrate (C–S–H) is not considered. This is based on the fact that the solubility of C–S–H is quite low compared with that of calcium hydroxide, i.e., the solubility of C–S–H, in the case of $6CaO \cdot 5SiO_2 \cdot 6H_2O$) $K_{sp} = [Ca^{2+}]^6[HSiO_3^-]^5[OH^-]^7 = 5.5 \times 10^{-49}$, and the solubility of calcium hydroxide, in the case of

$Ca(OH)_2$ $K_{sp} = [Ca^{2+}][CO_3^{2-}] = 4.14 \times 10^{-5}$ (Japan Concrete Institute 1993). It has to be noted, however, that these values are for a normal environment. Therefore, we understand that it is necessary to consider chemical reactions of C–S–H solution for the sake of predicting deterioration phenomena under severe environmental action and/or environmental action over quite a long time. The carbonation reaction of C–S–H will be included in the model in Section 4.2.

The rate of the reaction in Equation 4.14 can be expressed by the following differential equation, assuming that the reaction is of the first order with respect to $Ca^{2+}$ and $CO_3^{2-}$ concentrations as

$$Q_{CO_2} = \frac{\partial \left( C_{CaCO_3} \right)}{\partial t} = k[Ca^{2+}][CO_3^{2-}] \tag{4.15}$$

where $C_{CaCO_3}$ is the concentration of calcium carbonate [mol/l], and $k$ is the reaction rate coefficient [l/mol.sec]. The concentration of calcium carbonate per unit of time obtained by Equation 4.15 is equal to the consumption rate of carbonic acid, which is the sink term $Q_{CO2}$ in the mass balance Equation 4.2.

As already mentioned, Equation 4.15 is a differential equation determining the reaction rate of calcium ions and carbonate ions in solution. For this reaction to occur, molecular collisions of each ion are necessary, and the right-hand side of Equation 4.15 represents the rate of such collisions. Here, a molecular decomposition occurs to form a new product only if a molecule has energy greater than the activation energy. The ratio of molecules having a kinetic energy above this activation energy is determined by Boltzmann's distribution law, which shows strong dependency on temperature. In Equation 4.15, therefore, a temperature effect is involved in the reaction rate coefficient $k$. In other words, $k$ only represents the reaction rate at a certain temperature. Therefore, in order to evaluate the carbonation process at an arbitrary temperature, it is necessary to consider the temperature effect on rate reaction coefficient $k$. In this section, a unique coefficient is applied, i.e., coefficient $k$ is assumed to be constant ($k = 41.6$ [l/mol.sec]) using a value determined from several sensitivity analyses. An extended model for various temperature conditions based on Arrhenius's law of chemical reaction will be discussed in Section 4.2.

The discussion herein concludes the formulation of the carbonation reaction. In the following section, we will describe the modeling of the ionic equilibrium in order to obtain the reaction rate by Equation 4.15.

### 4.1.2.2 Ion equilibrium in solution

In the pore solutions of concrete, various ions coexist, i.e., calcium, aluminum, iron, magnesium, sodium, potassium, and others. Though sodium and potassium ions form carbonate salts, here we consider only calcium ions, which are comparatively abundant (in the case of ordinary cement, CaO: 65%, $Na_2O$: 0.3%, $K_2O$: 0.5%). Kobayashi (1991) reported, however, that sodium and potassium ions may cause pH fluctuations, and also a change in the carbonation rate. In addition, Matsumoto *et al.* (1998) pointed out that the pH value after carbonation would depend on the concentrations of these ions. Therefore, in order to accurately evaluate the pH profile, it would be necessary to consider these alkali ions as well as calcium ions. As a first approximation, however, only calcium ions are considered, and other ions will be integrated in future.

The dissociations of the ions considered in the model are shown by the following Equation 4.16. We consider the dissociation of water and carbonic acid, and the dissolution and dissociation of calcium hydroxide and calcium carbonate.

$$H_2O \leftrightarrow H^+ + OH^-$$

$$H_2CO_3 \leftrightarrow H^+ + HCO_3^- \leftrightarrow 2H^+ + CO_3^{2-}$$

$$Ca(OH)_2 \leftrightarrow Ca^{2+} + 2OH^-$$

$$CaCO_3 \leftrightarrow Ca^{2+} + CO_3^{2-}$$

(4.16)

Strictly speaking, the representation of a proton by $H^+$ might not be correct. Rather, the hydronium ion $H_3O^+$ is present in water and confers acidic properties on aqueous solutions. However, it is customary to use the symbol $H^+$ in place of $H_3O^+$, so $H^+$ will be used in the following discussion.

As shown in Equation 4.16, carbonation is an acid-base reaction, where cataions and anions act as a Brönsted acid and base, respectively. Furthermore, the solubility of precipitations is dependent on the pH of the pore solution. Therefore, in order to obtain each ionic concentration at an arbitrary stage, the authors first introduce an equation with respect to the concentration of protons $[H^+]$. Once the concentration of protons $[H^+]$ is known, each equilibrium condition can be calculated for a given pH value. For each ion, the following basic principles should be satisfied (Freiser and Fernando 1963): (1) the law of mass action, (2) the mass conservation law and (3) the proton balance.

First of all, let us consider the equilibrium reaction of carbonic acid. Based on the law of mass action, the corresponding equilibrium expressions should be satisfied as

$$K_w = [H^+][OH^-] \, , K_a = \frac{[H^+][HCO_3^-]}{[H_2CO_3]} \quad K_b = \frac{[H^+][CO_3^{2-}]}{[HCO_3^-]}$$

(4.17)

where $K_i$ is the equilibrium constant of concentration for each dissociation. We give these values as $K_w = 1.00 \times 10^{-14}$, $K_a = 1.00 \times 10^{-14}$, and $K_b = 4.79 \times 10^{-14}$ at 25°C, respectively.

Next, the mass conservation law is applied for the ions resulting from the dissolution of carbon dioxide and re-dissolution of calcium carbonate as (Figure 4.4)

$$C_0 + S_1 = [H_2CO_3] + [HCO_3^-] + [CO_3^{2-}] \tag{4.18}$$

where $C_0$ is the concentration of dissolved carbon dioxide [mol/l], which can be obtained from $\rho_d$ in Equation 4.2. $S_1$ is the solubility of calcium carbonate, which can be calculated using the solubility–product constant, discussed later.

Using Equation 4.17 and Equation 4.18, concentrations of $H_2CO_3$, $HCO_3^-$ and $CO_3^{2-}$ can be obtained as

$$[H_2CO_3] = \alpha_0 \cdot (C_0 + S_1) \qquad \alpha_0 = \frac{[H^+]^2}{[H^+]^2 + K_a[H^+] + K_aK_b}$$

$$[HCO_3^-] = \alpha_1 \cdot (C_0 + S_1) \qquad \alpha_1 = \frac{K_a[H^+]}{[H^+]^2 + K_a[H^+] + K_aK_b} \tag{4.19}$$

$$[CO_3^{2-}] = \alpha_2 \cdot (C_0 + S_1) \qquad \alpha_2 = \frac{K_aK_b}{[H^+]^2 + K_a[H^+] + K_aK_b}$$

As this shows, the concentrations of $H_2CO_3$, $HCO_3^-$ and $CO_3^{2-}$ are functions of $S_1$. It is therefore necessary to obtain the solubility of calcium carbonate. That solubility can be derived by the following relationship as

$$K_{sp}^1 = [Ca^{2+}][CO_3^{2-}] \tag{4.20}$$

where $K_{sp}^1$ is the solubility–product constant of the calcium carbonate ($= 4.7 \times 10^{-9}$ at 25°C, Freiser and Fernando 1963). Similarly, the solubility of calcium hydroxide can be calculated as

$$K_{sp}^2 = [Ca^{2+}][OH^-]^2 \tag{4.21}$$

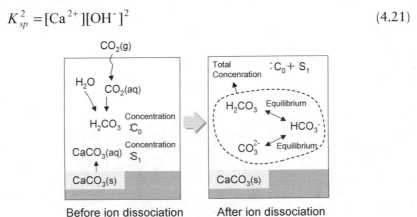

Before ion dissociation          After ion dissociation

*Figure 4.4* Mass balance and equilibrium conditions for carbonic acid.

where $K_{sp}^2$ is the solubility–product constant of the calcium hydroxide (= 5.5 × 10$^{-6}$ at 25°C, Freiser and Fernando 1963). In looking at the ion dissociations shown in Equation 4.20 and Equation 4.21, it is necessary to consider the common ion effect on the solubility. Namely, carbonate ions in Equation 4.20 come not only from the decomposition of calcium carbonate, but also from that of carbonic acid. As for the calcium ions in Equation 4.20 and Equation 4.21, a similar effect should be considered as well. Considering these common ion effects on each solubility, the solubility of calcium carbonate $S_1$ and that of calcium hydroxide $S_2$ can be substituted in Equations 4.19, 4.20 and 4.21 as

$$K_{sp}^1 = \left(S_1 + S_2\right)\cdot\alpha_2\left(C_0 + S_1\right) \tag{4.22}$$

$$K_{sp}^2 = \left(S_1 + S_2\right)\cdot[\text{OH}^-]^2 \tag{4.23}$$

Next, the law of mass balance is considered. Figure 4.5 shows the mass balance conditions in each system: dissolution of carbon dioxide, re-dissolution of calcium carbonate, and dissolution of calcium hydroxide. In order to distinguish the sources of ion dissociation, concentrations of ions are expressed with suitable notation, i.e., $[i]_d$, $[i]_s$ and $[i]_c$ are the concentrations of ions dissociated from the dissolution of $CO_2$ gas, calcium carbonate and calcium hydroxide, respectively. For example, the total concentration of carbonic acid $[H_2CO_3]$ shown in Equation 4.9 becomes the summation of $[H_2CO_3]_d$ from $CO_2$ gas and $[H_2CO_3]_s$ from $CaCO_3$.

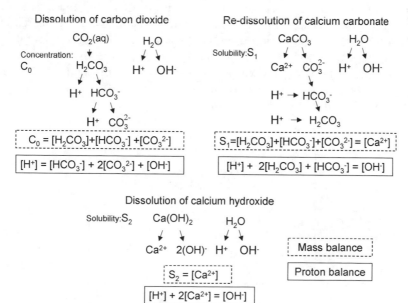

*Figure 4.5* Mass and proton balance equations in each system.

From the mass conservation conditions before and after ion dissociation, the following equations should be satisfied:

$$C_0 = [H_2CO_3]_d + [HCO_3^-]_d + [CO_3^{2-}]_d \tag{4.24}$$

$$S_1 = [H_2CO_3]_s + [HCO_3^-]_s + [CO_3^{2-}]_s = [Ca^{2+}]_s \tag{4.25}$$

$$S_2 = [Ca^{2+}]_c \tag{4.26}$$

In addition, the ions should satisfy the law of proton balance, in which the amount of donors is equal to that of accepters in terms of protons in the Brönsted–Lowry theory. The equation deduced by this law of proton balance is also shown in Figure 4.5. The overall proton balance equation can be obtained by summing up each equation as

$$[H^+] + 2[Ca^{2+}]_c + 2[H_2CO_3]_s + [HCO_3^-]_s = [OH^-] + [HCO_3^-]_c + 2[CO_3^{2-}]_c \tag{4.27}$$

From these equations describing ion equilibrium conditions, we finally obtain

$$[H^+] + 2(S_1 + S_2) = \frac{K_w}{[H^+]} + \alpha_1(C_0 + S_1) + 2\alpha_2(C_0 + S_1) \tag{4.28}$$

Equation 4.28 shows that the concentration of protons $[H^+]$ can be obtained at an arbitrary stage as an exact solution, given the concentrations of calcium hydroxide and carbonic acid before dissociation. After the concentration of protons is known, each individual ionic concentration can be calculated. As an example of a computation using the proposed method, the relationship between the pH value in solution and the existing ratio of carbonic acid, carbonic hydroxide ion and carbonate ion is shown in Figure 4.6. In the high pH range, carbonic ions are dominant, whereas carbonic hydroxide ions increase under low pH conditions.

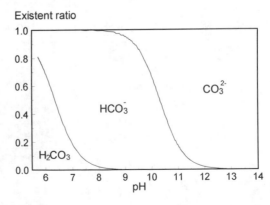

*Figure 4.6*  Relationship between equilibrium of carbonic acid and pH in solution.

### 4.1.2.3 Changes in micro-pore structures due to carbonation

It has been reported that the micro-pore structure of cementitious material may change due to carbonation (Japan Cement Association 1993; Houst and Wittmann 1994; Papadakis *et al.* 1991a). In this section, the authors take into account changes in porosity caused by carbonation by applying a simplified model. A universal model accounting for changes in micro-pore structure based on microscopic phenomena will be presented in Section 4.2. Here, it is assumed that the porosity distribution does not change, but that total porosity decreases as carbonation in progresses. Papadakis *et al.* (1991a) showed experimentally that the reduction in porosity of fully carbonated concrete is dependent on the water-to-cement ratio. This fact also coincides well with the experimental results of Houst and Wittmann (1994) as well as those of Ngala and Page (1997). Based on these experimental results, the reduction of fully carbonated concrete/mortar can be empirically formulated (Figure 4.7) as

$$\alpha_f = 0.67(W/C) + 0.27 \tag{4.29}$$

where $\alpha_f$ is the reduction in porosity when a large amount of $Ca(OH)_2$ is consumed by carbonation (fully carbonated specimen). Due to the lack of experimental results for a moderate level of carbonation (not fully carbonated), the porosity reduction factor for this level is simply modeled as a linear function of the remaining calcium hydroxide to its initial value, which also coincides with the experimental work of Saeki *et al.* (1991). The proposed bi-linear porosity reduction function is expressed as

$$\phi' = \alpha\phi \quad \begin{array}{ll} \alpha = \alpha_f & (R \le \beta) \\ \alpha - 1.25 \cdot R - 0.25 & (\beta < R \le 1) \end{array} \tag{4.30}$$

where $\phi$ is the porosity before carbonation, $\phi'$ is the porosity after carbonation, $\alpha$ is the reduction parameter in porosity, $R$ = degree of carbonation,

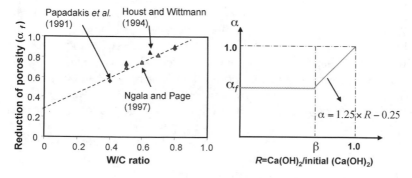

*Figure 4.7* Simplified porosity reduction model due to carbonation reaction.

which is expressed as the ratio of the amount of remaining $Ca(OH)_2$ to the initial amount of $Ca(OH)_2$. A parameter $\beta$ in Equation 4.30 is formulated as

$$\beta = 0.8\alpha_f + 0.2 \tag{4.31}$$

It has to be noted that the porosity reduction function was mainly derived from cases in which ordinary Portland cement was used. Past research has reported that the use of mineral admixtures, such as blast furnace slag and fly ash, may cause a change in the pore structure due to carbonation. This aspect remains a matter of future study for more precise prediction and wider applicability of the system.

Theoretically, a finer porosity results from the volume change of hydrates due to carbonation. The crystal volume of calcium carbonate is approximately 11.7% more than that of calcium hydroxide. The authors understand that it becomes possible to establish a more generalized treatment by implementing this volume change into the micro-pore structure model (Section 3.1 and Section 3.4). To this end, the simplified model has been enhanced, and this issue will be discussed in Section 4.2.2.2.

### 4.1.3 Verifications and numerical simulations

#### 4.1.3.1 Verification with accelerated carbonation tests

First, accelerated carbonation tests were used for verification. Experimental data obtained by Uomoto and Takada (1993) were used for this purpose. The data are for cylindrical specimens of radius 10 cm and height 20 cm. After two days of sealed curing, two curing conditions were specified: submerged in water for zero days, and cured in water for five days. After the prescribed curing periods, specimens were kept in a controlled chamber where the concentrations of $CO_2$ gas (1.0% and 10%), temperature (20°C) and relative humidity (55% RH) were kept constant. The-water-to-cement ratios of the specimens were 50%, 60% and 70%. The unit weight of water was held constant in the mix proportions. Results of the verifications are shown in Figure 4.8. All of the input values in the analysis corresponded to the experimental conditions. Analytical results show the relationship between exposure time and the concrete depth at which the pore water pH falls below 10.5, which is the phenolphthalein indicator point. The empirical formulae shown in the Figures were regressed with the square root equation of time (Kishitani 1963). The simulations are able to roughly predict the progress of carbonation for different $CO_2$ concentrations and water-to-powder ratios.

Next, the effect of ambient relative humidity on the progress of carbonation was verified using experimental data obtained by John et al. (1990). After one day of wet curing, specimens were stripped and then cured under standard curing conditions until the age of 28 days, before being kept at 50%

RH and 20°C for 10 days. In the accelerated test, specimens were exposed to a 10% concentration of $CO_2$ at a constant temperature of 25°C, and at three different ambient relative humidities: 50% RH, 65% RH and 80% RH. Two cases of water-to-cement ratio were tested (W/C = 65% and 55%). Figure 4.9 shows the results of the verification. The depth of carbonation for various ambient relative humidities can be predicted with reasonable accuracy by the proposed method.

### 4.1.3.2 *Distribution of pH, calcium hydroxide and calcium carbonate*

Figure 4.10 shows the distribution of pore water pH, $CO_2$, calcium hydroxide and calcium carbonate within concrete exposed to a $CO_2$ concentration of 3%. Two different water-to-powder ratios, W/C = 25% and 50%, were specified in the analysis, and these specimens were exposed to $CO_2$ gas after seven days of sealed curing. Consumption of calcium hydroxide, formation of calcium carbonate, and reduction of pH with time are estimated by the analysis. It is also shown that a higher resistance to carbonic acid action is achieved in the case of low W/C.

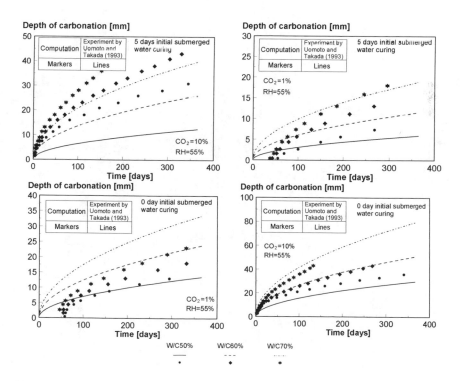

*Figure 4.8* Prediction of carbonation phenomena for different $CO_2$ concentrations, W/C and curing conditions.

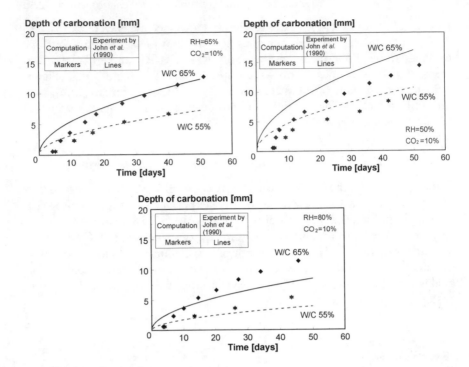

*Figure 4.9* Prediction of carbonation phenomena for different W/C and ambient relative humidity.

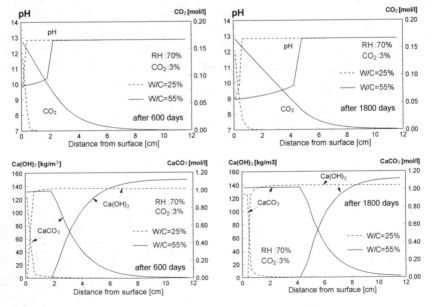

*Figure 4.10* Distribution of pH, calcium hydroxide and calcium carbonate under the action of carbonic acid.

## 4.2 Enhanced modeling of carbonation

For the purpose of establishing an evaluation method for the carbonation phenomenon, formulations are developed for the equilibrium of gaseous and dissolved carbon dioxide, their transport, ionic equilibriums and the carbonation reaction process in the previous section. The main feature is that the model is based on microphysical phenomena, enabling the pH profile of pore water and the degradation of the micro-pore structure to be predicted for arbitrary conditions. This contrasts with the conventional approach, where depth of carbonation is generally evaluated using an empirical formula. By means of the proposed methodology, it is possible to predict not only the carbonation phenomenon, but also the pH profile in pore solutions attacked by other acids, such as sulphuric acid and nitric acid. Through various numerical simulations, it is shown that the proposed modeling technique can roughly predict carbonation progress and pH fluctuations for different mix proportions, curing conditions and environmental conditions. In this section, further improvement and extended applicability will be challenged.

### *4.2.1 Applicability of the simplified modeling of Ca(OH)₂ carbonation*

#### *4.2.1.1 Verification of the carbonation model and sensitivity analysis*

As discussed in the preceding section, in order to express the carbonation phenomenon, the equilibrium of various ions associated with the carbonation reaction, the transport of dissolved and gaseous carbon dioxide, and the carbonation reaction itself have been modeled and implemented in the thermo-hygro system. The carbonation model is directly linked with characteristics of micro-pore structures, the state of internal moisture, the quantity of hydrated products, etc., which are calculated in each sub-system based on thermodynamic theory

Although the proposed simplified methodology has already shown its versatility and high applicability to carbonation progress for various cases, verifications have been carried out based mainly on short-term accelerated tests at normal temperature. Thus, in this section, the authors aim to improve the accuracy and to extend the applicable range of the simplified models, not only for controlled acceleration environments, but also for low concentrations such as the carbon dioxide concentration in the atmosphere, and for various temperature histories, taking into consideration the application of the thermodynamic model to actual structures.

Verification of the simplified modeling of carbonation was carried out using tests under a high-concentration accelerated environment (John *et al.* 1990) and long-term exposure in a natural environment (Yoda 2002). In the analysis, concrete depth at which computed pH of the pore water falls below 10.5, which is the phenolphthalein indicator point, was plotted against exposure time. Although the results of the simplified model, indicated by the

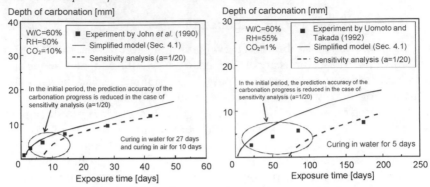

*Figure 4.11* Carbonation prediction in acceleration environment.

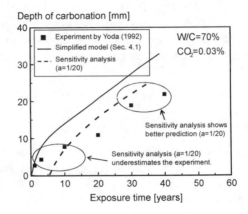

*Figure 4.12* Carbonation prediction in low concentration $CO_2$ environment.

solid lines in Figure 4.11 and Figure 4.12, can predict with good accuracy the accelerated experiments, the model tends to overestimate the long-term progress of carbonation in a low-concentration environment. It is also found that the difference between the measured values and predicted values tends to increase with the passage of time. Based on these results, it was decided to examine the sensitivity of the reaction rate coefficient that directly expresses the rate of progress of carbonation itself. The focus of the analytical investigation was whether it might be possible to improve the accuracy of the model over a wider range of conditions by simply changing the reaction rate coefficient. For discussion, Equation 4.15 is re-defined as follows.

$$R_{CH} = Q_{CO_2} = \frac{\partial C_{CaCO_3}}{\partial t} = a \cdot k \cdot \left[ Ca^{2+} \right] \cdot \left[ CO_3^{2-} \right] \tag{4.32}$$

where $R_{CH}$ is the reaction rate of calcium hydroxide [mol/m³.sec], and $a$ is the reaction rate reduction coefficient. The simplified model corresponds to $a = 1$.

Analysis was carried out with the reaction rate simply reduced by a factor of 20, in other words, with $a = 1/20$. The results are shown by the broken lines in the Figures. In this case, progress of carbonation is of course reduced, so the prediction accuracy is improved overall in a low-concentration environment, as shown in Figure 4.12. On the other hand, it can be seen that analytical results for the accelerated condition do not coincide with the experimental data. In particular, the progress of carbonation at the initial stage is not accurately predicted.

As it cannot be expected that the accuracy of the model can be improved by adjusting only the reaction rate coefficient, attention turned to interactions among the change in micro-pore structure due to carbonation, the moisture state in such pores and the transport of carbon dioxide through the porous medium. As stated previously, to represent the change in the micro-pore structure due to carbonation in the simplified model, the porosity only is reduced, without changing the distribution of pore structure (Equations 4.29–4.31). As shown in Figure 4.13, when the pore distribution is not changed after carbonation, the moisture equilibrium model estimates almost same degree of saturation under given relative humidity. This is because, from the thermodynamic point of view, the radius $r_c$ in which a vapor–liquid interface exists remains constant under constant relative humidity, which leads to almost the same degree of saturation of pores when the distribution

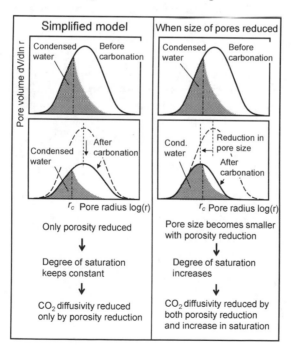

*Figure 4.13* Interdependency between characteristics of micro-pore structure, moisture state in the pores and progress of carbonation.

of pore structures does not change (Section 3.2 and Section 3.3). In other words, the ratio of liquid water, which is a barrier to carbon dioxide diffusion, is unchanged before and after carbonation, so the reduction in the diffusion coefficient after carbonation is expressed by only the reduction in porosity. On the other hand, it is known that in reality the distribution of micro-pore structures can be changed by carbonation. Referring to past studies in which the pore distribution after carbonation was measured by the mercury porosimetry method, there are many reports that fine pores increased, and pore radii overall became smaller (Saeki *et al.* 1991; Yoda and Yokomuro 1987).

In contrast to this, there are reports that pore structure becomes coarser by carbonation (Yoda 2002; Ohga and Nagataki 1988), but this trend is seen in cases where blast furnace slag or fly ash or other mixtures of materials are used. This is believed to be due to the different characters of the produced C–S–H gel. As a consequence of these overall study reports, in the present study, only ordinary Portland cement was used as a binder, and the authors set about a re-construction of the model assuming that after carbonation the pore radius distribution becomes finer.

If the relative humidity within the pores does not change before and after carbonation, the pore radius in which a vapor–liquid interface is created can be assumed to be constant, based on Kelvin–Laplace's equation. Thus, when the micro-pore distribution becomes finer, the proportion of pores in which condensed liquid water exists is increased due to the reduction in size of the pores, and as a result it is expected that the overall degree of saturation within the pores increases (Figure 4.13). In other words, even though the relative humidity remains the same, if the pore size distribution changes, the degree of saturation changes. If this hypothesis can be established, it means that the simplified model over-estimates the gaseous carbon dioxide diffusion after carbonation.

The above point is discussed in detail with reference to the carbon dioxide transport model. The flux of carbon dioxide $J_{CO_2}$ is expressed by the following Equation 4.33, taking both transport of gaseous and dissolved phases into account (Section 4.1.1.3).

$$J_{CO_2} = -\left(D_g \nabla \rho_g + D_d \nabla \rho_d\right)$$

$$D_g = \frac{\phi D_o^g}{\Omega} \frac{(1-S)^n}{1 - l_m / \{2(r_m - t_m)\}}$$

$$D_d = \frac{\phi D_o^d}{\Omega} S^n$$

(4.33)

If the pore size distribution becomes denser even though the porosity remains the same, the degree of saturation of the pores $S$ increases from the conditions of thermodynamic equilibrium. Therefore the diffusion coefficient of gaseous carbon dioxide $D_g$, which plays the leading role in carbon dioxide transport in accordance with Equation 4.33, is reduced. As a result, the total flux of carbon dioxide is reduced, which is considered to restrict the reaction rate.

The above mechanism was studied as a reason that the model overestimated the progress of carbonation in the long term, since the authors understand that it is very important to consider the coupling of geometric characteristics of pore structures, and the transport and reaction of moisture and carbon dioxide for improving the accuracy of the model. In other words, it was inferred that the progress of the carbonation reaction under various conditions was not governed by a single parameter, like the reaction rate coefficient, but that it was a problem in which various factors were mutually coupled in a non-linear manner. Thus, the model was re-constructed by returning again to the microscopic mechanisms, and combining them to predict the macroscopic phenomenon.

### 4.2.1.2 Changes in material characteristics due to carbonation and points for improvement in the model

To model the micro-pore structure after carbonation based on microscopic mechanisms, it is necessary to accurately determine the change in the characteristics of hydrates after carbonation reactions. Table 4.1 shows the results of measured porosity after carbonation in accelerated tests (Houst and Wittmann 1994). In the test, cement paste specimens were cured in water for six months, and then they were exposed to 85% carbon dioxide concentration. For comparison, the calculated values of the change in porosity are shown by assuming that all the calcium hydroxide was reacted and the same quantity of calcium carbonate was formed (Equation 4.14). In the calculation, the crystal volume of calcium carbonate is assumed to be approximately 11.7% more than that of calcium hydroxide. The mass of calcium hydroxide used in the calculation was obtained from the multi-component hydration model (Chapter 2). As shown in the Table, the actual measured reduction in porosity is significantly greater than the theoretical values for the reaction of calcium hydroxide only. In other words, looking at the change in porosity, it

Table 4.1 Change in porosity due to carbonation

| W/C (%) | Porosity (%) | | Change in porosity due to carbonation (%) | |
|---|---|---|---|---|
| | Non-carbonated part | Carbonated part | Measurement | Calculation |
| 40 | 24.1 | 13.4 | 10.7 | 1.7 |
| 50 | 32.3 | 23.0 | 9.3 | 1.6 |
| 80 | 48.0 | 42.3 | 5.7 | 1.2 |

is difficult to explain this actual change on the assumption that only calcium hydroxide contributes to the carbonation reaction.

Shirakawa *et al.* (2001) made cement paste specimens with different water-to-cement ratios, which were cured in water for specific periods (3, 7, 28 and 56 days). After a 28-day drying period, accelerated carbonation tests were carried out. Using thermal analysis, the quantity of calcium hydroxide just after curing and the quantity of calcium carbonate after the carbonation tests were measured. Although the molar ratio of calcium carbonate and calcium hydroxide varies slightly depending on the curing periods, as shown in Figure 4.14, regardless of the differences in water-to-cement ratio, a value greater than 2 was obtained. This result means that the actual generation of calcium carbonate was more than twice the amount indicated by the reaction in Equation 4.14, so it is inferred that substances other than calcium hydroxide are contributing to the reaction.

Figure 4.15 shows the measured values of carbon dioxide absorbed in accelerated carbonation tests (Kroone and Crook 1961) in which mortar specimens with a sand-to-cement ratio of 4.0, and a water-to-cement ratio of 0.64, were cast, removed from the mold after one day, and tested in a carbon dioxide concentration of 100% and relative humidity of 65%. The theoretical values of absorbed quantity of carbon dioxide for the case of only calcium hydroxide reacting are also shown. Normally, the quantity of calcium hydroxide generated in cement hydration reactions is about one-third of the cement mass (Kishitani 1963). If it is assumed that a quantity of carbon dioxide equivalent to the quantity of calcium hydroxide has all reacted, then the amount of carbon dioxide absorbed should be 20% of the mass of cement. In reality, however, the quantity of carbon dioxide absorbed by the hardened cement increased with exposure time, and exceeded 35% (Figure 4.15).

Taking the above into account, for enhancing the modeling of the carbonation, it is necessary to consider not only calcium hydroxide, but also other

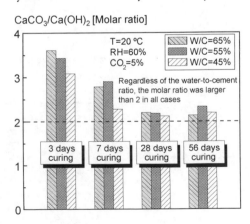

*Figure 4.14* Quantity of calcium carbonate generated.

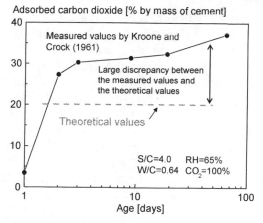

*Figure 4.15* Quantity of carbon dioxide absorbed.

hydrates like C–S–H gel. The authors understand that it becomes possible to construct a more generalized model based on microscopic mechanisms by considering the change of properties of the hydrates involved in this reaction.

## 4.2.2 Carbonation model including the C–S–H reactions

### 4.2.2.1 Equation for carbonation reaction rate at normal temperatures

Based on the discussion in the preceding section, the model (Equation 4.34) with the calcium hydroxide reaction was expanded to also include a C–S–H gel reaction. It was assumed that silica gel is generated by the following carbonation reaction.

$$(3CaO \cdot 2SiO_2 \cdot 3H_2O) + 3CO_2 \rightarrow 3CaCO_3 + 2SiO_2 \cdot xH_2O + (3-x)H_2O$$
$$(4.34)$$

Here $x$ is the equivalent mass of water contained in the silica gel. The density varies depending on the quantity of water contained, so the volume of the silica gel also varies according to the assumed quantity of water. Through the sensitivity analysis regarding the progress of carbonation and change in micro-pore structure due to the reaction, the value of $x$ was found to be 2.5, and the density of silica gel was taken to be 2,100 [kg/m³] (Sakai *et al.* 1999).

Here two points will be further explained. One concerns the generation of silica gel by the carbonation reaction of C–S–H gel, and the other concerns the chemical composition of silica gel. Funato *et al.* (1991) and Matsuzato *et al.* (1992) experimentally proved the generation of silica gel from C–S–H gel, using quantitative tests on carbonated test specimens. Also, Suzuki *et al.* (1989a) conducted a carbonation test by using synthesized C–S–H gel having a Ca/Si ratio of 1.15. They reported that in the initial stage of carbonation, the Ca/Si ratio reduced but there was no change in the structure. However, as a result of further carbonation, they obtained C–S–H gel with a Ca/Si ratio in the region 0.1 to 0.4 whose composition is close to that of silica gel. In other words, as a result of the reaction with carbon dioxide, calcium is eliminated from within the C–S–H gel, and the Ca/Si ratio gradually reduces, and ultimately it is thought that the C–S–H gel is separated into silica gel and calcium carbonate. This reaction process may vary according to the Ca/Si ratio of the C–S–H gel, the ambient temperature and humidity, and the concentration of carbon dioxide. At this moment, it is difficult to model the above process rigorously, so in the present study as the first preliminary approximation, the carbonation reaction of C–S–H gel was expressed at its simplest by Equation 4.34.

Other pioneering research that considered the C–S–H reaction system is the proposed model by Papadakis *et al.* (1991b). They hypothesized that C–S–H gel separated into calcium carbonate and silica gel (corresponding to $x = 3$ in Equation 4.34, which means no water produced). However, Shima *et al.* (1989) confirmed the phenomenon of the elimination of water from cement hydrates due to carbonation, from the relationship between the quantity of carbon dioxide obtained by elemental analysis of carbonated test specimens and the increase in mass after carbonation. It is difficult to rigorously determine the quantity of water eliminated during a reaction, so in the present study the parameter was determined from the model by means of sensitivity analysis. Specifically, $x$ was set to 2, 2.5 and 3, and it was determined that $x = 2.5$ gave the most appropriate analysis, in terms of porosity after carbonation, the quantity of produced calcium carbonate, and the change in the mass of the test specimens, etc.

Next, the reaction rate of C–S–H gel and carbon dioxide is discussed. Here, the following two assumptions are made. The first is that the reaction rate would be proportional to the quantity of C–S–H gel and the concentration of gaseous carbon dioxide. The second assumption is that the reaction rate would gradually reduce due to the calcium carbonate and silica gel generated by the reaction accumulating on the surface of the C–S–H gel. Accordingly, the following model was proposed.

$$R_{CSH} = k_{CSH} \cdot \exp(\gamma \cdot R) \cdot [CSH] \cdot [CO_2]$$
$$R = \frac{W_{carb-CSH}}{W_{all-CSH}} \tag{4.35}$$

where $R_{CSH}$ is the reaction rate of C–S–H gel [m³/mol.sec], $k_{CSH}$ is the reaction rate coefficient [m³/mol.sec], [CSH] is the molar density of C–S–H gel remaining in the hardened cement paste matrix [mol/m³], $R$ is the degree of carbonation of C–S–H gel, $W_{carb-CSH}$ is the mass of carbonated C–S–H gel [kg/m³], $W_{all-CSH}$ is the mass of all C–S–H gel [kg/m³], and $\gamma$ is the constant (= 20.7). The quantity of C–S–H is obtained from the total mass of hydration products $W_{hyd}$ [kg/m³], which is computed by the multi-component hydration model as

$$W_{hyd} = \alpha \cdot W_p + W_{chem} \tag{4.36}$$

where $\alpha$ is the degree of hydration, $W_{chem}$ is the quantity of bound water [kg/m³], and the quantity of binder $W_p$ [kg/m³] (Section 3.1 and Section 3.4). The mass of C–S–H gel is obtained by subtracting the mass of calcium hydroxide from the total mass of hydration products $W_{hyd}$. In the hydration model, the mass of the double salts such as ettringite and monosulfate can be calculated at an arbitrary stage in the reaction. Therefore, the most rigorous approach is to model the carbonation reaction process of each double salt as well as that of C–S–H gel, and to describe the overall phenomenon as the sum of these processes. However, in this book, as the first approximation and for simplicity, it was decided to classify all hydrates apart from calcium hydroxide as C–S–H. Therefore, in the model of C–S–H described in this sub-section, the carbonate reactions of ettringite and monosulfate, etc., are included in an averaged manner. The reaction rate coefficient in Equation 4.35 was determined to be $k_{CSH} = 1.0 \times 10^{-2}$ [m³/mol.sec] by carrying out systematic sensitivity analysis.

### 4.2.2.2 Calculation of change in mass and volume of hardened cement paste matrix due to carbonation

From the equation for the reaction rate of calcium hydroxide (Equation 4.32) and the equation for the reaction rate of C–S–H gel (Equation 4.35), the consumption and generation of the following equivalent masses can be obtained: calcium hydroxide and C–S–H gel consumed, calcium carbonate, silica gel and water generated. At an arbitrary stage during the carbonation, the mass of each of these can be calculated in accordance with the following Equation 4.37.

$$W_{CH}^n = W_{CH}^{n-1} + M_{CH} \cdot \left( C_{CH-hyd} - C_{CH-CA} \right)$$

$$W_{CSH}^n = W_{CSH}^{n-1} + M_{CSH} \cdot \left( C_{CSH-hyd} - C_{CSH-CA} \right)$$

$$W_{SH}^n = W_{SH}^{n-1} + M_{SH} \cdot C_{SH-CA} \tag{4.37}$$

$$W_{CA}^n = W_{CA}^{n-1} + M_{CA} \cdot C_{CA-CA}$$

where $W_i^n$, $W_i^{n-1}$ is the mass of substance $i$ in step $n$ and step $n$-1 [kg], $M_i$ is the molar mass [kg/mol], $C_i$ is the molar equivalent [mol]. The subscripts *CH*, *CSH*, *SH* and *CA* indicate calcium hydroxide, C–S–H gel, silica gel and calcium carbonate, respectively. *CH-hyd* and *CSH-hyd* indicate calcium hydroxide and C–S–H gel generated by hydration. Also, *CH-CA*, *CSH-CA*, *SH-CA* and *CA-CA* indicate calcium hydroxide, C–S–H gel, silica gel and calcium carbonate generated or consumed by carbonation. For example, as shown in the first equation of 4.37, the mass of calcium hydroxide in step $n$ is obtained by adding the difference between the mass of calcium hydroxide generated by hydration ($CH$-$hyd$) and the mass of calcium hydroxide consumed by carbonation ($CH$-$CA$) to the mass in the previous step $n$-1. When there is no further hydration reaction in the system, the mass of calcium hydroxide in step $n$ is simply obtained by subtracting the mass of consumed calcium hydroxide during a step from the mass in the previous step, i.e., $W_{CH}^{n} = W_{CH}^{n-1} - M_{CH} \cdot C_{CH-CA}$.

From Equation 4.14 and Equation 4.34, the molar equivalent of calcium carbonate and silica gel generated in each reaction are expressed by

$$\begin{cases} C_{CA-CA} = 3 \cdot C_{CSH-CA} + C_{CH-CA} \\ C_{SH-CA} = C_{CSH-CA} \end{cases} \tag{4.38}$$

It was assumed that the densities, which are necessary for calculating volumes, were as follows: silica gel 2,100 [kg/m³], calcium hydroxide 2,240 [kg/m³], C–S–H gel 2,440 [kg/m³], and calcium carbonate 2,720 [kg/m³] (Taylor 1997).

The change in volume of the hydrated products due to carbonation was obtained as the volumetric difference of the sum of calcium carbonate and silica gel generated minus the sum of calcium hydroxide and C–S–H gel consumed. Figure 4.16 shows the concept of the change in porosity in the proposed model. As shown in the Figure, besides the 11.7% increase in volume due to the reaction of calcium hydroxide, the proposed model takes into account the increase in volume of C–S–H gel due to the carbonation reaction. This increase is obtained as 38%, based on the assumed reaction Equation 4.38 and the density and molecular mass of each substance.

By inputting the calculated mass and volume of each substance after the reaction into a micro-pore structure formation model (Section 3.1), the porosity and pore distribution at an arbitrary stage can be obtained automatically. When we simulate the early-age development process, the model needs the mass and volume of calcium hydroxide and C–S–H gel generated

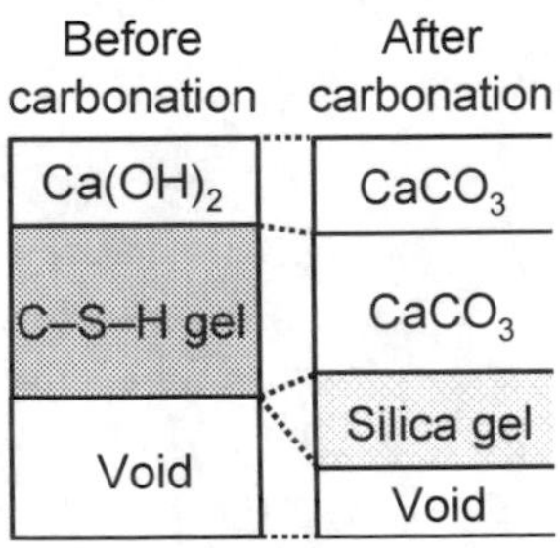

*Figure 4.16* Concept of change in quantity of porosity in the proposed model.

by hydration so that it can calculate the micro-pore structure formation at an arbitrary stage of early-age development. In this book, in addition to the progress of hydration, the mass and volume of new reaction products generated by carbonation are also taken into account in the microstructure development model. This is the big difference from the simplified model (Section 4.1.2.3) that empirically expresses the reduction in porosity corresponding to the quantity of calcium hydroxide consumed in the reaction.

### 4.2.2.3 Change in the specific surface area of C–S–H gel due to carbonation

From the model described in the preceding section, it is possible to deal with the change in porosity due to carbonation with a generalized methodology. On the other hand, the phenomenon of reduction in size of micro-pore distribution is well known. In the proposed model, the specific surface area of the gel increases due to carbonation, and it is assumed that as a result the micro-pore structure distribution is changed. Table 4.2 shows the results of measuring the specific surface area due to the carbonation reaction under conditions of differing relative humidities (Kim *et al.* 1995). The generation product of calcium hydroxide after carbonation is calcium carbonate, whereas the generation product of C–S–H gel after carbonation is inferred to be a mixture of a C–S–H gel with a lower Ca/Si molar ratio, silica gel and calcium carbonate. Compared with calcium hydroxide, the specific surface area of calcium silicate hydrates (mainly C–S–H gel) in the original test specimens is large. In addition, the proportional increase due to carbonation is large, and the specific surface area is larger by a factor of about two than it was prior to carbonation. In other words, as a result of the carbonation reaction, a finer gel is generated compared with the gel prior to carbonation. Therefore, it is considered that from the change in C–S–H gel density and specific surface area due to carbonation, the quantity of micro-pores in hardened cement is reduced overall, and micro-pores with large diameter are reduced, and micro-pores with small diameter are increased.

In the current pore structure development model, the specific surface area of the gel is taken to be 30 m²/g. Referring to the experimental results shown in Table 4.2, the relative change in the specific surface area of the gel due to carbonation is assumed to be a factor of two. Thus, in the proposed model,

*Table 4.2* Change in specific surface area due to carbonation (m²/g)

| | *Before carbonation* | After carbonation | | |
|---|---|---|---|---|
| *Test specimen* | | *RH 30%* | *RH 60%* | *RH 80%* |
| Ca(OH)₂ | 3.2 | 3.3 | 4.7 | 4.2 |
| C–S–H gel (Ca/Si = 1.23) | 6.0 | 8.2 | 10.9 | 18.9 |
| C–S–H gel (Ca/Si = 0.4) | 76.8 | 141.6 | 127.0 | 120.4 |

the specific surface area of the carbonated C–S–H gel is roughly assumed to be 60 m²/g, as a first primary approximation.

### 4.2.2.4 *Verification of the proposed model at constant room temperature*

#### 4.2.2.4.1 CHANGE IN MICRO-PORE STRUCTURE DUE TO CARBONATION

Figure 4.17 shows the analytical results for pore size distribution before and after carbonation in the case of a water-to-cement ratio of 60%. The simplified model (Section 4.1) just simulates the reduction of change in porosity without changing the pore-size distribution. In contrast, it can be seen that the current model calculates not only the reduction in porosity, but also the change in pore-size distribution. It is necessary to wait for further detailed research to determine whether the pore-size distribution calculated by the model is valid or not. However, the focus and interest of this chapter are not whether this distribution itself can be predicted with high accuracy. Rather, it is the improvement in the accuracy of the prediction of the internal moisture state and $CO_2$ diffusion, and the mutual dependency of each phenomenon in the carbonation reaction process. From the fact that various properties associated with carbonation can be rationally predicted as described later, it is inferred, although indirectly, that the pore-size distribution calculated by the model is generally appropriate.

Verification of the change in porosity due to carbonation was carried out using the experimental results of Houst and Wittmann (1994) and Ngala and Page (1997). The vertical axis of Figure 4.18 shows the ratio of the porosity after carbonation to that before carbonation. The size of the test specimens was approximately 1–3 mm, but the paper does not state when the measurements were made, so several assumptions were made for the analytical conditions. In other words, from the surface towards the depth direction, a

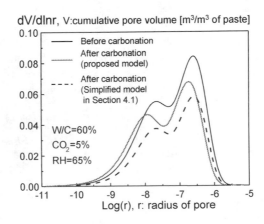

*Figure 4.17* Change in computed pore-size distribution due to carbonation.

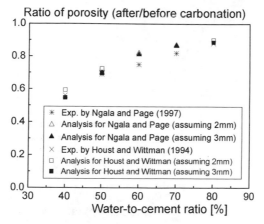

*Figure 4.18* Change in porosity due to carbonation.

different porosity is obtained as a calculated value at each position, so it is necessary to average all the values for comparison with the measured values. Here, the average values of porosity at the time when the depth of carbonation reached 2 mm or 3 mm from the surface are shown as the analytical results. As can be seen in the Figure, the measurements and the values calculated by the model display the same trends for different water-to-cement ratios (40% to 80%), and different carbon dioxide concentrations (5%, 85%). From this it is concluded that the newly proposed carbonation model is capable of reproducing well the actual changes in porosity.

4.2.2.4.2 VERIFICATION OF THE QUANTITY OF CALCIUM CARBONATE GENERATED BY CARBONATION

Figure 4.19 shows the results of a comparison of measured and computed values for the ratio of calcium carbonate and calcium hydroxide. Here, the measured values of Shirakawa *et al.* (2001) for 28 days curing in water are used. As shown in the Figure, the current model predicts well the quantity of calcium carbonate generated by carbonation for different water-to-cement ratios.

4.2.2.4.3 PROGRESS OF CARBONATION UNDER ACCELERATED AND NATURAL ENVIRONMENTS

Figure 4.20 and Figure 4.21 show a comparison of predictions and measurements of carbonation depth for exposure to high-concentration acceleration environments (John *et al.* 1990) and long-term natural environments (Yoda 2002), respectively. Under the natural environments the precise relative humidity is unclear, so in the analysis the average environment for Japan was assumed: a temperature of 20°C, and constant values of relative humidity

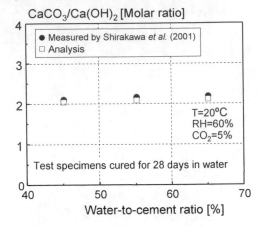

Figure 4.19  Quantity of calcium carbonate generated by carbonation.

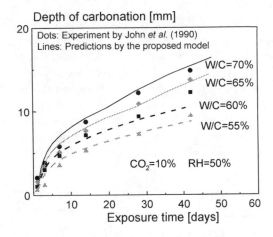

Figure 4.20  Progress of carbonation reaction in accelerated environment.

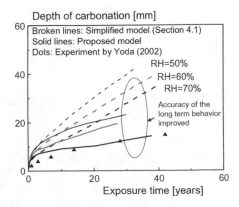

Figure 4.21  Progress of carbonation reaction in low concentration $CO_2$ environment.

of 50%, 60% and 70%. As shown in the Figures, there was no discrepancy between the measured values and the values predicted by the proposed model, in which the finer micro-pore structure and reduced porosity due to carbonation is appropriately expressed. In particular, the accuracy of the long-term behavior under natural environments was improved by expressing the increase in degree of saturation due to the smaller pore-size distribution after carbonation, and the resulting reduction in the diffusivity of gaseous carbon dioxide.

### 4.2.3 Modeling of carbonation reaction at arbitrary temperature conditions

Diffusion of carbon dioxide and the carbonate reaction in concrete is strongly dependent on the temperature. Generally it is known that the higher the temperature, the more carbonation is promoted (Kishitani 1963; John *et al.* 1990). On the other hand, the higher temperature accelerates the evaporation of moisture in concrete. Also, the solubility of calcium hydroxide and carbon dioxide reduce significantly as the temperature increases, which is known to have the effect of reducing the rate of the carbonation reaction. According to the research of Uomoto and Takada (1993), the rate of carbonation reaction is the highest in the range 30°–40°C, and tends to reduce in temperature environments higher than this. The carbonation rate is a function of many factors, such as the diffusivity and equilibrium of carbon dioxide, the solubility of calcium hydroxide, and the moisture state in pore structure, so at a glance it appears a very complex phenomenon. In this section, an investigation is carried out to make the model applicable to conditions of variable temperature. From an engineering viewpoint also, actual structures are subject to various temperature time histories, so it is an important task for the reaction model to take into consideration the temperature dependence.

#### 4.2.3.1 Temperature-dependent parameters for the carbonation reaction

The following relationship is proposed for the temperature-dependent diffusivities of carbon dioxide in water and in air, based on values in the literature (Lide 2004) (Figure 4.22 and Figure 4.23).

$$\ln\left(D^d / D_0^d\right) = -2.35 \times 10^3 \cdot \left(1/T - 1/T_0\right) - 3.00 \times 10^{-4} \tag{4.39}$$

$$\ln\left(D^g / D_0^g\right) = -6.04 \times 10^2 \cdot \left(1/T - 1/T_0\right) + 9.90 \times 10^{-3} \tag{4.40}$$

where $D^d$, $D_0^d$ represent the diffusivity of dissolved carbon dioxide at temperature $T$ [K], $T_0$ (= 293.15 [K]) [m²/s], and $D^g$, $D_0^g$ represent the diffusivity of gaseous carbon dioxide at temperature $T$ [K], $T_0$ (= 293.15 [K]) [m²/s].

The following equations are proposed for Henry's coefficient $H_{CO2}$, which describes the equilibrium relationship between the gaseous and liquid phases,

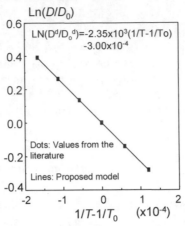

$$\text{Ln}(D^d/D_o^d) = -2.35 \times 10^3 (1/T - 1/To) - 3.00 \times 10^{-4}$$

*Figure 4.22* Relationship between diffusion coefficient of dissolved carbon dioxide and temperature.

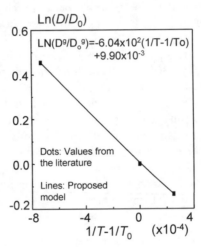

$$\text{Ln}(D^g/D_o^g) = -6.04 \times 10^2 (1/T - 1/To) + 9.90 \times 10^{-3}$$

*Figure 4.23* Relationship between diffusion coefficient of gaseous carbon dioxide and temperature.

the solubility product of the calcium hydroxide $K_{sp}$, and the equilibrium constant of concentration for the dissociation of water $K_w$, based on values in the literature (National Institute of Natural Sciences, 2002, Figures 4.24–4.26).

$$H_{CO_2} = -1.52 \times 10^2 \cdot T^3 + 1.65 \times 10^5 \cdot T^2 - 5.34 \times 10^7 \cdot T + 5.44 \times 10^9 \quad (4.41)$$

$$K_{sp} = 3.52 \times 10^{-2} \cdot \exp(-2.27 \times 10^{-2} \cdot T) \quad (4.42)$$

$$K_W = 5.88 \times 10^{-19} \cdot T^3 - 4.97 \times 10^{-16} \cdot T^2 + 1.40 \times 10^{-13} \cdot T - 1.32 \times 10^{-11}$$
$$(4.43)$$

Henry coefficient of
carbon dioxide [Pa/mol fraction]

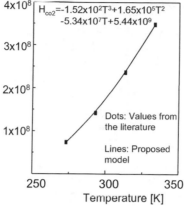

Figure 4.24   Relationship between Henry's coefficient and temperature.

Solubility-product of the
calcium hydroxide $K_{sp}$ [mmol³/l³]

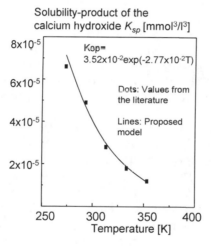

Figure 4.25   Relationship between solubility product constant of calcium hydroxide and temperature.

### 4.2.3.2  Carbonation reaction rate coefficients

As stated previously, the proposed model takes into consideration the carbonation reaction of the two substances, calcium hydroxide and C–S–H gel. Therefore, it is necessary to consider separately the effect of temperature on calcium hydroxide and on C–S–H gel. At the present stage, however, the reaction rates of these two substances in concrete at different temperatures are not well known.

The sensitivity analysis in the previous section focused on phenomena in a constant room temperature environment and, through such analysis, the rates of carbonation reaction were determined. These rates gave generally

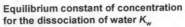

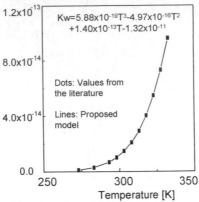

*Figure 4.26* Relationship between concentration equilibrium constant for water dissociation and temperature.

valid analyses that satisfied several measurement values, such as quantities of calcium hydroxide and calcium carbonate before and after carbonation, carbonation depth and change in porosity. In this section, reaction rate coefficients are determined by the same methodology. Here, it is assumed that the carbonation reaction rate coefficients for calcium hydroxide and C–S–H gel under various temperatures increase or decrease by the same ratio with regard to their respective referential temperatures. Using available experimental data by John *et al.* (1990) for accelerated carbonation carried out under a constant relative humidity (50%), and combinations of different temperatures (15°C, 22.5°C, 30°C and 37.5°C), as well as different water-to-cement ratios (55%, 60%, 65% and 70%), a temperature-dependent model for the carbonation reaction was constructed (Figure 4.27).

The analytical input – including mix proportion, curing and environmental conditions – was given according to the corresponding experimental conditions. Microphysical information such as temperature, degree of hydration, porosity and moisture distribution is automatically calculated as internal variables within the analysis system. Here, only the carbonation rate coefficients, which were the unknown parameters, were manually changed, to determine values that appropriately predicted the macroscopic carbonation depth. Through such back analysis, the following Equation 4.44 is proposed which gives the reaction rate coefficient in an arbitrary temperature environment as a ratio with respect to 20°C. It has to be noted here that, as a result, the proposed model is mathematically expressed in accordance with Arrhenius's law of chemical reaction (Figure 4.28).

$$\ln K_{ratio} = -1.51 \times 10^4 \cdot \frac{1}{T} + 5.14 \times 10^1 \tag{4.44}$$

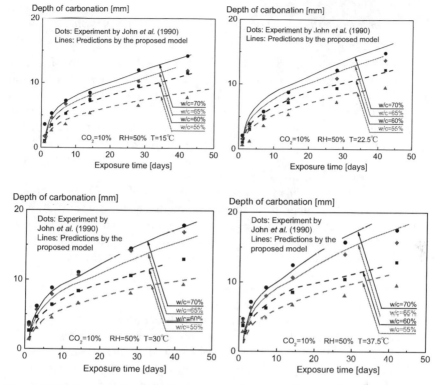

*Figure 4.27* Progress of carbonation under different temperature conditions.

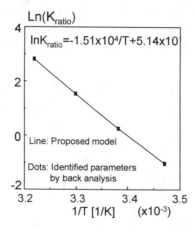

*Figure 4.28* Ratio of reaction rate coefficient at arbitrary temperatures.

where $K_{ratio}$ is the ratio of reaction rate coefficient at temperature $T$ [K] to that at referential temperature $T_0$ (= 293.15 [K]). As stated previously, Equation 4.44 is applied to both calcium hydroxide and to C–S–H gel.

### 4.2.3.3 *The carbonation phenomenon under various temperature conditions*

Figure 4.29 shows the analytical results for postulated carbonation acceleration tests with a water-to-cement ratio of 60%, relative humidity of 50%, and carbon dioxide concentration of 10%, at 10°C temperature intervals over the range 10°–50°C. As shown in the Figure, progress of the carbonation reaction was fastest at 30°C, and tended to become slower the further the temperature was from 30°C. It is considered that the temperature dependence of the carbonation reaction, which is known from previous research to peak at around 30°C, can be quantitatively predicted.

Generally, the mechanism by which the progress of carbonation shows a peak at a particular temperature is explained by the relationships of solubility and reaction rates to temperature canceling each other out. In other words, the solubility of carbon dioxide and calcium hydroxide is reduced by an increase in temperature, so the quantity available for reaction is reduced. On the other hand, as described by Arrhenius's law, the reaction rate coefficient itself becomes larger as the temperature rises. As a result of these two factors, at around 30°C the reaction rate is maximized. In the analysis using the proposed model, these factors are incorporated, so different rates of progress of carbonation were predicted according to the temperature conditions.

In this section, the carbonation reaction of C–S–H gel as well as that of calcium hydroxide was modeled for an arbitrary temperature environment based on the microscopic mechanisms. The change in mass and volume of hydration products due to carbonation was directly applied to a micro-pore structure formation model, and the change in the microscopic structure was reasonably predicted. Furthermore, by appropriately taking into account the interactions among temperature, moisture state in pores, diffusion of carbon

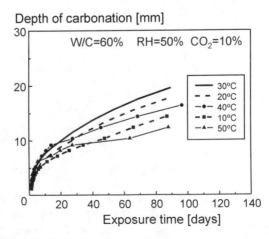

*Figure 4.29* Effect of temperature on carbonation progress.

dioxide and carbonation reaction, it is possible to make good predictions of the progress of the carbonation under an arbitrary temperature environment and an arbitrary concentration of carbon dioxide.

# 5   Calcium ion transport and leaching

The applicability of concrete as a solidifying barrier for the geological disposal of radioactive waste is being studied. Radioactive waste contains materials with long half-lives, so the period of stability required in design is several tens of thousands of years. This far exceeds the expected lifetime of the regular infrastructure. Calcium leaching over such an extremely long service life is one issue of crucial importance to be considered in the design and planning of such projects. The prediction of such phenomena over very long periods is difficult through experiments or accelerated tests alone. Experimental methods need to be combined with numerical analysis for more rational evaluations of performance, since methods of assessing barriers on a real-time scale are not appropriate. In feasibility studies as well as actual designs, numerical simulations that take a rational approach to account for mix proportion, types of constituent materials, structural dimensions and detailing are particularly useful. In other words, a theoretical scheme plays an essential role in linking accelerated tests with reality.

Degradation due to calcium leaching has been reported since the 1920s. Despite this, there have been few cases in which ordinary structures suffered from calcium leaching damage because the rate of degradation is relatively slow compared with carbonation, chemical attack and corrosive reactions. In a few cases, degradation caused by leaching has been observed in hydro-power installations and facilities related to water supply after long-term service of approximately 100 years. From a practical viewpoint, safety verifications related to calcium leaching from cement hydrates are unnecessary for cementitious composites in normal use. Consequently, the recent need to investigate long-term degradation caused by calcium leaching can be seen as a new challenge on the road to developing artificial safety barriers for nuclear waste depositories.

Cemented soils are new kinds of engineering materials that can be used to improve the ground, make use of surplus soil, and enhance soil-structure performance. Examples of such materials are cemented soil and gravel (CSG) (Hirose *et al.* 2001) and geosynthetic-reinforced cement treated backfills (Watanabe and Tateyama 2002). The cemented soils generally have high water-to-cement ratios and contain large voids, and are thought to deteriorate more quickly than ordinary concrete due to calcium leaching from cement hydrates. In Japan, cemented soils must be tested for hexavalent chromium leaching before they can be used, despite their low cement content per unit volume. Research to date has focused primarily on increasing the strength of cemented soils, and there has been little investigation of their long-term durability. Particularly needed is a study of the properties of cemented soils that have voids much larger than the pores in cement paste.

In this chapter, a new model of calcium leaching is proposed. The key issue is the analysis of a transient process in which the microstructure of cement hydrates varies in time and space. Calcium leaching generally results in a coarser micro-pore structure that allows easier mass transport. At the same time, the newly created voids may form new spaces in which additional cement hydration can take place. Calcium leaching and the evolving microstructure are strongly coupled with each other, and both need to be simulated in order to realize performance assessments of cement composites over extremely long periods of time. By integrating a calcium ion transport model with a statistical microstructural model of chemo-physics (Chapter 3), the aim is to achieve wide applicability to cover an arbitrary choice of mix proportion, constituent material characteristics and ambient conditions. Systematic experimental verification and validation of the proposed model are also reported. The authors expect that this study will lay a foundation for future heavy metal dissolution analysis from cemented materials, too.

## 5.1 Modeling of calcium leaching from cement hydrates

### 5.1.1 Scheme of thermodynamics

Calcium leaching from cement hydrates correlates strongly with the processes of cement hydration, microstructure formation and moisture distribution (Figure 5.1 and Figure 5.2). As these chemo-physical processes are highly time-dependent, the consequent microscopic and macroscopic properties of cement composites can be treated as variables. In this chapter, a model of calcium leaching is overlaid over this analytical scheme (Chapter 1, Figure 5.1). This enables the strong interaction between calcium leaching and the micro–macro solid features of concrete to be taken into account consistently (Figure 5.2).

### 5.1.2 Mass conservation of calcium

Momentum, energy and material flows must satisfy the conservation laws. As the governing equation for calcium, we have mass conservation Equation 5.1 as given below. This relates the total mass of calcium in the pore solution and the solid phase calcium in the system. It was originally formulated by Gérard *et al.* (2002) as

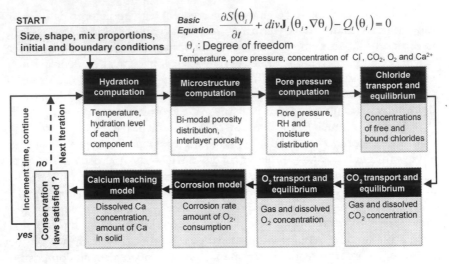

*Figure 5.1*   Overall scheme of *DuCOM* chemo-physical coupled system.

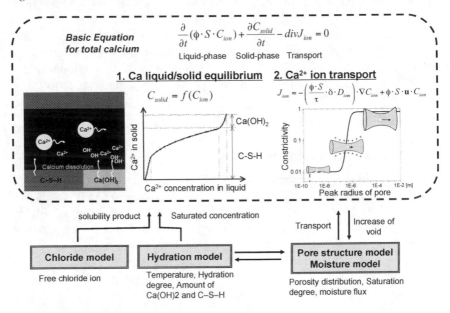

*Figure 5.2*   Calcium leaching model coupled with time-dependent material properties.

$$\frac{\partial}{\partial t}(\phi \cdot S \cdot C_{ion}) + \frac{\partial C_{solid}}{\partial t} - divJ_{ion} = 0 \tag{5.1}$$

where $\phi$ is porosity [m³/m³], $S$ is the degree of saturation, $C_{ion}$ is the molar concentration of calcium ions in the liquid phase [mmol/m³], $C_{solid}$ is the amount of calcium in the solid phase [mmol/m³], and $J_{ion}$ is the flux of calcium ions [mmol/m².s].

The first and second terms of Equation 5.1 represent the increments in the amount of calcium in the liquid and solid phases per unit of time and volume, respectively. The relation between calcium ion concentrations in the liquid and solid phases is given by the solid–liquid equilibrium model, as described later. Phase transformation (i.e., dissolution from solid to liquid phases or precipitation from liquid to solid phases) is implicitly expressed by assuming a quasi-equilibrated process. The third term of the equation represents the diverging mass flux. The calcium ion flux at a specific boundary surface is expressed by

$$q_S^{Ca} = -h_{Ca}\left(C_{ion} - C_{ion\_S}\right) \tag{5.2}$$

where $q_S{}^{Ca}$ represents the flux of calcium ions into the porous medium at the surface [mmol/m².s], $h_{Ca}$ is the surface calcium ion emissivity coefficient [m/s], and $C_{ion\_S}$ is the environmental concentration of calcium ions [mmol/l]. The value of the coefficient is assumed to be $1.0 \times 10^{-3}$ [m/s], the same as the surface chloride ion emissivity coefficient in the chloride model (Ishida and Maekawa 2000).

### 5.1.3 Solid–liquid equilibrium

As for the thermodynamic equilibrium of calcium in the solid and liquid phases, an isotherm correlation proposed by Buil *et al.* (1992) is installed with minor modifications for computational efficiency, as shown in Figure 5.3. As the original Buil model has a simple mathematical formulation with reasonable accuracy, it has been widely used by other researchers (e.g., Yokozeki *et al.* 2003). In this chapter, the original Buil model was slightly modified so as to uniquely determine the quantity of solid calcium from a given ion concentration in the pore solution. In this modification, the mathematical formulation adopted by Gérard *et al.* (2002) was referred to and the consistency of the model with the experimental data earned by Berner (1988) was maintained as

$$C_{Solid} = f(C_{ion}) = A \cdot \left\{C_{CSH} \cdot \left(C_{ion}/C_{satu}\right)^{1/3}\right\} + B$$

$$A = \begin{cases} -\dfrac{2}{x_1^3} C_{ion}{}^3 + \dfrac{3}{x_1^2} C_{ion}{}^2 & (0.0 \le C_{ion} \le x_1) \\ 1 & (x_1 < C_{ion}) \end{cases} \quad \cdots\cdots$$

$$B = \begin{cases} 0 & (0.0 \le C_{ion} \le x_2) \\ \dfrac{C_{CH}}{(C_{satu} - x_2)^3} \cdot (C_{ion} - x_2)^3 & (x_2 < C_{ion}) \end{cases} \quad \cdots\cdots \tag{5.3}$$

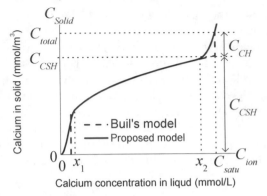

*Figure 5.3* Solid–liquid equilibrium model for calcium.

where $C_{CSH}$ is the amount of calcium in the solid phase of the C–S–H gel [mmol/m³], $C_{CH}$ is the amount of calcium in the solid phase of the calcium hydroxide [mmol/m³], $C_{satu}$ is the saturated liquid phase calcium ion concentration [mmol/l], $x_1$ is the liquid phase calcium ion concentration when the rapid transition of C–S–H gel into silica gel begins [mmol/l], and $x_2$ is the liquid phase calcium ion concentration when the calcium hydroxide has completely dissolved and the dissolution of C–S–H gel begins [mmol/l].

In this research, $x_1 = 3.0$ mmol/l and $x_2 = (C_{satu} - 0.7)$ mmol/l are adopted. Since calcium hydroxide has the highest solubility among the hydrates in the leaching process, it dissolves first. During dissolution, the calcium ion concentration of the solution gradually decreases to $x_2$ from $C_{satu}$. After the calcium hydroxide is lost, dissolution of C–S–H gel starts. As the dissolution of the C–S–H gel proceeds and the liquid phase calcium ion concentration falls below $x_1$, the C–S–H gel decomposes rapidly into silica gel. Here, the presence of alkali salts such as sodium and potassium is ignored. But, the presence of sodium chloride is implicitly taken into account through the change in solubility product, as described later, since it is known that sodium chloride accelerates calcium leaching (Imoto *et al.* 2004).

The parameters in the phase equilibrium (i.e., the amount of calcium and the saturated concentration of calcium ions) are defined not through experiments but by computation. All are calculated as time-dependent variables in order to take into account the influence of mix proportion, ambient conditions, hydration and degradation due to calcium leaching.

The total amount of calcium in the solid phase of the cementitious composite is determined from the chemical composition of the mixture. The amount of calcium in the fly ash and blast furnace slag is obtained from the CaO content, which is assumed to be present at a ratio of 4.4% (Japan Society of Civil Engineers 1988a) and 41.8% (Japan Society of Civil Engineers 1988b), respectively. The amount of calcium hydroxide in the cement paste is determined stoichiometrically using the following equations in the multi-component hydration model (Maekawa *et al.* 1999):

$$2C_3S + 6H \rightarrow C_3S_2H_3 + 3CH$$
$$2C_2S + 4H \rightarrow C_3S_2H_3 + CH \tag{5.4}$$
$$C_4AF + 2CH + 10H \rightarrow C_3AH_6 + C_3FH_6$$

When using fly ash and blast furnace slag as admixtures, calcium hydroxide is consumed in the reaction. Consumption is assumed to be 100% and 22% in the case of fly ash and blast furnace slag, respectively (Maekawa *et al.* 1999).

The concentration of saturated calcium ions in the liquid phase equates to the concentration of saturated calcium hydroxide in the pore solution. This is a simplified treatment disregarding the presence of coexisting ions such as sodium and potassium. By assuming that no other ions coexist with the calcium ions derived from the calcium hydroxide, the electroneutrality rule gives the concentration of saturated calcium ions from the solubility product as

$$C_{satu} = \frac{1}{2}\sqrt[3]{2K_{sp}} \tag{5.5}$$

where $C_{satu}$ is the concentration of saturated calcium ions [mmol/l], and $K_{sp}$ is the solubility product of calcium hydroxide [mmol³/l³]. In order to account for the effects of temperature and sodium chloride concentration, the solubility product is corrected by adding the contribution of sodium chloride to the change in solubility product, $\Delta K_{sp}{}^{Cl}$, to the solubility product including the effect of temperature, $K_{sp}{}^T$, as given by the following equation:

$$K_{sp} = K_{sp}^T + \Delta K_{sp}^{Cl} \tag{5.6}$$

where $K_{sp}{}^T$ is the solubility product of calcium hydroxide considering the effect of temperature [mmol³/l³], and $\Delta K_{sp}{}^{Cl}$ is a term for the effect of sodium chloride on the solubility product of calcium hydroxide [mmol³/l³]. For the temperature effect, the following approximation (Figure 5.4) is derived from the literature (CSJ 2004):

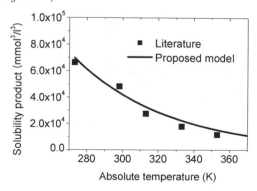

*Figure 5.4* Influence of temperature on solubility product.

$$K_{sp}^T = 0.0125 \times 10^9 \cdot e^{-0.019T} \tag{5.7}$$

where $T$ is absolute temperature [K]. In fact, far fewer data are available for deep discussion on the effect of sodium chloride. Then, in this study, the concentration of saturated calcium was measured for different sodium chloride solutions, and the term $\Delta K_{sp}^{Cl}$ was experimentally identified. Here, the authors tentatively propose a simple model having explicit formulae on the direct basis of the experiment as shown in Figure 5.5. It should be noted that there is still a possibility of different dissolution of calcium oxide in this experiment from that of calcium oxide in the real cement hydrate. In order to further discuss the ion interaction as the first stage, simple modeling is of great importance to conduct the system simulation of calcium leaching from cement composites and compare the derived phenomena with the reality in multiple views. Then, we have

$$\Delta K_{sp}^{Cl} = A' \cdot \left(5 \times 10^4 \cdot \sqrt[3]{Cl}\right)$$
$$A' = \begin{cases} -2Cl^3 + 3Cl^2 & (0.0 \le Cl \le 1.0) \\ 1 & (1.0 < Cl) \end{cases} \tag{5.8}$$

where $Cl$ is the concentration of sodium chloride in the pore solution [mass %]. The concentration of free chloride ions determined from the chloride ion transport model (Ishida and Maekawa 2000; Maekawa *et al.* 2003a) is taken to be the concentration of sodium chloride in the analysis system. Concerning the computational phase-equilibrium formulae, the sensitivity with respect to the perturbation of the chloride ion is intentionally set small to ensure computational stability of the highly non-linear range of the solubility product at the low chloride ion concentration as well as the chemo-physical accuracy.

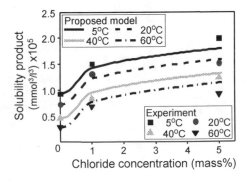

*Figure 5.5* Influence of chloride on solubility product.

### 5.1.4 Transport of calcium ion

By considering both diffusion and advection (Ishida and Maekawa 2000), the flux of calcium ions transported in a porous medium is written as (see Figure 5.2)

$$J_{ion} = -\left(\frac{\phi \cdot S}{\tau} \cdot \delta \cdot D_{ion}\right) \cdot \nabla C_{ion} + \phi \cdot S \cdot \mathbf{u} \cdot C_{ion} \tag{5.9}$$

where $\tau$ is tortuosity, $\delta$ is constrictivity, $D_{ion}$ is the diffusion coefficient of a calcium ion [m²/s], $\nabla^T = [\partial/\partial x \ \partial/\partial y \ \partial/\partial z]$ is the nabla operator, and $\mathbf{u}^T = [\ u^x \ u^y \ u^z\ ]$ is the velocity vector of a calcium ion transported by solution flow [m/s]. By treating the capillary and gel pores in the cement paste as ion transport pathways, the porosity denoted by $\phi$ is expressed by the following equation (Ishida and Maekawa 2000).

$$\phi = \phi_{cp} + \phi_{gl} \tag{5.10}$$

where $\phi_{cp}$ and $\phi_{gl}$ are the porosities of capillary and gel pores, respectively. Here, it is assumed that no ion is transported into the interlayer pores of the cement paste since the molecular-related size of the interlayer space is too small to allow substantial room of any ion.

The first term of Equation 5.9 represents the component of molecular diffusion, while the second is a component representing advection driven by the flow of condensed liquid water in the pore. The velocity of calcium ions related to advection is assumed to reflect the bulk motion of the condensed water, and it is determined from the transport model for the liquid phase (Chapter 3). The diffusion coefficient of calcium ions is expressed according to Einstein's theorem by

$$D_{ion} = R \cdot T \cdot \frac{\lambda_{ion}}{Z_{ion}^{\ 2} \cdot F^2} \tag{5.11}$$

where $R$ is the ideal gas constant [J/mol-K], $\lambda_{ion}$ is the molar conductivity of an ion [Sm²/mol], $Z_{ion}$ is the ion valence (= 2), and $F$ is Faraday's constant [C/mol]. Regarding the molar conductivity of an ion, $\lambda_{ion}$, temperature dependency is taken into account by Arrhenius's law (Yokozeki *et al.* 2003) as

$$\lambda_{ion} = \lambda_{ion\_25} \cdot \exp\left\{-1700\left(\frac{1}{T} - \frac{1}{298}\right)\right\} \tag{5.12}$$

where $\lambda_{ion\_25}$ is the molar conductivity of an ion at 25°C (Chemical Society of Japan 2004).

Ion transport in a porous medium like concrete is affected by the micro-pore structure. In this section, the effect of pore structure is expressed in terms of porosity, degree of saturation, tortuosity and constrictivity (Atkinson and

Nickerson 1984). Tortuosity is defined as a function of porosity, while constrictivity is a function of pore radius.

Tortuosity is formulated first. It expresses the increased length of actual ion transport pathways in accordance with the geometrical connectivity of pore spaces. Tortuosity is thought to be higher for cement paste with finer pores at lower water-to-cement (W/C) ratios and lower at larger W/C ratios or where coarse pores have been caused by leaching. The tortuosity of pores in the cement paste, $\tau_{paste}$, is assumed to be expressed in terms of effective porosity, $\phi_{paste}$, as an analogy with past research on chloride ion diffusion coefficients (Maekawa *et al.* 2003a; see Figure 5.6). This is because the tortuosity is an intrinsic characteristic expressing the geometrical property of pore structure and it can be assumed to be independent of the characteristics of ions.

$$\tau_{paste} = -1.5 \tanh \left\{ 8.0(\phi_{paste} - 0.25) \right\} + 2.5$$

$$\phi_{paste} = \frac{\phi_{cp} + \phi_{gl}}{V_{paste}}$$

$$(5.13)$$

where $\phi_{paste}$ is the effective porosity representing the sum of gel and capillary pores effective as ion transport pathways per unit volume of cement paste [$m^3/m^3$], and $V_{paste}$ is the unit volume ratio of cement paste [$m^3/m^3$]. Here, the equation was proposed as a function of porosity by modifying the original model with regard to the peak radius of capillary pores. In the modification, the relation between the porosity and the peak radius was examined by sensitivity analysis beforehand.

Next, constrictivity is formulated as an expression of the effect of pore size. Porosity and pore size distribution were taken into account in one past study of constrictivity (Van Brakel and Heertjes 1974). Pore dimensions are thought to be the main determinant of constrictivity (Atkinson and Nickerson 1984). A significant range of constrictivities has been reported, with a value of about 0.8 for ordinary porous materials (Van Brakel and Heertjes 1974; Petersen 1958) but of the order of about 0.01 for fine pores in cement paste (Atkinson and Nickerson 1984). In order to determine the

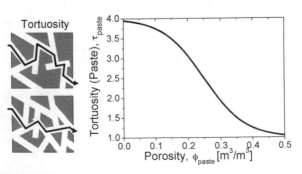

*Figure 5.6* Modeling of tortuosity for cement paste.

value of the constrictivity for the sound and deteriorated cement paste, the interaction between pore structure and ion transport needs to be discussed.

Pores in cement paste are widely distributed over the nanometer-to-micron scale and pores with varying dimensions are randomly connected. The average rate of ion transport is thought to be lower in fine pores because of interactions between the ions and pore walls in addition to the size factor. Electrical interactions between ions and pore walls have been cited as a reason for this (Sato *et al.* 1995). The pore walls in cement paste are positively charged, as are the calcium ions. Therefore, the effective dynamic pathway may be reduced in size due to electrostatic repulsion. This effect depends on the ion type, the ion concentration and other factors, but is largely dependent on pore size and ion radius. The smaller the pore size is, the more dominant the electrostatic effect (Sato *et al.* 1995). Accordingly, one fine pore in a connected series of pores within the cement hydrate can hinder ion transport strongly, so pore constrictivity becomes prominent in cement paste (Figure 5.7). The reduced transport resulting from pore connectivity and this interaction between ions and pore walls is expressed as constrictivity.

On the other hand, in the large pores of ordinary porous media such as soils and severely deteriorated cement paste, this sort of interaction becomes less but the constrictivity becomes larger. Based on the above discussion, the constrictivity of pores in a cement paste is defined with respect to the peak pore radius by Equation 5.14 (Figure 5.8). The equation gives a value of 0.01 for the fine pores in the sound cement paste, whereas it gives a value of 0.8 for the large pores in the totally deteriorated cement paste in the concrete composite. As a physical background of this treatment, the authors assume that the pore structure of the deteriorated cement paste would have similar characteristics to soils (see Section 5.4) as

$$\delta = 0.395 \tanh\left\{4\left(\log\left(r_{cp}^{peak}\right) + 6.2\right)\right\} + 0.405$$

$$\delta = 0 \text{ for } r_{cp}^{peak} \leq \frac{a_{Ca}}{2}. \tag{5.14}$$

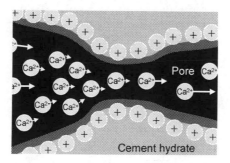

*Figure 5.7* Effect of electrostatic repulsion.

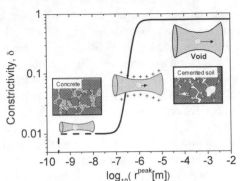

*Figure 5.8* Modeling of constrictivity.

where $r_{cp}^{peak}$ is the peak capillary pore radius [m] and $a_{Ca}$ is an ion diameter parameter for calcium ions = $0.6 \times 10^{-9}$ m. The peak pore radius in the Raleigh–Ritz distribution does not as a parameter provide a direct expression of the fine pores in which interactions between ions and pore walls are prominent, but it does determine the shape of the pore size distribution. As a first approximation, constrictivity is modeled in this simplified manner. A detailed study of statistical processing for the connectivity of pores is required in the future. The effect of ion concentration will also need to be taken into account (Sato *et al.* 1995).

### 5.1.5 Leaching-induced varying pore structures

It has been reported that the leaching of calcium leads to a coarse pore structure in the cement paste. This may accelerate calcium leaching. In this study, an increment in pore size and volume caused by the dissolution of calcium hydroxide and C–S–H gel is taken into account. That is, the change in pore structure and moisture re-distribution caused by calcium leaching is calculated by substituting the increased pore size and volume into the existing thermodynamic coupled analysis system (Chapter 3).

The increased pore volume resulting from dissolution of calcium hydroxide is expressed as a volumetric decrease in calcium hydroxide, $\Delta V_{CH}$, according to the following.

$$\Delta V_{CH} = \Delta m_{CH} / \rho_{CH}, \quad \Delta m_{CH} = \Delta C_{CH} M_{CH} \tag{5.15}$$

where $\Delta V_{CH}$ is the volumetric decrease in calcium hydroxide due to leaching [m³/m³], $\Delta m_{CH}$ is the decrease in mass of calcium hydroxide due to leaching [kg/m³], $\rho_{CH}$ is the density of calcium hydroxide (= $2.24 \times 10^3$ [kg/m³]), $\Delta C_{CH}$ is the decrease in solid-phase calcium in the calcium hydroxide [mmol/m³], and $M_{CH}$ is the molecular mass of calcium hydroxide (= $74.1 \times 10^{-6}$ [kg/mmol]).

The process of C–S–H gel dissolution is more complex than that of calcium hydroxide. In this study, the solid volume aside from that of the calcium hydrates and silicon dioxide is taken into account in the dissolution process. Then, the increase in pore volume due to dissolution of C–S–H gel is simply expressed by assuming a linear relation with the amount of leached calcium as

$$\Delta V_{CSH} = \frac{C_{CSH} - C_{solid}}{C_{CSH}} \cdot V_{CSH} \tag{5.16}$$

where $\Delta V_{CSH}$ is the volumetric change in C–S–H gel [$m^3/m^3$], which means the increase in porosity caused by dissolution of C–S–H gel. $V_{CSH}$ is the volume of C–S–H gel when dissolution is assumed to be zero. The volume $V_{CSH}$ is obtained from the following expression.

$$V_{CSH} = V_{unhyd} + V_{hyd} - V_{SiO_2} \tag{5.17}$$

where $V_{unhyd}$ is the volume of unhydrated powder [$m^3/m^3$], $V_{hyd}$ is the volume of hydrates [$m^3/m^3$], and $V_{SiO2}$ is the volume of silicon dioxide assumed not to be dissolved [$m^3/m^3$]. Thus, we have

$$V_{unhyd} - (1-\alpha)\frac{m_p}{\rho_p} \tag{5.18a}$$

$$V_{hyd} = \alpha m_p \left( \frac{1}{\rho_p} + \frac{\beta_p}{\rho_u} \right) \tag{5.18b}$$

$$V_{SiO_2} = \frac{Si \cdot M_{SiO_2}}{\rho_{SiO_2}} \tag{5.18c}$$

where $\alpha$ is the average hydration degree, $m_p$ is the mass of powder, $\rho_p$ is the density of the powder, $\rho_u$ is the density of the bound water in the hydrates (= $1.25 \times 10^3$ [$kg/m^3$]), $\beta_p$ is the mass of the bound water per mass of hydrated powder [$kg/kg$], $Si$ is the molecular amount of silicon [$mmol/m^3$], $M_{SiO2}$ is the molecular mass of silicon dioxide (= $60.1 \times 10^{-6}$ [$kg/mmol$]), and $\rho_{SiO2}$ is the density of silicon dioxide (= $2.20 \times 10^3$ [$kg/m^3$]). The amount of silicon in the cement is calculated from its mineral composition. The amount of silicon in fly ash and blast furnace slag is assumed to be 56.0% (Japan Society of Civil Engineers 1998a) and 33.6% (Japan Society of Civil Engineers 1998b), respectively. Unhydrated powder is not actually leached directly, since the leaching rate is much smaller than the hydration rate.

## 5.1.6 Computational discretization

In order to numerically solve Equation 5.1, we need to calculate incremental changes in porosity, degree of saturation and the number of calcium ions in the liquid and solid phases in an infinitesimal time. In the analytical scheme, all are calculated as time-dependent variables associated with ambient conditions, cement hydration and degradation due to leaching. These variables are interpolated between finite time steps to obtain a solution.

The increment in calcium ions in the liquid phase is discretized with respect to time as

$$\frac{\partial}{\partial t}(\phi \cdot S \cdot C_{ion}) = \frac{\phi^{n+1} \cdot S^{n+1} \cdot C_{ion}^{n+1} - \phi^{n} \cdot S^{n} \cdot C_{ion}^{n}}{t^{n+1} - t^{n}} \tag{5.19}$$

where $t^{n+1}$ and $t^n$ are the times [sec] at the $n+1$th step and $n$th step, respectively. In this notation, the superscript denotes the time step. So, for example, $\phi^{n+1}$ represents the porosity at the $n+1$th step. The porosity in Equation 5.19 is obtained from the microstructure model over time. The degree of saturation is also obtained from the moisture model.

Next, the increment in calcium ions in the solid phase is expressed in the form of a finite difference as

$$\frac{\partial C_{Solid}}{\partial t} = \frac{C_{Solid}^{n+1} - C_{Solid}^{n}}{t^{n+1} - t^{n}} = \frac{f^{n+1}(C_{ion}^{n+1}) - f^{n}(C_{ion}^{n})}{t^{n+1} - t^{n}} \tag{5.20}$$

When calculating the solid calcium associated with the varying amount of calcium ions, a simplified piecewise linearity of solid–liquid equilibrium is not adopted. Rather, the full non-linear relation is directly solved so as to maintain the accuracy of the numerical solution. That is because the strong coupling makes the isothermal solid–liquid equilibrium dynamically change with varying hydration, temperature and chloride concentration over time (Figure 5.9).

## 5.2 Analysis and verification

### 5.2.1 Immersion

The proposed model was verified by comparing with immersion experiments. Block-shaped specimens were submerged in deionized water to accelerate

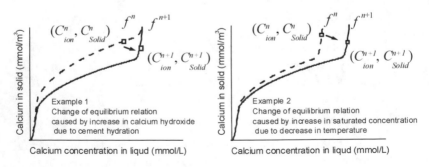

*Figure 5.9* Example of change in equilibrium relation.

calcium leaching. This method is able to quickly deliver more realistic results than immersion tests using cement powder.

In one experiment, the proposed model was used to simulate research by Watanabe *et al.* (2000). In their experiment, a cubic mortar specimen of W/C = 70% was prepared. All three dimensions were 40 mm. It was cured in sealed conditions for 42 days and then soaked in deionized water. The water-to-solid volume ratio was 50.0. The water was replaced with new deionized water every three months. After 12 months of immersion, the distribution of remaining calcium in the specimen was measured using energy dispersive X-ray spectroscopy (EDS). In the analysis, the concentration of calcium ions in the water was assumed be 2 mmol/l, to take into account the calcium remaining in the water. This value was determined based on the total amount of leached calcium ions. Here, hydroxide ions at a concentration of 4 mmol/l in the water are implicitly assumed to be the balanced anion of the cation (2 mmol/l calcium ions).

The experimental and simulated profiles of calcium distribution are shown in Figure 5.10. The simulation indicates slightly greater deterioration than observed in the experiment. Given that a higher calcium concentration in the water is inevitable near the specimen, this small discrepancy is thought to be acceptable. For a stricter estimate, it would be necessary to integrate a simulation of hydro-dynamics for the surrounding water into the model.

The proposed model was also tested against a sodium chloride experiment carried out by Imoto *et al.* (2004). Cubic paste specimens of W/C = 40% were prepared. All three dimensions were 40 mm. After casting, the specimens were cured at high humidity for one day followed by wetting for 27 days. The water-to-solid volume ratio was 50.0. Three kinds of water were used for immersion tests: deionized water, 0.28 mol/l sodium chloride solution, and 0.56 mol/l sodium chloride solution. The water was replaced every three months. After four years of immersion, the amount of residual calcium hydroxide and the total porosity were measured.

Figure 5.11 shows the experimental and simulated results for calcium hydroxide while Figure 5.12 gives the porosity results. Compared with

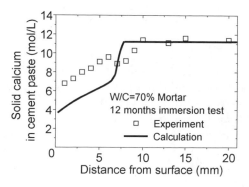

*Figure 5.10* Change in solid calcium due to leaching (immersion test).

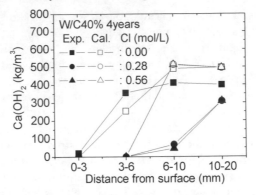

*Figure 5.11* Change in calcium hydroxide (immersion test).

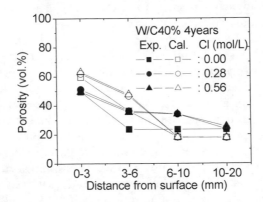

*Figure 5.12* Change in porosity (immersion test).

deionized water soaking, immersion in sodium chloride solution led to accelerated loss of calcium hydroxide and a rise in porosity. Further, the concentration of sodium chloride did not affect the results significantly in these cases. The model simulates these tendencies as a result of taking into account the effect of chloride ions on the solubility product.

### 5.2.2 Influence of temperature

The influence of temperature is considered in formulating the solubility product of calcium hydroxide and the diffusivity of calcium ions. Sensitivity to temperature is reviewed for the case of G50, which has a W/C ratio of 50%. The mix proportion is shown in Table 5.1.

First, the temperature sensitivity of the equilibrium relation is considered. In the simulation, after curing at 20°C for 80 days, the temperature was changed to 10°C, 20°C, 40°C and then 60°C in order of time under the sealed condition. The applied temperature history is shown in Figure 5.13a. As a sealed boundary is assumed, the total amount of calcium is constant. Figure

*Table 5.1* Mix proportion of concrete for analysis

| No. | W/P | Unit mass (kg/m³) | | | | |
|-----|-----|-----|-----|-----|-----|-----|
| | | W | C | FA | S | G |
| G25 | 25 | 190 | 760 | 0 | 600 | 775 |
| G50 | 50 | 190 | 380 | 0 | 700 | 988 |
| F50 | 50 | 190 | 304 | 76 | 700 | 950 |

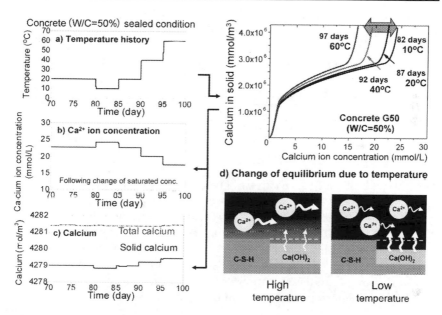

*Figure 5.13* Influence of temperature on phase change of calcium (a, b, c) and thermodynamic equilibrium (d).

5.13d shows sensitivity analysis with respect to the temperature-dependent solubility product of calcium hydroxide. Figure 5.13b shows the corresponding change in calcium ion concentration and Figure 5.13c is the response of total and solid calcium. When the temperature increases, the calcium ion concentration decreases with the fall in solubility product, resulting in increased solid calcium. The total amount of calcium remains unchanged with changing temperature.

Second, the influence of temperature on long-term calcium leaching is reviewed numerically. In this analysis, concrete specimens were exposed to water with a calcium ion concentration of 2 mmol/l at different temperatures after sealed curing at 20°C for 28 days. The analysis target was again G50 in Table 5.1. Figure 5.14 shows the results after 10,000 years of accelerated aging. The results indicate that an increase in temperature accelerates leaching. This increase causes a declining solubility product (Figure 5.13d) and rising

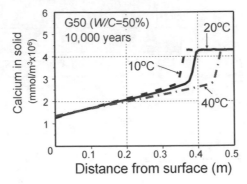

*Figure 5.14* Influence of temperature on long-term calcium leaching.

*Table 5.2* Influence of temperature on diffusivity

| Temperature | 10°C | 20°C | 40°C |
|---|---|---|---|
| $\dfrac{\delta}{\tau} D_{ion}$ (m$^2$/s) | 6.13E-12 | 7.79E-12 | 1.28E-11 |

diffusivity (Table 5.2), respectively. These changes have different effects on calcium leaching. Whereas the decrease in solubility product reduced calcium leaching, the increase in diffusivity accelerates the leaching. In the proposed equilibrium model, the effect of the reduced solubility product on equilibrium becomes less in the leaching process of C–S–H than in the process of calcium hydroxide (Figure 5.13d). Since the C–S–H leaching process plays a major role in the long-term deterioration of the solid, the effect of the higher diffusivity is thought to be significant in this analysis.

### 5.2.3 Influence of water-to-cement ratio

In order to investigate the influence that W/C has on long-term degradation caused by calcium leaching, sensitivity analysis was performed using the proposed model as well as a trial model. The targets of the analysis were G25 (W/C = 25%) and G50 (W/C = 50%). Both contained the same unit mass of water: 190 kg/m³. The mix proportions are shown in Table 5.1. In the analysis, the concrete specimens were exposed to water with 2 mmol/l of calcium ions after sealed curing for 28 days. The temperature was fixed at 20°C. In order to discuss interrelationships between increased porosity caused by calcium leaching and progress with hydration, additional calculations were performed using a trial model. In this sensitivity analysis of the trial model, the increase in porosity related to leaching is intentionally ignored in the hydration model. Porosity affects the amount of free water, which in turn influences the hydration rate. That is, the trial model does not take into account additional hydration associated with the change in pore structure caused by calcium leaching.

Figure 5.15 shows the analytical results after 10,000 years. Both the proposed model and the trial model show that a low W/C results in smaller deterioration depth. In the case of G50 (W/C = 50%), both models give almost the same results. In the case of G25 (W/C = 25%), however, the proposed model in which increasing porosity with leaching is taken into account exhibits less degradation compared with the trial model.

When the W/C is below 40%, such as with specimen G25, the initial mixing water is not enough to allow full hydration of the cement. As free water is short during the hydration of such low W/C concrete, unhydrated cement remains. Then, as leaching proceeds, the newly created pore spaces that result may act as free space for additional hydration. Consequently, unhydrated cement starts to react with existing water (Figure 5.16). This additional hydration causes the pore structure to become dense once again

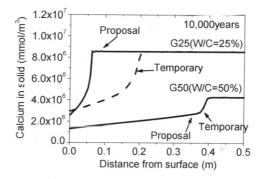

*Figure 5.15* Influence of W/C on long-term calcium leaching.

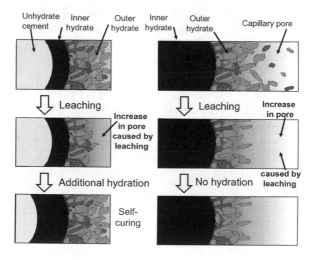

*Figure 5.16* Relationship between hydration and calcium leaching.

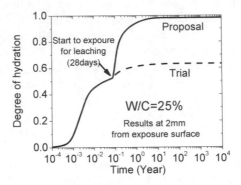

*Figure 5.17* Change in hydration degree due to leaching (W/C = 25%).

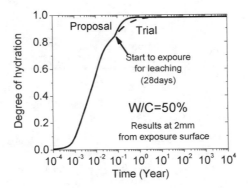

*Figure 5.18* Change in hydration degree due to leaching (W/C = 50%).

in a kind of self-curing system for low W/C concrete. In the case of normal or higher W/C, this self-curing is not observed since there is no unhydrated cement. The changes in degree of hydration illustrated in Figure 5.17 and Figure 5.18 clearly show the mechanism of self-curing as calcium leaching takes place. The proposed model calculates significant additional hydration of low W/C concrete as calcium leaching occurs. This self-curing system in low W/C concrete can be predicted only by a model that considers the strong coupling between hydration and changes to the pore structure. Whatever a further verification for the self-curing phenomenon is, the authors recognize that the analytical system and its predictions show the higher possibility of efficiently using low W/C concrete as a high-performance material for nuclear waste management.

### 5.2.4 Influence of fly ash

It has been reported that fly ash (FA) reduces degradation related to calcium leaching (Yokozeki 2004). In this section, the effect of FA on calcium leaching is investigated using the proposed model. The specimens subjected to analysis (G50 and F50) have a common mix proportion with *W/P* = 50% and a unit mass of water of 190 kg/m³. Two replacement ratios of FA (0% and 20%) were studied. The mix proportions are shown in Table 5.1.

The calculated distributions of solid calcium in the concrete after 10,000 years are shown in Figure 5.19. As this Figure makes clear, concrete with 20% FA has greater resistance to calcium leaching. This is for two reasons: first, it has a different calcium solid–liquid equilibrium and second, a different ultimate micro-pore structure. The use of FA leads to a different equilibrium relationship (Figure 5.20), since the pozzolanic reaction consumes calcium hydroxide, which has high solubility in pore water. The ratios of calcium hydroxide to total solid calcium were 32.4% in G50 (FA 0%) and 13.4% in F50 (FA 20%) in the sound part of the concrete after 10,000 years, respectively. In addition to the change in the equilibrium relationship, the micro-pore structure becomes denser with the use of FA. This trend can be seen in Figure 5.21. In the case of F50 with FA, the capillary pores disappear, resulting in lower diffusivity, as shown in Table 5.3. They show that 20% use of FA reduced the effective diffusion coefficient to almost half in this analysis.

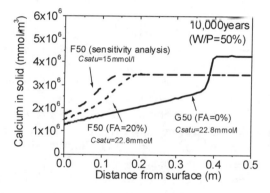

*Figure 5.19* Influence of fly ash on long-term calcium leaching.

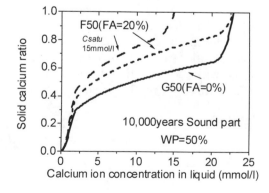

*Figure 5.20* Influence of fly ash on equilibrium relation.

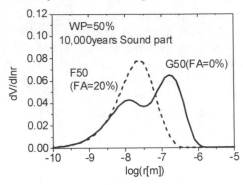

*Figure 5.21* Influence of fly ash on porosity distribution.

*Table 5.3* Influence of fly ash on diffusivity

| No. | G50 | F50 |
|---|---|---|
| $\frac{\delta}{\tau} D_{ion}$ (m²/s) | 7.79E-12 | 3.55E-12 |

These values of the coefficients almost agree with the experimental results measured by electrical potential methods (Horiuchi *et al.* 1998).

On the other hand, the coefficient of FA concrete is much smaller compared to the experimental results measured by diffusion cell tests (Uchikawa 1985). The effective diffusion coefficient of the concrete with 20% FA is about one-tenth of the coefficient of concrete without FA. The reason of the large decrease in the coefficients by the diffusion cell tests is thought to be the increase in the electrical interactions between ions and pore wall, because the electrical potential method may mask the internal electrical interaction due to the large external electric charge.

In order to investigate the adequacy of the effective diffusion coefficient in the proposed model, a comparison of deterioration depth between the experimental results by Yamamoto and Hironaga (2006) and analytical results by the proposed model was made. In the experiments, the specimens of cement paste were immersed in deionized water for 82 weeks after curing under 50°C for 91 days, in order to investigate the effect of FA on calcium leaching. The water-to-binder ratio was 40% and the replacement ration of FA was 30%. Table 5.4 shows the deterioration depth from the exposure surface in the analytical and experimental results. In the case of FA, the calculated

*Table 5.4* Deterioration depth after 82 weeks

| No. | Experiment (mm) | Calculation (mm) |
|---|---|---|
| OPC | 2.4 | 1.8 |
| OPC with 30% fly ash | 0.6 | 1.6 |

deterioration depth is much larger than the measured depth in the experiment. This indicates that the decrease in the effective diffusion coefficients calculated for FA concrete is underestimated. A detailed investigation of the effective diffusion coefficient is needed in future.

Finally, the effect of a decrease in calcium ion solubility in the equilibrium relation is investigated by using G50 and F50, since it is reported that solubility decreases in the presence of FA (Yokozeki 2004). Although the calculated effective diffusion coefficient for FA concrete in this analysis is larger than the reality, more qualitative investigation on the phase–equilibrium relation is thought to be possible. In this sensitivity analysis at 20°C, the saturated concentration of calcium ions in F50 was tentatively changed to 15.0 mmol/l based on past research, whereas the calculated value in the proposed model was 22.8 mmol/l (Figure 5.20). The result for residual calcium in the solid phase is shown in Figure 5.19. By decreasing the saturated concentration in the solid–liquid equilibrium, deterioration was reduced. The effect of decreasing the saturated concentration was, however, not as significant as that of changing the amount of calcium hydroxide and porosity, which are automatically taken into account in the proposed system.

## 5.3 Extended modeling of cemented soil

### 5.3.1 Scheme of thermodynamics

Soil and cementitious composites are the primary constituents of ground and infrastructures. Assessing the long-term serviceability of artificial cementitious composites coupled with natural soil foundations requires a unified physico-chemical model, especially when targeting intermediate materials such as cemented soils. In this section, state variables for pores in hardened cement hydrates (nanometer-to-micron in size) are integrated into thermodynamic state equilibrium equations that have governing formulae for voids (micron-to-millimeter in size) in soil particles. The multi-phase, multi-scale concrete models are extended to the geo-environment and used to assess mid- and long-term changes and fluctuations in the material properties of coupled soils and structures (Figure 5.22).

### 5.3.2 Extended pore structure model

A consistent method for analyzing natural geo-materials, cemented soils and concrete must be able to take into account the intrinsic characteristics of soils. One key issue is that the geometry of pores can vary greatly (Figure 5.23). Sandy soils have a skeleton of particles, with large airspaces (voids) dispersed throughout the material. In natural cemented soils, water and cement cannot fill up the airspaces. The connecting pores define the soil's mechanical properties and permeability.

Here, the analytical method incorporates a new microstructural component

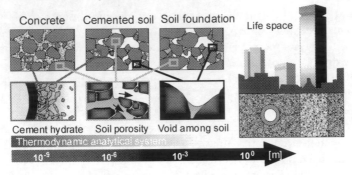

*Figure 5.22* Extended thermodynamic system to soil materials.

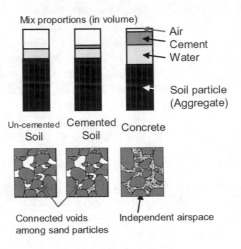

*Figure 5.23* Mix proportion and void in soil materials.

representing the large airspaces in cemented and natural soils (Figure 5.24). This means the expansion of the multi-scale analytical system (Chapter 3) into the large pores between micron and millimeter scales. The inputs are the pore volume and the average pore radius, which are set according to the target mix proportion and the properties of the sand. It is assumed that the micro-spaces in the large airspaces of the soil are occupied by water and do not provide space for cement hydrates to solidify. The existing model statistically expresses the general pore structure of cemented soil (Shimomura and Maekawa 1997). The distribution of nanometer-to-millimeter scale pores is represented by the following pore density function.

$$\phi(r) = \phi_{vd}V_{vd}(r) + \phi_{cp}V_{cp}(r) + \phi_{gl}V_{gl}(r) + \phi_{lr} \tag{5.21}$$

where $\phi_{vd}$ is the porosity of the voids between the soil particles [m³/m³], $V_{vd}(r)$ is functions that specify the size distribution of the voids, as given by the following equation (Chapter 3).

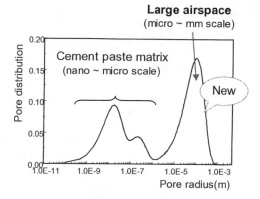

*Figure 5.24* Multi-scale pore structure model.

$$V_{vd}(r) = 1 - \exp(-B_{vd}r)$$
$$dV_{vd}(r) = B_{vd}r\exp(-B_{vd}r)d\ln r \tag{5.22}$$

where $B_{\iota}$ is a porosity distribution parameter [1/m], which represents the peak of porosity distribution on a logarithmic scale. For the voids between soil particles, a new characteristic value, $B_{vd}$, is incorporated to represent the peak pore radius of the soil. The geometric shape of the voids is assumed to be constant irrespective of hydration. In other words, $B_{vd}$ is a constant.

### 5.3.3 Extended moisture transport

The flux of liquid condensed water in continuous pores, denoted by $q_l$, can be determined by integrating the specific flux of liquid water over the entire pore structure using a random pore size distribution model (Chapter 3).

$$q_l = -\frac{\rho_l\phi^2}{50\eta}\left(\int_0^{r_c} rdV\right)^2 \nabla P \tag{5.23}$$

where $\phi$ is the total porosity (not including interlayer pores) [m³/m³], and $\eta$ is the viscosity of liquid water [Pa-s]. In order to consider the influence of pore size on moisture transport, the viscosity of water in capillary and gel pores of finer sizes is calculated as follows (Section 3.2.3.3).

$$\eta = \eta_0 \exp\left(\frac{G_e}{RT}\right) \tag{5.24}$$

where $\eta_0$ is the viscosity of liquid water under ideal conditions [Pa-s], and $G_e$ is the additional Gibbs' energy required for the flux of liquid water under non-ideal conditions [kcal/mol]. For the large-radius voids between soil particles, the additional Gibbs' energy, $G_e$, can be assumed to be zero, meaning that the intrinsic viscosity of bulk water is used.

## 5.4 Leaching model of calcium from cemented soil to soil foundation

### 5.4.1 Mass conservation

Momentum, energy and the mass flow of materials must satisfy the laws of conservation. The mass conservation equation explained in 5.1 is applied with small modification for considering the effect of soil materials.

$$\frac{\partial}{\partial t}(\phi \cdot S \cdot C_{ion}) + \frac{\partial C_{bound}}{\partial t} - div J_{ion} = 0 \tag{5.25}$$

where $S$ is the degree of saturation in the pore spaces, $C_{ion}$ is the molar concentration of calcium ions in the liquid phase [mmol/m³], $C_{bound}$ is the amount of bound calcium per unit volume [mmol/m³], and $J_{ion}$ is the flux of ions [mmol/ m².s]. The bound calcium, denoted by $C_{bound}$, is physico-chemically bound and not related to mass transport. For the cementitious composites, $C_{bound}$ is equilibrated with the calcium in the solid phase, $C_{solid}$, which is present mainly in the cement hydrates of C–S–H gel and calcium hydroxide (see Section 5.1.3). For ordinary soil materials, the amount of bound calcium is assumed to be zero. When the target soil material has the capacity to adsorb ions such as mineral clays, the adsorbed ions are treated as bound calcium. For bentonite, the binding of calcium ions, caused by the ion exchange of sodium ions in montmorillonite, is also taken into account.

### 5.4.2 Thermodynamic phase-equilibrium

As described in Section 5.1.3, an isotherm correlation rooted in the model described by Buil *et al.* (1992) is used to uniquely determine the amount of calcium in the solid phase from a given concentration of liquid phase calcium ion in cementitious materials.

### 5.4.3 Calcium transport

As described in Section 5.1.4, the flux of calcium ions transported in a porous medium takes the following form.

$$J_{ion} = -\left( \frac{\phi \cdot S}{\Omega_{ave}} \cdot \delta_{ave} \cdot D_{ion} \right) \cdot \nabla C_{ion} + \phi \cdot S \cdot \mathbf{u} \cdot C_{ion} \tag{5.26}$$

where $\Omega_{ave}$ is average tortuosity, $\delta_{ave}$ is average constrictivity, $D_{ion}$ is the diffusion coefficient of a calcium ion [m²/s], $\nabla^T = [\partial/\partial x \; \partial/\partial y \; \partial/\partial z]$ is the nabla operator, and $u^T = [\, u^x\ u^y\ u^z\, ]$ is the velocity vector of a calcium ion transported by a solution flow [m/s]. The capillary and gel pores in the cement paste and the voids between the soil particles are treated as ion transport pathways, and the authors assume that no ions are transported in the interlayer pores in the cement paste. Thus, porosity, $\phi$, can be expressed by the following equation.

$$\phi = \phi_{cp} + \phi_{gl} + \phi_{vd} \tag{5.27}$$

The transport properties of ions in porous materials, such as concrete and cemented and natural soils, depend on their pore structures. In this chapter, the pore structure is computationally characterized by porosity, degree of saturation, tortuosity and constrictivity (Atkinson and Nickerson 1984). Tortuosity is defined in terms of porosity, and constrictivity by the pore radius in the proposed multi-scale analytical system.

The tortuosity factor, $\Omega$, expresses the increased length of the actual ion transport pathway according to the tortuosity of the pores. For soil materials, the value depends largely on the soil type and the degree of compaction (Van Brakel and Heertjes 1974). This will be the case for cemented soil, as well. Thus, the tortuosity of the voids in the soil particles, $\Omega_{vd}$, is assumed to be formulated in the same manner as the porosity, $\phi_{vd}$, as given by Equation 5.28 (Figure 5.25), with reference to Van Brakel and Heertjes (1974),

$$\Omega_{vd} = -1.5\tanh\{4.0(\phi_{vd} - 0.15)\} + 2.5 \tag{5.28}$$

For materials such as cemented soils that contain both pores in hardened cement paste and voids between soil particles, the macroscopic average tortuosity is determined by simply assigning weights in proportion to the porosities in Equation 5.29.

$$\Omega_{ave} = \frac{(\phi_{cp} + \phi_{gl})\Omega_{paste} + \phi_{vd}\Omega_{vd}}{\phi_{cp} + \phi_{gl} + \phi_{vd}} \tag{5.29}$$

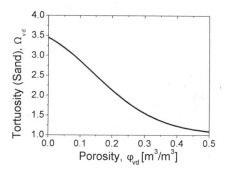

*Figure 5.25* Modeling of tortuosity for sand.

The constrictivity is then formulated in order to express the effect of the pore connectivity and the electric changes on the walls of the micro-pores during mass transport (Section 5.1.4). Equation 5.14 can consider the difference of pore size between cement paste and soils (Figure 5.8). Thus, the constrictivity of the pores in a cement paste and the voids of the soil particles are defined as a function of the peak pore radius by Equation 5.30.

$$\delta_i = 0.395 \tanh\left\{4\left(\log\left(r_i^{peak}\right) + 6.2\right)\right\} + 0.405$$

$$(5.30)$$

$$\delta_i = 0 \quad for \quad r_i^{peak} \leq \frac{a_{Ca}}{2}$$

where for parameter $r^{peak}$, the peak radius of a capillary pore is given in the case of hardened cement paste, whereas for the voids between soil particles, the peak radius of the voids is used. These values are determined from the inverse of the pore distribution parameters, $B_{cp}$ and $B_{vd}$, respectively. The peak pore radius, as a parameter in the Raleigh–Ritz distribution, does not directly express the fine pores in which interactions between the ions and the pore walls are prominent, but it does determine the shape of the pore size distribution (Figure 5.26). In this section, constrictivity is modeled in this simplified manner as a first approximation. In the future, a detailed study of statistical processing for pore connectivity will be necessary. In addition, the ion concentration is an influential parameter that also needs to be taken into account (Sato *et al.* 1995). For materials such as cemented soil that contain both cement paste pores and soil particle voids, the macroscopic average constrictivity is determined by simply assigning weights in proportion to the porosities.

$$\delta_{ave} = \frac{(\phi_{cp} + \phi_{gl})\delta_{paste} + \phi_{vd}\delta_{vd}}{\phi_{cp} + \phi_{gl} + \phi_{vd}}$$

$$(5.31)$$

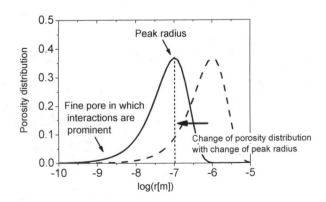

*Figure 5.26* Porosity distribution and peak radius.

where $\delta_{paste}$ is the constrictivity of the pores in the cement paste and $\delta_{vd}$ is the constrictivity of the voids between the soil particles.

## 5.5 General verification

### 5.5.1 Leaching of calcium from cemented soil

In order to compare calcium leaching from cement paste with that from cemented sand, cylindrical test specimens of cement paste and cemented sand were prepared and subjected to an immersion test. The water-to-cement ratio was 50% and the mix proportions are shown in Table 5.5. The amount of calcium ions leaching into the solution was measured. The cylindrical specimens were 50 mm in diameter and 100 mm in height. After sealed curing for seven days, the top and sides of the specimen surfaces (but not the bottom) were coated with rubber, and then immersed in deionized water at a liquid-to-solid ratio of 10 to 1 by volume (Figure 5.27). The water was stirred at intervals of a few days and replaced every 15 days. The conditions of the exposed surfaces before immersion are described in Figure 5.28. Large voids were visible in the cemented sand.

*Table 5.5* Mix proportions for cement paste and cemented soil

| No. | W/C (%) | Unit mass (kg/m³) | | | Voids (%) | Porosity at 7 days (%) |
|-----|---------|-------|--------|------|-----|----------------|
|     |         | Water | Cement | Sand |     |                |
| P50 | 50      | 612   | 1,223  | 0    | 0   | 32             |
| S50 | 50      | 79    | 157    | 1,706| 22  | 26             |

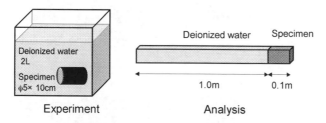

Experiment     Analysis

*Figure 5.27* Experiment and analysis for immersion test.

Cemented sand     Cement paste

*Figure 5.28* Condition of surface of specimen before immersion tests.

Figure 5.29 shows the amount of calcium that leached, as determined from the change in the concentration in the water. Figure 5.30 shows the cumulative amount of calcium that leached. The amount of calcium that leached became zero every 15 days when the water was replaced. The total amount of calcium that leached from the cemented sand was larger than the amount that leached from the cement paste, in spite of the fact that, at initial immersion, the cemented sand had smaller quantities of cement and a smaller volume of pores than the cement paste did (Table 5.5). The volume of pores is the sum of the volume of voids between the soil particles (determined during mixing) and the volume of gel and capillary pores in the cement paste at the age of seven days (as calculated in the analysis).

A simple one-dimensional analysis was carried out (Figure 5.27). Input data such as mix proportions and environmental conditions were based on the experiments. While the porosity distributions in the cement paste over time were calculated, the porosity of the voids among the soil particles was a fixed value determined from the input data. In order to reproduce the conditions of the immersion test, an analysis of calcium leaching, taking into account the concentration gradient in the water around the specimen and the decrease in the concentration gradient as ions leached into the water, was carried out by connecting the analytical elements representing the deionized water. To take into account the effects of three-dimensional diffusion in water

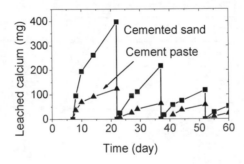

*Figure 5.29* Leached calcium from cement paste and cemented sand in experiments.

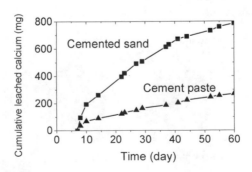

*Figure 5.30* Cumulative leached calcium in experiments.

as well as stirring, the diffusion coefficient of the calcium ions in the deionized water was set to a value ten times larger than the value in free water. The analysis did not take into account the replacement of the water.

Figure 5.31 shows the results of the analysis. The calculated results for still water without advection were compared first. When the cemented sand was assumed to have a fine pore structure comparable to that of the cement paste (peak pore radius: $10^{-7}$ m), the amount of calcium that leached from the cemented sand was much smaller than the amount that leached from cement paste. This is because the volume of pores and the cement content per unit volume are both smaller. On the other hand, in the analysis that considered the scale of the connected large voids in the cemented sand (peak pore radius: $10^{-4}$ m), the amount of calcium that leached was higher due to the greater diffusion coefficient. These analytical results demonstrate the absolute necessity of taking into account the characteristics of the pore structure.

Comparing the results of the analysis with those of the experiment, however, reveals that the amount of leached calcium was underestimated. The reasons are thought to be the effects of leached calcium from the sand grains and the effect of advection due to stirring. In a preliminary experiment in which the same quantity of sand alone was immersed for 15 days, about 80 mg calcium was leached. This means that sand ingredients can leach, but only to a limited extent. In order to maintain a constant concentration in the immersion solution, the solution was stirred. Because the pore structure of the cemented sand used in this section was very coarse, this stirring may have caused advection within the specimen, which may have accelerated the leaching of ions.

To look into this, a sensitivity analysis to determine the effect of stirring was carried out by applying a small pressure gradient to the inside of the test specimen in the outward direction, as shown in Figure 5.32. A water head difference of 0.1 mm per 100 mm of depth was applied. In the cement paste and the cemented sand, which was assumed to have fine pores, the water permeability was found to be low and there was practically no transport of

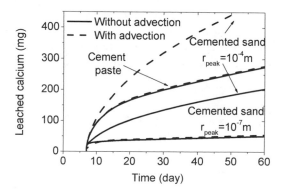

*Figure 5.31* Leached calcium from cement paste and cemented sand in analysis.

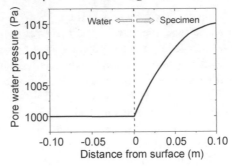

*Figure 5.32* Distribution of pore water pressure in sensitivity analysis.

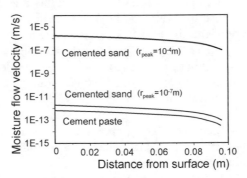

*Figure 5.33* Moisture flow velocity in sensitivity analysis.

water (Figure 5.33), so there would be no effect on the amount of calcium that leached (Figure 5.31). More specifically, under the small pressure gradient used to simulate stirring, there was practically no advection effect, and the transport of ions in the specimen was governed by diffusion alone. It should be noted that the cemented sand had high water permeability and showed significant transport of water (Figure 5.33), which resulted in a sharp and high-level increase in the amount of calcium that leached (Figure 5.31). This confirms the need to carefully examine the experimental conditions for their effects on advection, even though the effects may be minute. It is also vital to examine the durability of cemented soil under groundwater flow. To date, no firm conclusions have been drawn regarding any of these effects, although studies will continue.

## 5.5.2 Leaching of calcium from cementitious composites into soil foundation

The effects of the surrounding ground on the leaching of calcium from concrete were analyzed by applying the site structural survey proposed by Yokozeki *et al.* (2003) to an old underground concrete footing for a four-storied building constructed in 1929. Concrete samples were taken at 4 m below the ground surface and at about 70 cm below the groundwater level (Figure 5.34). The mix proportion of the concrete was estimated from the compressive strength. Table 5.6 shows the results of an examination of the samples. The amount of calcium and silicon were measured using an energy dispersive X-ray spectroscopy (EDS). The remaining calcium ratio for the solid phase was determined using two types of analyses: one analysis focused on the concrete itself, while the other examined the interaction of the concrete with the surrounding soil foundation.

In the analysis of the concrete alone, the concentration of calcium in the ground was taken as the boundary condition at the concrete's free surfaces. The specified value of 1.3 mmol/l was measured in the groundwater in the vicinity of the building. In the analysis of the interaction of the concrete and the surrounding ground, the soil elements were combined with the concrete elements. In consideration of the depth of ion diffusion, the thickness of the soil elements was set to 5.0 m. Since there were no data about the details of the properties of ground, it was assumed to be ordinary sandy ground without cement. Based on this assumption, the porosity and peak pore radius were defined as 35% and $1.0 \times 10^{-4}$ m, respectively. The effective diffusion coefficient of a calcium ion, including tortuosity and constrictivity, as determined from these set values, was calculated to be $5.9 \times 10^{-12}$ m²/s in the soil elements. Moisture transport of the groundwater was not a factor, since completely saturated states were assumed in analysis.

Figure 5.35 shows the ratio of measured and analytical distributions of the remaining solid-phase calcium for the underground structure. The ratio is calculated by dividing the amount of residual solid calcium by the average amount of solid calcium in the non-leached part. Figure 5.36 shows the

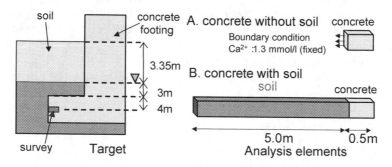

*Figure 5.34* Analytical target and elements.

*Table 5.6* Mix proportions for concrete

| No. | W/C (%) | Unit mass (kg/m³) | | | |
| --- | --- | --- | --- | --- | --- |
| | | Water | Cement | Sand | Gravel |
| D55 | 55 | 170 | 309 | 799 | 1,104 |

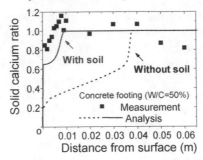

*Figure 5.35* Distribution of residual solid-phase calcium ratio.

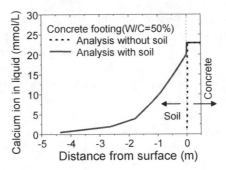

*Figure 5.36* Distribution of liquid-phase calcium ion.

distribution of liquid-phase calcium ion concentrations in the ground and structure (analyzed values). The coupled analysis of the interaction of the concrete with the surrounding soil foundation produced a degree of deterioration that was closer to the measured results than the one produced by the analysis of just the concrete (Figure 5.35). As can be seen in the coupled analysis, the concentration of calcium in the ground increased because of mass transport from inside the structure (Figure 5.36). The surrounding ground makes the concentration gradient small and reduces the rate of calcium leaching. The mass transport interaction is vital for rationally assessing long-term calcium leaching and the subsequent deterioration of cementitious composites in a foundation.

### 5.5.3 Leaching with ground adsorbing calcium ions

In order to analyze the effect that the surrounding ground had on leaching when the ground was treated as a boundary condition, a coupled analysis of the surrounding ground and the water environment was carried out (Figure 5.37). The subject of the analysis was cemented soil with a water-to-cement ratio of 100%, $\phi_{vd}$ of 40%, and a peak pore radius of $10^{-6}$ m. The mix proportions are listed in Table 5.7. After sealed curing for seven days, the specimens were exposed to deionized water and the ground environment (with and

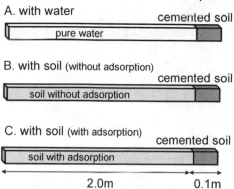

*Figure 5.37* Analysis elements.

*Table 5.7* Mix proportions for cemented soil

| No. | W/C (%) | Unit mass (kg/m³) | | | Voids (%) |
| --- | --- | --- | --- | --- | --- |
| | | Water | Cement | Sand | |
| B100 | 100 | 150 | 150 | 1,000 | 40.0 |

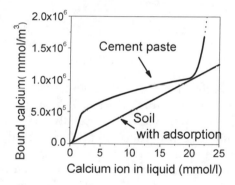

*Figure 5.38* Solid–liquid equilibrium relations.

without adsorption) (Figure 5.37). The ground was saturated, non-cemented soil with a water content of 60%, a $\phi_{vd}$ of 60%, and a peak pore radius of $10^{-6}$ m. Adsorption was taken into account by adopting a simple linear equilibrium adsorption curve, as shown in Figure 5.38. The amount of ions adsorbed was determined according to examples taken from past research (Usui *et al.* 2005). Figure 5.38 also shows the solid–liquid equilibrium curve for the sound portion of the cemented soil at the age of one year.

Figure 5.39 shows the distribution of solid-phase calcium remaining in the cemented soil one year after exposure, while Figure 5.40 shows the distribution of liquid-phase calcium ion concentrations. When adsorption was not taken into account, the cemented soil deteriorated to a lesser degree than in

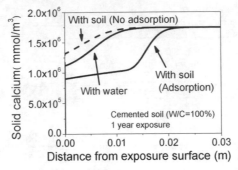

*Figure 5.39* Distribution of residual solid-phase calcium.

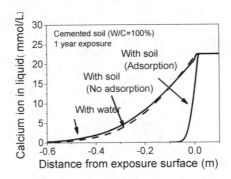

*Figure 5.40* Distribution of liquid-phase calcium ion.

the case where the specimen was in contact with water. This is because the volume of pores acting as ion pathways was reduced. When adsorption was taken into account, deterioration developed rapidly. This is because the pores in the ground maintained a low concentration of free calcium ions due to the adsorption of ions and the constantly high concentration gradient in the cemented soil. These results agree with past results: cemented soil exposed to clay ground deteriorates faster than cemented soil immersed in fresh water (Ikegami *et al.* 2004).

# 6 Chloride ion transport and corrosion

This chapter presents three items of importance for predicting chloride ion equilibrium, transport and steel corrosion in concrete. First, thermodynamic states of chloride are discussed with ionized, solid-phase and adsorbed chlorides, and state equations in micro-pores presented in Chapter 3 are formulated for ordinary Portland and mixed cement with pozzolans. Second, a chloride ion transport model is proposed for both sound and cracked concrete, which is linked with Chapter 9 and Chapter 10, along with the non-linear chloride-binding model in a thermodynamic coupled system. Third, a unified corrosion model is presented for steel reinforcement throughout the life of concrete structures based on an electro-chemical thermodynamic approach, and this model is coupled with structural performance assessment discussed in Chapters 8–10.

## 6.1 Chloride-binding capacity of cementitious materials

Chlorides present in hardened cement pastes are generally classified into chloride ions and bound chlorides (Figure 6.1, Maruya *et al.* 1998b). Chloride ions present in concrete are dissolved in the pore liquid and exist in a freely mobile form. Bound chlorides, on the other hand, apparently do not move at ordinary concentration gradients or in ordinary advective environments. Bound chlorides can be further classified into constituents of hydrates, as

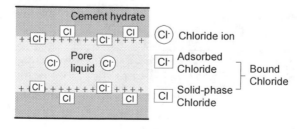

*Figure 6.1* Classification and definition of chlorides in hardened cement pastes (Maruya *et al.* 1998b).

Friedel's salt ($C_3A \cdot CaCl_2 \cdot 10H_2O$) and chlorides adsorbed into the pore walls. The total quantity of chlorides in a hardened cement paste is the sum of the chloride ions and bound chlorides. Hardened cement pastes with good binding capacity are expected to be relatively low in chloride ion content and highly resistant to steel corrosion. Accordingly, it is important to have a grasp of the relationship between chloride ions and bound chlorides in concrete under various conditions.

Chlorides bound as Friedel's salt (solid-phase chlorides) are thought to be dependent on the $Al_2O_3$ content in the binder and the quantity of monosulfate formed (Glass *et al.* 1997). Further, it is expected that the quantity of chlorides adsorbed into hydrates (adsorbed chlorides) will vary depending on the geometric characteristics of pore structure, electric charge of pore surface, etc. The objective of this section is to quantify the equilibrium of chloride ions and bound chlorides in hardened cement pastes. The specimens consist of cement and varying mineral compositions and gypsum contents as well as some made with cement and ground, granulated blast furnace slag and pozzolans, to examine the effects of admixtures.

### 6.1.1 Experimental methods

#### 6.1.1.1 Materials and mix proportions

The physical properties and chemical composition of the ordinary Portland cement (OPC), high early-strength Portland cement (HPC), low-heat Portland cement (LPC), ground granulated blast furnace slag (SG), fly ash (FA) and silica fume (SF) used in the experiment are listed in Table 6.1 with loss on ignition (LOI). The specific surface area of the silica fume in the table is not an actual measurement of that in the experiment but the average (published data) of products on the market.

*Table 6.1* Physical properties and chemical compositions of materials used in the tests

| Materials | Density $(g/cm^3)$ | Specific surface area $(cm^2/g)$ | LOI | $SiO_2$ | $Al_2O_3$ | $Fe_2O_3$ | CaO | MgO | $SO_3$ | Cl |
|---|---|---|---|---|---|---|---|---|---|---|
| OPC | 3.18 | 3,110 | 0.49 | 21.5 | 5.65 | 2.93 | 64.9 | 1.07 | 1.75 | 0.005 |
| HPC | 3.13 | 4,320 | 1.39 | 20.0 | 5.01 | 2.79 | 65.9 | 0.80 | 2.93 | 0.008 |
| LPC | 3.24 | 3,410 | 0.41 | 25.8 | 3.41 | 3.51 | 62.4 | 0.82 | 2.40 | 0.005 |
| SG | 2.91 | 4,150 | 0.01 | 34.31 | 14.33 | 0.25 | 43.0 | 5.66 | – | <0.001 |
| FA | 2.16 | 3,480 | 3.20 | 58.24 | 25.58 | 5.02 | 2.42 | 1.34 | – | <0.001 |
| SF | 2.25 | 200,000 | 1.13 | 95.8 | 0.62 | 0.68 | 0.18 | 0.46 | 0.10 | – |

*Table 6.2* Series of test specimens, materials and mix proportions used in the tests

| Test series | Content per unit volume of concrete (kg/m³) | | | |
|---|---|---|---|---|
| | Water | Cement | Gypsum | Fine aggregate |
| HC | 416.5 | 832.9 | – | 838.0 |
| NC | 419.0 | 838.0 | – | 838.0 |
| LC | 422.0 | 844.1 | – | 838.0 |
| NGS | 416.2 | 792.7 | 39.7 | 838.0 |
| NGL | 413.4 | 747.9 | 78.8 | 838.0 |
| Test series | Water | Cement (OPC) | Ground, granulated blast furnace slag | Fine aggregate |
| B20 | 416.0 | 665.7 | 166.4 | 838.0 |
| B40 | 413.1 | 495.7 | 330.5 | 838.0 |
| B60 | 410.2 | 328.2 | 492.2 | 838.0 |
| B80 | 407.3 | 162.9 | 651.8 | 838.0 |
| Test series | Water | Cement (OPC) | Fly ash | Fine aggregate |
| F20 | 404.3 | 646.8 | 161.7 | 838.0 |
| F40 | 390.5 | 468.6 | 312.4 | 838.0 |
| Test series | Water | Cement (OPC) | Silica fume | Fine aggregate |
| S20 | 406.1 | 649.7 | 162.4 | 838.0 |
| S40 | 393.9 | 472.7 | 315.1 | 838.0 |

The series of specimens, their materials and mix proportions are listed in Table 6.2. For the series using Portland cement without admixtures, specimens made with ordinary Portland cement (NC), high early-strength Portland cement (HC) and low-heat Portland cement (LC) were prepared. To examine the effects of gypsum content on the amount of bound chlorides, two experimental series (NGS and NGL) with varying gypsum contents were prepared. In this chapter, the total quantity of cement and gypsum is defined as the quantity of binder. Prefixes B, F and S added to the specimens in admixture series represent ground, granulated blast furnace slag, fly ash and silica fume, respectively. The number suffix on these experimental series indicates the admixture replacement ratio as a percentage by mass of total binder. Mortar specimens having a water-to-binder ratio of 50% were prepared using Japan Industrial Standard (JIS) standard sand (percentage of water adsorption: 0.42%; density: 2.64 g/cm³) for the fine aggregate and ion-exchanged water for mixing the cement pastes.

The properties of the fresh mortar specimens were almost the same except for the specimen with a 40% silica fume replacement ratio. Accordingly, 0.4% of an air-entraining and high-range water-reducing agent (as a percentage by mass of binder) was added to this specimen to adjust its properties.

### 6.1.1.2  *Method of preparing test specimens and curing conditions*

The method of mixing the cement pastes was in accordance with JIS R5201 10.4.3 and the mortar was mixed and placed in forms at 20°C. The dimensions of the prepared specimens were $\phi 50 \times 100$ mm. Forms were removed on the day following placement and the specimens were then cured under water for 28 days. For series F and S specimens, which included pozzolans, the specimens were cured under water for 91 days so that the pozzolanic reactions could proceed sufficiently. After completion of the curing, specimens were cut into slices and cubes for the various types of experiment as described later. To be more precise, sections 10 mm from the ends of a specimen were removed to eliminate the effects of segregation. From the remaining 80 mm-long section of the cylindrical specimens, six $\phi 50 \times 10$ mm slice specimens and about 250 cubic specimens 7 mm on a side were taken using a wet-type diamond cutter. The specimens were then immersed in the NaCl solution of the prescribed concentration described later.

### 6.1.1.3  *Test items and methods*

#### 6.1.1.3.1  DEFINITION AND DETERMINATION OF THE QUANTITIES OF CHLORIDE IONS AND BOUND CHLORIDES

Basically, the Japan Concrete Institute (JCI) method (JCI-SC4) was used to analyze the quantity of chlorides in the hardened cement pastes. Strictly speaking, however, the quantity of soluble chlorides obtained by this means is not equal to the total quantity of chloride ions present in the pore liquid. This is because soluble chlorides are extracted in warm water at 50°C, so they can be expected to include some chlorides that would be adsorbed at a normal temperature of 20°C and some of the solid-phase chlorides that take the form of Friedel's salt. For this reason, a comparison was made between the total quantity of chloride ions measured with a pore liquid extractor (Barneyback and Diamond 1981) and the quantity of soluble chlorides in a test specimen obtained under the same conditions, to formulate the relations of the two quantities for the respective experimental series (Maruya *et al.* 1998b). This allows for conversion of measurements obtained, using the relatively simple and easy JCI method, to the true quantity of chloride ions. The total quantity of chloride ions in a specimen was determined from the product of the concentration of chloride ions in the pore liquid and the volume of the pore liquid. The volume of pore liquid was defined as the volume of pores (3.2 nm to 320 μm in radius) measured using mercury intrusion porosimetry (MIP) in the specimen after immersion in the NaCl solution. More specifically, it

Table 6.3 Test items

| Test items | Shape of test specimens | Concentration of chloride ions in immersion solution | Period of immersion |
|---|---|---|---|
| Measurement of chloride ion concentration | | | |
| Measurement of quantity of chlorides by the JCI method | Sliced φ50 x 10 mm | 3 levels[1] (1%, 3%, 10%) | 2 periods (14, 28 days) |
| Determination of quantity of Friedel's salt | | | |
| Measurement of quantity of chlorides by the JCI method | Cubic 7 mm on a side | 1% | 56 days |
| | | 3% | 28, 56 days |
| | | 10% | 14, 28, 56 days |

[1] Five levels (1%, 3%, 6%, 10%, 15%) for test series NC and B20.

was assumed that the volume of spaces in which chloride ions may reside is equal to the volume of pores measured by MIP.

After the 10 mm test slices were immersed in NaCl solution (using three levels of concentration as listed in Table 6.3) for 14 and 28 days, the concentration of chloride ions in the pore liquid was measured, the quantity of total soluble chlorides was measured by the JCI method and the quantity of Friedel's salt was determined by powder X-ray diffractometry (XRD). Further, in order to accelerate the penetration of chloride ions from the exterior in a short period of time, cubic specimens with a large ratio of immersed area to volume were used in addition to the slice specimens for the measurement of the total quantity of chlorides and soluble chlorides. The concentrations of chloride ions in the solutions used for immersion of specimens and immersion periods are listed in Table 6.3.

6.1.1.3.2 DETERMINATION OF QUANTITY OF FRIEDEL'S SALT

The quantity of Friedel's salt was determined by the XRD internal standard method (referred to here as the XRD method) using synthesized Friedel's salt and its calibration curve. Methods of determining the quantity of calcium aluminate hydrates by the XRD method aiming at pastes of Portland cement and synthetic clinker minerals have been reported in several papers proposed by Inoue *et al.* (2002) and Sakai *et al.* (1998). However, there have been few examples of the XRD method being used for mortar specimens. Accordingly, the procedure used to determine the quantity of Friedel's salt by the XRD method is described below (Figure 6.2).

The standard substance of Friedel's salt was synthesized by hydrating the mixture of $C_3A$ and $CaCl_2$ with a stoichiometric ratio at a water-to-powder

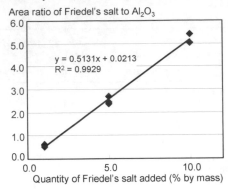

Area ratio of Friedel's salt to Al$_2$O$_3$

$y = 0.5131x + 0.0213$
$R^2 = 0.9929$

Quantity of Friedel's salt added (% by mass)

*Figure 6.2*  Calibration curve prepared by XRD internal standard method.

ratio of 2, and cured for seven days (Japan Cement Association 2001). The C$_3$A was prepared by firing the mixture of reagent-grade calcium carbonate and aluminum oxide. Ideally, the density and chemical composition of a sample for preparing the XRD method calibration curve should be close to those of the specimen to be measured (Japan Society for Analytical Chemistry 2005). For this experiment, standard sand at 50% as specified in the mix proportion was added so that the sample was under conditions as close as possible to the specimen after immersion in the NaCl solution. Calcium carbonate was used as a diluent to supply calcium and it was mixed with the synthesized Friedel's salt in prescribed quantities. For example, the quantity of Friedel's salt required to give 1% on a mass percentage basis on the calibration curve was obtained by mixing the standard sand, calcium carbonate and Friedel's salt to make a 50:49:1 mix of standard sand:calcium carbonate:Friedel's salt on a mass percentage basis. As an internal standard, 10% α-Al$_2$O$_3$ (by mass of sample) was added to the sample stated above. The standard sand, diluent, standard substance (Friedel's salt) and internal standard (α-Al$_2$O$_3$) were dried at reduced pressure using a water aspirator for seven days and pulverized before use.

The diffraction peak used for the determination of quantity was at $2\theta =$ 11.3° [002] for Friedel's salt and at $2\theta = 52.5$° [024] for α-Al$_2$O$_3$ (CuKα = 1.5405 Å). X-ray diffraction measurement conditions were as follows: tube voltage = 40 kV, tube current = 250 mA, scan rate = 0.2 deg/min and step size = 0.02 deg. The area of diffraction peak was determined by calculating an integrated intensity of the function after fitting of the peak profile using the least-squares method. The calibration curve was prepared from the ratio of diffraction peak area (of the standard substance to internal standard). In this case, if a diffraction peak of Kuzel's salt was present adjacent to that of Friedel's salt, the integrated intensities of the Friedel's salt and Kuzel's salt were separated by the addition of a function corresponding to the diffraction peak of Kuzel's salt and by function-fitting the diffraction peak. By this

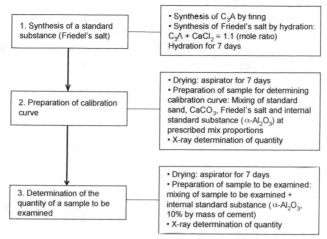

*Figure 6.3* Procedures for determining quantity of Friedel's salt.

means, only the integrated intensity of the Friedel's salt was used. As can be seen from Figure 6.3, the calibration curve shows good linearity.

Using the calibration curve, the quantity of Friedel's salt in specimens after immersion in the NaCl solution was determined as follows. A mortar specimen immersed in the solution for the prescribed period was pulverized with acetone to stop hydration, and after suction filtration the pulverized specimen was dried in a water aspirator for seven days. The specimen was further pulverized before measurement and 10% internal standard ($\alpha$-Al$_2$O$_3$) by mass of specimen was added to the pulverized material. The mass of the specimen used for determination of the quantity of Friedel's salt was standardized at 2.0 g. The ratio of diffraction peak area of the Friedel's salt to that of the internal standard, as obtained from X-ray diffraction, was calculated and the quantity of Friedel's salt in the mortar was then determined from the calibration curve.

6.1.1.3.3 METHOD OF CALCULATING QUANTITY OF CHLORIDES NORMALIZED BY MASS OF BINDER

The total quantity of chlorides and soluble chlorides determined by the JCI method and the quantity of Friedel's salt determined by the XRD method are given as a ratio to the mass of the sample (% by mass of sample). Although this measure is effective for evaluating the total quantity of chlorides in concrete, it does not provide a generalized unit suitable for various mixing conditions when evaluating differences in chloride binding capacity due to the type of cement and admixtures. For example, taking into account the fact that chlorides in hardened cement pastes are bound by the cement hydrates, thus in mixes with large cement content, the quantity of chlorides bound per mass of sample is high because the quantity of hydrates contributing to binding chlorides is large. For this reason, the measured chloride amounts were rearranged as a ratio to the mass of binder (% by mass of binder).

To convert the measured quantity of chlorides per mass of sample to the quantity of chlorides per mass of binder, the simplest method is to assume that the unit mass of binder, the unit mass of water and the unit mass of aggregate in the specimen are the same as in the original mix. However, as shown in Figure 6.4, the quantity of binder may vary even for an equivalent sample mass, since the quantity of water remaining in a sample after drying in a water aspirator (or D-dry) varies with drying conditions and the degree of hydration. To eliminate factors causing such errors wherever possible, corrections are made by the loss on ignition method in converting from the quantity of chlorides by mass of sample to that by mass of binder.

First, the ratio of a component in a specimen is determined after D-dry in use of the quantity of water determined by the loss on ignition method as

$$P_{Wn} = \frac{(P_S + P_B) \cdot I}{100 - I} \tag{6.1}$$

where $P_{Wn}$ is the ratio of mass of water (mainly bound water) contained in the sample determined by the loss on ignition, $P_S$ is the ratio of mass of fine aggregate, $P_B$ is the ratio of mass of binder, and $I$ is the measured loss on ignition (% by mass of sample). Because the quantity of fine aggregate is the same as the quantity of binder, $P_S$ and $P_B$ in Equation 6.1 are both set to unity.

From Equation 6.1, the quantity of chlorides per mass of sample obtained by the JCI method and XRD method can be converted into a quantity per mass of binder by using Equation 6.2.

$$Cl_{binder} = Cl_{sample} \cdot \frac{P_S + P_B + P_{Wn}}{P_B} \tag{6.2}$$

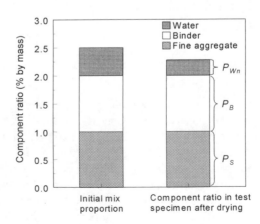

*Figure 6.4* Determination of component ratio of a component by loss on ignition method.

where $Cl_{binder}$ is the quantity of chlorides per mass of binder (% by mass of binder), and $Cl_{sample}$ is the quantity of chlorides per mass of sample (% by mass of sample).

Next, let us convert the concentration of chloride ions obtained with the pore liquid extractor. The concentration of chloride ions per mass of binder $C_{free}$ (% by mass of binder) is given by,

$$C_{free} = C_{Cl} \cdot M_{Cl} \cdot \frac{V_{ion}}{B} \cdot 100 \tag{6.3}$$

where $C_{Cl}$ is the concentration of chloride ions in the pore liquid [mol/m$^3$], $M_{Cl}$ is the atomic mass of chlorine [mol/m$^3$], $V_{ion}$ is the volume in which chloride ions are present [m$^3$/m$^3$], and $B$ is the mass of binder per unit volume of concrete [kg/m$^3$]. As the volume of pores (of radius 3.2 nm to 320 μm) measured by MIP is defined as the volume where chloride ions can reside, we have,

$$V = \frac{V_{ion}}{B + S + W_n} \tag{6.4}$$

where $V$ is the volume of pores (of radius 3.2 nm to 320 μm) per unit mass measured by MIP [m$^3$/kg], $S$ is the mass of fine aggregate per unit volume of concrete [kg/m$^3$], and $W_n$ is the ratio of mass of water (mainly bound water) contained in a sample determined by the loss on ignition [kg/m$^3$]. Finally, we have Equation 6.5 from Equations 6.3 and 6.4 as,

$$C_{free} = \frac{C_{Cl} \cdot M_{Cl} \cdot V}{B} (B + S + W_n) \cdot 100 \tag{6.5}$$

### 6.1.2 Chloride-binding capacity of mortar using Portland cement without admixtures

#### 6.1.2.1 Relationship between chloride ions and soluble chlorides

The concentration of chloride ions in the pore liquid is measured by potentiometric titration of a solution obtained with a pore liquid extractor using a silver nitrate solution. Figure 6.5 shows actual examples of measurements, i.e., the concentration of chloride ions in the pore liquid extracted from test specimens HC, NC and LC immersed in NaCl solutions with three levels of concentration for 28 days. NaCl solutions of 1%, 3% and 10% are equal to 173, 529 and 1,900 mol/l in chloride ion concentration. As this figure shows, the concentration of chloride ions present in hardened cement pastes is higher than the concentration in the immersion solution and varies with the type of cement. Similar tendencies were reported in the past (Maruya *et al.* 1998b, Someya *et al.* 1989). In addition, the lower the concentration of the NaCl immersion solution, the greater is the increment in the concentration of

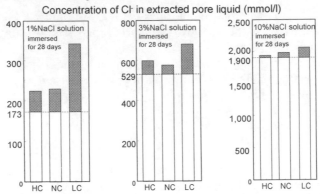

*Figure 6.5* Concentration of chloride ions in pore liquid extracted from test samples.

chloride ions in the pore liquid. This phenomenon is interpreted as: positively charged pore surfaces attract negative chloride ions, resulting in an increased concentration of chloride ions in the pore liquid (Maruya *et al.* 1998b). Accordingly, this result indicates that the same value cannot be assumed for the concentrations of chloride ions in hardened cement pastes and pore liquid even if the chloride ions are in equilibrium.

Figure 6.6 shows the relation of the quantity of chloride ions and that of soluble chlorides (expressed as percentage by mass of powder). The quantity of soluble chlorides is determined from Equation 6.2 and the quantity of chloride ions in a specimen is determined from Equation 6.5. For the experimental series with gypsum added, the quantity of powder is the sum of cement and gypsum. As shown in the figure, there is a greater quantity of soluble chlorides than the chloride ions in the specimens under the same conditions of immersion period and concentration. This relation of chloride ions and soluble chlorides is formulated by a power approximation of the experimental results within the range of the experimental conditions, i.e., for chloride ions up to 2.5% by mass of binder as

$$
\begin{array}{llll}
\text{HC} & C_{sol}=1.56 \cdot C_{free}^{\,0.64} & (R^2=0.996) & \\[2mm]
\text{NC} & C_{sol}=2.07 \cdot C_{free}^{\,0.55} & (R^2=0.983) & \\[2mm]
\text{NGS} & C_{sol}=1.70 \cdot C_{free}^{\,0.65} & (R^2=0.954) & (6.6) \\[2mm]
\text{NGL} & C_{sol}=1.37 \cdot C_{free}^{\,0.70} & (R^2=0.860) & \\[2mm]
\text{LC} & C_{sol}=1.67 \cdot C_{free}^{\,0.59} & (R^2=0.972) &
\end{array}
$$

where $C_{free}$ is the quantity of chloride ions [% by mass of binder] and $C_{sol}$ is the soluble chlorides [% by mass of binder]. From a physical viewpoint, the power function means that the soluble chlorides decrease as the chloride ions increase. To be more specific, the quantity of partially adsorbed and solid-

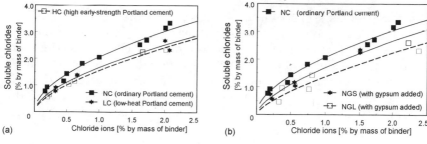

*Figure 6.6* Relationship between quantities of chloride ions and soluble chlorides in mortar using Portland cement without admixtures.

phase constituents extracted from specimens using warm water decreases relatively and the quantity of chloride ions and that of soluble chlorides gradually equalizes in the region of higher concentration of chloride ions.

### 6.1.2.2 Effects of quantity of gypsum

Figure 6.7 shows the relations of the quantities of chloride ions and bound chlorides in the series NC, NGS and NGL. The experimental results in the figure include those obtained by both the pore liquid extraction and the JCI methods. The quantity of soluble chlorides determined by the JCI method is converted to the quantity of chloride ions using Equation 6.6. Calculation results exceeding the applicable range of Equation 6.6, i.e., chloride ions exceeding 2.5% by mass of binder, are removed from the plot. The quantity of bound chlorides is determined by deducting the quantity of chloride ions from the measured total quantity of chlorides. As can be seen from the figure, the quantity of bound chlorides shifts greatly with the addition of gypsum, and chloride-binding capacity decreases markedly as more gypsum is added. To study the mechanism in more detail, the authors attempted to separate the bound chlorides into solid-phase chlorides and adsorbed ones.

The study was carried out by using the specimens immersed in NaCl solution for 14 days. The results obtained by the XRD method are shown in

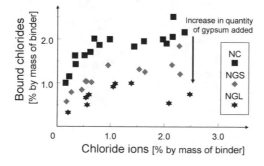

*Figure 6.7* Effects of quantity of added gypsum on chloride binding.

Figure 6.8. Only the experimental results at diffraction angles $2\theta = 10°–12°$ are extracted and plotted. As is clear from the series NC and NGS, the formed Friedel's salt tends to increase as the concentration of chloride ions increases. In addition, the diffraction peak intensity corresponding to Friedel's salt decreases as the gypsum increases. No Friedel's salt was observed in the series NGL, in which the greatest amount of gypsum was added. Figure 6.9 shows the results obtained by the XRD method (at $2\theta = 8.5–10.5°$) immediately

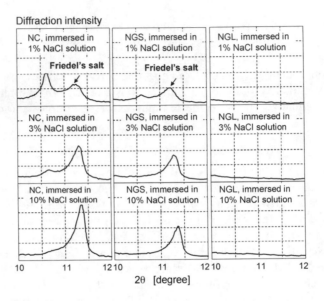

*Figure 6.8*  XRD peak (after immersion).

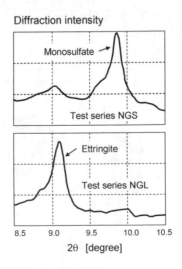

*Figure 6.9*  XRD peak (before immersion).

before the immersion of specimens in the NaCl solution. As is apparent from the figure, the addition of a large quantity of gypsum in the series NGL increases the formation of ettringite (the peak at $2\theta = 9.1°$) but does not cause a shift to mono-sulfate (the peak at $2\theta = 9.9°$). This is why the formation of Friedel's salt is not observed in the series NGL. A similar tendency was reported in cement pastes containing gypsum mixed with ground, granulated blast furnace slag (Kato *et al.* 2003).

A diffraction peak at $2\theta = 10.6°$ distinct from the diffraction peak corresponding to Friedel's salt is observed in the series NC and NGS at low NaCl concentrations (Figure 6.8). The same tendency was observed in the series described later where ground, granulated blast furnace slag was used (Figure 6.15). This substance identified at low NaCl concentrations is likely to be Kuzel's salt ($C_3A \cdot (0.5CaSO_4 \cdot 0.5CaCl_2) \cdot 10H_2O$), as pointed out in the past research. However, this diffraction peak is not included with the solid-phase chlorides in this study, because no method of determining the quantity of Kuzel's salt by the XRD method has yet been established (Glasser *et al.* 1999).

Figure 6.10 shows the relationships between the quantities of chloride ions, solid-phase chlorides and adsorbed chlorides for respective experimental series. The quantity of adsorbed chlorides is determined by deducting the quantity of solid-phase chlorides obtained by the XRD method from the total quantity of bound chlorides. As described earlier, the solid-phase chlorides indicate only Friedel's salt ($C_3A \cdot CaCl_2 \cdot 10H_2O$). Accordingly, compounds that bind chlorides in a different form, such as Kuzel's salt, may be included in the adsorbed chlorides.

The quantity of solid-phase chlorides varies greatly depending on the experimental series and decreases with the amount of gypsum added. On the other hand, no clear difference is observed in the amount of adsorbed chlorides among the various series of tests. Adsorption behavior was expected to vary with the specific surface area or charge state of the pore walls, but from a macroscopic viewpoint, no particular changes in adsorption behavior were

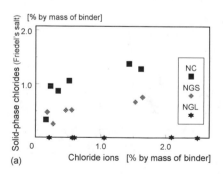

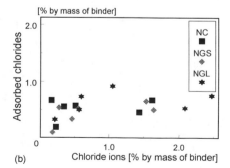

*Figure 6.10* Relationships between quantities of solid-phase chlorides and chloride ions (effects of addition of gypsum).

observed with the addition of gypsum. As noted previously, the difference of the chloride-binding capacity among the series NC, NGS and NGL is chiefly dependent on the quantity of solid-phase chlorides.

### 6.1.2.3 Effects of mineral composition

Next, the effects of mineral composition are discussed. Figure 6.11 shows the equilibrium between chloride ions and bound chlorides for the series NC (ordinary mortar), HC (high early-strength mortar) and LC (low-heat mortar). Quantities are calculated in the same way as described in the previous section. With regard to chloride-binding capacity, the ranking of mortar types from lowest to highest is NC, HC and LC. This is thought to reflect the $Al_2O_3$ content in the cement (Table 6.1).

Bound chlorides were again separated into solid-phase chlorides and adsorbed ones to investigate the chloride binding by solid-phase and adsorbed chlorides, as shown in Figure 6.12. The solid-phase chloride varies greatly according to mineral compositions. In particular, the quantity of formed Friedel's salt is small in the series made with low-heat cement. Mineral

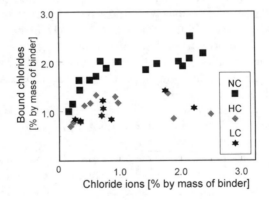

*Figure 6.11* Effects of differences in mineral composition on chloride binding.

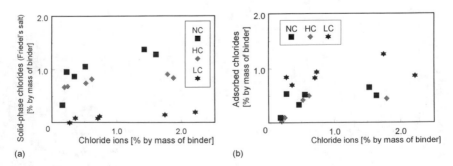

(a)   (b)

*Figure 6.12* Relationships between quantities of chloride ions and solid-phase chlorides and adsorbed chlorides (effects of mineral composition).

compositions estimated by Bogue's formula based on the chemical compositions in Table 6.1 are shown in Table 6.4. The estimated proportions of calcium aluminate, $C_3A$, in ordinary and low-heat cements are 10.0% and 3.1%, respectively. The quantities of solid-phase chlorides in the series made with ordinary and low-heat cements were expected to reflect the proportion of $C_3A$, but the discrepancy from actual measurements is larger than originally thought.

The chemical proportion of $C_3A$, based not on absolute quantity but in the form of mole ratio of gypsum to $C_3A$ ($CaSO_4/C_3A$), is also listed in Table 6.4. The estimated mole ratio for the series LC made with low-heat cement is 2.61, which is nearly the same as that for the series NGL (2.68) in which a large quantity of gypsum is added to ordinary cement. More specifically, the quantities of gypsum contained in both series are relatively high in relation to $C_3A$ and it is thought that, as a result, the quantity of Friedel's salt formed from mono-sulfate is lower. This hypothesis is considered applicable not only to the series LC and NGL but also to all other experimental cases. In other words, the quantity of formed Friedel's salt correlates with the mole ratio of $CaSO_4/C_3A$.

On the other hand, no noticeable difference was observed in the quantity of adsorbed chlorides with different mineral compositions. This is similar to the result discussed in the previous section. Although adsorption behavior was expected to vary with the pore structure of the hardened cement paste and the C/S ratio, it was clear that no particular change in the quantity of adsorbed chlorides took place within the range of conditions where only Portland cement was used as the binder.

Table 6.4 Mineral compositions estimated by Bogue's formula.

| Test series | Type of cement | Mineral composition estimated by Bogue's formula (%) | | | | $CaSO_4^{*1}$ | Converted quantity of total $CaSO_4^{*2}$ | $CaSO_4/$ $C_3A$ Mole ratio |
|---|---|---|---|---|---|---|---|---|
| | | $C_3S$ | $C_2S$ | $C_3A$ | $C_4AF$ | | | |
| NC | OPC | 53.5 | 21.3 | 10.0 | 8.9 | 3.0 | 3.0 | 0.59 |
| HC | HPC | 70.7 | 3.9 | 8.6 | 8.5 | 5.0 | 5.0 | 1.16 |
| LC | LPC | 23.1 | 56.4 | 3.1 | 10.7 | 4.1 | 4.1 | 2.61 |
| NGS | OPC | 51.0 | 23.0 | 9.5 | 8.5 | 2.8 | 6.6 | 1.58 |
| NGL | OPC | 48.4 | 19.2 | 9.1 | 8.1 | 2.7 | 10.2 | 2.68 |

*1 $SO_3$ in cement is counted as part of the total quantity of $CaSO_4$ (%).
*2 Total quantity of $CaSO_4$ in the binder, including the quantity of $CaSO_4$ added, as a percentage by mass of aggregate.

### 6.1.3 Chloride-binding capacity of mortar including admixtures

#### 6.1.3.1 Effects of ground, granulated blast furnace slag as cement replacement

In line with the method described in the previous section, the first step is to experimentally obtain the relation of the chloride ions and soluble chlorides for mortars using ground, granulated blast furnace slag (Figure 6.13) as

$$
\begin{array}{lll}
\text{B20} & C_{sol}=2.23 \cdot C_{free}^{0.53} & (R^2=0.977) \\
\\
\text{B40} & C_{sol}=2.17 \cdot C_{free}^{0.55} & (R^2=0.946) \\
& & & (6.7) \\
\text{B60} & C_{sol}=1.86 \cdot C_{free}^{0.56} & (R^2=0.971) \\
\\
\text{B80} & C_{sol}=1.60 \cdot C_{free}^{0.65} & (R^2=0.985)
\end{array}
$$

Figure 6.14 shows the relationship between chloride ion and bound chloride quantities for the series B using ground, granulated blast furnace slag (with a cement replacement ratio of 0%–80%). The specimens with low replacement ratios, such as B20 and B40, have a relatively high chloride-binding capacity, whereas the chloride-binding capacity of the specimen with a 60% replacement ratio is comparable to that of the series NC without replacement. The test specimen with an 80% replacement ratio is lower than that of all specimens using Portland cement without admixtures. A similar tendency in the relationship between the replacement ratio of ground

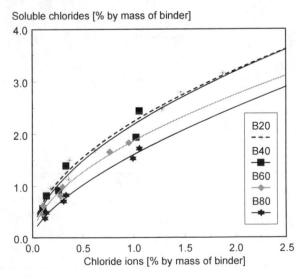

*Figure 6.13* Relationship between quantities of chloride ions and soluble chlorides (where ground, granulated blast furnace slag is used).

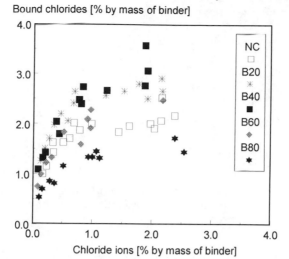

Bound chlorides [% by mass of binder]

Chloride ions [% by mass of binder]

*Figure 6.14* Chloride binding capacity of mortar with varying ratios of ground, granulated blast furnace replacement.

granulated blast furnace slag and chloride-binding capacity was reported by Kato *et al.* (2003). To study the mechanism in detail, bound chlorides were separated into solid-phase chlorides and adsorbed chlorides as in the previous section.

The study was carried out by using the case where specimens were immersed in NaCl solution for 14 days. The results obtained by the XRD method are shown in Figure 6.15. Results at a diffraction angle $2\theta = 10°$–$12°$ only are extracted. The common tendency among this series is that the formed Friedel's salt increases with increasing salt concentration in the solution, but the diffraction peak of Friedel's salt decreases as the replacement ratio exceeds 40%. To quantitatively analyze these phenomena, the

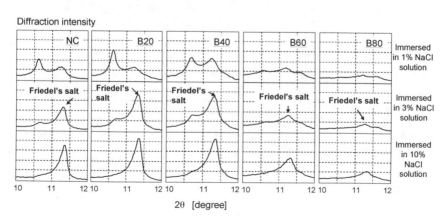

*Figure 6.15* XRD peak (test series using ground, granulated blast furnace slag).

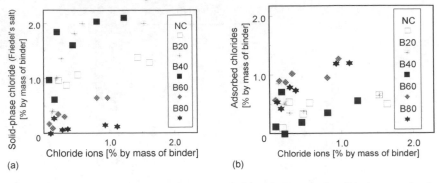

*Figure 6.16* Relation between quantities of chloride ions and solid-phase and adsorbed chlorides (effects of replacement of cement with ground, granulated blast furnace slag).

relationships between the quantities of chloride ions and solid-phase and adsorbed chlorides are plotted (Figure 6.16). As this figure makes clear, the quantities of solid-phase chlorides in the series B20 and B40 are markedly higher. The inferred reason for this is that, although the quantity of $Al_2O_3$ originating from Portland cement decreases as the replacement ratio is increased, the formed Friedel's salt increases through monosulfate formed from the reaction between ground, granulated blast furnace slag and gypsum or the reaction between ground, granulated blast furnace slag itself and chloride ions (Wanibushi *et al.* 1999). The other phenomenon is that the solid-phase chlorides decrease at replacement ratios of 60% or more. The detailed mechanism behind this is unknown, but possible factors are a change in the chemical composition of the pore liquid and a change in the calcium aluminate hydrates as the replacement ratio increases.

Next, attention is directed to the adsorbed chlorides. For the series with high replacement ratios, such as B60 and B80, the quantity of adsorbed chlorides was markedly higher. Although a simple comparison is not possible because the compositions of the formed C–S–H (particularly the C/S ratio) vary with the replacement ratio, a possible explanation is that the use of ground, granulated blast furnace slag increases specific surface area of the pore structure, resulting in a greater adsorbable area.

### 6.1.3.2 *Effects of pozzolans as cement replacement*

Figure 6.17 shows the relation of the quantities of chloride ions and soluble chlorides in the series using fly ash. Regression curves for the relation are given by

$$F20 \qquad C_{sol} = 1.95 \cdot C_{free}^{0.46} \qquad (R^2 = 0.999)$$
$$F40 \qquad C_{sol} = 1.84 \cdot C_{free}^{0.52} \qquad (R^2 = 0.985)$$

$$(6.8)$$

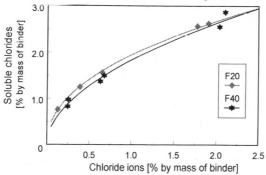

*Figure 6.17* Relation between quantities of chloride ions and soluble chlorides (where fly ash is used).

The relation of the chloride ions and bound chlorides is shown in Figure 6.18. As compared to the series using cement without admixtures, the amount of bound chlorides is slightly lower in the high-concentration region.

The bound chlorides were separated into solid-phase chlorides and adsorbed ones. The resulting relationships between quantities of chloride ions and solid-phase chlorides and adsorbed chlorides are shown in Figure 6.19. The quantity of solid-phase chlorides tends to be higher than in the series NC at a replacement ratio of 40%. This indicates that $Al_2O_3$ in the fly ash contributes greatly to the formation of Friedel's salt. Focusing on the quantity of adsorbed chlorides, the adsorption capacity of the series F20 is comparable to that of NC but is substantially lower at a replacement ratio of 40%. The reason for this decrease is that, although the pore structures become finer and the specific surface area increases with the inclusion of fly ash, the composition of the formed C–S–H, particularly the C/S ratio, decreases substantially, resulting in a decrease in adsorption capacity (Beaudoin *et al.* 1990).

Next, the mortars with part of the cement replaced with silica fume are studied. The relation of the quantities of chloride ions and soluble chlorides are shown in Figure 6.20 and formulated as

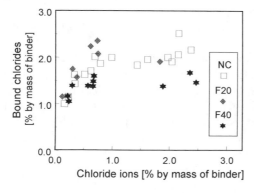

*Figure 6.18* Chloride binding capacity of mortar using fly ash.

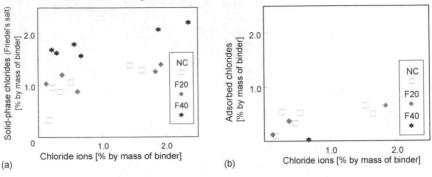

*Figure 6.19*  Relation between chloride ions and solid-phase chlorides and adsorbed chlorides (effects of replacement of cement with fly ash).

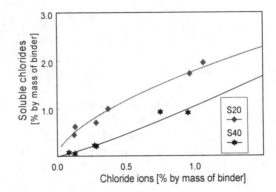

*Figure 6.20*  Relation between quantities of chloride ions and soluble chlorides (where silica fume is used).

$$
\begin{aligned}
\text{S20} \qquad & C_{sol} = 1.78 \cdot C_{free}^{0.61} \qquad (R^2 = 0.958) \\
& \qquad\qquad\qquad\qquad\qquad\qquad\qquad\qquad (6.9) \\
\text{S40} \qquad & C_{sol} = 1.04 \cdot C_{free}^{1.18} \qquad (R^2 = 0.943)
\end{aligned}
$$

Equation 6.9 is applicable up to a concentration of 1.5% of chloride ions (by mass of binder) because the quantity of chloride ions permeating the interior is low.

Figure 6.21 shows the equilibrium between chloride ions and bound chlorides when part of the cement is replaced with silica fume. Commonly, the silica fume replacement ratio is 5%–15% of the cement and the water-to-binder ratio is often kept low for the purpose of increasing strength and durability. In contrast to this standard practice, a high water-to-binder ratio and intentionally higher replacement ratio was used in the experimental mix proportions here because it proved very difficult to extract pore liquid from test specimens with a low water-to-binder ratio. This should not affect the most important objective of this research, which is to make clear the role of silica fume as a pozzolan in chloride binding. As is clear from Figure 6.21,

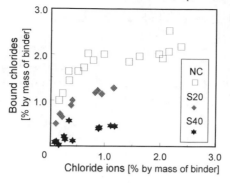

*Figure 6.21* Chloride binding capacity of mortar using silica fume.

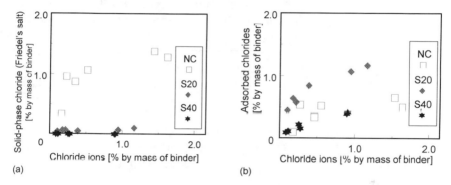

(a)    (b)

*Figure 6.22* Relation between quantities of chloride ions and solid-phase chlorides and adsorbed chlorides (effects of replacement of cement with silica fume).

the quantity of bound chlorides falls as the replacement ratio is increased. As before, the bound chlorides were separated into solid-phase chlorides and adsorbed ones. The relationships between quantities of chloride ions and solid-phase chlorides and adsorbed chlorides are shown in Figure 6.22. The detected amount of solid-phase chlorides was infinitesimal. This is attributed to the fact that the silica fume itself contains little $Al_2O_3$ and the pozzolanic reaction substantially reduces the concentration of $OH^-$ ions. As a result, Friedel's salt disappears. The quantity of adsorbed chlorides in the series NC is similar in tendency to that of S40, while in the series S20, it is higher. With regard to chloride-binding capacity, the observed macroscopic change in the quantity of adsorbed chlorides is thought to occur as a result of the combination of the positive effect of the finer micro-pore structure (an increase in surface area) and the negative effect of a decrease in the C/S ratio of hydrates caused by the addition of silica fume.

As discussed in the previous section, changes in the mineral composition and quantity of gypsum added do not greatly affect the quantity of adsorbed chlorides in mortar using Portland cement without admixtures. The primary factor that controls chloride-binding capacity in such cases is the quantity of formed Friedel's salt. For mortar including admixtures, the formed Friedel's salt is dependent on the chemical composition of a binder and, further, the properties of the formed hydrates vary, which leads to a large difference in the macroscopic quantity of adsorbed chlorides. This is one of the characteristic phenomena associated with the use of admixtures.

## 6.2  Modeling of chloride ion transport

### 6.2.1  Governing laws and equations

For structures exposed to sea-water and/or de-icing salt, chloride penetration is a common cause of deterioration of reinforced concrete. In order to predict the service life of such structures, it is necessary to quantify the chloride transport process in cementitious materials. It is well known that binding capacity and chloride diffusivity have a significant effect on the chloride diffusion process. In this section, the chloride-binding model describing the equilibrium between chloride ions and bound chlorides is also modeled as a non-linear function based on Langmuir's equation, referring to the experimental data in the preceding section.

Chloride transport in cementitious materials under usual conditions is an advective–diffusive phenomenon. In modeling, the advective transport due to bulk movement or pore solution phase is considered, as well as ionic diffusion due to concentration gradients. Mass balance for (moveable) chloride ions can be expressed as (Ishida and Maekawa 1999; Maekawa *et al.* 1996, 2003b)

$$\frac{\partial}{\partial t}(\phi S C_{Cl}) + div J_{Cl} - Q_{Cl} = 0, \tag{6.10}$$

where $\phi$ is porosity, $S$ is the degree of saturation of porous media, $C_{Cl}$ is the concentration of chloride ions in pore solution, and $J_{Cl}$ is the total flux of chloride ion. The first term in Equation 6.10 represents the rate of change in total amount of chloride ion per unit of time and volume, the second term is the flux of the chloride ion, and the third term, $Q_{Cl}$, is a sink term. Similar to the case of carbon dioxide transport (Chapter 4), only capillary and gel pores, which can act as transport paths, or locations for chemical reactions, are considered (the porosity $\phi$ in Equation 6.10 is the sum of the capillary and the gel pores). Figure 6.23 represents the summary of the governing equation for chloride transport.

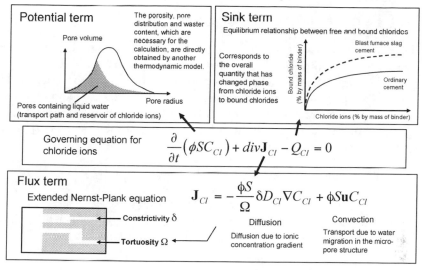

*Figure 6.23* Law of mass conservation for chloride ions.

## 6.2.2 Modeling of chloride diffusion coupled with the non-linear chloride-binding model

### 6.2.2.1 Chloride flux due to advection and diffusion

The flux of chloride ion through a porous body, taking both the diffusion and the convection, is expressed by,

$$J_{Cl} = -\frac{\phi S}{\Omega} \delta D_{Cl} \nabla C_{Cl} + \phi S \mathbf{u} C_{cl} \tag{6.11}$$

where $J_{Cl}$ is the flux of chloride ion (mol/m².s), $\Omega$ is tortuosity (a reduction factor in terms of complex micro-pore structure), $\delta$ is constrictivity (a reduction factor due to the interaction between pore structure and ion transport), $D_{Cl}$ is the diffusion coefficient of chloride ion in pore solution (m²/s), $C_{Cl}$ is the concentration of chloride ions in pore solution phase (mol/l), $u^T = [\, u^x \; u^y \; u^z \,]$ is the velocity vector of ions due to the bulk movement of pore solution phase (m/s), and $\phi$ is porosity of the porous medium (m³/m³).

In porous media, the diffusion coefficient is lower than that in the absence of a porous medium. Diffusion paths in concrete are constrained because pore structure is tortuous compared with diffusion paths in free liquid and path direction is not parallel to concentration gradient. Tortuosity is introduced to account for this complex micro-pore structure. This parameter expresses a reduction factor in terms of chloride penetration rate due to complexity of the micro-pore structure. The tortuosity model in Section 5.1.4 is applied to chloride ions as well. The value of tortuosity changes according to geometric characteristics of the pore structures and it is a function of porosity as

$$\Omega = -1.5\tanh\big(8.0\big(\phi_{paste} - 0.25\big)\big) + 2.5 \qquad (6.12)$$

where $\phi_{paste} = \phi_{cp} + \phi_{gl}$ is the total paste porosity (m³/m³), $\phi_{cp}$ is the capillary zone porosity (m³/m³), and $\phi_{gl}$ is the gel zone porosity (m³/m³).

Another parameter defined as constrictivity considers the effect of interaction between pore structure and ion transport. If the cross-section of a pore space segment is straight, then constrictivity becomes unity, whereas if the segments are restricted at certain points, then the value of constrictivity is less than unity as (Section 5.1.4)

$$\delta = 0.395\tanh\{4.0(\log(r_{cp}^{peak}) + 6.2)\} + 0.405 \qquad (6.13)$$

where $r_{cp}{}^{peak}$ is the peak radius of capillary pores [m].

Similar to Equation 5.11 in Section 5.1.4, the diffusion coefficient of chloride movement is expressed according to Einstein's theorem by

$$D_{Cl} = RT\frac{\lambda_{ion}}{Z_{Cl}^2 F^2}\left(1 + \frac{\partial\ln\gamma_{Cl}}{\partial\ln C_{Cl}}\right) \qquad (6.14)$$

where $R$ is the gas constant (8.314J/mol-K), $T$ is the absolute temperature (K), $Z_{Cl}$ is the electric charge of the chloride ion (=-1), $F$ is Faraday's constant (9.65E+4 C/mol), $\lambda_{ion}$ is ion conductivity (Sm²/mol), and $\gamma_{Cl}$ is molar conductivity of chloride ions. In order to obtain the correction term $\partial\ln\gamma_{Cl}/\partial\ln C_{Cl}$ in Debye–Hückel's theory, it is necessary to identify the anions and cations in equilibrium in the solution. However, the effect of this term on diffusion of chloride ions can be negligible when the concentration of chloride ions is around 3% NaCl (by mass), which is the main target of this study. Therefore, in the modeling, the correction term is ignored. Regarding the molar conductivity of an ion, $\lambda_{ion}$, temperature dependency is taken into account by Arrhenius's law (Yokozeki *et al.* 2003) as

$$\lambda_{ion} = \lambda_{25}\exp\left[-\frac{E_a}{R}\left(\frac{1}{T} - \frac{1}{298}\right)\right] \qquad (6.15)$$

where $\lambda_{25}$ is ion conductivity at 25°C (= $7.63\times10^{-3}$(Sm²/mol)), and $E_a$ is the activation energy for free pore fluid (= $17.6\times10^3$ (J/mol)).

### 6.2.2.2 Modeling of chloride binding

Chlorides in cementitious materials have free and bound components. The bound components exist in the form of chloro-aluminates and adsorbed phases on the pore walls, making them unavailable for free transport. In this chapter, by following the classification of chlorides as mentioned above, the relationship between chloride ions and bound chlorides under equilibrium condition is modeled as a Langmuir type equation based on experiments (Section 6.1).

$$C_b = \frac{\alpha.C_f}{1+4.0C_f} \tag{6.16}$$

where $C_b$ is the concentration of bound chlorides (% by mass of binder), and $C_f$ is the concentration of chloride ions (% by mass of binder). The binding capacity varies depending on the cement mineral composition, the types of admixtures and the replacement ratio, etc. These effects are expressed by parameter $\alpha$ in Equation 6.17. Based on experimental data presented in the preceding section, parameter $\alpha$ is formulated for ordinary Portland cement (OPC), blast furnace slag (SG) and fly ash (FA) as shown in Figure 6.24.

$$
\begin{aligned}
\alpha &= 11.8 && OPC \\
\alpha &= 34.0b^2 + 23.3b + 11.8 && (0 \le b \le 0.6) && SG \\
\alpha &= -15.5f^2 + 1.8f + 11.8 && (0 \le f \le 0.4) && FA
\end{aligned} \tag{6.17}
$$

where $b$ and $f$, are the replacement ratios by mass of blast furnace slag and fly ash, respectively.

### 6.2.2.3 Modeling of chloride flux at the surface

The surface flux of chloride ions on the boundary was modeled by taking into account the diffusion and quasi-adsorption flux (Maruya *et al.* 1992). It was experimentally known that the concentration of chlorides in concrete near exposure surfaces is higher than that of the submerged environment, as can clearly be seen in Figure 6.5. To simulate this phenomenon, Maruya *et al.* (1992) proposed the condensation model at the surface.

It considers the diffusive movement due to the concentration gradient and the quasi-adsorption phase by electro-magnetic attractive force between the positively charged pore walls and the negatively charged chloride ions. The flux of chloride ions through the boundary surface $q_{Cl}$ (mol/m².s) is described

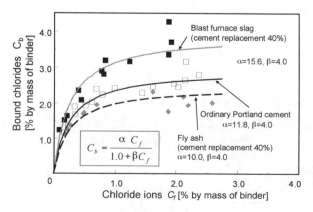

*Figure 6.24* Non-linear chloride binding model for various types of binders.

as the summation of the diffusive component $q_{diff}$ and the contribution of quasi-adsorption $q_{ads}$. It is expected that the quasi-adsorption flux will decrease as the chloride ions and adsorbed chlorides increase, because the migrating chlorides neutralize positive charges on the pore wall. The flux of quasi-adsorption is described by the following function, in which the flux decreases as a function of chlorides in the porous medium. Mathematically, the surface flux model can be expressed by,

$$q_{Cl} = q_{diff} + q_{ads}$$

$$q_{diff} = -E_{Cl}(C_{Cl} - C_s), \quad q_{ads} = k_{Cl}\left(\frac{C_{Cl}}{C_0}\right)^2 \exp(-1.15C_{Cl}) \tag{6.18}$$

where $C_{Cl}$ is the concentration of chloride ions at the exposure surface (mol/l), $C_s$ is the environmental concentration of chloride ions (mol/l), $E_{Cl}$ is the coefficient of chloride flux at the surface (= $1.0 \times 10^{-3}$), $C_0$ is the referential concentration of chloride ions (= $0.51$[mol/l]), and $k_{Cl}$ is a coefficient determined in accordance with the type of binder material. In the case of using admixtures such as blast furnace slag and fly ash, it has been reported that the contribution of quasi-adsorption decreases due to the change of electromagnetic state of the pore walls (Maruya *et al.* 1992). Here, the parameter $k_{Cl}$ has been formulated based on the results of chloride migration experiments and sensitivity analyses.

$$K_{Cl} = 1.5 \times 10^{-3} \qquad\qquad\qquad\qquad OPC$$

$$K_{Cl} = (1.7b^2 + 0.075) \cdot 2.0 \times 10^{-2} \quad (b \le 0.5) \quad SG$$

$$K_{Cl} = 1.0 \times 10^{-2} \qquad\qquad\qquad (b > 0.5) \quad SG \tag{6.19}$$

$$K_{Cl} = (4.4f^2 + 0.075) \cdot 2.0 \times 10^{-2} \quad (f \le 0.2) \quad FA$$

$$K_{Cl} = 0.5 \times 10^{-2} \qquad\qquad\qquad (f > 0.2) \quad FA$$

where $b$ and $f$ are the replacement ratios by mass of blast furnace slag and fly ash, respectively.

### 6.2.2.4 Verification of the chloride transport model

By using the chloride transport model described in the preceding sections, transport simulation was carried out. For verification, a prism-shaped mortar specimen ($5 \times 5 \times 10$ cm) cast with ordinary Portland cement, 50% of the water-to-cement ratio (W/C) and 52.1% of the volume ratio of aggregate was used. After 28 days of curing, the specimens were immersed in 3% NaCl solution by mass of water for 182 days, and the profile of total chloride content was measured by potentiometer technique. The experiment results are shown in Figure 6.25. In this analysis, in addition to the non-linear binding model given by Equation 6.16 and Equation 6.17, a simple chloride-binding model assuming a linear equilibrium relation of the free chloride and the bound one is introduced for comparison as

$$C_b = 2.5 C_f \tag{6.20}$$

Figure 6.25 shows the experiment and computational results by linear and non-linear chloride-binding models, whereas Figure 6.26 shows the analytical results in which the chloride-binding capacity was changed by altering the value of $\alpha$. In these two analyses, apart from the binding models, other data for calculation were exactly the same. A qualitative difference in trend is seen in Figure 6.25 between the non-linear binding model prediction and the experimental ones. In other words, the shape of the distribution obtained from the analysis by non-linear binding has two points of inflection. The surface concentration is relatively high, with a trend of sharp reduction in chloride concentration with depth. On the other hand, the analytical results using the linear binding model show a curve that is convex downwards,

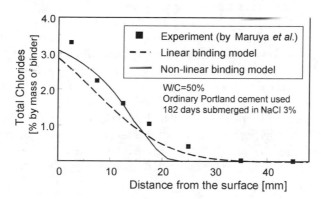

*Figure 6.25* Chloride distributions by linear and non-linear binding model.

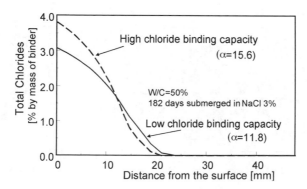

*Figure 6.26* Sensitivity analyses with high and low binding capacities.

which is consistent with Fick's law of diffusion and is close to the actual measured values. This kind of anomalous results by non-linear binding has been reported in past research by Nilson *et al.* (1994) as well.

Furthermore, Figure 6.26 shows the results of sensitivity analysis by altering the binding capacity. In the analysis, $\alpha$ in Equation 6.16 was changed from 11.8 (ordinary cement) to 15.6 (corresponding to 40% replacement with blast furnace slag). For hardened cement having a high binding capacity, although the depth of penetration was slightly reduced, the total quantity of chloride that penetrated into the cement increased enormously, with very high chloride concentration near the surface layer. Since it was not possible to logically explain the phenomenon of chloride transport with a non-linear binding model, it was decided to enhance the parameters associated with chloride diffusion and binding capacity.

### 6.2.2.5 Enhanced modeling of chloride diffusion

Based on the analytical results shown in Figure 6.25 and Figure 6.26, the authors propose a model that explicitly takes into account the interactions between bound chlorides and diffusion movement. In Section 5.1.4, a model of contrictivity for $Ca^{2+}$ (positive ion) is proposed. On the basis of this model, a constrictivity model for $Cl^-$ (negative ion) in which the effect of ion and pore wall interaction and the effect of bound chlorides on the diffusive movement of chloride ions is formulated as shown in Figure 6.27. Here, a new hypothesis was made that when bound chlorides increase, the diffusion rate of chloride ions moving through the pore water is reduced by the increase in the adsorbed component (adsorbed chlorides) on the pore. The physical image behind this hypothesis is shown in Figure 6.27. It is assumed that as the adsorbed chlorides on the pore wall increase, the chloride ions passing through the narrow pore spaces are acted upon electrically by the negatively

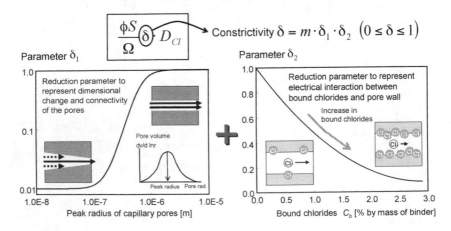

*Figure 6.27* Enhanced constrictivity parameter.

charged adsorbed chlorides, and as a result the diffusion rate would be reduced. An alternative explanation is that the rate of the surface diffusion component driven by the action of the positively charged wall surfaces decreases as the adsorbed chlorides increase. As a result, the apparent diffusion rate decreases. To define the model it was assumed that the constrictivity includes the effect of dimensional changes and connectivity of the pores, and the effect of electrical interaction with the wall surfaces of the pores, which was expressed as follows.

$$\delta = m \cdot \delta_1 \cdot \delta_2 \quad \left(m = 6.38, \text{ If } \delta \geq 1.0, \quad \delta = 1.0\right)$$

$$\delta_1 = 0.495 \tanh\left\{4.0\left(\log\left(r_{cp}^{peak}\right) + 6.2\right)\right\} + 0.505 \tag{6.21}$$

$$\delta_2 = 1.0 - 0.627C_b + 0.107C_b^2$$

where $\delta_1$ is the reduction parameter for dimensional change and connectivity of the pores, and $\delta_2$ is the reduction parameter for electrical interaction.

### 6.2.2.6 Experimental verification of enhanced chloride transport model

The analytical results by the enhanced chloride-binding model are verified by Maruya's experiment (1995). In this experiment, rectangular prism-shaped test specimens (5 × 5 × 10 cm) were made with a water-to-powder ratio of 50% (mass ratio), and a volume ratio of aggregate of 52.1% in all cases. After curing for 28 days, the specimens were immersed in 3% (mass ratio) NaCl solution, and the total chloride content was measured after 182 days and 365 days of constant immersion. Figure 6.28 shows the results for the ordinary Portland cement case. The analysis was carried out both with and without the proposed reduction factor for constrictivity, as shown in Equation 6.23. Figure 6.29 shows the results for 50% blast furnace slag and 20% fly ash replacement of the cement by percent of mass. As shown in Figure 6.28 and Figure 6.29, the proposed model predicted well for all

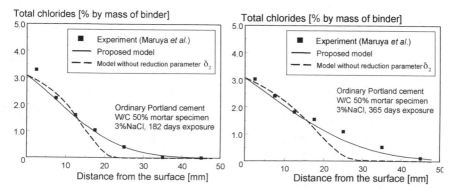

*Figure 6.28* Chloride profile in mortar specimen with ordinary Portland cement.

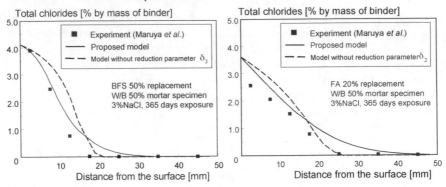

*Figure 6.29* Chloride profile in mortar specimen with admixtures.

cases. For the blast furnace slag case, the analysis has simulated nicely the low chloride penetration trend as compared with the case for ordinary Portland cement, due to the combined effect of fine micro-pore structure and greater chloride-binding capacity.

### 6.2.3 Modeling of chloride diffusivity in cracked concrete

#### 6.2.3.1 Modeling of cracking as a transport path

In the real environment, concrete structures are not always crack-free, and the formation of cracks increases the transport properties of concrete so that moisture along with chloride ions and oxygen easily penetrate and reach the reinforcing steel and speed up the initiation of steel corrosion in concrete. The chloride diffusion model, therefore, is extended for cracked concrete. Crack widths range from very small internal micro-cracks, to quite large cracks caused by unwanted interactions with the environment and external loading. In this section, the transport of chloride ions in cracked concrete is simulated by section large void spaces in a control volume to represent the crack and by proposing a model of chloride diffusivity through the cracked region. To simulate the chloride movement in the cracked path, the chloride diffusion phenomenon is separately defined for cracked and sound concrete.

Figure 6.30 shows a multi-scale pore size distribution model, which includes the large pores between micron and millimeter scales as well as capillary and gel pores. In Section 5.3.3, the large void component was implemented to represent the large airspaces in cemented and natural soils, whereas in this section this large void space model is adopted to represent cracked elements. To model a specific crack width, macro pore size distribution factor $B_{cr}$ is used. This factor considers the peak radius of pore size distribution, and hence is used to represent the crack width. In reality, a crack does not have a size distribution, unlike the void structure in soil or cemented soil. Specifically, soil consists of numerous voids randomly connected to each

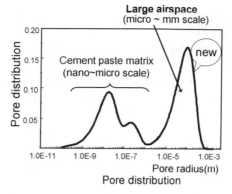

*Figure 6.30* Multi-scale pore structure model.

other, whereas a crack may have a single route. Although cracks are treated as a certain group of large voids, chloride diffusivity in the cracked part is formulated as a function of two predominant factors, i.e., volume fraction of cracks $\phi_{cr}$ and the parameter $B_{cr}$ representing crack width. In other words, for the sake of simplicity, the flux of chloride ions is not obtained by integrating the flow contributions of numerous pores, but two macroscopic factors (volume and averaged crack width) determine the apparent chloride diffusivity. Careful verification will be necessary to validate this methodology in future, however, it has to be noted also that, if such a modeling based on a smeared crack concept is valid, it would be very useful for large-scale computation, such as real-scale structural analysis.

Thus, the authors assume that in this modeling, cracks are modeled by two parameters, namely volume fraction of cracks in a control volume and average width of cracks, and, similar to Equation 5.21 for soil, the distribution of multi-scale pores including cracks is represented by the following function.

$$\phi(r) = \phi_{cr}V_{cr}(r) + \phi_{cp}V_{cp}(r) + \phi_{gl}V_{gl}(r) + \phi_{lr} \tag{6.22}$$

where $\phi_{cr}$ is the volume fraction of cracks [m³/m³], $V_{cr}(r)$ is a function that specifies the average size of cracks, as given by the following equation (Chapter 3).

$$V_{cr}(r) = 1 - \exp(-B_{cr}r) \tag{6.23}$$

where $B_{cr}$ is a porosity distribution parameter for cracks [1/m], which corresponds to average crack width $r^{cr}$.

By using these two parameters, single as well as multi-cracks can be modeled. Therefore, the diffusion of chloride ions in a wide crack is distinguished from that of multiple small cracks with the same gross volume.

### 6.2.3.2 Diffusivity of chloride ions in cracked concrete

To simulate chloride transport through cracked concrete in a comprehensive way, the diffusivity of chloride ions is formulated separately for the sound part of hardened concrete and cracks. For saturated conditions, the convective term in Equation 6.11 is negligible, thus the chloride flux can be formulated as

$$J_{Cl} = -\left( \frac{\phi S}{\Omega_{cp}} \delta_{cp} D_{Cl} + \zeta \frac{\phi_{cr} S_{cr}}{\Omega_{cr}} \delta_{cr} D_{Cl} \right) \nabla C_{Cl} \qquad (6.24)$$

where $S_{cr}$ is the degree of saturation of cracks, $\delta_{cp}$ and $\delta_{cr}$ are constrictivity parameters for cement paste and crack, $\Omega_{cp}$ and $\Omega_{cr}$ are tortuosity parameters for cement paste and crack, respectively. The parameters $\delta_{cp}$ and $\Omega_{cp}$ in Equation 6.24 correspond to $\delta$ and $\Omega$ in the original Equation 6.11.

When the width of crack is large enough (mm scale), it can be assumed that there is no effect of electrical interaction between chloride ions and the crack, unlike the case of capillary and gel pores (nm–μm scale) in hardened cement paste. As the width of crack decreases, however, such interaction will be dominant, which leads to apparent reduction of diffusive movement of chloride ions. As a first approximation, the relation between $\delta_{cr}$ and crack width is given as (Figure 6.31),

$$\delta_{cr} = 0.99 \tanh\{1.4 \times 10^4 \cdot r^{cr} (\log(r^{cr}) + 5.5)\} + 0.01 \qquad (6.25)$$

On the other hand, the tortuosity parameter is assumed to be unity for cracks, because a crack can be simplified as a straight transport path and be less complex than the micro-pore structure of hardened cement paste. Hence a value of 1.0 is given to the tortuosity $\Omega_{cr}$.

When $D_{Cl}$, the diffusion coefficient of chloride ion in solution (in free space), was used for a cracked part, it was found that the apparent movement of chloride ions in cracking was much underestimated. This is because mass transport in such bulk spaces is driven not only by concentration gradient,

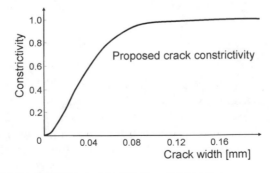

*Figure 6.31* Proposed constrictivity parameter for cracked concrete.

but also by convection current due to the small temperature gradient and/or small hydraulic pressure gradient, which are negligible in the case of mass transport in cementitious materials (Section 5.5.1). To take into account the transport of chloride ions due to the convection currents generated by temperature and/or small hydraulic pressure gradients in the crack space, the diffusion coefficient of the chloride ions in the void is set to a value of fifty times larger than the value in free water by introducing the convection current parameter $\zeta = 50$, as shown in Equation 6.24.

### 6.2.3.3 Verification of chloride diffusivity model in cracked concrete

The analytical results for chloride diffusivity are compared with the Kato *et al.* (2005), Win Pa Pa (2004a) and Win Pa Pa *et al.* (2004b) experiments. In the analysis, three-dimensional solid elements are used and the mesh layout is shown in Figure 6.32, which directly correlates to the experimental block used to determine the averaged chloride penetration profiles in and around the crack. For stability reasons, the nodes are restrained at the saturated pore pressure. The crack volume and the average crack width were set according to the generated crack.

In the case of the Kato *et al.* (2005) experiment, the specimens were prepared with a rectangular cross-section of 70 mm (W) × 120 mm (H), and their total length was 380 mm. A 0.2 mm-wide slit was introduced in the specimens. Three round steel bars of 10 mm in diameter were embedded longitudinally in the specimens, whose cover depths were fixed to 30 mm, 60 mm and 90 mm from the exposed surface. The initial chloride content was 0.01% by mass of cement. The compressive strengths of W/C = 0.39 and W/C = 0.55 concretes at cracking were 57.4 MPa and 41.0 MPa, respectively. In order to control the penetration of chloride ions into concrete, all surfaces except the bottom surface of the specimens were coated with epoxy.

After undergoing curing in water for 28 days, all the specimens were then subjected to accelerated penetration of chloride ions for 91 days through a wet test. In the wet test, specimens were submerged in a sodium chloride solution (3% NaCl). The environmental temperature was kept constant at 40°C for accelerating the transportation of chloride ions. After the

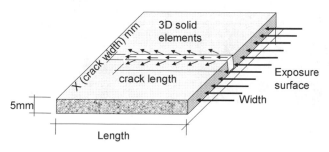

*Figure 6.32* FE Mesh layout for cracked part in the analysis.

accelerated penetration of chloride ions, concrete samples were taken from each specimen for measuring the chloride ion contents. The thickness of the cracked zone was 5 mm from each side of the crack face. Concrete samples were taken from the cracked zone at intervals of 10–20 mm and then milled into powder samples. After the powder samples were dissolved, the chloride content at each sampling point was measured, based on the JCI test method (JCI-SC4).

Figure 6.33 shows a two-dimensional profile of chloride penetration obtained by analysis and comparison with the experiment and analysis of the chloride profile in the cracked zone for 0.55 water-to-cement ratio concrete. Similarly, Figure 6.34 shows the corresponding results for 0.39 water-to-cement ratio concrete. The analytical results match up very nicely with the experimental data, which provide an independent verification of the proposed methodology to predict the chloride profile in cracked concrete.

Next, verification has been made for the chloride penetration model in cracked concrete by using experimental data by Win Pa Pa (2004a) and Win Pa Pa *et al.* (2004b). In this experiment, specimens were prepared as beam (prism) specimens of $100 \times 100 \times 400$ mm in size, which were reinforced with $2 \times \phi 10$ mm round bars at the tension side. The specimens were sealed in a plastic bag for the first 28 days at 20°C. Visible crack lengths ranged from

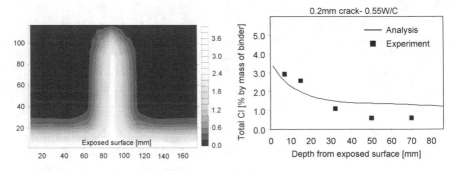

*Figure 6.33* Chloride penetration in cracked concrete (W/C = 0.55).

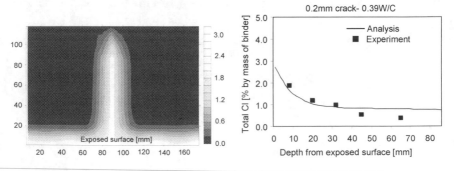

*Figure 6.34* Chloride penetration in cracked concrete (W/C = 0.39).

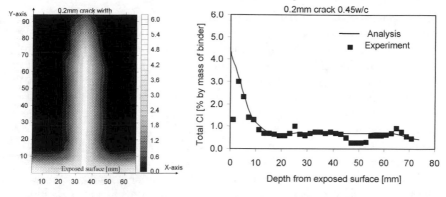

*Figure 6.35* Chloride profile in cracked concrete (one-month exposure, W/C = 0.45).

60 to 90 mm. After all preparations, the specimens were kept in the control room at 20°C and RH of 60% for pre-curing and waiting for exposure to 8% NaCl solution. The experimental set-up was started at 3 months of concrete age in the control environment of 20°C and RH 60%.

The specimens were laid in the NaCl solution trays of specified concentration for an exposure time of one month. The concentration values in this section were calculated as the average of values obtained at 2 mm depth intervals in X-direction on a fixed area of 76 × 76 mm by EPMA (electron probe micro analyzer). The concentration obtained in this experiment was that of the total content of chlorides penetrated both from the exposed and crack surface. Figure 6.35 shows the two-dimensional profile of the chloride penetration along and across the 0.2 mm crack. The analytical result showed good agreement with the experiment.

### 6.2.4 Coupled chloride and moisture transport under cyclic drying and wetting

Chloride ingress solely by pure diffusion is a rare occurrence in a marine environment, where civil structures are exposed to continuous wetting and drying by tidal and splash action. When structures are below the low tidal level or constantly submerged, the progress of corrosion will be much limited due to the lack of oxygen supply in the underwater environment. Therefore it is very important to consider cyclic wetting and drying conditions in numerical simulation of corrosion progress, which is the root cause of deterioration of concrete structures. In this zone, the rate of chloride penetration is maximized by convection during wetting, and exposure to sufficient atmospheric air brings an inward flow of chloride ions due to the high concentration gradient at the surface caused by depleting moisture during drying. In past research, simple and limited experimentation was carried out to model this harsh scenario using a seven-day alternate wetting and drying test. However, different parts of the same structure are exposed to varied durations of wetting and drying during the same tidal cycle by virtue of their position from the highest tidal level, i.e., the part of the structure close to the high tidal level is exposed to wetting for a very short time during the complete tidal period, while the part of the structure close to the low tidal level will be exposed to more wetting as compared to drying. Therefore it is very important to simulate the influence of various wetting and drying patterns on the chloride ingress. The objective in this section is to develop a model which can predict chloride transport under such turbulent exposures.

The tidal, splash and submerged zones are not the only important ones, as the atmospheric zone also has its significance, as a source of airborne chloride transport. Strong winds in the vicinity of the sea pick up small droplets of moisture from breaking waves and crash them into the civil infrastructure along with dissolved salt. The distance traveled by this mechanism depends on the topography of the area and wind speed. This problem requires the modeling of environmental actions, which are responsible for transportation of chloride ions through the air phase. So far, there are few studies dealing with this phenomenon (Swatekititham and Okamura 2006). The primary scope of this section is the material modeling, which deals with the simulation of chloride transport inside the concrete coupled with moisture migration under well-controlled laboratory environmental conditions. In future research, the authors will try to combine an environmental model to simulate the transportation of chloride ions through the air phase with the proposed material modeling, in order to realize the durability assessment under real, complex marine environmental actions.

### 6.2.4.1 Modeling of submerged environment

Since convection (transport of chloride ions with moisture flow in concrete) is one of the major causes of chloride transport in wetting and drying environments, hence in this section, emphasis is laid on the transport of moisture from the exposed surface towards the inner core. For moisture movement dominant in vapor form, the surface mass flux is given by,

$$q_s = -E_b(h - h_s) \tag{6.26}$$

where $E_b$ is the emissivity coefficient and a value of $5 \times 10^{-5}$ is given to the constant, $h$ is the relative humidity of the interior, and $h_s$ is the relative humidity of the surrounding environment. This equation is valid for the case where moisture transport takes place mainly in vapor form, such as exposure to drying or moisture adsorption.

In Equation 6.26, the environmental relative humidity $h_s$ as well as that of the interior can be correlated to the pore pressure head. The relationship between the pore pressure $P_l$ and the relative humidity $h$ is described by the following Kelvin's equation as (Section 3.2.2),

$$P_l = \frac{\rho RT}{M} \ln h \tag{6.27}$$

where $\rho$ is the density of liquid water, $M$ is the molecular mass of liquid, $R$ is the universal gas constant, and $T$ is the absolute temperature. Relative humidity in Equation 6.27, $h$ has a domain from 0 to 1.0, i.e., the surrounding humidity value ranges from 0% (completely dry air) to 100% (fully saturated air). As found in the formulation, from a theoretical viewpoint, both these extremes are singularity points in Kelvin's equation. Another important aspect of this equation is that it always gives a negative value of pore pressure for any value of relative humidity.

In a water-submerged environment, however, hydraulic pressure acts on the exposed surface, and moisture moves under the positive pore pressure head. This condition cannot be reproduced by Kelvin's equation, even though a value sufficiently close to 100% is given. In other words, completely saturated conditions can never be simulated, always ending up in suction pressure (negative pore pressure).

To distinguish the water-submerged conditions from the unsaturated humid environment, there are two options to define the boundary conditions in the analytical system. For water submergence, positive pressure equal to the hydraulic head of water is applied to the exposed surface, whereas for an unsaturated environment, the negative pore pressure computed by Kelvin's equation (Equation 6.27) is used. In the following verifications, submerged environments (completely saturated conditions) are reproduced by applying positive hydraulic pressure at the surface node of the elements so that the sorption flux of liquid water can be simulated in a more realistic manner, and the influence of different boundary conditions (such as the difference between a completely submerged condition and a high-humidity condition) on coupled moisture and chloride transport will be studied.

### 6.2.4.2 Experimental verification of the moisture conductivity model

#### 6.2.4.2.1 TEST METHOD

Cylindrical specimens of 50 mm diameter and 100 mm height were prepared. After casting, all the specimens were sealed and cured for 28 days at 20°C. Then the cylindrical specimens were coated with epoxy to make them watertight and a 4 mm slice was cut from the top surface with a concrete cutter. All the specimens were placed inside an environment control chamber at 60% relative humidity (RH) with 20°C and 40°C temperatures for 30 days. After this exposure, the respective specimens were exposed to a 99.5% relative humidity environment and water-submerged conditions for another 30 days.

During drying at 60% RH, wetting at 99.5% RH and water submergence, the weight loss/gain readings were taken regularly by an electronic weighing machine sensitive to two decimal places.

Ordinary Portland cement, regular tap water, coarse sand and a W/C of 50% and 35% were provided to all specimens. The mix proportion of cement-to-sand was kept at 1:2.25.

### 6.2.4.2.2 MOISTURE GAIN UNDER DIFFERENT BOUNDARY CONDITIONS

The boundary conditions in the modeling are simulated as outlined for the experiment. For the initial curing period, moisture flux in and out of the specimen is restricted, whereas heat flux is allowed during curing and environment exposure conditions. The humid wetting cycle is modeled by providing 99.5% RH (since the environmental chamber is capable of producing 99.5% RH) and the drying environment corresponds to 60% RH. To simulate the water-submerged case, hydraulic pressure is applied to the surface (boundary) elements.

Figure 6.36 and Figure 6.37 show the moisture gain at 20°C and 40°C.

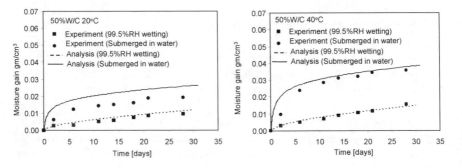

*Figure 6.36* Moisture gain under different wetting conditions at 20°C and 40°C (W/C = 50%).

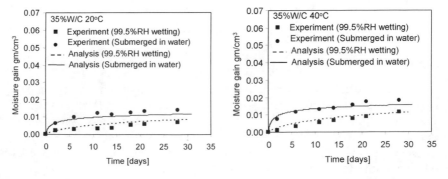

*Figure 6.37* Moisture gain under different wetting conditions at 20°C and 40°C (W/C = 35%).

In these results, comparisons are made between experiment and analysis for wetting at 99.5% RH and for a water-submerged environment. The analysis shows good results for both high- and low-water cement ratio concretes at 20°C and 40°C.

### 6.2.4.3 Chloride transport under alternate wetting and drying conditions

Alternate wetting and drying is a typical marine environment. In order to verify the applicability of the model under various boundary conditions, three alternate wetting and drying cycles were designed in a laboratory-controlled environment: 1% wet–99% dry, 10% wet–90% dry and 40% wet–60% dry cases. These exposure cycles are shown in Figure 6.38.

6.2.4.3.1 TEST METHOD

Ordinary Portland cement, regular tap water, coarse sand and a W/C of 50% were provided to all specimens. The mix proportion of cement to sand was kept at 1:2.25. Specimens consist of cylinders 50 mm diameter and 100 mm height. Curing was done for 28 days in sealed conditions at 20°C. After curing, the top 4 mm slice from the surface of all the specimens was removed to minimize the surface disturbances. Dry-cutting was used to slice the specimens.

For the determination of chloride profiles, specimens were tested after one, seven and 12 months exposure, by the potentiometer titration technique. In this experiment, a slicing method was used for the determination of chloride contents. The thickness of each slice was measured and specimen loss due to the blade thickness of the cutter was accounted for in measuring the depth of each slice from the surface. After this, the slices of respective specimens were ground and tested for chloride contents according to the relevant American Society for Testing and Materials (ASTM) standard C1152/C1152M-04 (for acid soluble chloride).

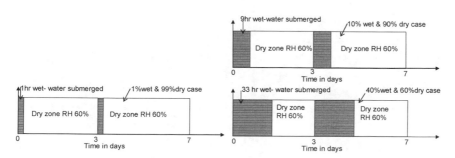

*Figure 6.38* Weekly exposure cycle for alternate wetting and drying experiment.

6.2.4.3.2 ANALYTICAL BOUNDARY CONDITIONS

The boundary conditions in the modeling were simulated as specified for the experiment. In the modeling, only one surface was exposed to the environment, so as to simulate the sealed specimen with one face exposed. For the initial curing period, moisture flux in and out of the specimen was restricted, whereas heat flux was allowed during curing and environment exposure conditions. The submerged wetting cycle was modeled by applying a hydraulic head of water as well as providing 99.5% RH, and the drying environment corresponds to 60% RH.

During the experiment, a water layer was observed on the surface of the vertically placed specimens during drying. For all specimens, the wet surface remained for four hours. Thus, this condition is modeled in the analysis by increasing the wetting duration by four hours, and subsequently decreasing the drying time by four hours.

The relative humidity at the surface becomes equal to the surrounding environment (60%) as it reaches to air-dry state. This condition remains stable throughout the drying period, and is simply modeled at 60% RH with moisture flux allowed at the surface.

6.2.4.3.3 ANALYSIS BY CONSIDERING 99.5% RH AS A WATER-SUBMERGED CONDITION

First, analysis is carried out by modeling the moisture migration by using Kelvin's equation. A value of 99.5% relative humidity was given as a boundary condition, since Kelvin's equation cannot deal with completely saturated conditions, namely 100% RH, which corresponds to the singularity point as already discussed in earlier sections.

Figure 6.39 shows the experimental and analytical chloride penetration profiles for 33 hours wetting case for 6% NaCl at 20°C, and the corresponding moisture gain and loss in several cycles. It can be seen that slight underestimation is seen in the analytical chloride profiles. The same trend

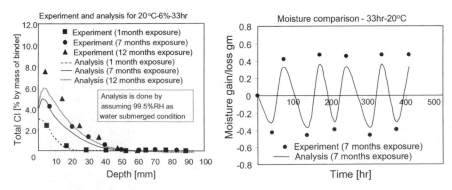

*Figure 6.39* Chloride profiles and moisture gain/loss for 33 hours wetting case (99.5% RH given).

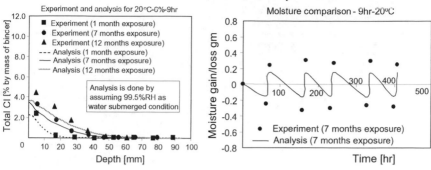

*Figure 6.40* Chloride profiles and moisture gain/loss for 9 hours wetting case (99.5% RH given).

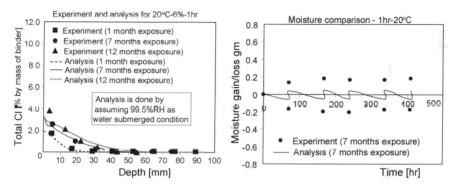

*Figure 6.41* Chloride profiles and moisture gain/loss for 1 hour wetting case (99.5% RH given).

of underestimation is observed in the corresponding moisture migration analysis. Similarly, Figure 6.40 and Figure 6.41 show chloride profiles and corresponding moisture gain/loss for nine hours wetting and one hour wetting, respectively. In these cases also, underestimations can be observed in the analytical chloride profiles and corresponding moisture migration results. This result suggests that there exists a strong coupling between moisture migration and chloride transport, and the underestimation can be attributed to the use of a simplified method of assuming 99.5% RH as water submergence. A more refined and precise way of treating this alternate wetting and drying environment is investigated by applying hydraulic pressure at the surface elements.

6.2.4.3.4 ANALYSIS BY SATURATED PORE PRESSURE

Application of saturated pore pressure is used to model alternate wetting and drying for 33-hour, nine-hour and one-hour wetting cycles. The experimental and analysis result shown in Figure 6.42 to Figure 6.44 corresponds to 33-hour, nine-hour and one-hour wetting cycles for 6% NaCl at 20°C. In this

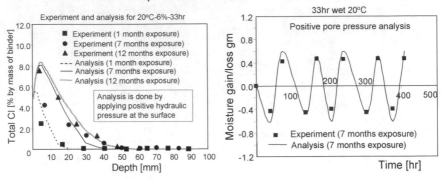

*Figure 6.42* Chloride profiles and moisture gain/loss for 33 hours wetting case (hydraulic pressure given).

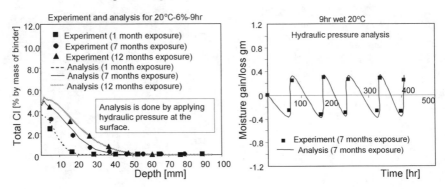

*Figure 6.43* Chloride profiles and moisture gain/loss for 9 hours wetting case (hydraulic pressure given).

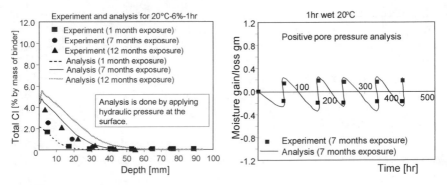

*Figure 6.44* Chloride profiles and moisture gain/loss for 1 hour wetting case (hydraulic pressure given).

section, analysis is done by the application of positive hydraulic pressure, which corresponds to a completely saturated environment. It may be seen that the underestimation by assuming 99.5% RH as water submergence is eliminated in this method. The moisture migration analysis also matches up nicely with the experimental data. For the highly eccentric case of one-hour wetting (Figure 6.44), there is a slight overestimation in the analysis for chloride transport, and the same trend can be seen in the corresponding moisture migration analysis in Figure 6.44. Although the model still needs more enhancement for extreme cases, overall behaviors are well simulated.

## 6.3 Modeling of steel corrosion in concrete

Proper design, selection of good material, good construction practice, and periodic monitoring, maintenance and repairs should all be considered to extend the service life of reinforced concrete structures in the most effective way. In order to accomplish such a task, it is essential to compute the lifetime durability performance of reinforced concrete structures for corrosion under the severe environmental actions of chloride and high temperature. This leads to the objective of this section, i.e., the investigation of severe environmental conditions, especially those involving the coupled effects of chloride and temperature on the corrosion of reinforcing steel in concrete, both by experimentation and using a modeling approach, to clarify the mechanisms involved therein. It was found that existing experimental data for the combined effect of chloride and temperature on the corrosion of reinforcement materials, especially in the higher range, is rather limited (Nishida 2005). Therefore this section consists of modeling and verification of the corrosion potential and rate of reinforcement steel in concrete, for the coupled effect of chloride and temperature, from low- to high-range variable environmental conditions.

### 6.3.1 General formulation

#### 6.3.1.1 Governing laws and equations

First, a general scheme of micro-cell-based corrosion is introduced, with reference to thermo-dynamic electro-chemistry. The electric potential of a corrosion cell is obtained from ambient temperature, pH in pore solution and partial pressure of oxide, which are calculated by other subroutines in the system. The half-cell potential can be expressed with Nernst's method, expressed in Equation 6.28 as

$$Fe(s) \rightarrow Fe^{2+}(aq) + e(Pt)$$
$$E_{Fe} = E_{Fe}^{\ominus} + (RT/z_{Fe}F)\ln h_{Fe^{2+}}$$
$$O_2(g) + 2H_2O(l) + 4e(Pt) = 4OH^-(aq)$$
$$E_{O_2} = E_{O_2}^{\ominus} + (RT/z_{O_2}F)\ln(P_{O_2}/P^{\ominus}) - 0.06pH$$

(6.28)

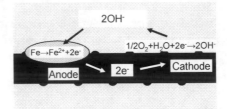

*Figure 6.45* Electro-chemical mechanism of steel corrosion in concrete.

where $E_{Fe}$ is the standard cell potential of Fe, anode (V, SHE), $E_{O2}$ is the standard cell potential of $O_2$, cathode (V, SHE), $E^{\ominus}_{Fe}$ is the standard cell potential of Fe at 25°C (= -0.44V,SHE), $E^{\ominus}_{O2}$ is the standard cell potential of $O_2$ at 25°C (= 0.40V,SHE), $z_{Fe}$ is the number of charge of Fe ions (= 2), $z_{O2}$ is the number of charge of $O_2$ (= 2), $F$ is Faraday's constant, and $P^{\ominus}$ is atmospheric pressure. Strictly speaking, the solution of other ions in pore water might affect the electric potential of a cell. However, it is difficult to consider the effect of ion solutions on half-cell potentials. Thus, we adopt the above equations, assuming ideal conditions.

For a physical explanation of the corrosion reaction of steel in concrete in relation to these equations, refer to Figure 6.45. Ionic current flows in the pore solution of wet concrete area, acting as an electrolyte from anode to cathode. At the same time, electric current flows by the movement of electrons on the surface of corroding rebar from cathode to anode.

### 6.3.1.2 *Incorporation of the Tafel diagram in the corrosion model*

The diagram proposed by Tafel (1906) is a useful tool for simulating the corrosion phenomenon of metals. From the electric potential and the formation of passive layers, electric current that involves chemical reaction can be calculated so that the conservation law of electric charge should be satisfied in a local area. Corrosion potential $E_{corr.}$ and corrosion current $i_{corr}$ can be obtained as the point of intersection of the two lines. Refer to Figure 6.46 for the diagrammatic explanation of the above discussion. Most of the metal corrosion occurs via electro-chemical reactions at the interface between the metal and an electrolyte solution. A thin film of moisture on a metal surface forms the electrolyte for atmospheric corrosion. Corrosion normally occurs at a rate determined by equilibrium between opposing electro-chemical reactions. First there is an anodic reaction, in which a metal is oxidized, releasing electrons into the metal. Then there is a cathodic reaction, in which a solution species (often $O_2$ or $H^+$) is reduced, removing electrons from the metal. When these two reactions are in equilibrium, the flow of electrons from each reaction is balanced, and no net electron flow (electrical current) occurs. The two reactions can take place on one metal or on two dissimilar metals (or metal sites) that are electrically connected.

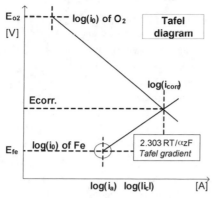

*Figure 6.46* Corrosion current and corrosion potential on Tafel diagram.

The relation of electric current and voltage for the anode and cathode can be expressed by Nernst's formula, in Equation 6.29, as

$$\eta^a = (2.303RT/0.5 \cdot z_{Fe}F)\log\!\left(i_a/i_0^a\right)$$
$$\eta^c = -(2.303RT/0.5 \cdot z_{O_2}F)\log\!\left(i_c/i_0^c\right)$$

(6.29)

where $\eta^a$ is the overvoltage at anode [V], $\eta^c$ is the overvoltage at cathode [V], $F$ is Faraday's constant, $i_a$ is the electric current density at anode [A/m²], $i_c$ is the electric current density at cathode [A/m²], $i_0^a$ is the equilibrium exchange electric current density at anode ($= 1.0 \times 10^{-5}$ at 20°C) [A/m²], and $i_0^c$ is the equilibrium exchange electric current density at cathode ($= 1.0 \times 10^{-10}$ at 20°C) [A/m²].

### 6.3.1.3 Corrosion in concrete and the effect of chloride

The existence of a passive layer reduces the corrosion's progress but when chlorides break down this passive layer, corrosion begins to set in. This phenomenon of breakdown of the passive layer by the intrusion of chloride ions is described by changing the anodic Tafel slope $b_a$ as a function of chloride ions in this model.

$$b_a = (2.303RT/0.5 \cdot z_{Fe}F) \cdot f_p$$

(6.30a)

$$f_p = \begin{cases} 1.0 \times 10^4 & \left(C_f \le 0.04\right) \\ 0.4/C_f & , \left(0.04 < C_f < 0.4\right) \\ 1.0 & \left(0.4 \le C_f\right) \end{cases}$$

(6.30b)

where $f_p$ is an anodic Tafel gradient factor, and $C_f$ is a concentration of chloride ions [% by mass of binder]. For an explanation of the above relation, refer to Figure 6.47a and b. In the model, the semi-empirical theoretical factor $f_p$ accounts for the change in anodic gradient. In theory, when the chloride concentration is zero or very low (numerically defined as being up to 0.04% chloride ions), the factor $f_p$ is assumed to be very large, approaching infinity ($f_p = 1.0 \times 10^4$), considering the rebar to be in perfect passive condition represented by a 90° perpendicular anodic line of the Tafel diagram, as shown in the Figure 6.47. As the chloride concentration increases, the passive layer starts breaking down – a process simulated by the decrease in $f_p$. This results in the fall of anodic gradient and a shift of the point of intersection between the anode and cathode lines towards the more negative corrosion potential. Consequently, the corrosion current continues to increase until the threshold chloride value is reached and the passive layer is completely destroyed.

When chloride ions enter into the system, the passive region which has been occupied by $Fe_3O_4$ until this time will begin to disappear. Also, the protective region of $Fe_2O_3$ will divide into two portions. One is passive and the other becomes somewhat unstable due to the localized destruction and pitting (West 1986). It has been reported by Broomfield (1997) that as chlorides increase around reinforcement, the current density of corrosion becomes larger due to the breakdown of passive film.

The vital issue here is to define the threshold value of chloride concentration for the initiation of corrosion. Various researchers (Byung *et al.* 2004; Glass *et al.* 1991; British Standard (BS) 1997; American Concrete Institute 2001; Page and Vennesland 1983) each have their own points of view, which show a degree of divergence among themselves, due to defininitional issues regarding the initiation point of corrosion. Some researchers consider that corrosion starts with the very first transfer of electrons between anode and cathode sites and the consequent initiation of a corrosion current. Others believe that the corrosion initiation is the point when a considerable amount of metal has been oxidized. And yet other researchers consider that corrosion has occurred when the structure shows a considerable fall in strength and

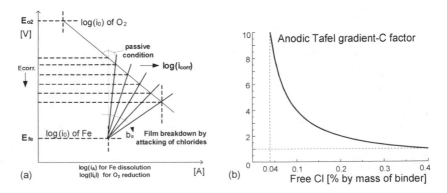

*Figure 6.47* Potential and current increment by chloride migration and breakdown of passive layer.

serviceability. In this model, the threshold chloride concentration is fixed as 0.4% chloride ions by mass of binder. At the threshold chloride value, the passive layer is completely destroyed, thus the factor $f_p$ is assumed to become 1.0 and the gradient of anodic reaction in the Tafel diagram becomes equal to the actual Tafel gradient as given by Nernst's equation already explained above. Since the model is based on the amount of free chloride existing in the pore water, conversion of chloride ions to bound chlorides and vice versa is done using the chloride-binding model as Equation 6.7.

### 6.3.1.4 Effective corrosion current reduction factor $A_{fr}$ for saturated area of concrete

Corrosion is an electro-chemical phenomenon which requires electrolytes for the occurrence of a corrosion reaction. Therefore, it is necessary to calculate the effective corrosion current, $i_{corr\_eff}$, in the system with reference to the saturated area of concrete only as

$$
\begin{aligned}
i_{corr\_eff} &= i_{corr} A_{fr} \\
A_{fr} &= \left( \phi_{cp} S_{cp} + \phi_{gl} S_{gl} \right) \cdot \left( 1.0 - V_{agg} \right)
\end{aligned}
\tag{6.31}
$$

where $\phi_{cp}$ is the capillary porosity, $S_{cp}$ is the saturation of capillary pores, $\phi_{gl}$ is the gel porosity, $S_{gl}$ is the saturation of gel pores, and $V_{agg}$ is the volume fraction of aggregate.

As the hydration proceeds, new products are formed and deposited in the large, water-filled spaces known as capillary pores. Moreover, the gel products also contain interstitial spaces, called gel pores, which are at least one order of magnitude smaller than the large voids of capillary pores. Consider Figure 6.48 below for further clarification of the function modeled in Equation 6.31. Both theoretically and numerically, the saturated area depends on capillary porosity, gel porosity and their respective degree of saturation in the aggregate-free volume of concrete, which is a material heterogeneous in nature. Thus it is essential to estimate the saturated area of concrete.

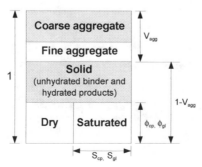

*Figure 6.48* Schematic presentation of saturated area in concrete.

### 6.3.2 Experimentation

#### 6.3.2.1 Materials and mix proportions

Ordinary Portland cement (OPC) as per JIS R5210 specifications was used throughout this research. Natural river sand passed through JIS A1102 sieve No. 4 (4.75-mm openings) was used as fine aggregate for all concrete mixes. Its density and water absorption were 2.65 g/cm$^3$ and 2.21%, respectively. Crushed sandstone with a maximum size of 20 mm was used as coarse aggregate with a density of 2.70 g/cm$^3$ and water absorption 0.59%. It was retained on the sieve No. 4 (4.75-mm openings) and cleaned before being used. Deformed round carbon steel bars 13 mm in diameter were used as reinforcing steel. The surface of the steel bar was polished using sandpaper No. 200. Finally, steel bar was degreased with acetone just prior to being placed in the mold. The W/C ratio was kept at 0.45, with an air content of 3.5±1%. Mix proportions are shown in the following Table 6.5.

#### 6.3.2.2 Specimen preparation and experiment scheme

The specimen is derived from the previous research survey by Byung *et al.* (2004), with necessary modifications as explained below in order to take care of factors previously overlooked. A schematic diagram and original picture are given of the prismatic specimen (100 × 100 × 200 mm) with two 13-mm diameter deformed mild steel bars, with appropriate spacers to hold the bars firmly in the required position (one bar completely embedded and other coming out from both faces, each of which has a clear cover of 15 mm), cast in steel molds (Figure 6.49).

*Table 6.5* Mix proportions

| Specimen | W/C | Binder (kg/m³) | Water (kg/m³) | Sand (kg/m³) | Gravel (kg/m³) |
|---|---|---|---|---|---|
| Concrete | 0.45 | 386 | 174 | 629 | 1,122 |
| Mortar | 0.45 | 686 | 309 | 1,118 | ----- |

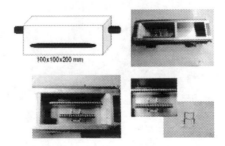

*Figure 6.49* Schematic diagram of the specimen.

The reason for using two steel bars is to make it possible to measure corrosion potential and corrosion mass loss using the same specimen, which is an original feature of this experiment scheme. The bar emerging out from the two sides can be used for the corrosion potential measurement only, since the edges are not embedded in the concrete and are not under chloride attack. Therefore, in order to find the mass loss using the same specimen a separate steel bar was embedded completely into the concrete. This was done to obtain more reliable and accurate results. The clear cover from the surface of the steel to the surface of the concrete was kept at 15 mm, because the measured half-cell potential values at the specimen's surface could be considered as the actual values at a steel surface, if cover depth is within 20 mm (Uomoto 2000).

The test consists of 24 specimens having a total chloride concentration varying from 0% to 10% (NaCl: 0%–16.5% by mass of binder) in mixing water, consisting of three sets for 20°C, 40°C and 60°C temperature conditions and 60% relative humidity, in environment control chambers. All the specimens were allowed to set and harden in the mold for a day in controlled, sealed curing conditions at 20°C before being de-molded. Half-cell potentials were measured with two days interval for all specimens using a copper–copper sulphate reference electrode (CSE) in accordance with standard specifications which can be found elsewhere (ASTM 1999). For further illustration of the measurement procedure adopted in this chapter, consider Figure 6.50. A standard voltmeter with 0.0001V accuracy is connected with the reinforced concrete specimens and the standard electrode through the specified wire in order to make a half-cell potential measuring circuit.

Perhaps one of the most important points in the measurement of half-cell potential is to make the concrete surface wet enough before taking the measurement, so that the resistivity of the concrete is reduced to such an extent which does not affect experiment measurement results. If the measured value of half-cell potential does not stop fluctuating during the measurement, it means that the surface of the concrete is not wet enough and the resistivity of concrete is hindering the formation of proper contact between the electrode and the concrete electrolyte.

Finally, the specimens were split along the position of the steel in the

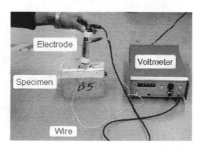

*Figure 6.50* Half-cell potential measurement assembly.

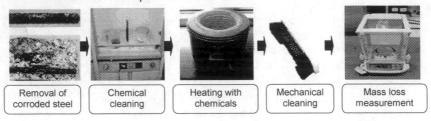

| Removal of corroded steel | Chemical cleaning | Heating with chemicals | Mechanical cleaning | Mass loss measurement |

*Figure 6.51* Gravimetric weight loss determination.

concrete and the steel bars were removed carefully from the concrete. After a photographic evaluation of the corroded steel reinforcement bars, the corrosion mass loss was determined by the gravimetric method. This was done by chemical cleaning (ASTM 2002) of the corrosion products (Figure 6.51). Care must be taken that the base metal does not dissolve along with the corrosion products during this cleaning treatment.

### 6.3.2.3 Experiment results and discussions

Figure 6.52a–c and Figure 6.53a–c show results of half-cell potential and corrosion rate measurements for ordinary Portland cement concrete mixed with 0.0%, 0.025%, 0.25%, 0.6%, 1.0%, 1.82%, 3.65%, 6.0%, 8.0% and 10.0% chloride by mass of binder and three temperature conditions of 20°C, 40°C and 60°C, respectively. The half-cell potential values for various cases have a tendency to increase in early age and then to approach certain less variable values – though not strictly constant ones – later on. In general,

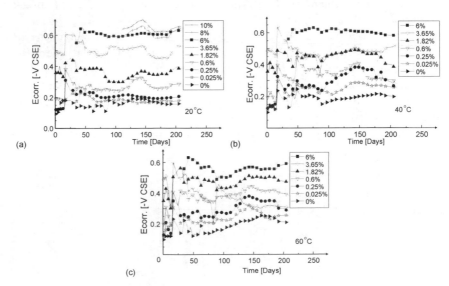

*Figure 6.52* Variation of half-cell potential with age, chlorides and temperature (OPC, W/C = 0.45).

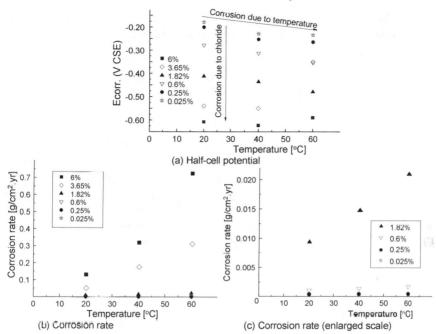

*Figure 6.53* Combined effect of chloride and temperature on steel corrosion.

the corrosion potential and rate show a non-linear rise with the increase of chloride content and temperature (Figure 6.52 and Figure 6.53). The reason for the variations in the half-cell potential values throughout the timespan is largely due to the fact that corrosion is a non-uniform phenomenon. Additionally, the fact that concrete is heterogeneous in nature makes for even more variability in the values.

Averaged treatment has been adopted for picking up a certain value corresponding to a specific chloride and temperature condition for comparison with the model analysis or any other investigations in the coming pages of this section. Averaged values have been taken for the entire time period of measurement of half-cell potentials except the first 30 days of measurement. This is because a suitable method of wetting the surface of specimens in order to achieve proper contact between the reinforced concrete and the electrode before taking half-cell potential measurement could not be achieved in the first month.

It has to be noted here that, in the case of specimens having a high concentration of chlorides, a reduction in the corrosion potential values (the opposite trend) was observed as temperature increased. A possible reason could be the occurrence of corrosion cracking generated by a high concentration of chloride and high temperature, which could result in the loss of moisture needed for the corrosion process and discontinuity of the specimen materials between the reinforcement steel and the concrete or the crack face. Therefore, in establishing an enhanced corrosion model, care has been taken to avoid the influence of this cracking effect on the experiment results. Details will be discussed later in this section.

### 6.3.3 The effect of chloride

#### 6.3.3.1 Comparison of experiment results and model analysis

From the comparison of experiment results and model analysis (Figure 6.54 and Table 6.6), it can be seen that the model underestimates especially at higher chloride concentrations. Thus the model needs improvement.

It should be kept in mind here that the corrosion model gives the value of corrosion potential, $E_{corr}$, in SHE (standard hydrogen electrode) or NHE (normal hydrogen electrode) units. These are the standard units used in all electro-chemical formulations. SHE reference electrodes, however, do not

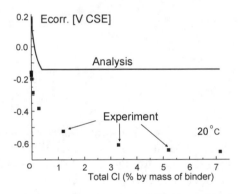

*Figure 6.54* Comparison of experiment and analytical models for different chloride concentrations.

*Table 6.6* Variation in corrosion rate with chloride content (T = 20°C)

| Total Cl⁻ | Free Cl⁻ | Corrosion rate (g/cm².yr) | |
|---|---|---|---|
| *% mass of cement* | *% mass of cement* | *Analysis* | *Experiment* |
| 0.000 | 0.000 | 2.282E-06 | 1.192E-04 |
| 0.025 | 0.002 | 2.282E-06 | 3.006E-04 |
| 0.250 | 0.020 | 2.282E-06 | 4.508E-04 |
| 0.600 | 0.058 | 1.926E-05 | 1.112E-03 |
| 1.820 | 0.270 | 4.995E-04 | 9.438E-03 |
| 3.650 | 1.200 | 1.164E-03 | 5.440E-02 |
| 6.000 | 3.300 | 1.164E-03 | 1.330E-01 |
| 8.000 | 5.200 | 1.164E-03 | 2.938E-01 |
| 10.00 | 7.160 | 1.164E-03 | 4.325E-01 |

lend themselves easily to laboratory experiments. Also, the type of reference electrodes for half-cell potential measurement varies, depending on certain experimental and field conditions. The most common type of reference electrodes used in field and laboratory works are CSE (copper–copper sulphate electrode) reference electrodes. Therefore, in order to compare the model analysis with experiment results which are in CSE units, it is necessary to use the standard conversion factors as

$$CSE(V)=SHE(V)+0.316V \tag{6.32}$$

### 6.3.3.2 *Enhancement of the chloride-induced corrosion model*

#### 6.3.3.2.1 CORROSION IN CONCRETE AND THE EFFECT OF CHLORIDE

Although most corrosion takes place in water, corrosion in non-aqueous systems is not unknown. When it comes to the modeling of corrosion in concrete – as with that which occurs in steel reinforcement embedded in concrete under the effect of severe environmental conditions, such as chloride attack – then the situation becomes even more complex. In such cases there is a need to introduce some semi-empirical equations which logically satisfy the existing corrosion science laws and principles as well as being verified by the experiments. Indeed, the target of any research is to obtain results relevant not merely to laboratory specimens in a controlled environment but to the real infrastructure in the everyday world. Therefore, theory and practice need somehow to be combined via an empirical-theoretical fusion approach.

In general, there are three types of electro-chemical corrosion reaction in nature, i.e., anodic control, cathodic control and resistivity control. It was reported that the presence of chloride ions leads to an anodic control reaction. In this type of reaction, the primary controlling phenomenon is chloride-induced corrosion initiation at the anode site controlled by the anodic Tafel gradient, anodic potential and anodic current (Glass *et al.* 1991).

Therefore, in this section the modeling has been accomplished by varying the anodic parameters. In this corrosion model, the phenomenon of passive layer breaking and the progress of corrosion are described by varying the anodic Tafel gradient and anodic reaction potential as a function of chloride concentration.

#### 6.3.3.2.2 ANODIC POTENTIAL FACTOR $F_{Cl}$

As already discussed, the model underestimates the amount of corrosion in comparison to the experimental results. The reason for this underestimation is thought to lie in the present model's neglect of the effect of chloride on the anodic potential. Therefore, to address this shortcoming and based on the fact that the chloride-induced corrosion reaction is anodically controlled (Glass *et al.* 1991), a semi-empirical factor $F_{Cl}$ is introduced into the system in order to initiate anodic potential variation as a function of chloride content as

$$E_{\mathrm{Fe}} = E_{\mathrm{Fe}}^{\ominus} + \left(RT/z_{\mathrm{Fe}}F\right)\ln h_{\mathrm{Fe}^{2+}} \cdot F_{Cl} \tag{6.33}$$

The factor $F_{Cl}$ shifts the anodic curve diagonally downwards with the increase in the chloride content, thus moving the point of intersection of cathode and anode polarization curves towards the more negative potential and higher corrosion current direction. Consider Figure 6.55 for further illustration of the working of $F_{Cl}$.

In the absence of chloride, the factor $F_{Cl}$ is assumed to be unity and the anodic potential becomes equal to the original anodic potential of the Tafel diagram given by Nernst's equation as already explained. As the chloride attack sets in, the factor $F_{Cl}$ is assumed to start increasing more than 1.0 with the increase in chloride concentration.

The cathodic gradient $b_c$ varies according to the nature of the chemical reaction and the availability of oxygen near the surface of the steel bar embedded in the concrete. In the corrosion model under discussion, the amount of corrosion can be limited when oxygen is in short supply. Coupled with the $O_2$ transport model, this effect can be simulated by the *DuCOM* corrosion model. Thus, considering a free flow of oxygen for the corrosion reaction of steel in chloride-contaminated concrete, $b_c$ has been semi-empirically determined as 0.14 by sensitivity analysis and has been incorporated in the corrosion model. Lastly, using Faraday's law, the electric current of corrosion is converted to the rate of steel reinforcement corrosion.

Figure 6.56 shows a summary of the enhanced model, in which both the variable anodic potential model and the Tafel slope model are implemented. In the following sections, parameter identifications will be tackled based on the experimental data.

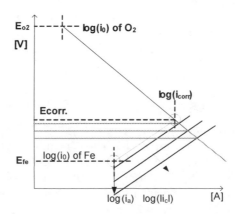

*Figure 6.55* Anodic potential variation as a function of chloride content.

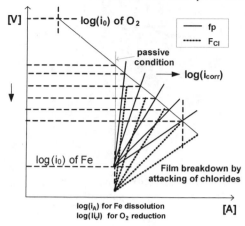

*Figure 6.56* Enhanced modeling of chloride-induced corrosion.

6.3.3.2.3 ENHANCED EFFECTIVE CORROSION CURRENT REDUCTION FACTOR $A_{FR}$ FOR A SATURATED AREA OF CONCRETE

Concrete is a non-homogeneous material. Due to the effects of gravity, any deformation of the steel bars, the size and dimensions of aggregate particles and so on, a thin gap is created between the concrete and the steel. This gap is filled by paste only, due to the paste's fineness. Thus it was assumed that the effect of the volume of coarse aggregate on corrosion is not significant since the area around the steel bar is mostly surrounded by fine material only. Furthermore, from visual observations it was noted that the interfacial transition zone (ITZ) around the steel rebar in concrete does not contain coarse aggregate and is primarily composed of cement paste and possibly some fine aggregate.

In order to verify this fact, a mortar specimen was cast and its corrosion potential values were compared with those of the concrete specimen to find the effect of the volume of coarse aggregate on corrosion. The corrosion potential values obtained in the case of mortar and concrete specimens show the same average values, which can be seen in Figure 6.57a. This fact was further corroborated by measuring the corrosion weight loss and resulting corrosion rate in the two cases. As shown in Figure 6.57b, the corrosion rate of the mortar specimen was just a little more than the corrosion rate of the concrete specimen.

Thus from the experiment results discussed above, it was found that the effect of the volume of coarse aggregate on corrosion is insignificant, since the ITZ around the steel rebar in concrete does not contain coarse aggregate but is instead primarily composed of cement paste and possibly some fine aggregate, depending on its nature, fineness, modulus and grain size distribution (Raja 2006). Thus, Equation 6.31 in Section 6.3.1 is modified as

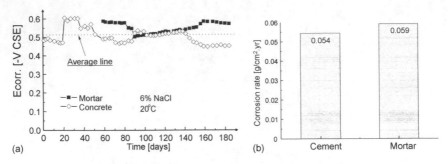

(a)    (b)

*Figure 6.57* Corrosion potential and corrosion rate of mortar and concrete specimens with 6% NaCl and 20°C temperature conditions.

$$i_{corr\_eff} = i_{corr} A_{fr}$$
$$A_{fr} = \left(\phi_{cp} S_{cp} + \phi_{gl} S_{gl}\right) \cdot \left(1.0 - \alpha_{Vf} \cdot V_f\right) \tag{6.34}$$

where $V_f$ is the volume fraction of fine aggregate, and $\alpha_{Vf}$ is the fraction of fine aggregate present in the vicinity of the reinforcing bar. Consider Figure 6.58 for a clarification of this enhancement in comparison to Figure 6.48. In this new model, factor $\alpha_{Vf}$ representing the percentage of fine aggregate present in the vicinity of the rebar varies from 0 to 1.0 depending on parameters such as the fineness modulus of sand, particle size distribution, the profile of fine aggregate, etc. This variation will need to be further investigated in future research. In this chapter, meanwhile, the value is tentatively taken as 1.0.

### 6.3.3.3 Parameter identification and comparison of corrosion potential and corrosion rate models with experiment results for the effect of chloride

The experiment results were compared with the model analysis in order to fix parameters for the anodic Tafel gradient factor $f_p$ and anodic potential

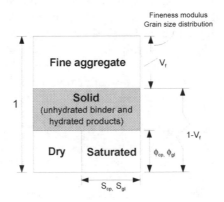

*Figure 6.58* Schematic representation of saturated area in the interfacial transition zone around steel reinforcement.

factor $F_{Cl}$ based on the semi-empirical approach discussed in detail in the previous sections. The numerical formulations are shown in the following equations.

$$b_a = \left(2.303RT/0.5 \cdot z_{Fe}F\right) \cdot f_p$$
$$f_p = 3.17 \times 10^{-2} \cdot C_f^{-1.04}$$

(6.35)

$$E_{Fe} = E_{Fe}^{\Theta} + \left(RT/z_{Fe}F\right)\ln h_{Fe^{2+}} \cdot F_{Cl}$$
$$F_{Cl} = \frac{1.2}{C_f + 1} + 3 \times 10^{-3} \cdot \ln\left(1.0 + 1 \times 10^4 \cdot C_f\right) + 1.0$$

(6.36)

The parameter $f_p$ has been fixed with the experiment results such that its value becomes infinity and 1.0 at zero and threshold chloride concentrations, respectively (Figure 6.59). In the light of past research (Byung *et al.* 2004; BS 1997; ACI 2001; Page and Vennesland 1983), the threshold chloride value has been fixed as 0.4% of total chlorides by mass of binder.

The parameter $F_{Cl}$ has been fixed such that, in the absence of chloride, the factor is equal to one. As the chloride attack becomes active, factor $F_{Cl}$ starts increasing more than 1.0 with the increase in chloride concentration in a non-linear path, as shown in the Figure 6.60.

After fixing the parameters as discussed above, the corrosion model shows good agreement with the experiment results (Figure 6.61) for the effect of chloride on the corrosion of steel reinforcement embedded in concrete for both corrosion potential and corrosion current, thus providing evidence for the efficiency and accuracy of the model. The model, though simple, predicts well and takes into account influential parameters involved in the process of corrosion in reinforced concrete structures.

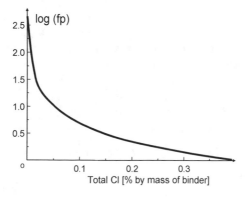

*Figure 6.59* Anodic Tafel gradient factor $f_p$.

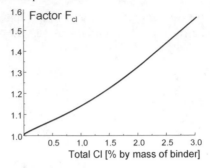

*Figure 6.60* Anodic potential factor $F_{Cl}$.

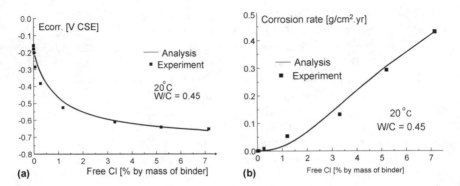

*Figure 6.61* Corrosion potential and corrosion rate for different chloride concentrations.

### 6.3.4 *The effect of temperature*

#### 6.3.4.1 *Comparison of experiment results and model analysis*

The effect of temperature on corrosion progress is taken into account in the original Nernst equations, as temperature is one of the variables in these equations (Section 6.3.1). Although the model does account for the variation in temperature as far as the calculation of electric potential is concerned, the model was primarily designed for constant normal temperature conditions for this calculation. Consequently, from the comparison of experimental results and model analysis (Figure 6.62), it can be seen that the model underestimates at high temperature conditions by more than 20°C and improvement is therefore required.

#### 6.3.4.2 *Theoretical discussion about the dependency of corrosion rate on temperature*

The model works satisfactorily for normal temperature conditions of 20°C (Figure 6.61). In order to extend the model for variable temperature

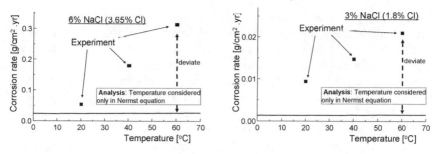

*Figure 6.62* Corrosion rate at different temperature conditions.

conditions, Tafel's equation derived from Arrhenius's law has been modeled for estimating temperature-induced corrosion in reinforced concrete structures.

Since the corrosion of steel bars in concrete is an electro-chemical process, it is generally said that the electro-chemical reaction is accelerated due to temperature. Therefore it is considered that the corrosion rate of a steel bar embedded in concrete rises as temperature increases. However, research dealing with these matters is as yet rather thin. Therefore, the objective of this enhancement is to evaluate the temperature dependency of corrosion rates for steel bars embedded in concrete affected by chloride.

In general, the chemical reaction rate is theoretically illustrated using Arrhenius's equation as

$$A = k \cdot \exp(- \Delta E_a / RT) \tag{6.37}$$

where $A$ is the reaction rate, $k$ is the frequency factor, $\Delta E_a$ is the activation energy, $R$ is the gas constant, and $T$ is absolute temperature. Equation 6.37 can be transformed into logarithmic form as

$$\ln A = -\left( \frac{\Delta E_a}{R} \right) \cdot \frac{1}{T} + \ln k \tag{6.38}$$

From Equation 6.38, it is apparent that the logarithm of the reaction rate ($\ln A$) is proportional to the reciprocal of the absolute temperature ($1/T$). And the diagram illustrating the relationship between the logarithm of the reaction rate and the reciprocal of the absolute temperature is called Arrhenius's plot.

### 6.3.4.3 The electro-chemical temperature dependency of corrosion

In the corrosion model under discussion, Nernst equations are used to calculate the respective half-cell potential values on anode and cathode sites, as explained in the previous sections (refer to Equation 6.28). This is done to deal with the equilibrium conditions concerning $E$, pH, concentrations of ions, partial pressures of gases, and so on, so that the model can predict the effect of various parameters related to the corrosion of steel in concrete. No doubt these Nernst equations in the model allow for a better understanding of the stable phase and reactions occurring in the E–pH diagram, or Pourbaix's diagram or Tafel's diagram (Denny 1996). However, with respect to temperature, the model only calculates potentials and slopes. Thus when Tafel slope $b_a$ in the model (Equation 6.29 and Equation 6.30) increases with higher temperature, it causes a reduction in the corrosion current, $i_{corr}$ (Figure 6.46). This is the reason why temperature in the induced corrosion model is underestimated.

Further, it was found that in Equation 6.29, not only slope $b_a$ but also the exchange electric current density at anode $i_0^a$ increases with higher temperature – another fact which needs to be incorporated in the model. From past research (Vaccaro *et al.* 1997; Piron 1991), it was found that the effect of temperature is much higher on $i_0^a$ than on $b_a$. As a result, even though the slope $b_a$ increases with temperature causing a decrease in $i_{corr}$, the increase in $i_0^a$ is much higher. This results in an overall increase in the corrosion current. So far in the model, a standard constant value of $i_0^a$ was used and this was set at $1.0 \times 10^{-5}$ A/m². Such a value is satisfactory enough when a constant normal temperature model of 20°C is assumed. But when it is intended to extend the model for variable temperature conditions, then it is necessary to account for the effect of temperature by interpreting the original Arrhenius's law as

$$i_{0,T}^a = \left(i_0^a\right)_\infty \cdot \exp\left(- \Delta E_a / RT\right) \tag{6.39}$$

where $i_{0,T}^a$ is the equilibrium exchange anodic current density at temperature $T$, and $(i_0^a)_\infty$ is the reference equilibrium exchange anodic current density at infinite temperature (an imaginary situation).

It is not possible to get the value of $i_0^a$ directly from experiment results alone. Therefore, it was decided to back-calculate the values of $i_0^a$ at variable temperature by using sensitivity analysis. The referential temperature has been set at 20°C and the standard value of $i_0^a = 1.0 \times 10^{-5}$ A/m², used as a constant value in the original model, has now been set as the referential value at 20°C in the enhanced model. With reference to the previous literature review (Maekawa *et al.* 1999), Equation 6.40 can be narrated for the setting of referential values as

$$i_{0,T}^a = i_{0,Ts}^a \cdot \exp\left( - \frac{\Delta E_a}{R} \left( \frac{1}{T} - \frac{1}{T_s} \right) \right) \tag{6.40}$$

Here, the value of referential current $i_{0,Ts}^a$ corresponds to the current at 293[K] (= $1.0 \times 10^{-5}$ A/m²). This enhanced model derived from Arrhenius's law gives the direct relation between the current, $i_0^a$, and any arbitrary temperature, $T$.

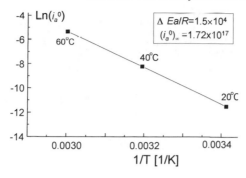

*Figure 6.63* Arrhenius's plot for referential $i_a{}^0$ at 20°C, 40°C and 60°C.

The back-calculation is executed based on the corrosion current and potential. Corrosion potential is obtained as a direct experimental measurement while corrosion current is obtained by using Faraday's law from the gravimetric mass loss measurement in the experiment. Then by using Nernst's equation (Equation 6.30) in the model and by comparing experimental results with analysis, the respective values of anodic current $i_0{}^a$ at 40°C and 60°C are back-calculated so that the model analysis and experiment results become coherent on an averaged basis for all the available cases of variable chloride and temperature conditions. The anodic current values $i_0{}^a$ obtained as above were plotted against the inverse of absolute temperature to obtain Arrhenius's plot. It was observed that this plot came out to be a perfect straight line (Figure 6.63). This proves that Arrhenius's law is applicable and the proposed enhancement method is valid. The activation energy can now be obtained as the rate of increase of $i_0{}^a$ with temperature following Arrhenius's law. From Arrhenius's plot (Figure 6.63), the activation energy of reaction came out to be '$\Delta E_a/R = 1.5 \times 10^4$'.

### 6.3.4.4 Comparison of the enhanced model and experiment results

When the results of the enhanced, extended model for the effect of temperature on the corrosion of steel in concrete are compared with the experiment data (Figure 6.64 and Figure 6.65) they show good averaged agreement with the experiment results for the coupled effect of chloride and temperature on the corrosion of reinforced concrete structures.

It is interesting and worth discussing here that $i_0{}^a$ is the only variable in this formulation and all other parameters are not changed manually but obtained automatically from the same original equations in the corrosion model. Furthermore, the standard value of $i_0{}^a$ at 20°C is also kept the same in the enhanced model as it was in the original model. It is observed that Arrhenius's plot (Figure 6.63) came out to be almost linear, resulting in a unique value

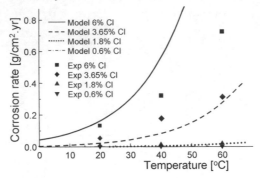

*Figure 6.64* Comparison of enhanced temperature model for the estimation of corrosion rate under variable chloride environment.

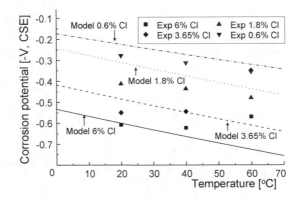

*Figure 6.65* Comparison of enhanced temperature model for the estimation of corrosion potential under variable chloride environment.

of activation energy. This proves that Arrhenius's law is applicable and the proposed enhancement methodology is justified.

This fusion of empirical and analytical methodology for anodic current estimation is rather new and has not been adopted in previous research. Yet, it is interesting to note here that the overall Arrhenius's plot method used for parameter identification is still in accordance to what has been done in the past (Nishida *et al.* 2005).

# 7 Time-dependent mechanics of cement concrete composites

The time-dependent deformation of concrete, including shrinkage and creep, may cause cracks and has much to do with the durability performance of structural concrete. This chapter presents the multi-scale linkage of micro-pore-based thermodynamics with the macroscopic, time-dependent deformation of concrete. Thus, this chapter will offer a "bridge" between material science and structural mechanics. First, some experimental facts are demonstrated to explain how the micro-pore moisture may influence the macroscopic mechanics. Second, the multi-scale solidification modeling of concrete creep and shrinkage is proposed, based on the thermodynamic states of micro-pore moisture with different scales. Lastly, the versatility of the multi-scale approach is shown with regard to the moisture interaction of porous aggregate and cement hydrate inside concrete composites in view of multi-scale modeling.

## 7.1 Moisture in micro-pores and time dependency

From a historical review, the dominant theories of the drying shrinkage mechanism are pore pressure due to water surface tension under high or moderate relative humidity and an increase of the surface energy of gel under low relative humidity (Ramachandran *et al.* 1981; Wittmann 1982). Further, viscous flow of capillary and gel water under permanent loading has been deemed as the main mechanism of creep (Neville 1970). Obeying these concepts, properties of liquid inside micro-pores must control the mechanical behaviors of concrete. In this section, experimental studies are introduced to check these theories in shrinkage and creep focusing on the liquid properties inside cementitious micro-pores. After replacing pore water with various liquids, macroscopic shrinkage and creep behaviors are examined.

### 7.1.1 Testing program of liquid control in micro-pores

Prism specimens of cement paste (40 × 40 × 160 mm) were used for shrinkage tests, while the cylindrical specimens of radius 50 mm and height

*Table 7.1* Mixing proportion

| W/C (%) | Volume ratio of limestone powder (%) | Water (kg/m³) | Cement (kg/m³) | Limestone powder (kg/m³) | Super- plasticizer (kg/m³) |
|---|---|---|---|---|---|
| 35 | - | 518 | 1,483 | - | 1.48 (0.1%) |
| 60 | 15 | 549 | 916 | 405 | - |

Cement: Ordinary Portland Cement; specific gravity is 3.15, Blaine specific surface area is 3450(cm²/g)
Lime stone powder; specific gravity is 2.70, Blaine specific surface area is 7400(cm²/g), exchanged as inert powder, Superplasticizer: Poly carboxylate-based

100 mm were made for creep tests. Two mix proportions having different water-to-cement ratios (0.35 and 0.60) by weight were prepared. The mixing proportions are shown in Table 7.1. One day after casting, the contact chips were directly buried in specimens for shrinkage tests to measure strains. Then, specimens were cured in water for 28 days at the room temperature of 20°C.

### 7.1.1.1 Test series for drying shrinkage

First, let us examine the solid surface energy concept in terms of drying shrinkage. It has been reported that the drying shrinkage of concrete under severe dry states is caused by the increased surface energy of gel particles (Ramachandran *et al.* 1981; Wittmann 1982). But, the energy is decreased by the adsorption of liquids. Suppose that concrete is oven-dried and then submerged in various liquids. Here, recovery of shrinkage during wetting was studied. After moist curing of 28 days, the specimens were oven-dried at 105°C. Then, the specimens were completely soaked in different liquids, i.e., water, methanol, ethanol and lubricating oil, at the constant temperature of 20°C. Here, the molecule sizes of the liquids are not remarkably different (about 3Å for water, 4Å for methanol, 5Å for ethanol and 8~10Å for lubricating oil). During the wetting phase, strain and weight of the specimens were measured with time. The degree of saturation is expressed by

$$Degree\ of\ Saturation(\%) = \frac{(W_l - W_d)}{(W_w - W_d)} \cdot \frac{\rho_w}{\rho_l} \times 100 \tag{7.1}$$

where $W_w$ is the weight of the specimen after 28 days moist curing, $W_d$ is the weight of the specimen after oven-drying, $W_l$ is the weight of the specimen during penetration of liquid, $\rho_w$ is the density of water, and $\rho_l$ is the density of each liquid.

Second, the contribution of pore pressure to drying shrinkage is to be examined. When the shape of pore structure is assumed to be cylindrical, the pressure is governed by Laplace's equation as (Section 3.2.2.1)

$$P_G - P_L = \frac{2\pi r \cdot \gamma_{LG} \cos\theta}{\pi r^2} = \frac{2\gamma_{LG} \cos\theta}{r} \qquad (7.2)$$

where $P_G$ is the pressure of gas, $P_L$ is the pressure of liquid, $r$ is the pore radius, $\gamma_{LG}$ is liquid–gas interfacial tension (when gas is inert air, it corresponds to the surface tension of liquid), and $\theta$ is the contact angle of liquid on the surface of the solid. From a theoretical viewpoint, any liquid used in the experiment can cause pore pressure according to Equation 7.2. It is assumed that those different surface tensions may lead to different pore pressures, which results in various drying shrinkage patterns. In the experiment, shrinkage strain was measured after the complete recovery of liquid. Vacuum drying was applied in order to prevent vapor in the air from permeating the specimens soaked in methanol and ethanol.

Third, a mechanism of drying shrinkage under high relative humidity should be verified. Specimens submerged in water after oven-drying were used in order to eliminate the effect of micro-cracking on the total deformation. It is intended only to focus on the effect of pore pressure. The specimens were dried under 60% and 80% RH at a constant temperature (20°C). After about three weeks of drying, the specimens were soaked in ethanol and the recovery of shrinkage strain was measured with time. It can be assumed that the radius of pores in which meniscus forms increases due to the penetration of ethanol, and pore pressure may finally disappear, since ethanol can dissolve in water. In this case, the recovery of shrinkage may correspond to the shrinkage strain caused by pore pressure.

### 7.1.1.2 Creep test procedure

After replacing the water in concrete with other liquids, creep tests were carried out to verify the seepage theory, which has been regarded as a dominant mechanism of creep under low-stress levels (Wittmann 1982; Neville 1970). From a theoretical viewpoint, the viscous flow of liquids under external load should occur regardless of the kinds of liquid stored in micro-pores. It is expected that the rate of creep would be strongly influenced by the viscosity of liquid.

After moist curing for 28 days, the specimens were oven-dried at 105°C until the weight loss was over. Next, the specimens were soaked in ethanol, lubricating oil and water at room temperature (about 25°C). Before the ethanol and water soaking, dried specimens were kept in a wrapped plastic cup which was then half-filled with the liquid in order to expel the air smoothly from pores. As oil penetration is slower due to its high viscosity, vapor in the air may infiltrate the specimens prior to penetration by the oil. To avoid this, dried specimens were put in the container under vacuum drying before it was half-filled with non-volatile oil. Compression creep tests were conducted after the penetration of each liquid was finished. The setting temperature was 25°C during the tests. Here, discussion will focus on sealed creep. Asamoto and Ishida (2003) reported the details of their experiments.

### 7.1.2 *The effect of liquid in micro-pores on drying shrinkage*

When the specimens were oven-dried and soaked in several liquids, the strain and degree of saturation of the specimens were measured as shown in Figure 7.1. For the specimens submerged in water, saturation was almost fully re-covered. The other liquids also exhibited recovery of saturation. Therefore, it can be said that capillary suction exists for non-water liquid as well as for water, although in the case of non-water liquid the recovery is only partial, i.e., about 62% for a water-to-cement ratio (W/C) of 35% and about 75% for W/C 60%. It is assumed that any liquid can penetrate into the capillary pores, since the pores' size is two or three orders of magnitude larger than the molecular size of liquid, which may be idealized as a condensed continuum. This being the case, the experimental difference of sorption may attribute to the sorption into gel and interlayer pores. In order to verify this assumption, micro-pore volumes of the cement paste after 28 days of moist curing were calculated according to the multi-scale micro-pore development model

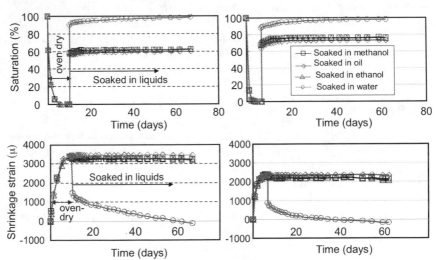

*Figure 7.1* Degree of saturation and shrinkage during oven-drying and soaking in liquids (left: W/C = 0.35, right: W/C = 0.60).

*Table 7.2* Experimental result of saturation after soaking in each liquid for 50 days

|           | *Water* | *Lubricating oil* | *Methanol* | *Ethanol* |
|-----------|---------|-------------------|------------|-----------|
| W/C: 35%  | 98.6%   | 59.6%             | 62.0%      | 62.0%     |
| W/C: 60%  | 98.2%   | 73.3%             | 76.3%      | 76.1%     |

*Table 7.3* Pore structure computed by analysis

|  | Porosity of pore radius $r \geq 3$ nm (mainly capillary pore) | Porosity of pore radius $r < 3$ nm (mainly gel pore and interlayer pore) |
|---|---|---|
| W/C: 35% | 62.1% | 37.9% |
| W/C: 60% | 75.3% | 24.7% |

in Chapter 3, as shown in Table 7.2 and Table 7.3. Recovery of the liquids other than water in the experiment mostly corresponds to the volume of pores larger than 3 nm. According to the results, it can be proved that liquids other than water can be absorbed only in the macro-pores, which consist mainly of capillary pores.

Next, let us discuss the variation of strain after soaking in each liquid. If drying shrinkage is caused solely by the increased surface energy due to desorption of adsorbed water, then shrinkage of the completely dry state would be zero due to the decreased surface energy, independently to the type of liquid adsorbed in gel. However, recovery of the shrinkage was observed only when the specimens were submerged in water. Water can penetrate into pores that are shrunken or even fully closed due to oven-drying, which recovers pore space, while other liquids cannot do this, resulting in an un-recoverable pore structure (Figure 7.2). Moreover, it was reported that the space between crystals of hydrated cement could be altered by adsorption and partial dehydration of bound water (Taylor 1972). It can be confirmed from the experiment that the special characteristics of water in extremely fine pores of less than several nanometers or in crystals could vary the surface energy of gel particles and control drying shrinkage under low ambient relative humidity.

Figure 7.3 shows the relation of the shrinkage strain and the saturation of the vacuum-dried specimen after soaking in methanol, ethanol and water. In the case of liquids other than water, only pore pressure can affect the shrinkage mechanism because no gel/interlayer water can exist in very

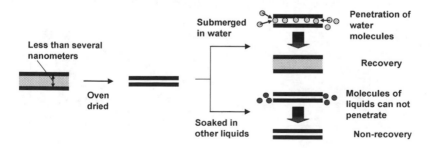

*Figure 7.2* Schematic representation of liquid vapor and adsorption mechanism in gel or interlayer pores.

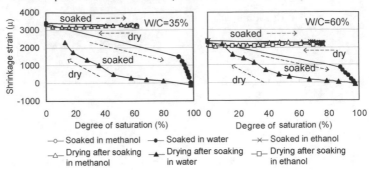

*Figure 7.3* Relation between shrinkage and saturation during penetration of liquids and vacuum drying.

minute pores and bound water. However, shrinkage occurs only when water penetrates into nano-scale pores and crystals. From the experimental results, two hypotheses are possible. One is that the pore pressure may not greatly influence drying shrinkage. The other is that the pore pressure itself becomes very small in the case of liquids other than water, because the liquid surface tension is relatively small and its contact angles on the gel surface may be larger than those of water. In order to verify these hypotheses, the influence of pore pressure was experimentally investigated.

When the specimens were soaked in ethanol after drying under 60% and 80% RH, their shrinkage recovered strongly in the case of both specified water-to-cement ratios and dry conditions (Figure 7.4). As mentioned for the testing procedure, recovery of the shrinkage may correspond to the shrinkage strain caused by pore pressure, because ethanol cannot penetrate into interlayer pores and crystallizations. It can be concluded that the contribution of pore pressure on drying shrinkage is relatively large and constitutes an indispensable component for modeling drying shrinkage. In addition, it is suggested from this experiment that other liquids may have a large contact angle on the gel surface, resulting in a very low pore pressure. This would also fit with the fact that the adsorption of non-water liquids is small.

Some non-recoverable strain was observed. Plastic strain caused by

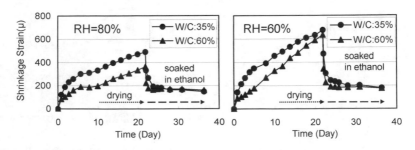

*Figure 7.4* Variation of shrinkage strain when soaked in ethanol after drying.

cracking on the surface would not have much influence because the oven-dried specimens already showed cracks before the drying. Moreover, pore pressure disappears after soaking in ethanol. So non-recoverable shrinkage is probably caused by the disappearance of the interlayer and bound water. Some interlayer water and bound water can evaporate under normal temperature and humidity (Chapter 3). Besides, evaporated interlayer water and bound water are unrecoverable if the specimen is soaked in ethanol. So drying shrinkage under high or moderate relative humidity is affected mainly by the pore pressure as well as the partial vaporization of interlayer water and bound water. It will be necessary to show the linkage between the moisture state in micro-pores and the modeling of shrinkage and time-dependent deformation.

### 7.1.3 The effect of liquid in micro-pores on creep

#### 7.1.3.1 Strength and stiffness

Table 7.4 shows compressive strength and stiffness. The oven-drying process causes a similar magnitude of stiffness reduction to normal curing. This may imply that some micro-cracking damage remains in the specimens. Regarding strength, the oven-dried specimen shows the most, although water intake after drying dramatically reduces it. The intake of non-water liquids brings about a strength similar to that obtained by normal moist curing. It seems that water penetrating inside micro-pores unavailable to oil or ethanol might reduce compressive strength. This fact cannot be well explained solely by reference to Griffith's crack theory (Hori 1962; Ohgishi *et al.* 1983), although that theory does offer a reasonable explanation for the case of the oven-dried specimen alone.

#### 7.1.3.2 Creep

The creep deformation of the specimens in Table 7.4 is shown in Figure 7.5 with creep of the non-oven-dried one after 28 days moist curing for comparison. The sustained load in each creep test was set to be 20% of compressive strength. The load was determined differently according to the type of liquid, since the compressive strength depends on penetration of liquids. Then, the elastic strains were different on a case-by-case basis. Creep can be seen only for the specimen with water.

It has been reported by many researchers that creep takes place due to the viscous flowing of water in concrete pores. It could be predicted from this mechanism that the higher the liquid viscosity, the slower the creep. However, creep was not observed when lubricating oil and ethanol are found inside the pores. Indeed, creep occurs only when water fills the micro-pores. Seepage of liquids may be influenced by the space where they exist, i.e., liquid water in macro-pores can flow smoothly and be stabilized in a short time. On the other hand, it can take a long period for the water molecules in very fine pores

*Table 7.4* Result of compression test

| Specimen | W/C | Compression strength (MPa) | Young's modulus (GPa) | Degree of saturation (%) |
|---|---|---|---|---|
| Normal | 35% | 64.5 | 23.6 | - |
| | 60% | 43.0 | 14.2 | |
| Oven-dried | 35% | 71.0 | 19.1 | - |
| | 60% | 49.7 | 11.0 | |
| Dry-water | 35% | 33.6 | 18.7 | 99.7 |
| | 60% | 22.6 | 12.1 | 100.1 |
| Dry-oil | 35% | 50.1 | 18.2 | 62.9 |
| | 60% | 43.4 | 11.1 | 78.4 |
| Dry-ethanol | 35% | 55.0 | 20.2 | 60.1 |
| | 60% | 40.6 | 9.7 | 77.2 |

After moist curing for 28 days, Normal: compression test soon; Oven-dried: oven-dried: compression test; Dry-water: oven-dried: submerged in water: compression test; Dry-oil: oven-dried: soaked in oil: compression test; Dry-ethanol: oven-dried: soaked in ethanol: compression test.

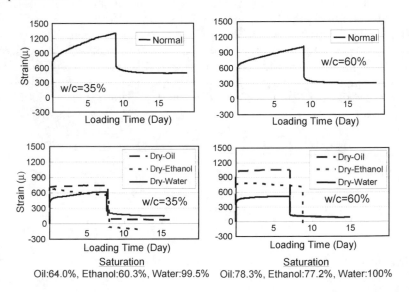

*Figure 7.5* Creep test result when exchanged water inside for various liquids.

to move and be stabilized. Water can infiltrate micro-pores whose sizes are as small as several nanometers, whereas other liquids can only penetrate into macro-pores, including capillary pores. Therefore, liquids could flow and be stabilized in the elastic zone in the case of lubricating oil and ethanol, which can exist only in macro-pores, and for such cases no creep was observed. Yet creep may occur with water, because moisture in the micro-pores can move to be stabilized very slowly. It is thought that water in large capillary pores can flow for a few hours or days while it may take several weeks or even years for water to flow in gel or interlayer pores of micro-scale size.

Creep recovery is not complete after unloading (Wittmann 1982; Neville 1970) because water in micro-pores recovers quite slowly or incompletely. Thus, it can be considered that the recoverable seepage theory is reasonable for capillary pores although it cannot be applied to gel and interlayer pores. Creep strain is unrecoverable for oil because the elastic strain at unloading is smaller than that at loading. This is attributed to the progressive hydration from water slightly infiltrated in the air. As evidence, the elastic strain in the case of water at unloading was also smaller than that at loading. Figure 7.6 shows the creep component of removed elasticity. Creep deformation has completely recovered in the case of oil, while it has not for the case of water. This result corresponds to the hypothesis as well.

Free shrinkage strain is shown in Figure 7.7. It only occurs in the case of the non-oven-dried specimen with a low water-to-cement ratio, because autogenous drying occurs by self-desiccation. In spite of sealing, swelling was

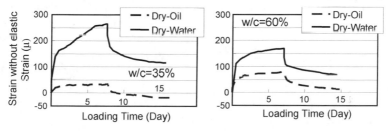

*Figure 7.6* In-elastic part of creep deformation.

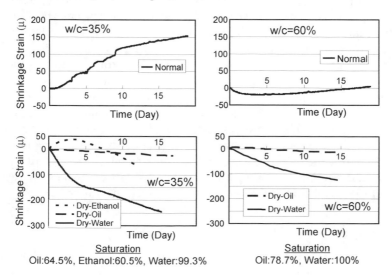

*Figure 7.7* Shrinkage strain of non-loaded specimens.

observed in the specimen submerged in water after oven-drying, due to the recovery possibility inherent in the gel and interlayer water. In the case of the specimen submerged in water after oven-drying, water penetrated into the macro-pores at first and the saturation apparently became about 100%. So liquid water in macro-pores may be distributed to micro-pores. Measured creep strain of the loaded specimen would be lower than the actual strain, because the same phenomenon may also occur during loading. Moreover, swelling was also observed in the case of ethanol and oil. The infiltration of vapor due to imperfect sealing might be a cause of swelling. However, it could not influence the creep, because the swelling strain was very small, at about 20–30 μ.

## 7.2  Solidification model: long-term deformation

Studies on the mechanisms of creep, drying and autogenous shrinkage of cement paste have revealed (Japan Society of Civil Engineers 2000; Murata and Okada 1980; Powers 1968a; Yonekura 1994; Mabrouk *et al.* 2002; Neville 1970; Lokhorst and Breugel 1997; Zhu *et al.* 2004) that the microstructure of cement paste and the state of moisture in it are greatly related to the deformation of cement paste, as stated in Section 7.1. As micro-pore sizes range from $10^{-10}$ m to $10^{-3}$ m, the thermodynamic states of moisture in micro-pores of different sizes are not unique, and the effects of moisture on the deformation of cement paste may vary accordingly. Past studies and hypotheses of microscopic mechanisms, which have been verified by taking a comprehensive view of macroscopic phenomena, constitute the basis of concrete mechanics. But the knowledge gained has not been well integrated. As micro-pore structure may differ widely depending on such factors as mix proportions, materials used and curing conditions, it has been difficult to quantitatively express in generalized form the macroscopic relationships among stresses, time and strains on the basis of microscopic models.

In the multi-scale platform, the micro-pore structural geometry and the thermodynamic states of moisture can be simply quantified as shown in Chapter 3. Thus, in this chapter, an attempt is made to derive a macroscopic material constitutive law by applying the following knowledge to micro-pores of corresponding sizes and integrating it to the microscopic models, such as capillary tension theory and seepage hypothesis, which are considered applicable to condensed water stored in pores ranging in size from $10^{-7}$ to $10^{-4}$ m, and disjoining pressure and surface tension theory, which are related to micro-pores of similar sizes to water molecules ($10^{-10}$ m to $10^{-9}$ m).

### 7.2.1  Concept of solidification

The key feature of the proposed model is its generality that can determine deformation characteristics based on the thermodynamic state variables of moisture in microstructure, such as temperature, relative humidity, degree of saturation, porosity distribution and degree of hydration. Here, the authors propose a method for deriving a space-averaged constitutive law according to the moisture distribution in microstructure, by modeling the deformation of gel particles constituting cement paste and the deformation characteristics of capillary pores composed of gel particles separately, and, at the same time, quantifying changes in capillary tension and solid surface energy of gel particles by desorption of adsorbed water. The moisture state variables can be calculated by the model in Chapter 3.

It is necessary to incorporate a transient response model for the hydration process into the multi-scale scheme for relating microscopic mechanisms and macroscopic deformation characteristics. Here, the concept of solidification theory (Bazant and Prasannan 1989) is accepted. The growth of cement paste with hydration is expressed as a formation of new clusters around previously solidified products (Mabrouk *et al.* 2002; Zhu *et al.* 2004). By using hysteretic variables to memorize the different state variables of individual clusters at the time of solidification, obviously complex hysteretic properties are quantified. At this stage, it is assumed as the first approximation that the mechanical properties of individual clusters are identical. The authors believe that in the cases where microscopic mechanical behavior varies depending on the type of admixture or the process of hydration, it is possible to achieve generality by giving appropriate mechanical properties to each cluster.

### 7.2.1.1 Two-phase (aggregate–cement paste) model

Assuming that cement paste is the matrix and that the aggregate distributed in cement paste is elastic, the system of a three-dimensional constitutive law at the most macroscopic level is presented first. The distributed aggregate particles have no shear resistance. However, since the volumetric stiffness of cement paste caused by volumetric changes is sufficiently high, a two-phase model is defined first on the basis of the following indices related to volumetric and deviatoric components.

$$\sigma_0 = (\sigma_{xx} + \sigma_{yy} + \sigma_{zz})/3 \qquad (7.3)$$

$$\varepsilon_0 = (\varepsilon_{xx} + \varepsilon_{yy} + \varepsilon_{zz})/3 \qquad (7.4)$$

$$S_{ij} = \sigma_{ij} - \sigma_0 \delta_{ij} \qquad (7.5)$$

$$e_{ij} = \varepsilon_{ij} - \varepsilon_0 \delta_{ij} \qquad (7.6)$$

where $\sigma_o$ is volumetric stress, $S_{ij}$ is the deviatoric stress tensor, $\sigma_{ij}$ is the total stress tensor, $\varepsilon_o$ is volumetric strain, $e_{ij}$ is the deviatoric strain tensor, $\varepsilon_{ij}$ is the total strain tensor, and $\delta_{ij}$ is Kronecker delta.

Equation 7.7 and Equation 7.8 can be derived by modeling concrete as a simple two-phase system composed of aggregate and cement paste, as shown in Figure 7.8, and taking into account the equilibrium of volumetric

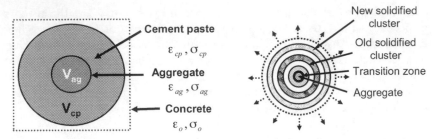

*Figure 7.8* Schematic representation of aggregate, cement paste and solidification process.

components and compatibility conditions for volume changes. The aggregate is assumed to be linearly elastic, as expressed in Equation 7.9, and a volumetric constitutive equation of cement paste, Equation 7.10, is described after Section 7.2.2.

$$\sigma_0 = V_{ag}\sigma_{ag} + V_{cp}\sigma_{cp} \tag{7.7}$$

$$\varepsilon_0 = V_{ag}\varepsilon_{ag} + V_{cp}\varepsilon_{cp} \tag{7.8}$$

$$\varepsilon_{ag} = \sigma_{ag}/(3K_{ag}) \tag{7.9}$$

$$\varepsilon_{cp} = f(\sigma_{cp}) \tag{7.10}$$

where $\sigma_{ag}$ and $\sigma_{cp}$ are volumetric stresses in aggregate and cement paste, $\varepsilon_{ag}$ and $\varepsilon_{cp}$ are volumetric strains in aggregate and cement paste, $V_{ag}$ and $V_{cp}$ are the unit volume of aggregate and cement paste, and $K_{ag}$ is the volumetric stiffness of the aggregate (formulated in Section 7.2.4).

Equations 7.7–7.10 include five variables, namely: $\sigma_0$, $\sigma_{ag}$, $\sigma_{cp}$, $\varepsilon_{ag}$ and $\varepsilon_{cp}$. For completeness, it is necessary to add a restraint condition related to the shear stiffness of the cement paste denoted by $G_{cp}$. If the cement paste, like a viscous fluid, has no shear stiffness, then the uniform pressure condition $\sigma_{ag} = \sigma_{cp}$ must hold true. Conversely, if shear stiffness of the cement paste is infinitely large so as to inhibit changes in shape, the aggregate and the cement paste will retain the same deformations and $\varepsilon_{ag} = \varepsilon_{cp}$. In reality, shear stiffness of the cement paste lies in two extremes. Generalizing this method gives the following restraint condition.

$$\left(\frac{\sigma_{ag} - \sigma_{cp}}{f(G_{cp})}\right) + (\varepsilon_{ag} - \varepsilon_{cp}) = 0 \tag{7.11}$$

The function $f(G_{cp})$ is the shear stiffness of cement paste. Benveniste (1987) proposed $f(G_{cp}) = 4G_{cp}$, and this value is also used in the modeling here.

Generally, shear stiffness, due to the contact between distributed aggregate particles, can be ignored. It can be assumed, therefore, that the shear stiffness of a two-phase system is roughly equal to the shear stiffness of cement paste used as the matrix. It is necessary, however, to take into consideration that a low-stiffness (30–50% of the stiffness of the cement paste) transition zone is formed at the interface between aggregate and cement paste, and that as aggregate density increases, the degree of restraint on free rotation of aggregate particles increases sharply.

Figure 7.9 shows a finite element analysis of shear deformation for two-phase modeling of suspended aggregate and cement paste. Figure 7.10 shows shear deformation modes of a two-phase system. Figure 7.11 shows the relationship between the volume of aggregate, $V_{ag}$, and the shear strain ratio, $\gamma_{cp}/\gamma$, determined through the analysis. Because the contact between aggregate particles is ignored, $V_{ag}$ is assumed to range between 0 and 0.75, for the purposes of sensitivity analysis. Of the two strains, $\gamma_{cp}$ is the average of local shear strains at different locations in the cement paste in the analytical region and $\gamma$ is the average shear strain in the system including suspended aggregate particles.

It has been clearly shown that $\gamma_{cp}/\gamma$ is a function of the volume of aggregate,

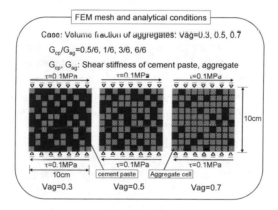

*Figure 7.9*  Two-phase modeling of suspended aggregates.

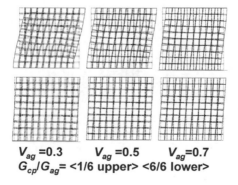

$$V_{ag}=0.3 \qquad V_{ag}=0.5 \qquad V_{ag}=0.7$$
$$G_{cp}/G_{ag}= <1/6 \text{ upper}> <6/6 \text{ lower}>$$

*Figure 7.10*  Local and global modes of shear with particles.

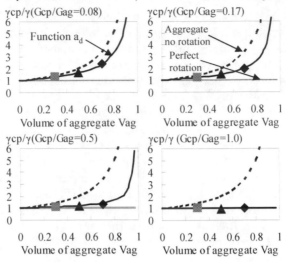

*Figure 7.11* Local shear strain of paste matrix among aggregates.

$V_{ag}$, and the shear stiffness ratio, $G_{cp}/G_{ag}$. As the volume of aggregate, $V_{ag}$, increases, the rotation of aggregate particles decreases and $\gamma_{cp}/\gamma$ increases. Conversely, as the shear stiffness ratio of cement paste and aggregate, $G_{cp}/G_{ag}$, increases, the consistency of cement paste and aggregate increases and the shear strain ratio $\gamma_{cp}/\gamma$ decreases. According to Figure 7.11, the calculated values lie between the mode in which aggregate particles do not rotate and the mode in which they rotate under perfect freedom. According to scientific sensitivity analysis, the relation between the intensity of aggregate particle dispersion, $x_d$, and $\gamma_{cp}/\gamma$ can be expressed with $G_{cp}/G_{ag}$ and $V_{ag}$. Here, this relation is employed as the shear interdependence model at the macroscopic level.

$$\frac{\gamma_{cp}}{\gamma} = 1 + g(x_d) \cdot \left( \frac{\gamma_{cp}}{\gamma_{max}} - 1 \right), \quad \frac{\gamma_{cp}}{\gamma_{max}} = \frac{1}{1 - V_{ag}}$$

$$g(x_d) = 1 \qquad\qquad\qquad x_d \leq 0.01$$

$$g(x_d) = -0.21 \ln(x_d) \qquad 0.01 < x_d \leq 1.0$$

(7.12a)

$$x_d = \frac{G_{cp}}{G_{ag}} \cdot \left( V_{cp} \cdot \frac{G_{cp}}{G_{ag}} + V_{ag} \right) \quad 0.0 \leq x_d \leq 1.0$$

(7.12b)

where $\gamma_{cp}$ is the shear strain in cement paste, $\gamma$ is the shear strain in concrete, $G_{cp}$ and $G_{ag}$ are the shear stiffness of cement paste and aggregate (described in Section 7.2.4), and $\gamma_{cp}/\gamma_{max}$ is the shear strain ratio generalized by shear strain of the mode in which aggregate particles do not rotate. In cases where the shear stiffness of the cement paste is higher than that of the aggregate, $x_d$ may take a value of 1.0 or more. This study, however, deals mainly with

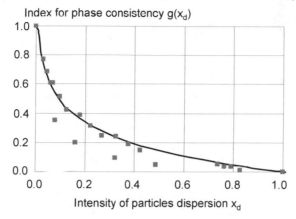

*Figure 7.12* Local and global modes of shear with particles.

the case in which ordinary natural aggregates are used, so Equation 7.12 is applied within the range of $0.0 \le x_d \le 1.0$. The index g, which represents the consistency of the two-phase system, is strongly correlated with the intensity of particle dispersion $x_d$ (Figure 7.12).

If it is assumed that deviatoric stresses in a two-phase system are not transferred by the contact between aggregate particles, they can be expressed simply by a function of shear stress and shear strain of the cement paste (Equation 7.13).

$$S_{ij} = (S_{ij})_{cp} = f((e_{ij})_{cp}),$$

$$(e_{ij})_{cp} = h(x_d) \cdot e_{ij}, \quad h(x_d) = z_d(a_d + b_d)$$

$$a_d = g(x_d) \cdot (\frac{1}{1 - V_{ag}} - 1) + 1, \quad b_d = \frac{1}{1 - V_{ag}} \tag{7.13}$$

$$z_d = 0.5(1 + \exp(-10 x_d)), (if \ x_d < 0.1, then \ x_d = 0.1)$$

where $(S_{ij})_{cp}$ is the deviatoric stress tensor for the cement paste, and $(e_{ij})_{cp}$ is the deviatoric strain tensor for the cement paste. Equation 7.13, however, is not applicable to a system in which the transfer of shear stresses by the contact between aggregate particles cannot be ignored, such as pre-packed concrete.

The function $h$, which is the ratio between cement paste shear strain and concrete shear strain, is assumed to lie between the value $a_d$, corresponding to an analytical solution obtained when perfect consistency between the aggregate and the cement paste is assumed, and the value $b_d$, corresponding to the state in which the aggregate particles in the cement matrix are free from rotation. The coefficient $z_d$ is a parameter reflecting the shear strain properties of young cement paste.

### 7.2.1.2 *Solidification of cement paste*

For the microstructural development model of cement paste followed by hydration, this study uses the solidification theory that assumes that fictitious clusters are gradually added to existing cement paste. Increment of stiffness due to hydration is given in relation to increment of cluster thickness, which is represented by increment of the volume of cement paste, and it is assumed that the properties of clusters do not vary with time (Figure 7.8). However, since clusters are formed at different ages, history variables for different clusters are treated as mutually independent variables. The hardening cement paste is expressed by the formation of these clusters.

These assumptions make it logical to relate the formation of clusters to the degree of hydration. The degree of hydration, $\Psi(t)$, is defined in terms of the ratio between the amount of heat generated by the hydrated cement, $Q(t)$, and the total amount of heat generated, $Q_\infty$, which is peculiar to the type and quantity of cement mineral.

$$\Psi(t) = Q(t)/Q_\infty \tag{7.14}$$

The degree of hydration is a dimensionless variable ranging from 0 to 1, and the thickness of the fictitious cluster is assumed to be a dimensionless variable that is equal to the increment in degree of hydration. The total amount of fictitious clusters and the incremental amount of clusters shown in Figure 7.13 can be calculated if the degree of hydration and the increment in the degree of hydration are known.

The stress carried by the cement paste is expressed by the sum of the stress ($\sigma_{ly}$) carried by each cluster. As a new cluster is formed, the initial stress in the newly formed cluster must be zero, even if the strain in the cement paste is not zero. In this case, stresses in other clusters are not zero. Consequently, the state of stress in each cluster is a function of strain at the present time $t$,

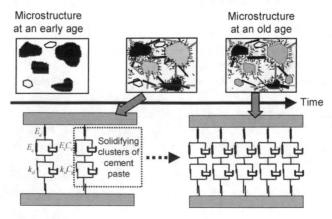

*Figure 7.13* Solidifying clusters of cement paste and aging.

strain at the time $t'$ at which the cluster is formed, and the thermodynamic state.

Figure 7.13 illustrates the formation of fictitious clusters and changes in the microstructure of cement paste. It is assumed that the strains of cement paste are the same as strains of all clusters regardless of their formed ages, and that the sum of cluster stresses is equal to the stress of the cement paste, although stresses in different clusters vary. This solidification theory can be expressed by the volumetric and deviatoric stress components in the form of generalized convolution integral equations.

$$\sigma'_{cp}(t) = \int_{t'=0}^{t} \sigma_{ly}(t,t')d\psi(t') \tag{7.15}$$

$$(S_{ij})_{cp}(t) = \int_{t'=0}^{t} S'_{ij}(t,t')d\psi(t') \tag{7.16}$$

where $\sigma_{ly}(t, t')$ is the volumetric stress in the cluster and $S_{ij}(t, t')$ is the deviatoric stress tensor in the cluster, and the relations $\sigma_{ly}(t', t') = 0$ and $S_{ij}(t', t') = 0$ must hold true.

### 7.2.2 Simplified rheological model for fictitious cluster of cement paste

In this section, at the microstructure level, a constitutive law that holds among microstructure, moisture thermodynamic state variables and cement paste deformation is formulated in a way similar to that of the microstructure formation model. The pore structures under consideration are classified into three categories and a characteristic model is defined for each type of microstructure as follows.

* *Interlayer pores and interlayer water:* C–S–H gel in hydration products has a layer structure similar to that of tobermorite, and the pore spaces between layers are called interlayer pores and contain one layer of water molecules. Feldman and Sereda (1970) state that the volume of cement paste changes as the interlayer water enters or exits the C–S–H gel layers, and that volume changes occurring at a relative humidity lower than 35% are mostly dependent on this mechanism. Interlayer water is not likely to be moved easily by applied stress alone.
* *Gel pores and gel water:* Hydration products form gel particles including interlayer pores, which have peculiar structures and characteristic configurations. Gel particles are highly hydrophilic porous media. These pores in the gel particles are referred to as gel pores. Water in gel pores is thought to be condensed water that can move under applied stress.
* *Capillary pores and capillary water:* Micrometer-scale pores that are not filled with gel particles are called capillary pores. Capillary water can move relatively easily.

Now, with the aim of deriving a constitutive law applicable to both short-term (up to 10 days) time-dependent deformation and long-term (years) deformation, an attempt is made to divide the water in cement paste into capillary water, gel water and interlayer water and to model reversible and irreversible deformation components caused by each type of water separately.

Capillary pores are relatively large ($10^{-8}$ to $10^{-6}$ m), so water in capillary pores is thought to be able to move relatively fast and reversibly under sustained stress. Moisture motion within capillary pores is likely to be assumed as a component governing short-term creep or creep at the earlier age that accounts for the greatest percentage of time-dependent deformation. As the hydration proceeds, hydration products gradually fill capillary pores so that they decrease in both volume and size. The rate of change in creep decreases as the pores become smaller.

Moisture motion within gel pores ($10^{-9}$ to $10^{-8}$ m) is thought to move relatively slowly and to constitute a deformation component that exists over a long period. Gel pore surfaces have a lot of surface energy, so deformed gel pores do not recover easily. It is thought, therefore, that irreversibility is a characteristic of the modeling of this part.

Since the interlayer distance ($10^{-10}$ m) of interlayer pores is within the range of Van der Waal's force interaction, dissipation of interlayer water is likely to be accompanied by instant closure of the pores. The driving force that causes this moisture motion is related mainly to temperature and to the density of vapor in equilibrium, and stress-induced percolation is likely to occur only under very high pressure. Moisture motion within interlayer pores, therefore, might be the prime cause of volume changes that occur under severe environmental conditions such as low relative humidity and high temperature.

On the basis of these hypotheses concerning moisture motion within microstructure and deformation, the rheological model as shown in Figure 7.14 has been constructed. The formulation related to the volumetric part is shown as follows. Tensor variables for the deviatoric component are also formulated in the same manner. The total strain in each cluster $\varepsilon_{ly}$ is expressed as the sum of instantaneous elastic strain $\varepsilon_e$ in the gel particle and capillary pore structures, visco-elastic strain $\varepsilon_c$ in the capillary pore structures, visco-plastic strain $\varepsilon_g$ in the gel particle structures, and instantaneous plastic strain $\varepsilon_l$ related to the interlayer pore structures.

$$\varepsilon_{ly} = \varepsilon_e + \varepsilon_c + \varepsilon_g + \varepsilon_l \tag{7.17}$$

where $\varepsilon_{ly}$ corresponds to the increment of strain that occurs after general cluster formation. As new clusters are formed, their initial strains are equal to the total strain at the stage incorporated into the parallel system, so as to correspond to zero initial stress. The volumetric strain $\varepsilon_{ly}$ in the cluster at time $t$, therefore, can be expressed as the difference between the volumetric strain $\varepsilon_{cp}$ in cement paste at time $t$ and that at the time $t'$ of general cluster formation.

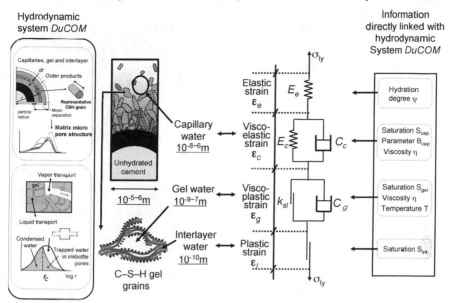

*Figure 7.14* Rheological model for each cluster linked with thermo-hydro states of moisture in micro-pores.

$$\varepsilon_{ly}(t) = \varepsilon_{cp}(t) - \varepsilon_{cp}(t') \tag{7.18}$$

### 7.2.2.1 Elastic model of gel particles and capillary pore structures

The spring in Figure 7.14 represents elastic deformation of newly formed gel particles and the larger-scale capillary pore structures composed of these gel particles. An increment of stiffness due to hydration is expressed in relation to cluster formation.

$$\sigma_{ly} = E_e \cdot \varepsilon_e \tag{7.19}$$

where $\sigma_{ly}$ is volumetric stress in the cluster, $E_e$ is volumetric stiffness of the elastic spring, and $\varepsilon_e$ is instantaneous elastic strain. The volumetric stiffness $E_e$ of the elastic part represents the stiffness of the solid part corresponding to the newly formed cluster and can be expressed as follows.

$$E_e = \frac{dK_{cp}}{d\psi} \tag{7.20}$$

where $K_{cp}$ is the volumetric stiffness of the cement paste (see Section 7.2.4).

### 7.2.2.2 Visco-elastic model of capillary pore structures

The visco-elastic Kelvin chain shown in Figure 7.14 represents the reversible deformation in the time-dependent deformation of the cluster. It is thought to be the deformation component of the capillary pore structure formed by gel particles, and the viscous dashpot is related to the motion of water in the capillary pores.

$$\sigma_{ly} = E_c \cdot \varepsilon_c + E_c C_c \frac{d\varepsilon_c}{dt} \tag{7.21}$$

where $E_c$ is the volumetric stiffness of the elastic spring, $\varepsilon_c$ is the visco-elastic strain, and $C_c$ is the viscous coefficient (1/day) of the dashpot. The volumetric stiffness $E_c$ of the visco-elastic part is used to decide the limiting value of time-delayed deformation of the capillary pore structures and is expressed as

$$E_c = a_{ec} \cdot E_e \cdot f_{ec}(S_{cap})$$
$$a_{ec} = 3, \quad f_{ec} = \exp(0.69(S_{cap} - 1)) \tag{7.22}$$

where $a_{ec}$ is a constant, and $S_{cap}$ is the degree of saturation of capillary pores at time $t$. The ratio between the limiting value of delayed reversible deformation and the instantaneous elastic deformation is estimated to be about one-quarter or one-half (Neville 1959; Roll 1964; Rusch *et al.* 1983; Yue and Taerwe 1992), so one-third is assumed as the first approximation ($a_{ec} = 3$).

The function $f_{ec}$ indicates the influence of the moisture equilibrium state in capillary pores on the volumetric stiffness $E_c$. As the degree of saturation of capillary pores decreases, the amount of water which is considered to carry stress decreases so that delayed limiting deformation increases. In fact, it has been shown that for cement paste specimens (W/C = 35%, 60%), the elastic modulus of specimens subjected to 28 days water curing is 20% greater than that of specimens oven-dried after 28 days water curing. In other words, the value of the function $f_{ec}$ at the degree of saturation of 0% is roughly equal to one-half of the value at full saturation. The elastic stiffness of dried concrete is generally known be lower than that of wet concrete. On the basis of the ideas about the degree of saturation and the volumetric stiffness $E_c$, the function $f_{ec}$ has been formulated as shown in Equation 7.22 (Figure 7.15).

The coefficient $C_c$ related to the viscosity of the visco-elastic part has been formulated, according to the state variables of capillary pores (degree of saturation, pore structure, viscosity of water) provided by the thermodynamic system, as follows.

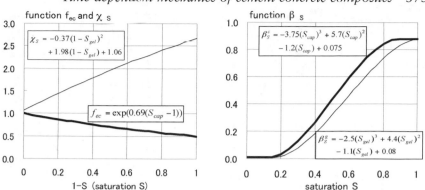

*Figure 7.15* Effect of moisture states in capillary and gel pores on visco-plasticity.

$$C_c = a_{cc} \cdot \beta_S^c(S_{cap}) \cdot \beta_T(\eta(T)) \cdot \beta_r^c(B_{cap})$$

$$a_{cc} = 4.3 \times 10^3, \ \beta_T = 10 + \eta(T), \tag{7.23}$$

$$\beta_r^c = B_{cap}/(10^6) \ (if \ \beta_r^c < 1, \ \beta_r^c = 1),$$

where $a_{cc}$ is a constant (1/day), $\eta$ is the non-dimensional viscosity of liquid water under non-ideal conditions (Section 3.2.3.3), which is normalized by the standard viscosity of water at T=295K under ideal conditions, and $B_{cap}$ is the distribution parameter that determines the shape of the function for the capillary pore structures at time $t$, where $1/B_{cap}$ corresponds to the representative radius of capillary pores.

As the viscosity $\eta$ of the liquid water in the micro-pores increases, mobility of the liquid water decreases and the viscous coefficient of the dashpot increases. As gel volume increases with the progress of hydration, capillary pores become smaller (the representative radius $1/B_{cap}$ of capillary pores decreases), and the viscous coefficient of the dashpot also increases. If capillary pores are completely filled with condensed water, percolation into adjacent pores does not occur easily. The model is designed, therefore, so that the viscous coefficient of the dashpot also increases as the degree of saturation rises. Figure 7.15 and Figure 7.16 show the sensitivity of these thermodynamic state variables expressed in terms of the functions $\beta_s^c$ and $\beta_r^c$. According to analytical results, in oven-dry conditions, time-dependent deformation comes to an end instantly, apparent stiffness decreases, and delayed strain after the instantaneous strain becomes zero. These results agree with observed phenomena.

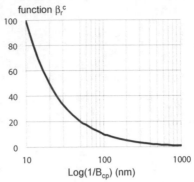

function $\beta_r^c$

*Figure 7.16* Effect of capillary pore size.

### 7.2.2.3 Visco-elastic model of gel pore structures

The delayed plastic deformation part shown in Figure 7.14 consists of a plastic slider and a dashpot. These are deformation elements related to the motion of water in gel pores and can be expressed as

$$\frac{d\varepsilon_g}{dt} = \frac{1}{C_g} \cdot (\varepsilon_{g\lim} - \varepsilon_g)$$

(7.24)

where $\varepsilon_g$ is visco-plastic strain, $\varepsilon_{glim}$ is the limiting value of visco-plastic deformation and $C_g$ is the viscous coefficient (1/day) of the dashpot. As in the case of the visco-elastic part, the viscous coefficient $C_g$ has been formulated to reflect the material properties according to the water motion characteristics (degree of saturation and viscosity of water) of gel pores from the thermodynamic variables as

$$C_g = a_{cg} \cdot \beta_S^g(S_{gel}) \cdot \beta_T(\eta(T)), \quad a_{cg} = 0.9$$

(7.25)

where $a_{cg}$ is a constant (1/day), and $S_{gel}$ is the degree of saturation in gel pores at time $t$. The theoretical approach concerning the degree of saturation $S_{gel}$ and the viscosity $\eta$ of liquid water in the micro-pores is basically the same as the approach concerning the visco-elastic part related to capillary pore structures. However, since the rate of change in creep caused by gel water in the micro-spaces is lower than that of the deformation of capillary pore structures, a relatively large value is used for the constant $a_{cg}(a_{cg} = 0.9, a_{cc} = 4.3 \times 10^{-3})$.

Figure 7.15 shows the function $\beta_S^g$. As the dissipation of moisture from saturated gel pores caused by drying proceeds and the unsaturated region expands, the motion and percolation of moisture can occur with increasing ease. This phenomenon can be explained by viscosity reduction of the dashpot caused by drying.

It is assumed that the limiting value $\varepsilon_{g\lim}$ of visco-plastic strain $\varepsilon_g$ is related to the stress acting on the cluster, the degree of saturation in gel pores, and temperature. It is generally known from the results of creep experiments in underwater conditions that, if gel pores are saturated with water, creep deformation is small even under large stresses. Conversely, in drying conditions, water in gel pores will be dissipated so as to increase the space for gel deformation, and the limiting value of plastic deformation is likely to become greater.

As temperature rises, the Gibbs' energy of gel water will increase, thereby facilitating the motion of water molecules in gel pores. This is a factor increasing the plastic deformation of gel pore structures. On the basis of these considerations, the model expressed in Equation 7.26 has been employed.

$$\varepsilon_{g\lim} = \varepsilon_{g\lim}^{linear}(\sigma_{ly}, S_{gel}, T) \cdot \chi_{non}(\varepsilon_{g\lim}^{linear})$$

$$\varepsilon_{g\lim}^{linear} = \left(\frac{\sigma_{ly}}{E_e}\right) \cdot \chi_1 \cdot \chi_S(S_{gel}) \cdot \chi_T(T),$$

$$\chi_1 = 2, \qquad\qquad (7.26)$$

$$\chi_S = -0.37(1 - S_{gel})^2 + 1.98(1 - S_{gel}) + 1.06,$$

$$\chi_{non} = \exp(0.2 \cdot (abs(\varepsilon_{g\lim}^{linear})/(2.0 \times 10^{-3}) - 1))$$

$$(\chi_{non} = 1 \; if \; \chi_{non} < 1)$$

In Equation 7.26, $\varepsilon_{g\lim}$ is the limiting value of visco-plastic strain due to the stress $\sigma_{ly}$ acting on the cluster. In this modeling, the non-linear function as shown in Figure 7.17a is used. In the low-stress region, the relationship between $\varepsilon_{g\lim}$ and $\sigma_{ly}$ is almost linear, represented by the function $\varepsilon_{g\lim}^{linear}$. Judging from specimen results formerly finished, the ratio between the limiting value of irreversible deformation and instantaneous deformation ranges between about 1 and 4. As a first approximation, the limiting value of irreversible deformation is assumed to be twice the instantaneous deformation ($=\sigma_{ly}/E_e$). The non-linear part due to the applied stress is formulated using the function $\chi_{non}$.

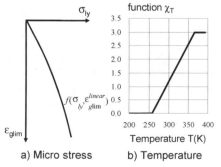

a) Micro stress          b) Temperature

*Figure 7.17* Influential factors on visco-plastic potential.

The function $\chi_{non}$ expresses that the relationship between stress and plastic deformation in the high-strain region is in the non-linear range. By referring to the relationship between stress and strain for cement paste alone, a function has been defined so that the relationship begins to be substantially non-linear when compressive strain exceeds about 2,000 µ. Non-linear states do not occur in the case of basic creep free from drying, or shrinkage strain due solely to drying. If, however, a number of factors such as high temperature, drying and sustained stress are combined, a non-linear plastic model of gel particles may influence analytical results.

The function $\chi_S$ shown in Equation 7.26 expresses the influence of the degree of saturation in gel pores on the limiting value $\varepsilon_{glim}$ of plastic deformation (Figure 7.15) and is based on a physical concept similar to the one in the case of the deformation of capillary pores. The function $\chi_T$ shown in Equation 7.26 has been derived by referring to the experimentally verified fact that as the Gibbs' free energy of water molecules increases with temperature, plastic slip involving gel pores is accelerated (Figure 7.17b).

### 7.2.2.4 Plastic model of interlayer pore structures

The deformation component composed of a slider alone shown in Figure 7.14 is related to the motion of water in interlayer pores. For simplification, it is assumed that deformation of interlayer pores is coupled only with thermodynamic state variables, such as the motion and dissipation of interlayer water, and is not related to the applied stress.

$$\frac{d\varepsilon_l}{dt} = a_{int} \cdot \frac{dS_{int}}{dt}, \quad a_{int} = 5 \times 10^{-3} \tag{7.27}$$

where $\varepsilon_l$ is plastic strain, $a_{int}$ is a constant, and $S_{int}$ is the degree of saturation in interlayer pores. This modeling becomes significant in cases where water is depleted because of severe drying but, under normal environmental conditions, the contribution of this component to overall deformation is small. For this reason, a much simplified formula has been used.

### 7.2.3 Coupling with inherent driving force for shrinkage

A number of hypotheses (Feldman and Sereda 1970; Shimomura and Maekawa 1997; Ishida *et al.* 1997; Soroka 1976; Powers 1968b; Wittmann 1982; Architecture Institute of Japan 2003) have been put forward concerning driving forces for volume changes in cement paste (under zero stress). Referring to existing knowledge, driving forces for volume changes are quantified by explicitly incorporating different phases of liquid water (condensed water, adsorbed water) existing in micro-pores. As shown in Figure 7.18, capillary tension acting on condensed water stored in relatively large micro-pores is considered as a driving force for shrinkage acting on gel and

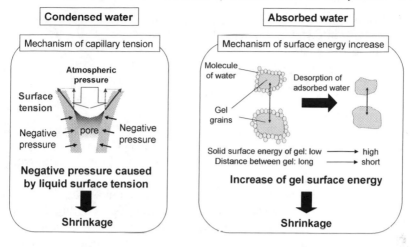

*Figure 7.18* Inherent driving forces of shrinkage.

capillary pore structures of cement paste. Also, the increment of solid surface energy of gel particles caused by the desorption of adsorbed water on the pore surfaces is taken into account. In the latter mechanism, the molecular interaction forces (Van der Waal's forces) of water molecules and solid molecules are considered as driving forces, so the latter mechanism is associated with smaller micro-pores than is the former mechanism.

This study attempts, therefore, to separate the inherent driving force for shrinkage of cement paste into the part related to condensed water and the part related to adsorbed water, and to model them respectively

$$\sigma_s = \sigma_{sc} + \sigma_{sd} \tag{7.28}$$

where $\sigma_s$ is the total shrinkage stress applied, $\sigma_{sc}$ is the shrinkage stress related to the decrease in pressure of condensed water (capillary pressure) and $\sigma_{sd}$ is the shrinkage stress related to the desorption of adsorbed water (disjoining pressure). The following sections discuss the modeling of these two parts.

### 7.2.3.1 Shrinkage stress related to condensed water

Shrinkage stress related to condensed water is driven mainly by capillary tension in the capillary phenomenon of condensed water. Volumetric shrinkage of concrete in the high-humidity range is thought to be caused mainly by this mechanism. The average shrinkage stress acting on cement paste can be evaluated by Equation 7.29

$$\sigma_{sc} = \beta \cdot P \tag{7.29}$$

where $\sigma_{sc}$ is the shrinkage stress related to decrease in pressure of condensed water, $P$ is the capillary tension in condensed water of cement paste (Pa) and $\beta$

is an effective coefficient for the volume acted upon by capillary tension. The capillary tension $P$ can be calculated, using Laplace's equation and Kelvin's equation, from Equation 7.30.

$$P = -\frac{\rho \cdot R \cdot T}{M} \ln h \qquad (7.30)$$

where $\rho$ is the density ($kg/m^3$) of liquid water, $R$ is the universal gas constant ($J/mol \cdot K$), $T$ is the absolute temperature (K), $M$ is the molecular mass of liquid water (kg/mol), and $h$ is the relative humidity of the vapor in equilibrium with condensed water in the micro-pores.

The effective coefficient $\beta$ for the volume acted upon by capillary tension is related to the moisture state stored in cement paste. In this multi-scale modeling, the average degree of saturation of capillary and gel pores evaluated by Equation 7.31 is used as an effective indicator, because it is thought to be the contact ratio of a solid surface with condensed water.

$$\beta = \frac{\phi_{cap} \cdot S_{cap} + \phi_{gel} \cdot S_{gel}}{\phi_{cap} + \phi_{gel}} \qquad (7.31)$$

where $S_{gel}$ is the degree of saturation of gel pores, $\phi_{gel}$ is the gel porosity, $S_{cap}$ is the degree of saturation of capillary pores, and $\phi_{cap}$ is the capillary porosity. These variables can be obtained according to Chapter 3.

### 7.2.3.2 Shrinkage stress related to adsorbed water

It has been pointed out that as drying proceeds, C–S–H solid surfaces have increasing opportunities to come into contact with one another so that increases in the solid surface energy of gel particles caused by the desorption of adsorbed water induce volume changes in gel particles. If solid surfaces are covered with adsorbed water, the adsorbed water may be regarded as part of the surface layers of gel particles. In such cases, the adsorbed water is subjected to surface tension, and the surface forces of gel particles are reduced. As adsorption proceeds, therefore, cement paste expands microscopically. If relative humidity becomes lower than about 30%, the surface areas of gel particles that are not covered with adsorbed water increase and the surface tension, too, increases, so that binding forces arise among gel particles and result in solid shrinkage. This study, therefore, uses the simplest physics model in which shrinkage stress due to bonding forces is expressed by multiplying the unit surface tension of gel particles by the effective specific area that is not covered with adsorbed water.

$$\sigma_{sd} = S_{sd} \cdot \gamma \qquad (7.32)$$

where $\sigma_{sd}$ is the shrinkage stress related to adsorbed water, $\gamma$ is the surface tension of the gel particle (of the order of 350 mN/m (Hori 1962)) and $S_{sd}$ is the distribution density function for surface tension (1/m).

The distribution density function $S_{sd}$ for surface tension of micro-pore surfaces can be calculated by the micro-pore surface area and thermodynamic states of the adsorbed water layer, and its magnitude can be expressed by

$$S_{sd} = f(h)S_{pore} \tag{7.33}$$

where $h$ is the relative humidity in the micro-pores, and $S_{pore}$ is the specific surface area of the micro-pore surfaces (1/m, per unit volume).

An attempt is made to calculate the function $f(h)$ from the equilibrium state of the adsorbed water layer. In the case where there is no adsorbed water on the micro-pore surfaces and the surface area acted upon by the surface tension of gel particles is maximized, $f(h) = 1$ is defined. In the case where adsorbed water molecules trapped on the micro-pore surfaces are in a saturated state, $f(h) = 0$ is defined. Real phenomena that occur fall between these extreme cases, so $f(h)$ ranges from 0 to 1. The function $f(h)$, which expresses the ratio of the micro-pore surface area that is free from adsorbed water, can be calculated, according to BET theory (Figure 7.19) based on statistical mechanics, as follows.

$$f(h) = \frac{V_m - V_1}{V_m} = (V_m - \frac{k_1 V_m h}{1 - k_2 h + k_1 h})/V_m \tag{7.34a}$$

$$f(h) = \frac{1 - k_2 h}{1 - k_2 h + k_1 h} \tag{7.34b}$$

where $V_m$ is the volume of a completely saturated adsorbed water layer, $V_1$ is the volume of the first adsorbed layer nearest to the micro-pore surface, $h$ is the relative humidity in the micro-pores, and $k_1$ and $k_2$ are constants of proportionality. As in the microstructure formation modeling, $k_1 = 15$ and $k_2 = 1/h_m$ are assumed (Section 3.2.2.2). The variable $h_m$ is the relative

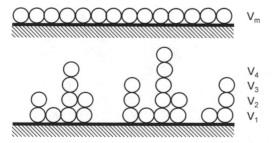

$V_m$: Volume of water fully adsorbed on a planar surface
$V_1, V_2, V_3, V_4$: volume of water adsorbed in each layer

*Figure 7.19* Moisture adsorption model by BET theory.

humidity required for completely filling the micro-pores with liquid water. This is another state variable that is given through sequential and parallel computations (Chapter 1 and Chapter 3)

Judging from the fact that the sizes of gel pores are one order smaller than those of capillary pores and that Van der Waal's forces are inversely six orders proportional to distance, it may be safe to think that the specific surface area $S_{pore}$ of micro-pore surfaces is roughly equal to the specific surface area of gel particles.

$$S_{pore} = S_{pore}^{gel} \tag{7.35}$$

where $S_{pore}^{gel}$ is the specific surface area (1/m) of gel pores.

The total volumetric stress $\sigma_{cp}$ occurring in the cement paste, therefore, can be calculated as

$$\sigma_{cp} = \sigma'_{cp} + \sigma_s \tag{7.36}$$

where $\sigma_{cp}$ is the total volumetric stress occurring in the cement paste, $\sigma_{cp}'$ is the volumetric stress occurring in the skeleton of cement paste, and $\sigma_s$ is the shrinkage stress occurring in the microstructure of cement paste.

Finally, the coupling of the inherent shrinkage forces and the stress occurring in the cement paste can be expressed in the form of (Zhu *et al.* 2004),

$$\sigma_{cp} = \sigma'_{cp} + \sigma_s = D \cdot \varepsilon_{cp} + (C + \sigma_s) \tag{7.37}$$

Figure 7.20 shows the relationships among total shrinkage forces acting in the cement paste, the contribution of condensed water, and the stress related to adsorbed water. As shown, in the relatively high-humidity range, the decrease in pressure of condensed water drives the volumetric shrinkage of the cement paste. In the low-humidity range, the interlayer bonding forces

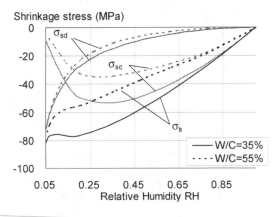

*Figure 7.20* Inherent driving forces in terms of relative humidity.

of micro-pores predominate. It can also be seen that the lower the water-to-cement ratio of the cement paste, the greater the driving forces for shrinkage become. However, since the stiffness of the cement paste skeleton with a low water-to-cement ratio also increases, the value of shrinkage does not increase in proportion to the driving forces for shrinkage.

### 7.2.4 *Instantaneous elasticity of cement paste and aggregate*

For modeling concrete as a two-phase system consisting of aggregate and cement paste, it is necessary to give the elastic modulus of aggregate and cement paste accurately. The elastic modulus of aggregate is formulated in Equation 7.38 (Kawakami 1991). This formula is applicable to natural aggregates. The relationship between the elastic modulus and the specific gravity is shown in Figure 7.21.

$$E_{ag} = (23.5\gamma_{ag} - 57.8) \cdot 10^4 \ (N/mm^2) \tag{7.38}$$

where $E_{ag}$ is the elastic modulus of aggregate, and $\gamma_{ag}$ is the specific gravity of aggregate. The volumetric and shear stiffness of aggregate can be calculated from Equation 7.39:

$$K_{ag} = \frac{E_{ag}}{3(1 - 2\upsilon_{ag})}, \quad G_{ag} = \frac{E_{ag}}{2(1 + \upsilon_{ag})} \tag{7.39}$$

where $K_{ag}$ and $G_{ag}$ are the volumetric stiffness and shear stiffness of aggregate, and $\upsilon_{ag}$ is Poisson's ratio of aggregate ($= 0.2$).

The elastic modulus model of cement paste has been calculated using Equation 7.40, based on the elastic modulus for cement paste experimentally obtained from Asamoto (2006), Kawasumi (1992) and Kiyohara *et al.* (1999, 2002) (see Figure 7.22).

$$E_{ag} = (23.5\gamma_{ag} - 57.8) \cdot 10^4 \ (N/mm^2) \tag{7.40}$$

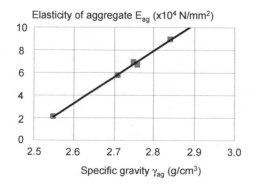

*Figure 7.21* Modulus of elasticity of normal weight aggregate.

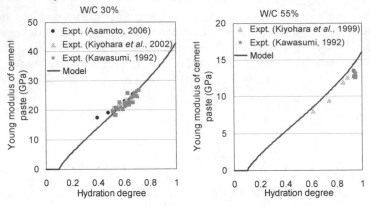

*Figure 7.22* Modulus of elasticity of cement paste.

where $E_{cp}$ is the elastic modulus of cement paste, and $f_{cp}$ is the compressive strength of cement paste. For the compressive strength $f_{cp}$ of cement paste, a formula expressed by a function of the water-to-cement ratio $W/C$ and the degree of hydration $\Psi$ has been proposed on the basis of experimental data for cement paste by Asamoto (2006), Kawasumi (1992) and Kiyohara *et al.* (1999, 2002):

$$f_{cp} = 820 \cdot \exp(-0.5 \cdot W/C) \cdot f(\psi) \quad (\text{N/mm}^2)$$

$$f(\psi) = \begin{cases} 1.50 - \sqrt{2.47 - 2.22\psi} & \psi \geq 0.1 \\ 0.0 & \psi < 0.1 \end{cases}$$

(7.41)

From Equation 7.42, the volumetric stiffness and shear stiffness of cement paste can be calculated as

$$K_{cp} = \frac{E_{cp}}{3(1 - 2\upsilon_{cp})}, \quad G_{cp} = \frac{E_{cp}}{2(1 + \upsilon_{cp})}$$

(7.42)

where $K_{cp}$ and $G_{cp}$ are the volumetric stiffness and shear stiffness of cement paste, and $\upsilon_{cp}$ is Poisson's ratio of cement paste only.

Unlike Poisson's ratio of cement paste only, Poisson's ratio of cement paste phase in the concrete is likely to be influenced by the aggregate particles distributed in the concrete. If the volume ratio of aggregate is high, the deformation of the cement paste in the direction perpendicular to the applied load is restrained by the effect of aggregate particles, with their stiffness being different from that of cement paste. This has been reproduced through two-dimensional finite element analysis, and the scheme, along with the mesh and the analytical conditions used, is shown in Figure 7.23. Figure 7.24 shows the relationship between the volume ratio of aggregate, $V_{ag}$, and $\upsilon_{cp}/(\upsilon_{cp})_{pure}$. The effective Poisson's ratio $\upsilon_{cp}$ of cement paste phase in the concrete is

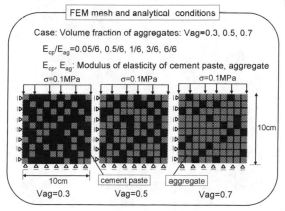

*Figure 7.23* Finite element dispersed aggregates in paste matrix for computing effective Poisson's ratio.

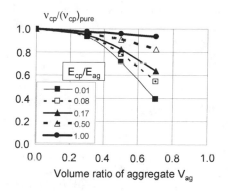

*Figure 7.24* Effective Poisson's ratio in terms of volume ratio of aggregate.

derived by averaging apparent Poisson's ratio (calculated value) in different locations over the entire analytical domain. The variable $(v_{cp})_{pure}$ is defined as Poisson's ratio of cement paste only that is free from local restraint due to aggregate particles.

The ratio $v_{cp}/(v_{cp})_{pure}$ is a function of the volume ratio of aggregate, $V_{ag}$, and the elastic modulus ratio $E_{cp}/E_{ag}$. This can be clearly seen from Figure 7.24, too. In the case the elastic modulus ratio $E_{cp}/E_{ag}$ is low, as the volume ratio of aggregate $V_{ag}$ increases, lateral deformation of the cement paste is restrained by the aggregate and $v_{cp}/(v_{cp})_{pure}$ decreases. As the elastic modulus ratio $E_{cp}/E_{ag}$ of the cement paste and the aggregate increases, the integrity degree of the cement paste and the aggregate rises so that $v_{cp}/(v_{cp})_{pure}$ increases. On the basis of systematic sensitivity analysis, the relationship among $v_{cp}/(v_{cp})_{pure}$, $E_{cp}/E_{ag}$ and $V_{ag}$ has been formulated as shown in Equation 7.43. Since it is assumed that usage of natural aggregate and the volume ratio of aggregate at which contact between aggregate particles occurs can be ignored, the sensitivity analysis ranges are $E_{cp}/E_{ag}$= 0–1 and $V_{ag}$= 0–0.75.

$$v_{cp} = (1 - y(x_p)) \cdot (v_{cp})_{pure}$$

$$y(x_p) = 0.98(1 - x_p)^{1.8} + 0.02 \quad (y = 0.02 \text{ if } x_p > 1), \tag{7.43}$$

$$x_p = V_{cp} + (E_{cp}/E_{ag}) \cdot V_{ag}$$

where $(v_{cp})_{pure}$ is Poisson's ratio of cement paste only (assumed to be equal to 0.25). The index $x_p$ is defined as the intrinsic volume of cement paste converted from concrete, and the index $y$ represents the confinement of the cement paste phase from the aggregate phase. The larger the intrinsic volume of cement paste, the smaller becomes the confinement of the cement paste phase from the dispersed aggregate phase. In fact, the two-dimensional analytical results show that there is a strong correlation between the intrinsic volume of cement paste and the phase confinement, as shown in Figure 7.25.

The stiffness model described above has been used to verify the elastic modulus of early-age concrete. Figure 7.26 compares the model with Westman's (1999) experimental data.

### 7.2.5 Verification

In this section, the applicability of the constitutive model is verified comprehensively through comparison between experimental data and analytical results for different combinations of time, stress and strain with different mix proportions and under various ambient conditions. Figure 7.27 illustrates the relationship between microstructure formation, thermodynamic state variables of moisture (temperature, relative humidity and the degree of saturation in micro-pores, porosity distribution, adsorbed molecular layer thickness, the degree of hydration of cement) and each element model. Thermodynamic state variables are provided by the analytical program *DuCOM* (Chapter

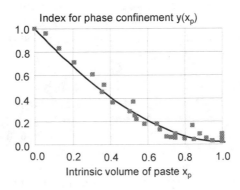

*Figure 7.25* Sensitivity index for effective Poisson's ratio in terms of volume ratio of aggregates.

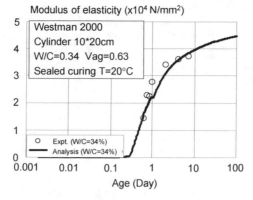

*Figure 7.26* Verification of concrete initial stiffness model.

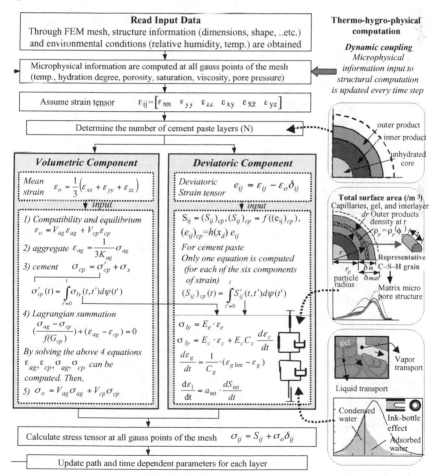

*Figure 7.27* Scheme of modeling of micro-pore-based physics linked with thermo-hydro states of moisture.

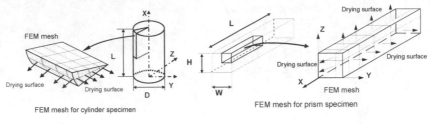

FEM mesh for cylinder specimen

FEM mesh for prism specimen

*Figure 7.28* Mesh of representative.

1). In drying conditions, different states of moisture exist at the surface and center of the specimen. The specimen, therefore, was divided into finite elements, thermodynamic state variables at different locations of the specimen were calculated, and a structural response analysis was conducted taking into account the moisture conditions in different types of micro-pores. The standard mesh used for finite discretization is shown in Figure 7.28. Although a finer mesh is used in the case of larger specimens, the prismatic or cylindrical mesh geometry is similar to that shown in Figure 7.28 for all cases. With the proposed analytical model, it is not necessary to distinguish between different types of time-dependent deformation such as basic creep, drying creep, drying shrinkage and autogenous shrinkage. Instead, the proposed generalized model provides all information through calculation.

### 7.2.5.1  *Autogenous shrinkage: stress-free sealed condition*

As hydration causes the consumption of water and as volume changes of hydration products lead to the formation of unsaturated micro-pores, the resultant decreases in relative humidity in the micro-pores cause self-desiccation and shrinkage in a relatively short period. Factors greatly affecting analytical results are the driving forces for shrinkage dependent on the moisture state and the short-term deformation capacity of the capillary pore structures. Boundary conditions have been defined so that moisture movement does not occur at the surface of the specimen and surface stress becomes zero, and the relationship between time and deformation has been determined. Autogenous shrinkage in Figure 7.29a–c shows good agreement between the experimental data and analytical results for relatively small strains occurring over short periods, but the difference between experimental data and analytical results increases with the passage of time. This tendency is thought to be due to calculation inaccuracy of a decrease in relative humidity caused by self-desiccation and the degree of saturation. This means that in numerical analysis, the relative humidity resulting from the shortage of free water at the late stages of the hydration process might have been underestimated. This is currently under study with focus on the time dependence (Iwata and Ishida 2003) of the moisture state. It is believed that the accuracy of shrinkage prediction can be improved by refining the moisture model.

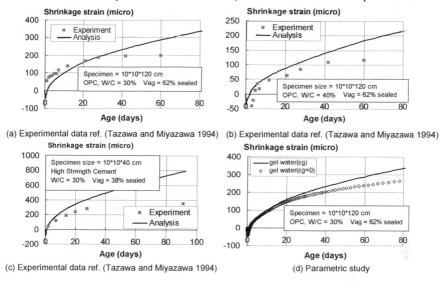

(a) Experimental data ref. (Tazawa and Miyazawa 1994)  (b) Experimental data ref. (Tazawa and Miyazawa 1994)

(c) Experimental data ref. (Tazawa and Miyazawa 1994)  (d) Parametric study

*Figure 7.29* Verification of autogenous shrinkage.

Figure 7.29d also shows the results of sensitivity analysis in which the deformation of gel particles is assumed to be zero. As shown, the influence of the moisture state in the gel pores on short-term autogenous shrinkage is small, and the decrease in relative humidity in the capillary pores is considered to be the primary driving force for autogenous shrinkage.

## 7.2.5.2 Drying shrinkage: stress-free drying condition

As the dissipation of moisture caused by drying proceeds, the decrease in pressure of condensed water can be the driving force for shrinkage that causes deformation of the cement paste skeleton. The relationship between deformation and time under the boundary conditions of zero stress and constant humidity at the surface of the specimen has been calculated. Since drying shrinkage continues for a relatively long period, this is the case where the modeling of the plastic deformation of the gel pores, which occurs over a long period, greatly influences analytical results. In contrast, behavior at the early stage of drying is dependent strongly on the deformation model of the capillary pore structures.

Figure 7.30 shows short-term drying shrinkage results. Drying shrinkage varies with the age at drying, the relative humidity and the shape of the specimen. If relative humidity is low and the age at drying is early, then the value of drying shrinkage is large. This tendency has been captured by the analytical model with fair accuracy. Figure 7.31 and Figure 7.32 compare analytical results and experimental data concerning long-term drying shrinkage. The analytical results show that the value of drying shrinkage increases along with the increase of unit water content, the decrease of aggregate ratio and relative

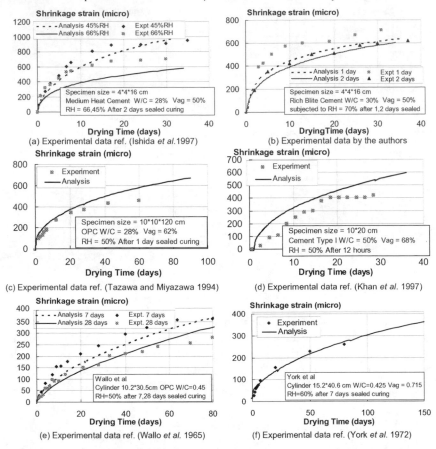

*Figure 7.30* Verification of short-term drying shrinkage.

humidity. In the conditions of different specimen size, mix proportion, curing and environment, although the tendency of the experimental data has been predicted with fair accuracy by the analysis method used, there is still room for improvement. It is necessary to verify and refine the moisture state in the micro-pores and microstructure model as well as the mechanical model of hydration products, both individually and comprehensively.

### 7.2.5.3　Basic creep: constant-stress sealed condition

No water dissipation at the specimen surface and constant uniaxial compressive stress were assumed as boundary conditions, and analytical results concerning time and strain were compared with experimental data. It is obtained from the analysis that all micro-pores are in a saturated condition if the water-to-cement ratio is 50% or more. Consequently, basic creep at the early stage of loading and at young age is greatly influenced by the deformation

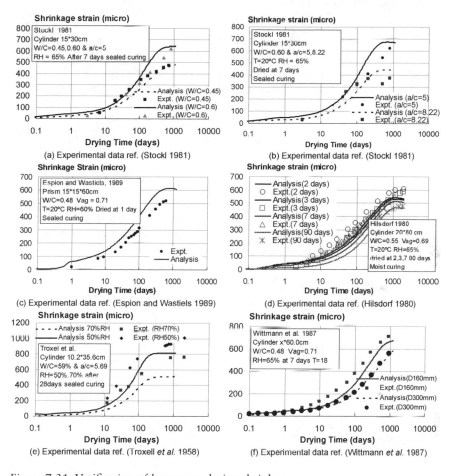

*Figure 7.31* Verification of long-term drying shrinkage.

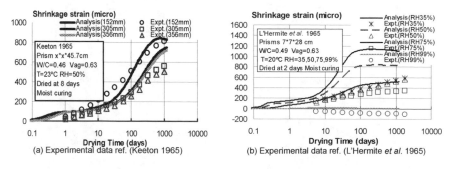

*Figure 7.32* Verification of long-term drying shrinkage with parametric study.

model of capillary pore structures in the saturated state, and long-term basic creep is dependent solely on the deformation model of gel pore structures. In cases where the water-to-cement ratio is low, since self-desiccation occurs even if water dissipation does not occur, driving forces for shrinkage act on the specimen in addition to external stresses, and the deformation model of gel pore structures in an unsaturated state influences analytical results.

Figure 7.33a shows short-term basic creep results and Figures 7.33b–7.34f show long-term creep results. The water-to-cement ratio to be verified ranges from about 30% to 60%, and basic creep behavior is reproduced with fair accuracy for mix proportions including those at which self-desiccation occurs. If the water-to-cement ratio is low, hydration gradually progresses even during creep loading and the moisture content in micro-pores continues to decrease, so plastic characteristics of the gel pore structures are influenced gradually. Hydration is also accompanied by cluster formation. With a combination of these mechanism models, different types of creep deformation have been predicted with fair accuracy. Accuracy of the ultimate value of long-term behavior, however, requires further study although creep rates at the early stage of loading have been predicted with reasonable accuracy. Under the environmental conditions considered in this study, calculation accuracy of the elastic modulus of aggregate and cement paste is one of the factors affecting the prediction of long-term creep. Relatively old experimental data are used for verification, so there is a need for study on the degree of agreement between them and elastic modulus model formulas derived

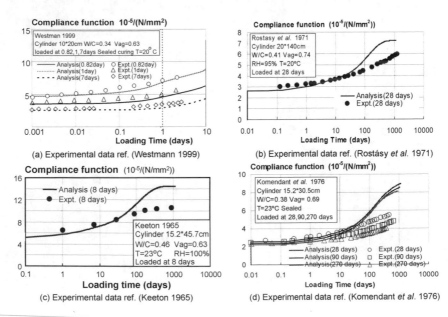

*Figure 7.33* Verification of basic creep (1).

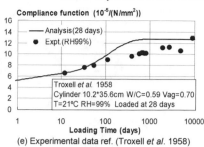

(e) Experimental data ref. (Troxell *et al.* 1958)

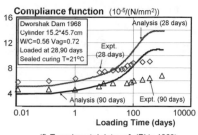

(f) Experimental data ref. (Pirtz 1968)

*Figure 7.34* Verification of basic creep (2).

from more recent material experimental data. This is an area that requires further study.

### 7.2.5.4 Drying creep: constant-stress dry condition

Assuming the humidity was constant at the surface of the specimen and water dissipation was allowed, the relationship between time and strain under constant stress and under zero stress would be determined respectively, and what is left after the former subtracts the latter was taken as drying creep. An unsaturated state of micro-pores is often observed in the case of basic creep of concrete with a low water-to-cement ratio. In the case of drying creep, this unsaturated state can occur at any water-to-cement ratio. Therefore, dependence of the deformation models for the gel and capillary pore structures on the moisture condition can be verified. Because of the influence of both drying and sustained stress, the non-linearity of plastic strain in cement paste can also be verified and simulated, as shown Figure 7.35f.

Figure 7.35 shows experimental data and analytical results. Figure 7.35a shows short-term drying creep or early-age behavior, and Figures 7.35b–7.35e show long-term drying creep. It is generally known that apparent drying creep is greater than basic creep. The analytical model described in this section has been used to calculate the macroscopic time-dependent deformation of concrete from microscopic mechanisms in cement paste without distinguishing between basic creep and drying creep. As a result, it has been confirmed that observed behaviors have been calculated with fair accuracy.

## 7.3 Multi-chemo physics of creep for cement hydrates: advanced solidification

### 7.3.1 Kinematics of gel water

First, some parameters relating to moisture migration in pores are re-examined with the aim of upgrading the basic model in Section 7.2 in view of long-term deformation. A slow, irreversible creep is associated with the motion of

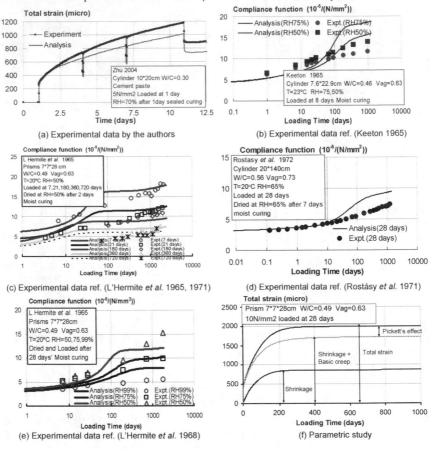

*Figure 7.35* Verification of drying creep.

water in gel pores, while a relatively rapid reversible deformation denoted "delayed elasticity" is affected by the viscous flow of water in capillary pores. In the model, a Kelvin chain consisting of an elastic spring and a dashpot is applied to the capillary component and a plastic slider with attached dashpot is assigned to the gel, as shown in Figure 7.14 (see Equations 7.21–7.26) in Section 7.2.2.1 and Section 7.2.2.2).

First, let us focus on dashpot parameter $a_{cg}$ of gel water. As shown in Figure 7.36a, the observed basic creep deformation under sealed conditions is not completed even after more than 1,000 days of loading, while the analytical creep comes close to convergence before 1,000 days. Thus, $a_{cg}$ was simply set to 10 or 100 times the original value. The same tendency was observed in the case of drying shrinkage and drying creep, too.

As shown in Figure 7.37, the firm dashpot for the gel component provides a more reasonable result for long-term creep than the original one does, especially when it is set to 10 times the original $a_{cg}$. However, on the contrary, it

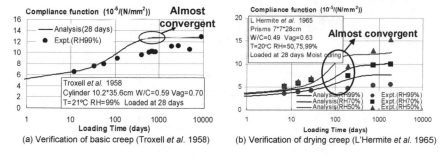

(a) Verification of basic creep (Troxell *et al.* 1958)

(b) Verification of drying creep (L'Hermite *et al.* 1965)

*Figure 7.36* Verification of current model in long-term time-dependent deformation.

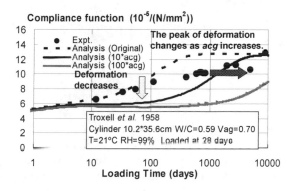

*Figure 7.37* Re-examination of $a_{cg}$ in the case of basic creep.

gives smaller deformation from 10 to 100 days than observed experimentally. Even with the higher $a_{cg}$, the creep versus logarithmic time curve exhibits a sinusoidal shape, while the observed result is almost linear. As no moisture evaporation is allowed for in the basic creep analysis, the functions of $\beta_s{}^c$, $\beta_s{}^g$ are analytically kept constant. Further, pore structures do not significantly change. This means stable $\beta_r{}^c$, because the sustained stress was applied after 28 days when hydration was almost complete. Thus, even if the formulations regarding moisture states and pore-structure formation are re-examined, long-term creep cannot be suitably modeled, and smooth creep in logarithmic time cannot be obtained simply by altering $a_{cg}$.

Next, we direct our attention at the capillary component, using a smaller $a_{cc}$ to accelerate the short-term creep in the analysis while keeping $a_{cg}$ at 10 times its original value. When the dashpot of the capillary component is reduced to 0.1 or 0.01 times its original size, the creep at 0–10 days just after loading rises but no substantial change occurs in the 10–100 days field (see Figure 7.38). In fact, visco-elastic strain plays a comparatively small role in overall creep. This indicates that it may be hard to reproduce long-term creep in a reasonable way solely by changing parameters $a_{cc}$ and $a_{cg}$. To obtain more accurate short- and long-term creep at the same time, more detailed modeling of the gel component is required.

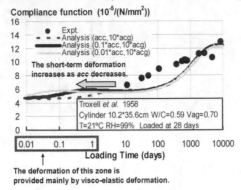

*Figure 7.38* Re-examination of $a_{cc}$ in the case of basic creep.

## 7.3.2 Visco-plasticity of gel water

To refine the creep model based on moisture kinematics in gel pores, the definition of micro-pore spaces needs to be reviewed again. Figure 7.39 shows the scale definition of gel and capillary pores (Bazant 1988; Daimon *et al.* 1977; Maekawa *et al.* 1999; Wittmann 1982). Capillary pores are defined in *DuCOM* as large voids where precipitation of new C–S–H gel grains is allowed at any temperature if condensed water exists. The large geometrical gaps among C–S–H gel grains are generally categorized as capillary pores. The gel pores are thermodynamically defined as spaces where new hydrated product may conditionally precipitate simply with a rise in temperature (Powers 1968b; Nakarai *et al.* 2005). Geometrically, the internal pores of the grains and a small fraction of the inter-particle spaces among the grains

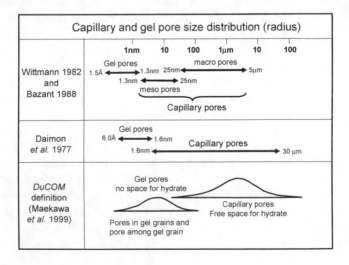

*Figure 7.39* Definition of pore size.

are thermodynamically equivalent to the gel pores defined in this modeling. The size distribution of these gel pores is expressed by a single Raleigh–Ritz function in terms of moisture hydrodynamics.

In the original model, the mechanistic properties of the gel component are also expressed as an average of the internal pores and inter-particle micro-spaces. However, in reality, the physical chemistry state of moisture in the extremely tiny spaces within C–S–H gel grains is thought to be much different from that of water in the inter-particle spaces. In such fine pores, water behaves almost as part of the solid. Figure 7.40 represents a schematic of the moisture migration rate in each pore. The seepage of moisture in macro-pores, such as capillary pores, would cause a relatively short-term, time-dependent deformation, whereas the water migration in fine pores, such as gel and interlayer pores, may lead to a long-term gradual deformation.

In this modeling, we let gel pores S denote the internal pore spaces within a C–S–H gel grain and gel pores L the inter-particle spaces where new hydrate grains can form conditionally. According to previous research, gel pores L are considered geometrically to consist of relatively large pores 3 nm or more in diameter (Bazant 1988; Daimon *et al.* 1977; Nakarai *et al.* 2005; Wittmann 1982). On the other hand, gel pores S within C–S–H gel grains are assumed to be less than 3 nm. Since water molecules in gel pores S are assumed to be firmly confined by Van der Waal's and Coulomb forces from the walls of the C–S–H solid, it may take quite a long time for the moisture in gel pores S to reach an equilibrated state under sustained stress.

It has been experimentally suggested that water in fine pores, which non-water liquids cannot infiltrate, causes long-term, time-dependent deformation (Asamoto and Ishida 2003). Thus, in order to classify the seepage of gel pores S and L, the visco-plastic component of the previous rheological model for each cluster is simply divided into two parts, one for relatively short-term behavior and the other for long-term behavior. The divided visco-plastic model is specified by gel pore size. Based on previous research (Bazant 1988; Neville 1970, 1995; Wittmann 1982), it is assumed that the mechanical model for long-term deformation is related to the moisture state in gel pores S, while

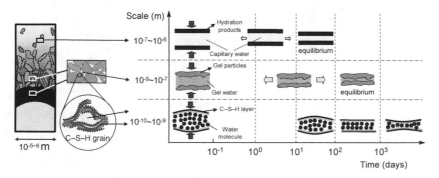

*Figure 7.40* Schematic image of moisture motion in each pore.

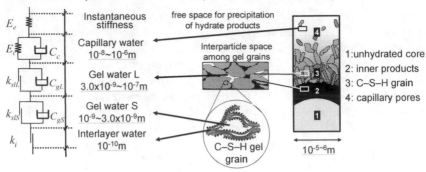

*Figure 7.41* Schematic image of proposed rheological model for each layer.

the short-term one depends on the moisture kinematics of gel pores L. A schematic image of this situation is given in Figure 7.41.

The dashpot viscosity and plastic strain limit are tailored for association with pore saturation as well. It is assumed that dry, empty pores may rapidly reach converged plasticity while liquid water-filled pores may exhibit slow viscous flow. As creep deformation in the submerged state is smaller than when dry, and since lower humidity causes greater creep, larger ultimate plastic strain is assumed under severer drying conditions in the model. A summary of dashpot viscosity and convergent plastic strain for each gel pore is given in Figure 7.42.

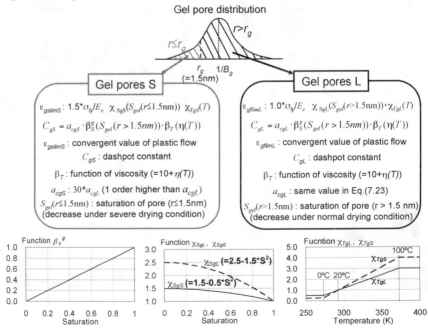

*Figure 7.42* Model functions of gel pores S and L.

Parameter $\beta_S{}^g$ ($0.0 \le \beta_S{}^g \le 1.0$), which is related to the rate of moisture migration, is simply assumed to be linear with the degree of saturation of gel pores, as shown in Figure 7.42. Here, the same function is adopted for both gel pores S and L. According to Kelvin's equation, the moisture in pores of size less than 3 nm can evaporate only under extremely low humidity (RH < 20%) conditions. Thus, the saturation of gel pores S generally takes a value of 100% under moderate ambient conditions and function $\beta_S{}^g$ for moisture in gel pores S remains 1.0. The value of $\beta_L{}^g$ for gel pores L takes a smaller value than in the original model because the distribution of gel pores L does not include the internal space in the C–S–H gels, so the saturation of gel pores L is less than it was originally. In addition, to represent the strong confinement of water molecules in gel pores S resulting from force interactions with the C–S–H solids, the dashpot viscosity parameter $a_{cgS}$ for gel pores S is about one order higher than $a_{cgL}$ for gel pores L. The original value of $a_{cg}$ (=0.864) is assigned to gel pores L and the $a_{cgS}$ of gel pores S is set at 30 times the value in the original model.

The limit plastic strains for gel pores S and L are determined from the total creep limit at infinite time. The convergent strain of visco-elasticity is assumed to be 0.3 times the instantaneous elastic strain in consideration of observed reversible creep (Neville 1959, Roll 1964; Rusch *et al.* 1983; Yue and Taerwe 1992). The total ultimate plasticity for gel pores S and L was formerly set at 2.5 times the instantaneous elasticity under the saturated condition, because the creep coefficient ranges from 1 to 6, with 2.5 the typical value (Bazant 1988; Wittmann 1982). Here, the ultimate creep coefficient under the saturated condition is assumed to be 2.8 at 20°C.

Functions $\chi_{SgS}$ and $\chi_{SgL}$ represent the effect of pore saturation on the creep plastic limit of gel and capillary spaces. From parametric studies, it is assumed that total ultimate plastic strain finally reaches 4.75 times the elastic strain at the time when moisture in the gel pores is lost. As the ultimate plastic creep for each unsaturated state is barely identifiable from experiments, simple functions are assigned as a first approximation such that higher humidity (50–100%) is more sensitive than lower values (0–50%), according to experimental findings of shrinkage (L'Hermite *et al.* 1965; Verbeck 1958). The function related to the visco-elastic limit for capillary pores is modified slightly as well. Details on the formulae for the proposed and original models are summarized in Figure 7.58.

The plasticity of gel pores induced by seepage of gel water may be accelerated at elevated temperatures, since the kinematic energy of the moisture is amplified. Thus, functions $\chi_{TgS}$ and $\chi_{TgL}$ are adopted such that plastic strain arises at higher temperatures. Since basic creep appears to increase almost linearly with temperature up to 100°C (Arthanari and Yu 1967; England and Ross 1962), linear functions were adopted as shown in Figure 7.42.

The original model simply assumes that the creep rate of gel pores is proportional to the creep potential evolved at infinite time (Equation 7.24). The creep potential is also simply assumed from the updated creep plastic strain.

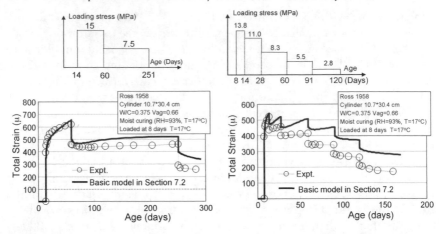

*Figure 7.43* Comparison between analysis and experiment (Ross 1958) under decreasing stress.

To verify the stress path dependency on creep rate, the authors analytically reproduced the special stress paths shown in Figure 7.43. When the applied stress is partly released, creep evolution is suppressed as the simulation predicts. However, actual creep evolution suppressed further than the prediction, as shown in Figure 7.43. In contrast, the analysis was verified for a history of increasing stress as being of reasonable accuracy. Thus, it is necessary to consider the path-dependent creep rate for more versatile modeling. Here, the concept of equivalent creep denoted by $\varepsilon_{g,eq}$ is introduced in a manner similar to the theorem of plasticity as

$$\frac{d\varepsilon_g}{dt} = \frac{1}{C_g}(\varepsilon_{g\lim} - \varepsilon_{g,eq})$$

$$\varepsilon_{g,eq} = \begin{cases} \max\left(\dfrac{W}{\sigma_{ly}}, \varepsilon_g\right) & (when \ \sigma_{ly} > 0) \\[2ex] \min\left(\dfrac{W}{\sigma_{ly}}, \varepsilon_g\right) & (when \ \sigma_{ly} \le 0) \end{cases}$$

$$W = \int \sigma_{ly} d\varepsilon_g$$

(7.44)

where $\varepsilon_g$ is the visco-plastic strain, $\varepsilon_{g\lim}$ is the limit value of plastic deformation, $\varepsilon_{g,eq}$ is equivalent plastic strain, $C_g$ is the parameter representing viscosity of the dashpot, and $\sigma_{ly}$ is a volumetric stress applied to clusters. When the applied stress is kept constant, the updated plastic creep strain is equal to the equivalent one. The influence of equivalent plastic strain is summarized in Figure 7.44.

The formulation of Equation 7.44 is adapted to both components, gel pores S and L. Although the same mathematical models are set down for

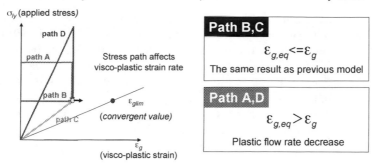

*Figure 7.44* Visco-plastic strain model depending on stress history.

the moisture kinematics of gel pores S and L, the dashpot parameters and saturation are associated with the characteristics of the moisture in each pore and result in different plastic flows.

### 7.3.3 Instantaneous plasticity associated with interlayer water

Interlayer pores are close to water molecules in size. Thus moisture loss from the pores would cause the interlayer space to close due to the strong surface energy of C–S–H solids, resulting in instantaneous volumetric contraction. We then describe the instantaneous plasticity related to loss of interlayer water in terms of the saturation of interlayer pores, as shown in Equation 7.27.

This component $\varepsilon_l$ is not in flux under normal ambient conditions and has never been directly verified. Recently, Ishida *et al.* (2007) reported the experimental finding that some of the interlayer water could disperse from tobermorite sheets even at 40–60°C, regardless of ambient relative humidity. In order to include hot climates and environments within the model scope, the plasticity dependent on interlayer water migration is required. In this research, the removal of only interlayer water was experimentally tackled.

Prismatic mortar specimens measuring $4 \times 4 \times 16$ cm were used. The mix proportion is shown in Table 7.5. Before the tests, 90 days of moist curing was carried out. After curing, the specimens were oven-dried (at 105°C) for three different periods: no drying, two days and three days. As the amount of interlayer water loss may depend on oven-drying time, each specimen was assumed to have a different degree of interlayer water saturation. The specimens were then infiltrated with ethanol to extract liquid water from the capillary and gel pores. It can be assumed that specimens retain only inter-layer water after extraction. This assumption has been verified in previous

*Table 7.5* Mix proportion of mortar (kg/m³)

| W/C | Water | Cement | Sand | CaCO₃ powder |
|------|-------|--------|------|--------------|
| 0.65 | 336 | 517 | 912 | 405 |

Cement: ordinary Portland cement.

research (Ishida *et al.* 2007) through the fact that the amount of extracted liquid water almost corresponds to the analytical result given by *DuCOM* for various water-to-cement ratio cases. After extracting all the liquid water from the capillary and gel pores, the specimens were oven-dried again. At that time, any drying shrinkage is caused by the evaporation of interlayer water, since the liquid water was previously replaced with ethanol and ethanol in capillary and gel pores cannot result in shrinkage (Asamoto and Ishida 2003). The experimental results are summarized in Table 7.6. The linear relationship between interlayer water saturation and corresponding shrinkage is clearly shown in Figure 7.45.

Through this experimental study, it was verified that the instantaneous plasticity of interlayer pores is directly associated with saturation, as formulated in Equation 7.27 and given in more generalized form as

$$\frac{d\varepsilon_l}{dt} = E_l \cdot \phi_{int} \frac{dS_{int}}{dt} \tag{7.45}$$

where $\varepsilon_l$ is the plastic strain of each cluster, $\phi_{int}$ is the porosity of interlayer pores, $S_{int}$ is the saturation of interlayer pores, and $E_l (= 1.22 \times 10^4 [\mu])$ is the plastic strain per unit of porosity. According to calculations using *DuCOM*, normal concrete with a water-to-cement ratio of 50% has a porosity of about 0.1 after 28 days of moist curing. When the interlayer water is completely

*Table 7.6* Loss of interlayer water and shrinkage due to disappearance of interlayer water

| Oven-dried | No drying | 2 days | 3 days |
|---|---|---|---|
| Loss ratio of interlayer water[*] | 0.0349 | 0.00833 | 0.00524 |
| Shrinkage strain [μ] | 1023 | 264 | 155 |

[*]　Loss ratio of interlayer water: loss of interlayer water (g)/specimen weight before oven-drying (g).

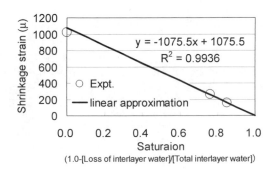

*Figure 7.45* Relation between saturation of interlayer water and shrinkage by loss of interlayer water.

lost, as here, the resultant instantaneous plasticity reaches approximately 1200 μ.

### 7.3.4 Coupling with moisture model

By coupling the moisture dynamics with the proposed time-dependent deformation model, the instantaneous plastic strain caused by the loss of interlayer water is computed in the dry state at elevated temperatures of 40–60°C. In the computation, the dispersion of moisture trapped by the ink-bottle effect promotes the varying moisture in capillary and gel pores because the dashpot viscosity and creep limit depend on pore saturation. This is reasonable, since greater drying creep under higher temperatures is experimentally observed.

Regarding drying shrinkage, however, it has been reported that higher temperatures lead to greater shrinkage (Mihashi and Numao 1992; Numao and Mihashi 1991), while Asamoto and Ishida (2005) and Cebeci *et al.* (1989) have reported that ultimate drying shrinkage at high temperatures differs little from that at normal temperatures. To look into this, drying shrinkage at elevated temperatures is investigated with the proposed model, comparing results with and without use of the enhanced moisture model.

Figure 7.46 shows the analytical results for drying shrinkage and moisture loss at a variety of temperatures. When dispersion of moisture held by the ink-bottle effect is not taken into account, drying shrinkage at higher

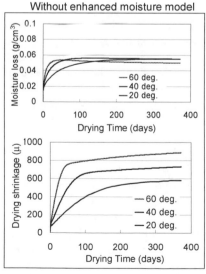

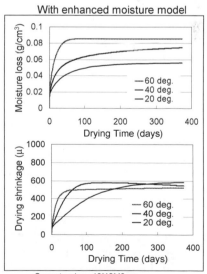

W/C=0.528, W=177,C=335, S=814, G=916 (kg/m³)

Concrete prisms 10*10*40 cm
28 days moist curing at 20°C
Dried at 28 days (RH=60%, 60, 40, 20°C)
Thermal coefficient is assumed to be 10μ/°C.

*Figure 7.46* Case study of drying shrinkage and moisture loss at different temperature.

temperatures is greater. However, against reality, moisture loss is little different at each temperature and the weight loss at 60°C is less than that at 20°C and 40°C. This is because the hydration reaction accelerates at elevated temperatures and ambient moisture is adsorbed by the hydration reaction. On the other hand, with the new moisture model, ultimate shrinkage at elevated temperatures is almost the same as at normal temperatures or smaller than that at normal temperature, and both shrinkage and moisture loss coincide with reality. This demonstrates that it is more rational and realistic to couple the proposed mechanical model with the new moisture model.

Let us examine this apparently weak dependence on temperature from a micro-physics viewpoint. The multi-scale and multi-chemo physics modeling yields the stress-free shrinkage as shown in Equations 7.29–7.31 of Section 7.2.3.1.

The diffusion of water trapped in ink-bottle-shaped pores decreases the saturation in capillary and gel pores and results in a smaller value for parameter β in Equation 7.29. The decline in β provides the decrease in shrinkage stress. According to Equation 7.29, pore pressure is less affected by absolute temperature. With the enhanced moisture model, the amount of drying shrinkage at high temperatures is not significantly larger than that at normal temperatures for a given relative humidity.

### 7.3.5 Verification

#### 7.3.5.1 Time dependence at normal temperatures

Drying shrinkage is first discussed as shown in Figure 7.47. There is no significant difference between model predictions and reality. Although the plastic flow of gel pores S provides steady shrinkage, overall shrinkage is almost convergent after 1,000 days because the viscous flow provoked by moisture migration is quite slow under small shrinkage stresses. This tendency corresponds to reality. It is concluded that the proposed model can simulate drying shrinkage reasonably at these temperatures, just like the original model (Section 7.2).

Regarding basic creep with no drying, slight differences between analysis and experiment are observed, as for instantaneous stiffness. Most of the experiments used for the verification were conducted in the 1960s–1970s, so the exact properties of the cement and aggregates cannot be ascertained. They appear to be slightly different from current ones. Consequently, in the verification process, the elastic stiffness of cement hydrates in Equation 7.20 was inversely adjusted from the experimental instantaneous stiffness, and the resulting time-dependent strain was compared. Figure 7.48 shows the experimental data alongside the analytical results. The installation of the new gel component results in a more realistic prediction over the long term. In the case of low W/C, the analysis overestimates the experimental results (by Komendant *et al.*, 1976 and 1978). One of the reasons can be attributed

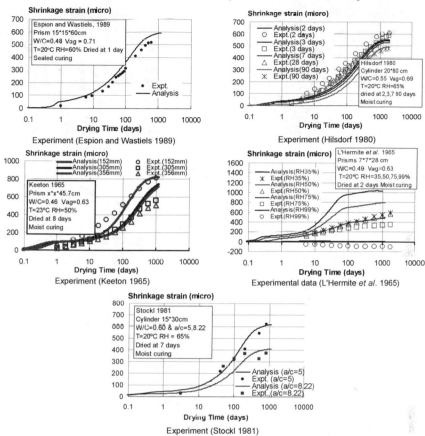

Figure 7.47 Verification of drying shrinkage.

to the accuracy of the moisture equilibrium model, which simulates the drop of internal relative humidity due to self-desiccation. In this analysis, the computed relative humidity inside pores reaches about 73% at 1,000 days. Nakarai *et al.* (2005) reported that the pore humidity predicted by the model under sealed conditions is slightly smaller than the experiment, especially in the case of a low water-to-cement ratio. The authors understand that a comprehensive study to improve other thermodynamic models as well as the time-dependent constitutive model should enhance overall accuracy of the system.

The initial stiffness of the drying creep cases was inversely identified from the experimental value in a similar way. A comparison between analysis and experiment is shown in Figure 7.49. The proposed model simulates slow continuous creep reasonably through the gradual viscous flow of moisture in the fine gel pores S. However, creep evolution is as smooth as in the experimental observations, especially when the ambient relative humidity is as

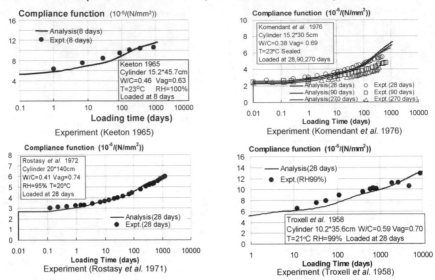

Figure 7.48 Verification of basic creep.

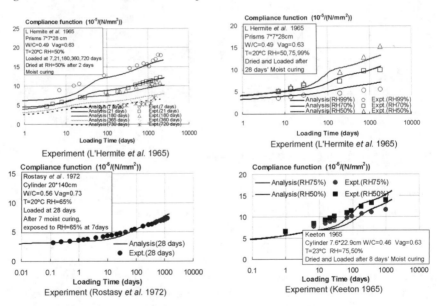

Figure 7.49 Verification of drying creep.

low as 50%. It will be necessary to improve the modeling of drying creep rate at low humidity in future.

Let us focus separately on the long-term total deformation (creep and shrinkage together) and on shrinkage alone under drying conditions. Figure 7.50 compares analytical results obtained by the proposed model and

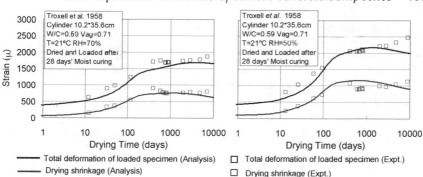

*Figure 7.50* Verification of long-term deformation (experiment: Troxell *et al*. 1958).

available experimental data. In the experiments, recovery of shrinkage was observed from 400 days to 1,000 days in both high and low relative humidity cases. Our analytical result also reproduces the recovery of time-dependent deformation. This is because moisture trapped by the ink-bottle effect disperses gradually and the internal stress decreases as stated in Equation 7.29, even while the pore pressure remains the same at vapor–liquid equilibrium. Figure 7.51 shows the mechanism of internal stress reduction. This experimental finding enables us to verify our assumption that extremely gradual dispersion of trapped water can occur even at room temperature. However, the recovery given by the analysis is more continuous than that in the experiment, especially in the case of $RH = 50\%$. This can be explained as follows.

Carbonation shrinkage can occur under long-term drying and it has been reported to cause an ongoing volume change (Verbeck 1958; Wittmann 1982) that lasts a long time. It is plausible that carbonation shrinkage may exceed the rate of recovery related to dispersion of the entrapped water, so the observed time-dependent deformation does not recover. Since the current analytical model does not include the effects of carbonation shrinkage, a further investigation involving integration of a carbonation model will be necessary.

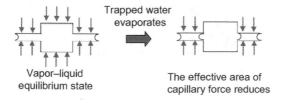

*Figure 7.51* Mechanism for internal stress decrease.

### 7.3.5.2 Time dependence at high temperatures

First, we consider basic creep at high temperatures. As moisture cannot escape to the outside, the accelerated moisture kinematics at high temperatures can be examined from a viewpoint of elevated Gibbs' energy. Next, drying shrinkage and creep at high temperatures are investigated. As discussed in Section 7.3.4, accelerated and specific moisture evaporation at high temperatures provides for a rational prediction of the time dependency of cementitious materials.

Basic creep accelerates greatly at high temperatures. In the model, the effect of temperature is represented by parameters $\chi_{TgS}$ and $\chi_{TgL}$ in regard to Gibbs' energy, as shown in Figure 7.42. The new component associated with moisture in gel pores S also provides reasonable results for long-term basic creep at high temperatures, as shown in Figure 7.52. However, the analysis underestimates short-term creep as temperature increases. This may be attributed to the temperature sensitivity function $\beta(T)$ in Equation 7.23 being common to both capillary pores, gel pores L and gel pores S (see Figure 7.42). It is plausible that the rate of moisture migration in each pore at elevated temperatures might be different. It might be possible in the future to upgrade the model to include the temperature sensitivity of the transport coefficient in each type of pore from micrometer to angstrom level.

Figure 7.53 compares experimental data (Asamoto and Ishida 2005) with analytical results of drying shrinkage at high temperatures, while Figure 7.54 represents the corresponding moisture loss. When dispersion of the trapped water is taken into account, the analytical results show reasonable tendencies of both drying shrinkage and moisture loss.

However, the analysis shows a recovery of drying shrinkage in the case of a high water-to-cement ratio. This can be explained by the same mechanism as shown in Figure 7.51. Moisture evaporation is much greater at high temperatures and these results in the vapor–liquid phase quickly reaching an equilibrium state. Then, if the trapped water in ink-bottle-shaped pores disperses, the internal stress can decrease according to Equation 7.29. There has been no experimental evidence that drying shrinkage recovers at high temperatures.

Currently, the decrease in internal stress resulting from gradual dispersion of water trapped in geometrically complex pores is able to model the ultimate drying shrinkage at high temperatures reasonably. The apparent drying shrinkage recovery in the analysis is related to mix proportion or drying conditions, and does not occur in some cases as shown in Figure 7.46. Moreover, other factors such as carbonation may also affect observed shrinkage. Therefore, a more explicit explanation can be expected if the simulation is integrated with other microphysics models before further experimental verification.

Figure 7.55 compares analyses with experiments of drying creep. In the case of a short-term experiment by Gross (1973, 1975), the analytical result

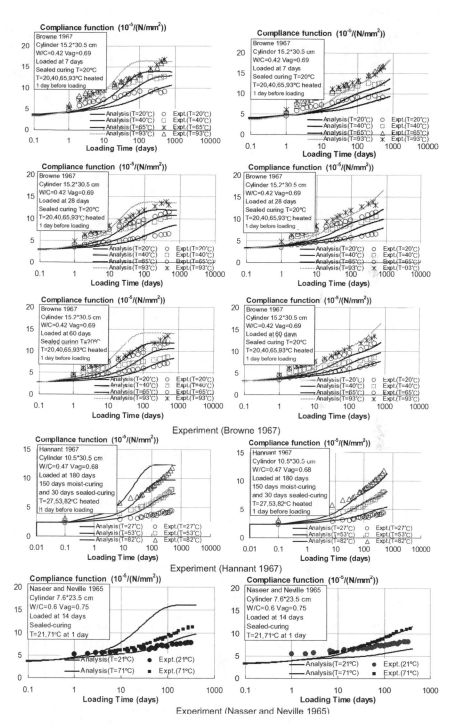

*Figure 7.52* Verification of basic creep under high temperature (left: basic model, right: enhanced model).

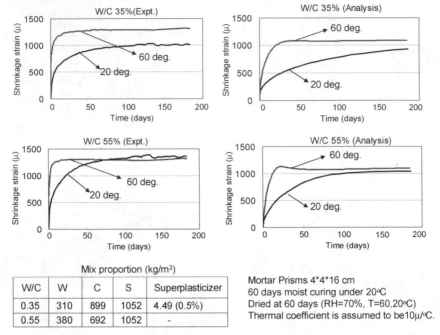

| Mix proportion (kg/m³) | | | | |
| W/C | W | C | S | Superplasticizer |
| --- | --- | --- | --- | --- |
| 0.35 | 310 | 899 | 1052 | 4.49 (0.5%) |
| 0.55 | 380 | 692 | 1052 | - |

Mortar Prisms 4*4*16 cm
60 days moist curing under 20°C
Dried at 60 days (RH=70%, T=60,20°C)
Thermal coefficient is assumed to be10μ/°C.

*Figure 7.53* Verification of drying shrinkage at high temperature (experiment: Asamoto and Ishida 2005)

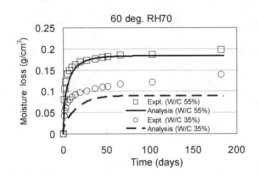

*Figure 7.54* Verification of moisture loss at high temperature. (experiment: Asamoto and Ishida 2005)

provides a reasonable prediction. However, the analysis overestimates the experimental results obtained by Arthanari and Yu (1967), and the computed creep rate appears to be greater than the experimental one. Here, the reported biaxial test data were transformed into equivalent uniaxial values assuming a Poisson's ratio of 0.2. Further, the ambient relative humidity of neither experiment was mentioned in the original paper and we simply assumed the same value as in Bazant's estimation (Bazant and Panula 1979). Thus, detailed investigation is somewhat difficult, although the overall behavior is

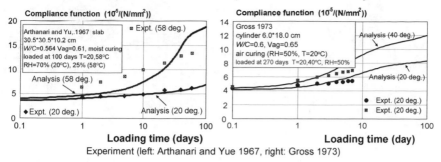

Experiment (left: Arthanari and Yue 1967, right: Gross 1973)

*Figure 7.55* Verification of drying creep under high temperature.

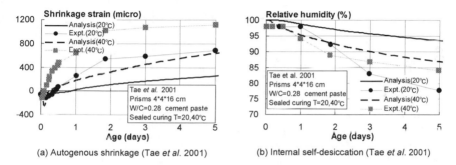

(a) Autogenous shrinkage (Tae *et al.* 2001)   (b) Internal self-desiccation (Tae *et al.* 2001)

*Figure 7.56* Verification of autogenous shrinkage and internal humidity under high temperature.

correctly generated by the model. It can be concluded that the temperature dependence of drying creep can be reproduced in outline by this analysis.

Figure 7.56 verifies the effect of temperature on autogenous shrinkage. This is a multi-factor problem involving the temperature dependence on the hydration rate, pore water equilibrium, moisture motion, and the adsorbed water model, which are taken into account when analyzing moisture state. The analytical model enables the tracking of phenomena. Model improvement is underway in connection with characteristics of moisture equilibrium and its movement in a high-temperature environment. Temperature dependence on moisture state and its effect on associated mechanical behavior require further comprehensive study.

### 7.3.5.3 *Creep deformation under decreasing stress*

Creep deformation under decreasing stress is investigated. Figure 7.57 and Figure 7.58 compare analytical results with experimental data under decreasing stress. As compared with the analytical results given by the previous model, the proposed model gives a better simulation of basic creep deformation under decreasing stress because of the introduction of equivalent strain. However, since this is a very limited comparison with experimental data,

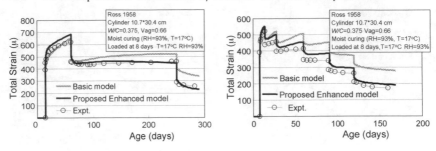

*Figure 7.57* Comparison between analysis and experiment (Ross 1958) under decreasing stress.

| | Basic model for Gel | Enhanced model for Gel pores L | Enhanced model for Gel pores S |
|---|---|---|---|
| Mathematical formulation | $\dfrac{d\varepsilon_g}{dt} = \dfrac{1}{C_g}(\varepsilon_{glim} - \varepsilon_g)$ | $\dfrac{d\varepsilon_g}{dt} = \dfrac{1}{C_g}(\varepsilon_{glim} - \varepsilon_{g.eq})$ $\varepsilon_{g.eq} = \begin{cases} \max\left(\dfrac{W_g}{\sigma_{ly}}, \varepsilon_g\right) & (when\ \sigma_{ly} > 0) \\ \min\left(\dfrac{W_g}{\sigma_{ly}}, \varepsilon_g\right) & (when\ \sigma_{ly} < 0) \end{cases}$ $W_g = \int \sigma_{ly} d\varepsilon_g$ | $\dfrac{d\varepsilon_g}{dt} = \dfrac{1}{C_g}(\varepsilon_{glim} - \varepsilon_{g.eq})$ $\varepsilon_{g.eq} = \begin{cases} \max\left(\dfrac{W_g}{\sigma_{ly}}, \varepsilon_g\right) & (when\ \sigma_{ly} > 0) \\ \min\left(\dfrac{W_g}{\sigma_{ly}}, \varepsilon_g\right) & (when\ \sigma_{ly} < 0) \end{cases}$ $W_g = \int \sigma_{ly} d\varepsilon_g$ |
| Dashpot constant | $C_g = a_{cg} \cdot \beta_S^g (S_{gel}) \cdot \beta_T (\eta(T))$ $a_{cg} = 0.864$ | $C_{gL} = a_{cgl} \cdot \beta_S^g (S_{gel}(r > 1.5nm)) \cdot \beta_T (\eta(T))$ $a_{cgl} = 0.864$ | $C_{gS} = a_{cgS} \cdot \beta_S^g (S_{gel}(r \le 1.5nm)) \cdot \beta_T (\eta(T))$ $a_{cgS} = 25.9$ |
| Limit value of visco-plastic strain | $\varepsilon_{glim} = \varepsilon_{glim}^{linear}\left(\sigma_{ly}, S_{gel}, T\right)$ $\cdot \chi_{non}\left(\varepsilon_{glim}^{linear}\right)$ $\varepsilon_{glim}^{linear} = 2.0 \cdot \left(\dfrac{\sigma_{ly}}{E_e}\right) \cdot \chi_S\left(S_{gel}\right)$ $\cdot \chi_T(T)$ | $\varepsilon_{glimL} = \varepsilon_{glimL}^{linear}\left(\sigma_{ly}, S_{gel}(r > 1.5nm), T\right)$ $\cdot \chi_{non}\left(\varepsilon_{glimL}^{linear}\right)$ $\varepsilon_{glimL}^{linear} = 1.0 \cdot \left(\dfrac{\sigma_{ly}}{E_e}\right) \cdot \chi_{SgL}\left(S_{gel}(r > 1.5nm)\right)$ $\cdot \chi_{TgL}(T)$ | $\varepsilon_{glimL} = \varepsilon_{glimL}^{linear}\left(\sigma_{ly}, S_{gel}(r \le 1.5nm), T\right)$ $\cdot \chi_{non}\left(\varepsilon_{glimL}^{linear}\right)$ $\varepsilon_{glimL}^{linear} = 1.5 \cdot \left(\dfrac{\sigma_{ly}}{E_e}\right) \cdot \chi_{SgS}\left(S_{gel}(r \le 1.5nm)\right)$ $\cdot \chi_{TgS}(T)$ |
| Elasticity | $f_{ec} = \exp(0.69(S_{cap} - 1))$ | $f_{ec} = 0.5(1 + S_{cap}^2)$ | |

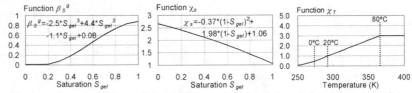

where, $S_{gel}$: degree of saturation of gel pores, $S_{gel}(r > 1.5nm)$: degree of saturation of gel pore-L, $S_{gel}(r \le 1.5nm)$: degree of saturation of gel pore-S, $\beta_T = 10 + \eta(T)$, $\chi_{non} = \exp[0.2 \cdot abs(\varepsilon_{glim}^{linear})/2.0 \times 10^{-3} - 1.0)]$.

*Figure 7.58* Comparison of visco-plastic component between basic and enhanced models.

further verification of the concept is required through comparisons with experimental results.

## 7.4 Mechanical nature of aggregates

### 7.4.1 *Extreme aggregates: facts and analysis*

Aggregate properties strongly affect the time-dependent mechanical properties of concrete (Troxell *et al.* 1958; Snowdon and Edwards 1962; Hansen and Nielsen 1965; Neville 1970, 1995; Arai *et al.* 1999, 2000; Tatematsu *et al.* 2001; Imamoto *et al.* 2006). Especially, Goto and Fujiwara (1979) investigated concrete shrinkage with many sorts of aggregates and concluded that the intrinsic shrinkage of aggregate is significant. Tatematsu *et al.* (2001) and Imamoto *et al.* (2006) also reached the same conclusions. Concrete shrinkage is affected, too, by the elastic modulus and the water absorption of aggregates. It has been reported that the lower elasticity of lightweight aggregate can lead to greater shrinkage of concrete compared to that of normal-weight aggregate (Neville 1995; Japan Society of Civil Engineers 2002). It is also possible that the loss of moisture from porous aggregates partially compensates for the drying of cement hydrates and reduces shrinkage of lightweight concrete (Japan Society of Civil Engineers 2002). These opposed features can be attributed to the mutual interaction of several mechanisms. In this section, the aim is to quantify the effect of aggregate properties, i.e., water absorption, elasticity and intrinsic shrinkage, on concrete shrinkage.

Table 7.7 shows the mix proportion which was reported to have extremely large shrinkage (Figure 1.36), as shown in Figure 7.59 (JSCE Concrete Committee 2005). The aggregates for mix A are thought to be problematic, while the ones for mix B are ordinary. The sand in mix C is standard but the gravel is the same as that in mix A. The specimens for autogenous shrinkage were sealed immediately after casting and kept at a constant temperature of 20°C. In the case of drying shrinkage, specimens were subjected to 60% relative humidity at 20°C after seven days of moist curing.

One of the aggregate properties thought to be possibly responsible for the

*Table 7.7* Mix proportions in committee test (kg/m³) (Japan Society of Civil Engineers Concrete Committee 2005)

|       | Water | Cement | Sand 1 |    | Sand 2 |    | Gravel |    |
|-------|-------|--------|--------|----|--------|----|--------|----|
| Mix A | 172   | 453    | 482    | S1 | 119    | S2 | 1,053  | G1 |
| Mix B | 172   | 453    | 544    | S3 | 61     | S4 | 1,065  | G2 |
| Mix C | 172   | 453    |        |    |        |    | 1,053  | G1 |

S1, S2: similar sand to that used in PRC bridge concrete; S3, S4: normal sand; G1: similar gravel to that used in PRC bridge concrete; G2: normal gravel.

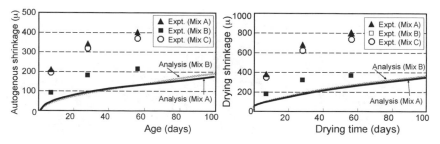

*Figure 7.59* Computed results and experimental results of autogenous shrinkage and drying shrinkage.

large shrinkage is the smaller elastic stiffness. The effect of aggregate stiffness on the macroscopic concrete shrinkage can be roughly estimated. An empirical formula of elastic stiffness based on the average density of sand and gravel is denoted by Equation 7.38 in Section 7.2.4. Accordingly, the aggregate Young's modulus of mix A is estimated as 31.6 GPa. In the simulation, a 10% lower value of estimated modulus, 28.4 GPa, was adapted, considering that the aggregates for mix A may be softer than the normal. In the case of mix B, Young's modulus obtained from Equation 7.38, 37.8 GPa, was used because the aggregates for mix B are standard ones. In simulation, the transport of moisture in aggregate pores is taken into account by considering aggregate absorption (Mix A: 1.86%, Mix B: 1.09%). The moisture model for the aggregates will be described in the detail in Section 7.4.2.

The experimental and computed results of autogenous and drying shrinkage are shown in Figure 7.59. In the case of concrete with normal aggregates, the model simulates shrinkage reasonably well, although the computed results are slightly lower than the observed ones. But both the observed autogenous and drying shrinkage of mix A concrete are much larger than those calculated by the model. As shown in Figure 7.60, a very low stiffness of aggregates (1.0 GPa) may bring the computed shrinkage close to reality. But such an extremely low stiffness is only likely to be possible with special, artificial lightweight aggregate.

Next, let us check the stiffness of concrete. The elastic stiffness of normal-strength concrete in Japan can be estimated by (Okamura 2000)

$$E_c = 8500 f'_c{}^{1/3} \tag{7.46}$$

where $E_c$ is Young's modulus of concrete [MPa], and $f'_c$ is cylinder compressive strength [MPa].

The compressive strength of concrete for mix A at 28 days of age is 68.3 MPa. Corresponding stiffness is estimated to be 34.7 GPa by Equation 7.46. This is close to the computed one by the two-phase model based on the normal stiffness aggregates as shown in Figure 7.61. Thus, it seems that the assumption of low aggregate stiffness offers little consistent explanation of both concrete shrinkage and stiffness.

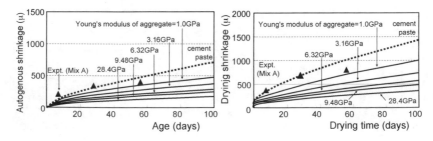

*Figure 7.60* Computed results of shrinkage with aggregates of varying Young's modulus.

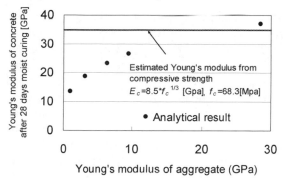

*Figure 7.61* Computed results of Young's modulus of concrete with aggregates of varying Young's modulus.

## 7.4.2 Aggregate stiffness and moisture loss

The influence of aggregate stiffness on the time-dependent deformation of concrete has been investigated comprehensively by experimental studies on the behavior of concrete containing extremely stiff or soft materials, such as steel, glass and rubber, in place of normal aggregate (Upendra 1964; Tazawa *et al.* 1992; Ono and Shimomura 2005). However, since the shape and density of steel, glass and rubber particles are quite different from those of sand and gravel particles, it is not clear that theories derived from shrinkage tests of concretes containing these alternative materials can be applied to concrete with normal aggregates. Further, the effect of moisture transport between aggregate and hardening cement paste on the time dependency of concrete needs to be given consideration. To fully study the varying shrinkage with different types of aggregate, it is essential to clarify the influence of each individual aggregate property on concrete shrinkage and to identify the combined effect of all the properties. In this section, computed results obtained by the multi-scale constitutive model are compared with experimental results, with the focus on the contribution of aggregate stiffness and absorption to concrete shrinkage, and then the model is verified quantitatively.

### 7.4.2.1 Summary of experimental program

The experimental results shown in Figure 7.59 suggest that gravel properties can significantly affect concrete shrinkage, since the different gravel types in specimens of mix B and C lead to a large difference in shrinkage, while the shrinkage of the mix A and B specimens made with the same gravel but different sands is small. Given these results, the shrinkage behavior of concrete made with different gravels (normal weight aggregate and artificial, lightweight aggregate) was studied for examining the effect of stiffness and water migration. The lightweight gravel had significantly lower Young's modulus and higher water adsorption than the normal gravel. Throughout this discussion, normal-weight aggregate (natural aggregate as used in general-purpose concrete) is defined as normal aggregate (gravel or sand).

The material properties of the normal and lightweight aggregates are shown in Table 7.8. The lightweight gravel coated with epoxy resin on the surface was also used in one of the experimental sets in order to prevent moisture migration from the gravel. Table 7.9 represents the mix proportions used in this experimental program. The type of sand and the volume ratio of aggregate were the same in every mix. The specimens cast using normal gravel are referred to as OG (ordinary gravel), the ones containing lightweight gravel are named LG (lightweight gravel) and specimens with epoxy-coated lightweight gravel are identified as ELG (epoxy-coated light-weight gravel). For example, OG30 stands for a specimen with W/C = 30% made with normal aggregate. In the case of a low water-to-cement ratio, a super-plasticizer was added to avoid the problem of insufficient compaction. The lightweight gravel for LG specimens was submerged in water and then used in saturated surface-dry condition, while the gravel surface for ELG specimens was manually epoxy-coated after oven-drying. Figure 7.62 shows the epoxy-coated gravel used in the ELG specimens. When prepared epoxy-coated gravel was immersed in water, the weight increase was limited to around 3% in spite of the high absorption of the uncoated gravel (27.5%). This test demonstrates that epoxy coating is an effective way of preventing moisture migration into the gravel. The gravel for ELG specimens was mixed manually to avoid damaging the epoxy-coated surface, while OG and LG concretes were mixed in a twin-shaft mixer. In order to exclude the influence of these different mixing methods on shrinkage behavior, the mortar for ELG

*Table 7.8* Aggregate properties

|  | *Density (g/cm³)** | *Absorption (%)* |
| --- | --- | --- |
| Sand (crushed sand) | 2.59 | 1.55 |
| Gravel (crushed stone) | 2.66 | 0.45 |
| Artificial lightweight gravel | 0.45 | 27.5 |

(*) Density of artificial lightweight gravel is in oven-dried condition. Density of other aggregates is in saturated surface-dry condition.

*Table 7.9* Mix proportion (kg/m³)

| *Specimen* | *W/C* | *Water* | *Cement* | *Sand* | *Gravel* | *SP* |
| --- | --- | --- | --- | --- | --- | --- |
| OG30 | 30 | 160 | 533 | 686 | 1,039 | 3.73 |
| LG30 | 30 | 160 | 533 | 686 | 176 | 3.73 |
| OG50 | 50 | 201 | 403 | 686 | 1,039 | - |
| LG50 | 50 | 201 | 403 | 686 | 176 | - |

Mix proportions of specimens ELG30 and ELG50 are the same as specimens LG30 and LG50, respectively.

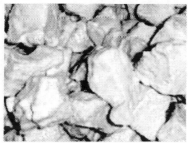

*Figure 7.62* Epoxy-coated artificial lightweight gravel.

specimens was first mixed in the twin-shaft mixer and then the gravel was manually distributed in the mix.

Prismatic specimens measuring 10 × 10 × 40 cm were used in the experiment. The form was removed one day after casting and then the specimens were cured in water at 20±2°C for six days. After this period of moist curing, specimens were exposed to 60±5% relative humidity at 20±2°C. Strains were measured with contact gauges having 5 μ resolution. All the results in drying shrinkage tests were obtained from the average of two specimens per situation.

The objective of the experiment is to evaluate the influence of aggregate stiffness on concrete shrinkage while eliminating variations in pore humidity and aggregate shrinkage by epoxy coating the lightweight aggregate. It was expected that the epoxy coating would prevent moisture transport between aggregate and cement paste solid while maintaining the same gravel stiffness. It was, however, necessary to investigate whether the epoxy-coated interface between gravel and cement paste affects shrinkage. To study the effect of the epoxy coating, shrinkage tests were conducted using normal and epoxy-coated normal gravels. The influence of the epoxy interface on shrinkage was directly examined in this test because there is very little water migration with non-epoxy-coated normal gravel due to its low water absorption, as shown in Table 7.8.

A specimen with a mix proportion of OG30 was used, since the influence on autogenous shrinkage can also be investigated with a low water-to-cement ratio. Cylindrical specimens were sealed after casting and autogenous shrinkage was measured at 20±2°C until 14 days of age. After 14 days, the surface sealing was removed and the specimens were dried at 60±5% relative humidity and 20±2°C. Strain was measured using a mold gauge. Figure 7.63 shows the results. The difference between the specimen with normal gravel and the specimen made with epoxy-coated normal gravel in autogenous shrinkage was around 30 μ at 14 days of age. Even after being subjected to drying conditions, this difference increased by just 5 μ (for a total shrinkage differential of around 35 μ). According to this experimental result, the epoxy coating on the gravel surface has no significant effect on concrete shrinkage.

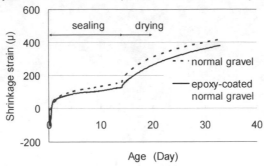

*Figure 7.63* Experimental study of the influence of epoxy-coated interface between cement paste and gravel on shrinkage.

The compressive strength and Young's modulus of each specimen at seven days of age are shown in Table 7.10. The mechanical properties of specimens ELG and LG were rather similar. The test results indicate that the epoxy coating on the gravel surface does not significantly affect bonding between gravel and hardened cement paste. It is concluded that coating the gravel surface with epoxy resin is an effective method of eliminating moisture migration in the gravel, while still maintaining other gravel properties such as stiffness and bonding with the cement paste.

### 7.4.2.2 *Analytical conditions and moisture loss of lightweight concrete*

In the multi-scale constitutive model, the effect of moisture migration between aggregate and cement paste on concrete behavior as well as aggregate Young's modulus can be taken into account by considering the average properties of sand and gravel, including density, particle size, volume ratio and absorption. In this section, the model for representing the aggregate roles in concrete behavior is summarized.

As explained in Section 7.2.1.1, concrete is modeled as a two-phase composite of an aging cement paste and an elastic aggregate without explicit separation of sand and gravel. Thus Young's modulus of the aggregate is idealized as an average value for sand and gravel. In the case of OG specimens, Young's modulus of the aggregate is computed to be 40.8 GPa, from Equation 7.38. Since the average density of the sand and lightweight gravel for LG and ELG specimens is out of range for Equation 7.38, Young's modulus of the aggregate was determined to be 2.0 GPa, based on experimental

*Table 7.10* Results of compressive test at seven days of age

|  | OG30 | LG30 | ELG30 | OG50 | LG50 | ELG50 |
|---|---|---|---|---|---|---|
| Compressive strength (MPa) | 62.8 | 18.8 | 18.1 | 23.7 | 7.5 | 12.9 |
| Young's modulus (GPa) | 34.3 | 14.0 | 15.5 | 23.4 | 9.6 | 8.8 |

values of concrete Young's modulus. Young's modulus of hardening cement paste is calculated according to the water-to-cement ratio and hydration degree. Figure 7.64 shows the experimental and computed results. The two-phase composite model is able to simulate reasonably well the different elastic stiffness of concretes with different aggregate elasticities, for both low and high water-to-cement ratios.

Moisture transport between aggregate and cement paste is idealized in Section 3.5.2. In order to determine the unknown parameter $\beta$ in Equations 3.125 and 3.126, parametric studies were conducted using the moisture transport model between cement paste and aggregate. In this analysis, the authors focused on obtaining an approximate value of $\beta$ by matching to the experimental results. And the moisture migration in aggregate pores depends on the aggregate isotherm that represents the relationship between pore humidity and moisture saturation. Here, the isotherm of normal aggregate proposed by Nakarai and Ishida (2003) based on experiments with normal sand was adapted. In the case of concrete with normal sand and lightweight gravel, it is supposed that the lightweight gravel is almost fully dried of moisture at relatively high levels of humidity due to the large pores, so the average saturation of the aggregate depends on the sand's saturation at relatively lower humidity levels. The proposed isotherm is shown in Figure 7.65.

The experimental conditions adopted in the investigation of parameter $\beta$

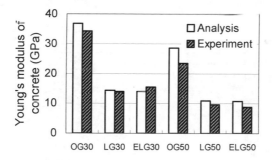

*Figure 7.64* Young's modulus of each concrete at seven days of age.

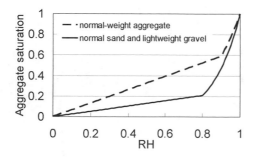

*Figure 7.65* Isotherm between pore humidity and aggregate saturation (Nakarai and Ishida 2003).

are summarized in Table 7.11 (Kokubu *et al.* 1969). The parametric numerical results of moisture loss obtained by varying parameter β are shown in Figure 7.66. With a larger value of β, the model is able to simulate moisture loss at the beginning of drying more reasonably, even though it is a slight improvement. Figure 7.67 shows the variations in cement paste and aggregate pore humidity in the model for cases where β is more than 0.01. When β is more than 0.1, moisture moves rapidly between cement paste and

*Table 7.11* Mix proportion of lightweight and normal weight concrete and experimental outline (Kokubu *et al.* 1969)

| Specimen | Unit water content (kg/m³) | Sand | | | Gravel | | |
|---|---|---|---|---|---|---|---|
| | | Type | Absorption (%) | Density (g/cm³) | Type | Absorption (%) | Density (g/cm³) |
| NN | 158 | River sand (N) | 1.9 | 2.63 | River gravel (N) | 1.1 | 2.66 |
| ON | 159 | River sand (N) | 1.9 | 2.63 | Artificial lightweight gravel (O) | 10.5 | 1.50 |
| XN | 162 | River sand (N) | 1.9 | 2.63 | Artificial lightweight gravel (X) | 16.1 | 1.46 |

Water-to-cement ratio: 0.45; sand volume ratio: 0.4, specimen size: 10 × 10 × 42 cm prism; drying conditions: specimen exposed to RH = 60% at 21°C after 7 days of moist curing.

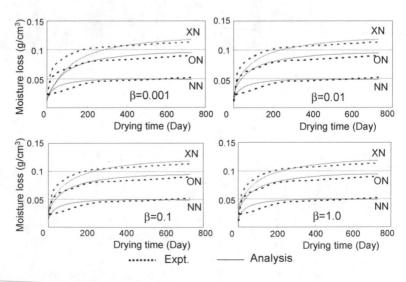

*Figure 7.66* Parametric study of parameter β.

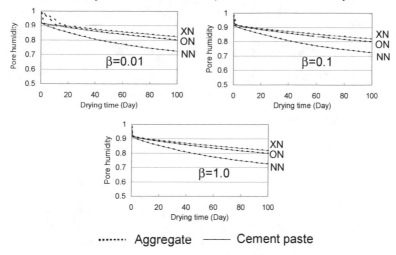

Figure 7.67 Pore humidity with varying parameter.

aggregate and the pore humidity of the cement paste almost corresponds to that of the aggregate during drying. Currently, the authors do not have a clear microscopic understanding of the possibility of this phenomenon under actual drying conditions. Thus, β was specified simply as 1.0 because the computed results with this value for moisture loss have better agreement with the experiment.

### 7.4.2.3 Influence of aggregate properties on concrete shrinkage

Figure 7.68 and Figure 7.69 show the computed results and experimental results for concrete drying shrinkage with different aggregates. In the simulation, the absorption of epoxy-coated gravel was assumed to be 0.0%, while the absorption values shown in Table 7.8 were input for the case of other aggregates.

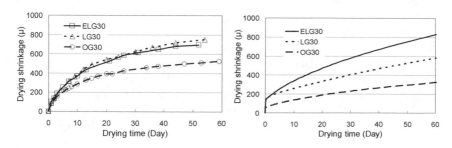

Figure 7.68 Experimental results and computed results of drying shrinkage (W/C = 30%).

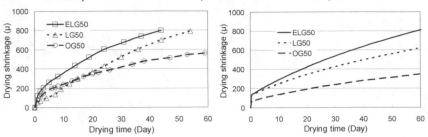

*Figure 7.69* Experimental results and computed results of drying shrinkage (W/C = 50%).

First, the influence of moisture in gravel pores on concrete shrinkage is discussed. The water in the gravel pores of LG specimens is gradually transferred into the cement paste matrix and then diffused outside. Since the pore humidity in the concrete of LG specimens is larger than that of ELG specimens, due to the diffusion of moisture from the gravel into cement paste at the beginning of drying, the drying shrinkage of LG specimens should be smaller than that of ELG specimens. In the case of high water-to-cement ratios, this speculation can be verified because both the computed result and the experimental result indicate larger shrinkage of ELG specimens than LG specimens. On the other hand, for low water-to-cement ratios, the experiment and analysis give different results. In the numerical simulation, moisture from the cement paste takes a longer time to migrate to the outside from the cement paste matrix, due to the dense pore distribution in comparison with the high water-to-cement ratios. Since moisture supplied from porous gravel remains in the paste matrix and therefore maintains pore humidity, the difference in shrinkage between LG specimens and ELG specimens with low water-to-cement ratios is larger than that at high water-to-cement ratios in the calculation. In the experiment, however, moisture in the gravel pores has little effect on shrinkage behavior. One possible reason for the difference is that, in the experiment, moisture from the porous gravel diffuses into the cement paste due to self-desiccation during curing and then may not migrate at drying. Since this assumption has not been verified yet, the authors recognize that further investigation is necessary.

Next, the influence of aggregate Young's modulus on drying shrinkage is investigated. The shrinkage of OG specimens made with stiff normal gravel is compared with that of ELG specimens containing soft epoxy-coated gravel, assuming, as noted previously, that moisture in the normal gravel does not affect drying shrinkage due to its low absorption. In the experiment, the shrinkage of ELG specimens with softer gravel is larger than that of OG specimens. The computed result shows the same tendency. In the case of epoxy-coated porous lightweight aggregate with a significantly low elastic modulus, the influence of aggregate stiffness on concrete shrinkage can be exactly observed because moisture migration is quite small. It is concluded that Young's modulus of the aggregate has an effect on concrete shrinkage and that the model simulates this effect reasonably well.

### 7.4.3 Modeling of intrinsic shrinkage of aggregates

Based on the discussion in the previous section, it is inferred that the multi-scale constitutive model is able to simulate concrete shrinkage reasonably well in consideration of aggregate elastic modulus and absorption. Thus, it is suggested that the large concrete shrinkage mentioned in Section 7.4.1 may be caused not by the low Young's modulus of the aggregate but by other aggregate properties. It has been reported that aggregate shrinkage as well as elastic modulus may influence the volumetric change of concrete (Snowdon and Edwards 1962; Hansen and Nielsen 1965; Goto and Fujiwara 1979; Tatematsu *et al.* 2001; Imamoto *et al.* 2006). In this section, an aggregate shrinkage model is proposed and implemented in the multi-scale constitutive model (Section 7.2 and Section 7.3). The proposed model is then verified by comparison with experimental results.

The volumetric stresses and strains for a two-phase system are given with respect to the component values as explained in Section 7.2.1. (See Equation 7.7 and Equation 7.8). For the purpose of considering aggregate shrinkage due to drying, the shrinkage strain is added to the volumetric strain of the aggregate arising from the volumetric stress as follows.

$$\varepsilon_{ag} = \frac{1}{3K_{ag}}\sigma_{ag} + \varepsilon_{ag}^{sh} \qquad (7.47)$$

where $K_{ag}$ is the volumetric stiffness of the aggregate, and $\varepsilon_{ag}^{sh}$ is the strain of the aggregate due to its shrinkage mechanism.

The drying shrinkage of aggregate is likely to be dependent on moisture states in the aggregate pores based on mechanisms such as capillary pressure and also on the increase in solid surface energy as well as in the concrete. Goto and Fujiwara (1979) presented the relationship between moisture

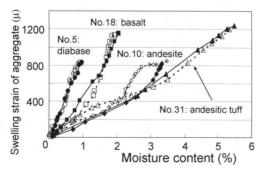

*Figure 7.70* Length change of normal aggregates due to absorption and drying (Goto and Fujiwara 1979).

content and shrinkage of normal aggregate for a situation where aggregate absorbed moisture after oven-drying and then dried again, as given in Figure 7.70. Figure 7.71 shows the relationship between the ratio of $\varepsilon_{ag}^{sh}$ (drying shrinkage of aggregate) to $\varepsilon_{agmax}^{sh}$ (maximum shrinkage of aggregate after oven-drying) and degree of saturation $S_{ag}$ (= moisture content/maximum moisture content) obtained from the experiment by Goto and Fujiwara (1979). Even for different types of aggregate with distinct values of maximum shrinkage strain, the relationship is almost linear through both the drying and wetting processes. Currently, there is no valid experimental evidence that verifies or contradicts this relationship based on a microscopic approach using capillary pressure, solid surface energy or similar. Once the implementation of a simple aggregate shrinkage model according to the experimental results and the respective investigation of each aggregate property are noted here, the shrinkage of normal aggregate is idealized as

$$\varepsilon_{ag}^{sh} = \varepsilon_{ag\,max}^{sh} \cdot (1.0 - S_{ag}/0.95) \quad S_{ag} \le 0.95$$

$$\varepsilon_{ag}^{sh} = 0.0 \qquad\qquad\qquad\qquad S_{ag} > 0.95 \tag{7.48}$$

The saturation $S_{ag}$ can be computed in the model from the pore humidity as obtained using the moisture transport model and the isotherm, as discussed in the previous section. Shrinkage up to a saturation of 0.95 is assumed to be 0 µ, because the moisture in large pores diffuses first and shrinkage due to changes in pore pressure and solid surface energy related to moisture states in large pores will be small.

The large concrete shrinkage as discussed in Section 7.4.1 is simulated once again, using the proposed aggregate shrinkage model. The isotherm between pore humidity and aggregate saturation was that proposed by Nakarai and Ishida (2003), as shown in Figure 7.65. The maximum drying shrinkage

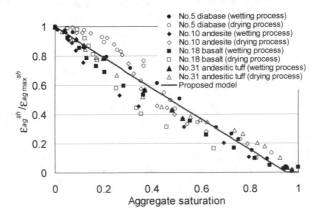

*Figure 7.71* Relationship of aggregate shrinkage and its saturation based on Goto and Fujiwara experiment.

$\varepsilon_{agmax}{}^{sh}$ is likely to be extremely large so it was set at 1,400 µ, the largest value of maximum shrinkage reported in previous research (Goto and Fujiwara 1979). The average elastic modulus of the sand and gravel was assumed to have the value of 28.4 GPa.

Figure 7.72 shows the computed results of autogenous shrinkage and drying shrinkage and Figure 7.73 gives the calculated aggregate shrinkage strain during drying. Both autogenous shrinkage and drying shrinkage are significantly larger when simulated in consideration of aggregate shrinkage and the computed results appear reasonable. Further, there is a remarkable amount of aggregate shrinkage: 600 µ. These analytical results strongly suggest that the contribution of aggregate shrinkage to concrete shrinkage is considerable and that significantly greater autogenous shrinkage and drying shrinkage can result from aggregate properties alone. Since various types of aggregate are used in cementitious composite materials, it is important to gain a correct understanding of the shrinkage properties of each aggregate during the design process. The model proposed here is further investigated by comparing the computed results with observed shrinkage of concretes made with various aggregates.

In order to evaluate the shrinkage of the aggregate itself, some researchers have focused on the specific surface area of the aggregate. Imamoto *et al.* (2006) reported that the amount of drying shrinkage of concrete is strongly related to the specific surface area of the sand and gravel. Goto and Fujiwara (1979) proposed the following equation, associating maximum aggregate

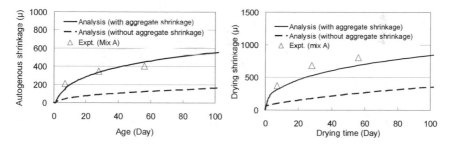

*Figure 7.72* Computed results considering aggregate shrinkage.

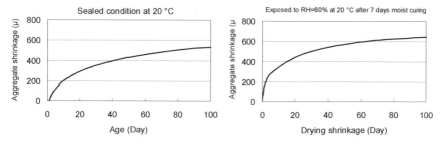

*Figure 7.73* Aggregate shrinkage in sealed conditions and drying conditions.

shrinkage strain $\varepsilon_{agmax}^{sh}$ with specific aggregate surface area $SA_{ag}$, assuming that the main mechanism of aggregate shrinkage is an increase in solid surface energy due to desorption of adsorbed water.

$$\varepsilon_{ag\,max}^{sh} = k \cdot SA_{ag} \tag{7.49}$$

As shown in Figure 7.74, an almost linear relationship is obtained experimentally between maximum aggregate shrinkage with oven-drying and specific surface area. Since the value of $k$ was not reported in the paper (Goto and Fujiwara 1979), $k$ was determined to be $136 \times 10^{-6}$ by the method of least squares. According to Equations 7.47–7.49, aggregate shrinkage can be computed in the model if the specific surface area of the aggregate is known.

Using the above methodology, computed results were compared with the shrinkage tests carried out by Imamoto *et al.* (2006). The mix proportion of concrete used in the experiments is shown in Table 7.12. The specimen size was $10 \times 10 \times 40$ cm and the specimens were exposed to RH = 60% at 20°C after seven days of moist curing at 20°C. From experimental data for specific surface area, as shown in Figure 7.75, the maximum aggregate shrinkage was determined using Equation 7.49: the maximum shrinkage for

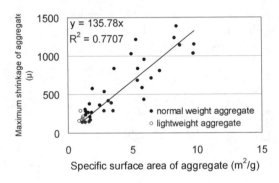

*Figure 7.74* Relation between specific surface area and maximum shrinkage of aggregate (Goto and Fujiwara 1979).

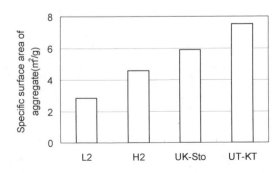

*Figure 7.75* Specific surface area of aggregate (Imamoto *et al.* 2006).

*Table 7.12* Concrete mix proportions (kg/m³) (Imamoto *et al.* 2006)

| Specimen | W | C | S1 | S2 | G |
|---|---|---|---|---|---|
| UT-KT | 180 | 360 | 517 | 222 | 1,006 |
| UK-STo | 180 | 360 | 522 | 224 | 1,006 |
| H2 | 180 | 360 | 517 | 222 | 980 |
| L2 | 180 | 360 | 517 | 222 | 1,010 |

S1 (UT-KT,H2,L2): sea sand; density 2.48g/cm³; absorption 2.50%, S1 (UK-Sto): sea sand; density 2.55g/cm³; absorption 1.61%, S2 (UT-KT,H2,L2): hard sandstone sand; density 2.61g/cm³; absorption 1.73%, S2 (UK-STo): limestone sand; density 2.58g/cm³; absorption 1.98%, G (UT-KT,UK-STo): hard sandstone gravel; density 2.66g/cm³; absorption 0.95%, G (H2): hard sandstone gravel; density 2.60g/cm³; absorption 0.73%, G (L2): limestone gavel; density 2.69g/cm³; absorption 0.41%.

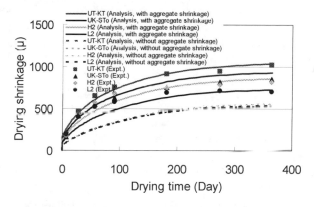

*Figure 7.76* Computed results of drying shrinkage considering aggregate shrinkage.

specimens UT-KT, UK-STo, H2 and L2 was 1,021 µ, 805 µ, 624 µ and 387 µ, respectively. Young's modulus of the aggregate used in specimens UT-KT, UK-STo, H2 and L2 was 33.1GPa, 37.2GPa, 25.1GPa and 37.0GPa, respectively, based on Equation 7.38.

Figure 7.76 shows the experimental results and the computed results with and without consideration of aggregate shrinkage. When aggregate shrinkage is included in the analytical model, the varying concrete shrinkage behavior when different types of aggregate are used is modeled more reasonably than when aggregate shrinkage is ignored. It is concluded, therefore, that the multi-scale constitutive model is able to acceptably predict concrete shrinkage in consideration of aggregate properties if density, water absorption and specific surface area of the aggregate are known.

# 8 Time-dependent mechanics of structural concrete

This chapter first presents a new time-dependent constitutive model which encompasses both near- and post-peak regions in concrete compression. In order to contribute towards a better evaluation of structural collapse under extreme loads, a coupled plastic–damaging law is presented and applied to non-linear collapse analysis in line with the multi-scale concept, and its applicability is verified. Repetition of higher stresses apparently causes progressive damage, which is also rate-dependent on strain and stress. Second, this chapter discusses the separate time-dependent cumulative non-linearity and the effect of repetition of the strain path on the overall damage evolution of concrete in compression. Third, this transient non-linear analysis is also proposed as a way of predicting the long-term deformation and safety performance of structural concrete with and without cracking, and a mechanistic creep constitutive model for pre- and post-cracking tension stiffness is presented. The effect of drying shrinkage is integrated into the predictive scheme using the thermo-hydro physics of porous media discussed in Chapter 3 and Chapter 7, and a simple equivalent method of analysis is discussed for the practical durability performance assessment of structural concrete. Careful verification of the model is carried out with respect to the creep deflection of reinforced concrete (RC) beams and slabs subjected to multi-axial flexure. Three-dimensional fiber and plate and shell elements are used for the space discretization of the analysis domain.

## 8.1 Elasto-plastic and fracturing model in compression: short-term non-linearity

In the scheme of performance-based design, demands for the simulation of structural collapse are increasing because of a need for a sounder description of global safety factors. Currently, research efforts are being made in the area of post-peak analyses of structural concrete, and non-linearity of softened compression is regarded as one of the key issues. Here, time dependency turns out to be predominant when stress levels exceed 70% of the specific uniaxial strength of concrete, and instability may occur in the form of creep rupture within a shorter stress period (Rusch 1960). This tendency is greatly accelerated beyond the peak. Current advanced computational mechanics enables us to simulate capacity as well as transient states until complete collapse. Thus, enhanced material modeling directly contributes to the upgrading of post-peak analyses and their reliability.

As loading rates of static experiments in laboratories greatly differ from those of dynamic experiments in the case of earthquakes, Okamura and Maekawa (1991) took into account time-dependency in their static/dynamic non-linear structural analyses under higher stresses by simply factorizing the plastic evolution law for concrete (Maekawa *et al.* 2003b). As ultimate limit states of reinforced concrete are comparatively less time dependent in practice, simplified approaches have been practically acceptable. As far as post-peak analysis is concerned, however, the rate effect is becoming significant and the authors empathize that explicit consideration of time-dependent non-linearity of compression softening is required in collapse analyses as well as long-term durability simulation.

Here, it should be noted that concrete in collapsing structures undergoes varying strain or stress rates. Even if the external displacement rate were kept constant, quite a high strain rate would be provoked around localized zones of deformation. On the contrary, the neighboring volumes next to the localized zone reveal rapid recovery of deformation. Thus, the scope of research has to be the short-term time dependency on both loading and unloading paths. This implies that the stress-strain model under a constant rate of loading cannot meet the requirement for collapse analyses. Thus, a new versatile time-dependent constitutive model, which encompasses both near- and post-peak regions in concrete compression, is proposed in this section.

Tabata and Maekawa (1984) and Song *et al.* (1991) investigated combined creep–relaxation hysteresis close to the uniaxial strength and extracted both short-term plastic and damage evolutions. They concluded that the combined elasto-plastic and damage concept could be applied to pre- and post-peak, time-dependent mechanics. This section basically goes along with this mechanics and further extends the applicability to generic three-dimensional confined conditions to fulfil the requirement of post-peak analyses. Concrete is a heterogeneous cohesive-frictional material that is highly pressure-sensitive and in which compressive stress transfer is accomplished to a great extent by frictional forces after the peak (Pallewatta *et al.* 1995; Sheikh 1982; Scott *et al.* 1982). There are two well-known facts: first, that confinement is very effective for improving post-peak responses, and second, that time dependency becomes significant in the post-peak region. However, a lack of combined knowledge prevents us from proceeding to post-peak analyses in time domains. Here, the authors primarily intend to formulate the one-dimensional, high-rate transient non-linearity of concrete compression.

Experimental works were conducted on the basis of uniaxial compressive loading to concrete (El-Kashif and Maekawa 2004a). Various confining pressures were applied by using steel rings embedded in advance ($\rho_s$ = steel

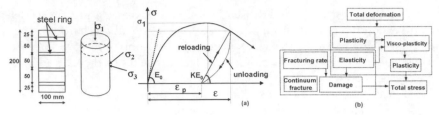

*Figure 8.1* Concept elasto-plastic and damaging model: (a) definition (b) scheme of formulation.

ratio). The specimens and their details are shown Figure 8.1. The axial capacity of concrete with lateral confinement is spatially averaged over the whole section (Irawan and Maekawa 1994). As a circular section was used, local stress over a section can be assumed uniform due to axial symmetry. The applied lateral stress can be calculated according to $\frac{1}{2}\rho_s f_y$ (Mander *et al.* 1988) by assuming full yield of lateral steel (denoted $f_y$) close to and after peak capacity. The measured strain is the special averaged strain over the length of the specimens. Thus, the compressive localized zone is included inside this characteristic length.

### 8.1.1 Basic concept extended to time dependency

The elasto-plastic and damaging model (Figure 8.1) is selected as the mechanical basis for expressing non-linearity. This is schematically expressed as assembled parallel elements consisting of elasto-plastic sliders. Here, broken elements are represented as damaged ones in past mechanics, and the total stress is taken as an integral of the internal stresses of remaining elements. According to the analogy shown in Figure 8.2a, the intrinsic stress intensity applied to each non-damaged component is directly proportional to elastic strain. Then, the time–elasticity path is thought to be a logical operator to rule both plastic and damaging growths.

Maekawa *et al.* (2003b) presented evolution laws of plasticity and damage with respect to elastic strain increment under high-rate loading. In order to further obtain their time dependency, the authors applied three patterns of total stress–strain hysteresis in experiment as shown in Figure 8.2, where A and B denote different mechanical states specified on stress–strain space. Rapid stress release from these two states may result in plastic strain increment and variation of the fracture parameter. The elastic strain also varies slightly from state A to state B, even if the total stress is kept constant. Here, the average elastic strain between state A and state B can be thought to represent the average internal stress intensity of active components during this transient condition. Thus, we can have a set of evolution ($d\varepsilon_p/dt$, $dK/dt$) by the form of finite time difference and corresponding averaged state variables ($\varepsilon_p$, $K$, $\varepsilon_e$) from an experiment where total compression is sustained. Under the relaxation path for keeping the total strain constant, we may get hold of

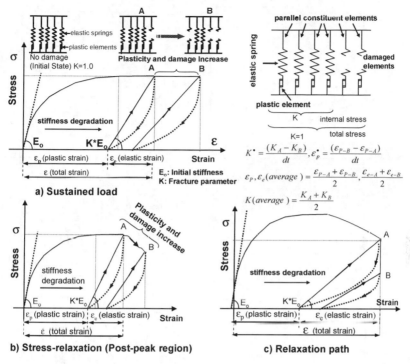

*Figure 8.2* Schematic representation of elasto-plastic and damaging concept and extraction of time-dependent evolution of plasticity and damaging.

evolved transient non-linearity and subsequent state variables in the same manner. Creep and relaxation paths were applied to concrete with different magnitudes of confinement. The authors included their combinations, in which total strain occurs but the total stress is softened in the post-peak region. If there were a unique relation between non-linearity rates and the state variables in spite of various time–stress–strain histories, it could serve as a rate-type constitutive law of generality. Here, attention is mainly directed to this distinctiveness.

Figure 8.3a shows the extracted rates of plasticity and fracturing ($d\varepsilon_p/dt$, $dK/dt$) from various loading paths under different intrinsic stress–strain levels of ($\varepsilon_p$, $K$, $\varepsilon_e$), as shown in Figure 8.11 under unconfined compression. It is clearly shown that the plastic rate under similar elastic strain levels is sharply reduced when the absolute plastic strain evolves, and that the plastic rate depends to a high degree on the magnitude of the elastic strain. Just before creep failure, an increase in progressive total strain was experienced. With this, the elasticity was simultaneously amplified due to simultaneously evolving damage even though the total stress level was kept unchanged. This trend is qualitatively similar to long-term creep under lower stress states where continuum damage is hardly seen. However, the observed transient plastic flow is thought to be caused by shear slides along micro-cracks (Maekawa

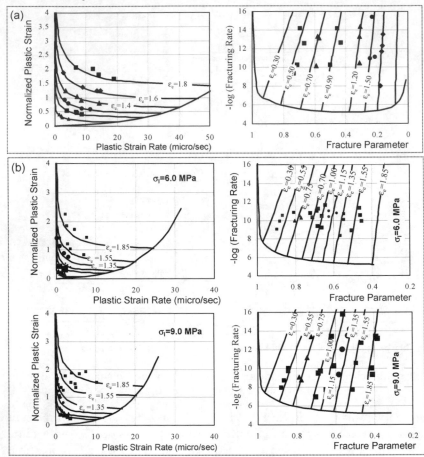

*Figure 8.3* Extracted plastic and damage rates: (a) unconfined, (b) confined concrete
(dot points = experimental data, solid lines = analytical model)

*et al.* 2003b), and its rate is significantly higher than the creep flow rate associated with C–S–H gel grains and moisture dynamics under lower stress states (Chapter 7). The larger elasticity is developed, the higher the rate of plasticity produced. This is also analogous to the long-term creep properties of concrete in nature but rather non-linear with respect to the magnitude of stress or elastic strains. These experimentally observed behaviors do not contradict the elasto-plastic and damage concepts.

The damage evolution is not seen under low stress states of less than 30% of the specific uniaxial compressive strength, but it becomes significant under near- or post-peak conditions. The fracturing rate is rapidly reduced when fracture itself occurs under a similar magnitude of elasticity that represents the internal stress intensity. Damage (reduction of elasticity) is attributed to the growth of distributed micro-cracks associated with deteriorated capacity of elastic shear strain energy (Maekawa *et al.* 2003b). In fact, it takes time for crack propagation to occur in a finite domain, and it can be seen in Figure 8.3a that the rate of fracturing is dependent on internal stress intensity in pre- and post-peak states.

The extracted data from experiments of confined concrete (Figure 8.12 and Figure 8.13) was processed and summarized as shown in Figure 8.3b. As all lateral steel rings (Figure 8.1) yield just before and after the peak strength of each specimen when non-linearity evolves, lateral confining stress is computed by solving lateral equilibrium as $\frac{1}{2}\rho_s f_y$. On unloading and reloading paths, lateral steel rings return to elasticity. Here, a reduced amount of non-linear evolution was observed. It is recognized that the damaging rate is reduced in absolute terms according to the increase in confining pressure. This means that higher confinement stabilizes transitional damaging of concrete and seems reasonable for composite frictional materials. Although the effect of confinement is seen in time-dependent plasticity, it is comparatively small. This means that the growth of plastic deformation is not well restrained by confinement when the stress is applied near the capacity, or when strain crosses the capacity threshold. In the following chapter, non-linear evolution laws will be formulated for computational mechanics.

### 8.1.2 Model formulation

#### 8.1.2.1 Potential term of non-linearity

The basic constitutive equations to express the elasto-plastic and damaging concepts (see Figure 8.2) and the general series of Taylor's state variables (plastic strain and fracture parameter) can be reduced to

$$\varepsilon = \varepsilon_e + \varepsilon_p , \quad \sigma = E_o \varepsilon_e K \tag{8.1}$$

$$d\varepsilon_p = \left(\frac{\partial \varepsilon_p}{\partial t}\right) dt + \left(\frac{\partial \varepsilon_p}{\partial \varepsilon_e}\right) d\varepsilon_e \ ; \ dK = \left(\frac{\partial K}{\partial t}\right) dt + \left(\frac{\partial K}{\partial \varepsilon_e}\right) d\varepsilon_e \tag{8.2}$$

where $K$ is the fracture parameter, $\varepsilon_e$ is the elastic strain, $\varepsilon_p$ is the plastic strain, $\sigma$ is stress, t is time, and $E_o$ is the initial elastic modulus.

For simplicity of formulation, the above stress and strain are defined as normalized stress and strain using the specific uniaxial compressive strength and the corresponding peak strain, respectively. Similar to the theory of plasticity, it was experimentally established that plastic and damage potentials $(F_p, F_k)$ exist and create an envelope surface described by $(F_p, F_k) = 0$, where the derivatives with respect to the elastic strain in Equation 8.2 reveal nonzero. As non-linearity also occurs by satisfying conditions $dF_p = 0$ and $dF_k = 0$, we have

$$\left(\frac{\partial \varepsilon_p}{\partial \varepsilon_e}\right) = 0 \quad \text{when } F_p > 0 \ , \quad \left(\frac{\partial \varepsilon_p}{\partial \varepsilon_e}\right) = -(\partial F_p/\partial \varepsilon_e)/(\partial F_p/\partial \varepsilon_p) \quad \text{when } F_p = 0 \tag{8.3}$$

$$\left(\frac{\partial K}{\partial \varepsilon_e}\right) = 0 \quad \text{when } F_k > 0, \quad \left(\frac{\partial K}{\partial \varepsilon_e}\right) = -(\partial F_k / \partial \varepsilon_e)/(\partial F_k / \partial K) \quad \text{when } F_k = 0 \quad (8.4)$$

The potential functions were experimentally formulated for unconfined normal strength concrete (Maekawa *et al.* 2003b) as

$$F_p = \varepsilon_p - 0.038\left(\exp\left(\frac{\varepsilon_e}{0.55}\right) - 1\right) \tag{8.5}$$

$$F_k = K - \exp\{-0.73\beta(1 - \exp(-1.25\beta))\}, \quad \beta = -\frac{1}{0.35}\left(\ln\left(1 - \frac{7\varepsilon_{ec}}{20}\right)\right) \tag{8.6}$$

where $\varepsilon_{ec} = \varepsilon_e$ in case of the unconfined condition. Figure 8.4a and Figure 8.5a illustrate the plastic and fracturing potentials, respectively (Maekawa *et al.* 2003b), derived from experiments under higher loading rates. Here, the time-dependent plasticity and fracture are thought to be negligible.

It is experimentally known that confining pressure elevates uniaxial strength (Richart *et al.* 1928) and ductility. This effect was successfully explained only by the suppressed damage evolution denoted by $F_k$, as illustrated in Figure 8.5d. It was also revealed that, on the contrary, plastic evolution of concrete is not affected by confinement (Maekawa *et al.* 2003a). Here, the authors extend the original damage potential of concrete (Equation 8.6) to more generic states under lateral confinement by using confinement parameter $\gamma$,

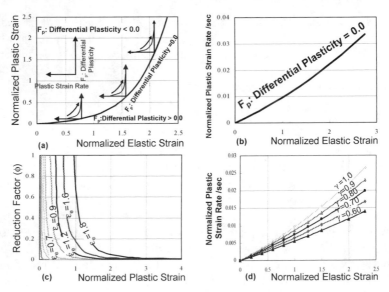

*Figure 8.4* Formulation scheme of time-dependent plasticity: (a) sensitivity of updated plasticity and elasticity on plastic flow rate, (b) flow rate on potential envelope, (c) decay of flow rate, (d) effect of confinement on flow rate on potential envelope.

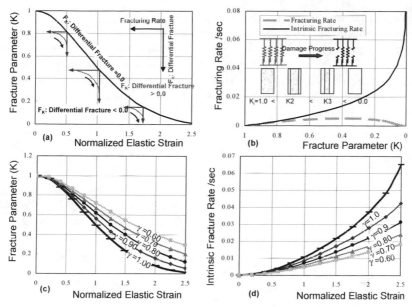

*Figure 8.5* Formulation scheme of time dependent damage: (a) sensitivity of updated damage and elasticity on damage evolution, (b) damage rate on potential envelope, (c) effect of confinement on damage evolution on potential envelope, (d) effect of confinement on damage rate on potential envelope.

defined by Equation 8.7. The unconfined condition corresponds to unity of $\gamma$ and higher confinement reduces this index to zero as an extreme confinement. The following strain modification factor for Equation 8.6 is presented so that the total stress–strain relation matches the reality of confined concrete as

$$\varepsilon_{ec} = \psi \cdot \varepsilon_e, \quad \psi = \gamma^a, \quad \gamma = \frac{f_c}{f_c + 4 \cdot \sigma_1} \tag{8.7}$$

$$a = Y^{0.4(1 - \exp(-5Y))}, \quad Y = \frac{\varepsilon_e}{3.25 - 2.65\gamma} \tag{8.8}$$

where $f_c$ is uniaxial compressive strength, and $\sigma_1$ is lateral confining pressure. In the following section, formulation of non-linear derivatives with respect to time will be discussed.

### 8.1.2.2 Plastic rate function

Figure 8.4a schematically shows the instantaneous plasticity envelope ($F_p = 0$) and the rate of plasticity on the ($\varepsilon_e$, $\varepsilon_p$) plane. It is natural to assume that the rate of plasticity reaches its maximum on the instantaneous plastic potential envelope, and that the specific plastic rate on the envelope increases in accordance with the magnitude of elasticity, as illustrated in Figure 8.3b. As described in Section 8.1.1 and shown in Figure 8.4a, the rate of plasticity decays as the state of active elasto-plastic components represented by ($\varepsilon_e$, $\varepsilon_p$) moves away from the envelope. Thus, the authors propose the following mathematical form

$$\frac{\partial \varepsilon_p}{\partial t} = \phi \left( \frac{\partial \varepsilon_p}{\partial t} \right)_b , \left( \frac{\partial \varepsilon_p}{\partial t} \right)_b = 0.034 \left( \exp\left( \frac{\varepsilon_{ep}}{4} \right) - 1 \right) \tag{8.9a}$$

$$\phi = \exp\left( -6 \left( \frac{F_p^{0.6}}{\varepsilon_e^{1.2}} \right) \right) , \quad \varepsilon_{ep} = \gamma \, \varepsilon_e \tag{8.9b}$$

where $(\partial \varepsilon_p / \partial t)_b$ means the referential rate defined on the plastic potential envelope, $\phi$ indicates the reduction factor in terms of plastic evolution as shown in Figure 8.4c, and $\varepsilon_{ep}$ means the equivalent elastic strain corresponding to the confining pressure level.

The values computed by Equation 8.9 are overlaid on Figure 8.3a and Figure 8.3b for both plain and confined concrete. The plastic rate function can be uniquely specified by the elastic strain to represent the intrinsic stress applied to the parallel components. If total stress is used for the plastic rate function, a unique relation of $d\varepsilon_p/dt$ and $\sigma$ cannot be found since at least two plastic strains may exist in the pre- and post-peak regions corresponding to the total stress.

### 8.1.2.3 Damaging rate function

Figure 8.6a indicates the tendency of the fracturing rate on the ($\varepsilon_e$, $K$) space. Similar to the case of plasticity, the fracturing rate decreases according to the continuum damage evolution under sustained elastic strains. For a rational formulation, it is useful to clarify the physical image of continuum fracturing represented by $K$. Song introduced fictitious non-uniformity of parallel elasto-plastic components as a source of damage (Song *et al.* 1991) and explained the instantaneous evolution of fracturing. This concept is implemented for the potential term in Equation 8.4, but it does not cover the delayed fracturing term denoted by $dK/dt$.

Delayed fracturing is thought to be associated with micro-crack propagation. Newly created micro-cracks can develop just inside the remaining non-damaged volume that is denoted by $K$. Thus, even though the probability of delayed fracturing would be common among individual components, the averaged fracturing rate of the assembly of remaining components will be proportional to $K$. In fact, the fracturing rate will be nil when the

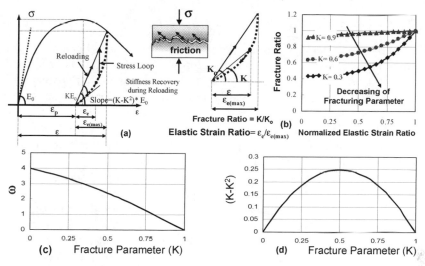

*Figure 8.6* Formulation scheme of internal loop: non-linear stress loop, (b) secant stiffness ratio, (c) non-linear order in terms of elastic strain, (d) stiffness ratio of internal stress loop ($\alpha$ in Equation 8.12).

fracture parameter converges to zero, as shown in Figure 8.6b. Then, the following formulae based upon this imaginary micro-fracture are introduced as

$$\frac{\partial K}{\partial t} = \left(\frac{\partial K}{\partial t}\right)_b \exp\left(\xi\left(\frac{K}{K - F_k} - 1\right)\right) \quad , \quad \xi = 45\left[\psi^{-0.5(1 - \exp(-5\varepsilon_e))}\right] \quad (8.10a)$$

$$\left(\frac{\partial K}{\partial t}\right)_b = \left(\frac{\partial K}{\partial t}\right)_n (K - F_k) \quad , \quad \left(\frac{\partial K}{\partial t}\right)_n = 0.015 \cdot \log(K - F_k) \quad (8.10b)$$

where $(dK/dt)_b$ represents the referential fracturing rate on the envelope ($F_k = 0$) on which instantaneous fracturing may occur with respect to the increment of elasticity.

In order to formulate $(dK/dt)_b$, the intrinsic fracturing rate to indicate the delayed evolution of micro-cracks per unit of active volume is given in the same manner as plasticity. As stated above, this rate is factored by the fracture parameter $(K - F_k)$ in order to consider the remaining volume where new micro-defects can develop. As the potential $F_k$ is a function of confinement, $(dK/dt)_b$ is also associated with the intensity of confinement, as discussed in Section 8.1.1 and shown in Figure 8.5d. The computed rates are shown with experiments in Figure 8.3a and Figure 8.3b.

### 8.1.2.4 Computational model for the internal loop

The simple combination of elasto-plastic and damage results in partial linearity in unloading–reloading paths without any hysteresis loop. Although this simplicity does not reflect the reality as shown in Figure 8.6, it is acceptable for structural analysis to verify some limit states of capacity. For seismic analysis, however, the partial linearity of the stress–strain relation leads to lower energy absorption under repeated cycles of loads. In the case of post-peak analyses, tangential unloading stiffness at high-stress states substantially affects the computed intensity of strain localization. Thus, this section focuses on the formulation of hysteretic non-linearity in unloading–reloading paths where instantaneous plasticity and damaging do not occur.

Figure 8.6 shows the transition of secant stiffness normalized by the fracture parameter for different damage levels and confinement. Generally, the unloading loop tends to deviate from the linear line specified by the fracture parameter according to damage evolution ($0.5<K<1.0$). This deviation is suppressed by the presence of lateral confinement. By referring to the computational model for cracked concrete, the authors incorporate the following fictitious loop stress into Equation 8.1 as

$$\sigma = E_o K \varepsilon_e + \sigma_{loop}, \quad \sigma_{loop} = E_{loop} \varepsilon_e \tag{8.11}$$

$$E_{loop} = -\alpha \cdot E_O \left( 1 - \left( \frac{\varepsilon_e}{\varepsilon_{e(max)}} \right)^{\omega} \right) \tag{8.12a}$$

$$\omega = 4 - 2.4K - 1.6K^2, \quad \alpha = (K - K^2) \tag{8.12b}$$

where the value of $\alpha$ is defined as the stiffness ratio of the internal loop at zero stress, as shown in Figure 8.6. The applicability of Equation 8.12 to

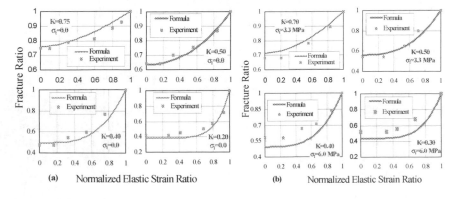

*Figure 8.7* Secant stiffness and elastic strain relation: (a) unconfined (b) confined concrete.

unloading paths can be seen in Figure 8.7. The reloading paths are assumed to come up linearly to the past maximum elastic strain denoted by $\varepsilon_{e(max)}$.

### 8.1.2.5 Extreme boundary of fracture

Concrete finally comes near to the assembly of aggregates which are mutually broken away and the cementing performance of the paste matrix is lost, due to full micro-crack propagation. But, this extreme state still retains residual stress with friction when lateral confinement is maintained, as shown in Figure 8.8. The residual capacity denoted by $\sigma_{lim}$ is thought to be equilibrated with the lateral confining pressure and frictional stress transfer along the shear band of aggregate assembly. As the fracture parameter corresponding to this extreme state of damage is thought to be lower bound of the confined concrete, we have

$$K_{lim} = \frac{\sigma_{lim}}{2E_o\varepsilon_{elim}} \tag{8.13a}$$

$$\sigma_{lim} = \sigma_l \frac{\sin(\theta) - \mu \cdot \cos(\theta)}{\cos(\theta) + \mu \cdot \sin(\theta)}, \quad \theta = \frac{\pi}{4}\left(2 - e^{(-\sigma_l/6.5)}\right), \quad \varepsilon_{elim} = 2.85.(\gamma^{-0.5}) \tag{8.13b}$$

where $\mu$ is the coefficient of friction for concrete and it equals 0.6, $\sigma_l$ is the lateral confining stress in MPa, $\theta$ is the directional angle of the shear fracturing band, obtained by equilibrium conditions in terms of confinement, and $\varepsilon_{elim}$ is defined as normalized elastic strain corresponding to the fracture limit. It is empirically obtained by Equation 8.13b.

It should be noted that this extreme boundary of the fracture parameter does not play any substantial role in the computed results of structural concrete but it does serve to head off any accidental numerical crush or divergence of the non-linear iterative computation.

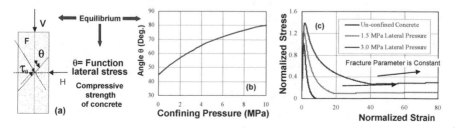

*Figure 8.8* Damage limit state of confined concrete.

### 8.1.3 Model verification

The differential form of the total stress–strain relation is derived by simultaneously solving Equations (1)–(13), and the total stress–strain can be integrated under any strain–stress history. Here, the experimental verification is performed under different histories for pre- and post-peak regions. It is clearly stated that the proposed modeling is applicable to concrete with a referential volume size of approximately 20–30 cm. This size is equivalent to that of standardized cylindrical specimens for strength, and, as reported by Lettsrisakulrat *et al.* (2000), the actual size of compression localization in RC structures is almost 20–30 cm. For consistency in experimental verification, a similar volume, on which the space-averaged strains are defined, is selected in this chapter. If the proposed modeling is to be used for finite elements whose size differs a good deal from the referential size of 20–30 cm, then the space-averaged constitutive model would need to be adjusted so as to give consistency of fracture energy and ultimate limit displacement in the shear band.

#### 8.1.3.1 Creep rupture

It is known that delayed creep failure may take place in compression when the applied sustained stress exceeds approximately 70% of the uniaxial compressive strength. Figure 8.9 shows the computed short-term creep under sustained stresses and experimental facts of unconfined concrete (specimen size: 10 cm diameter and 30 cm height), as reported by Rusch (1960). The creep rupture is computationally distinguished as a single point where the rates of total and elastic strains attain infinity but the plastic rate is negligible. This means that elasticity at this single point lies on the damaging envelope on the $(\varepsilon_e, K)$-plane mathematically indicated by $F_k(\varepsilon_e, K) = 0$, and instantaneous evolution of damage occurs at the same time. Here, the increase in intrinsic stress (represented by elastic strain) over the living elements is counterbalanced by the progressive fracturing of the continuum.

The ultimate strain and elapsed time until creep rupture are well simulated and the sensitivity of stress level is adequately predicted by the model. The computed elapsed time until the creep rupture is influenced mainly by the model of time-dependent damage, and the computed ultimate creep strain is largely affected by the formulated rate of plastic flow. Both models are expected to be validated.

#### 8.1.3.2 Monotonic stress–strain relation under constant strain rate

It is well known that the stress–strain relation is not unique but rather dependent on strain rate. In fact, the stress or strain rate is clearly specified in the standardized test for compressive strength of concrete. Figure 8.10 shows the computed stress–strain relation under different rates of strain and

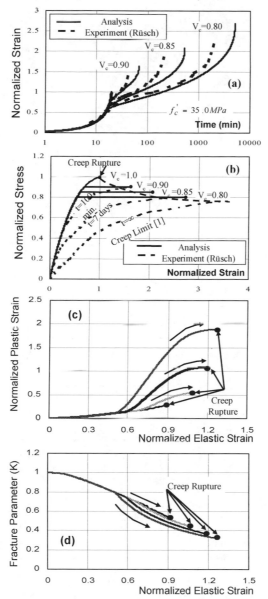

*Figure 8.9* Simulation of creep rupture: (a) creep strain in progress, (b) ultimate rupture strain and elapsed time, (c) elasto-plastic paths, (d) elasto-damaging paths.

comparison with experimental results as reported by Rusch (1960). Here, the strain is normalized by the peak strain corresponding to the specified compressive strength, as stated above. It is clearly found that the computed peak strength and its strain depend on loading rates. When the lower rate of strain

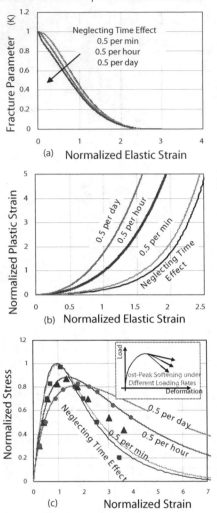

*Figure 8.10* Stress–strain relation under constant strain rates (a) elasto-plastic paths, (b) elasto-damaging paths, (c) normalized stress versus strain relation (dot points = experiments, lines = analytical model).

is assumed, the apparent strength decreases and the peak strain increases. This behavioral simulation quantitatively coincides with reality.

Figure 8.10 also indicates the internal variables of this simulation. The lower rate of loading causes large plasticity to a great extent and damaging occurs. Thus, the strength decay under lower rates of loading is mainly attributed to time-dependent damage, and the apparent reduced stiffness is brought about by progressive time-dependent plasticity. Reasonable agreement is seen in these loading paths. The tangential stiffness on the descending portion of stress–strain relations has much to do with computed results of compressive localization. The softened compressive stiffness can be recognized as time-dependent and is well simulated.

### 8.1.3.3 Generalized stress–strain paths

Figure 8.11 shows the short-term non-linear creep paths close to and beyond the uniaxial capacity of unconfined concrete. The greater progress of the total strain is seen especially in the post-peak zone but a comparatively smaller evolution of plasticity can be observed. The unloading stiffness varies over time and it drastically drops in the softening condition. It must be noted that time-dependent plasticity and fracturing continue to occur even though the unloading paths are enforced to concrete, and this leads to some non-linearity, shown in cyclic hysteresis under highly damaged conditions.

The relaxation paths in pre- and post-peak zones were also focused for validation of modeling, as shown in Figure 8.11. The computed and experimentally obtained elapsed times of stress relaxation are compared at each total strain level and good agreement is observed. The generalized loading paths of combined creep and relaxation are also checked in Figure 8.11. The simultaneous evolution of plasticity and damaging can be seen in the experimentally obtained stress–strain relation and the computational model fairly predicts this coupled behavior.

The experimental verification is performed for different levels of confining pressure. Figure 8.12 shows the influence of confining pressure on the overall stress–strain relation. The strength gain by confinement (3.3MPa) can be well simulated. It is rooted in suppressed evolution of damaging. Furthermore, the strain kinematics under higher stresses can be stabilized as shown in Figure 8.12a. Under the confined situation, stress can be well maintained for softened concrete even after the peak capacity, and the stabilized rate of creep deformation can be simulated as shown in Figure 8.12b. However, when the plastic and damage non-linearity is first experienced, singularity of creep rupture may occur in the post-peak softening in both analyses and experiments.

The non-linear creep deformation for more highly confined concrete under sustained stresses is shown in Figure 8.13. The non-linear creep strains were measured and simulated before and after peak capacity. Before the peak, the creep strain rate is greatly restricted by applying higher confinement when we compare these figures under the same stresses. But when the applied stress is set close to the elevated capacity by confinement, the creep rate is not small in practice. According to the confinement, unloading–reloading stiffness hardly decreases at all and the mechanical behavior looks like perfect elasto-plasticity. This means that damage evolution is effectively restrained by three-dimensional confinement. As a result, we have a low rate of time-dependent deformation even in post-peak conditions. The coupled elasto-plastic and fracturing model reasonably represents the behavioral simulation of materials with higher reliability.

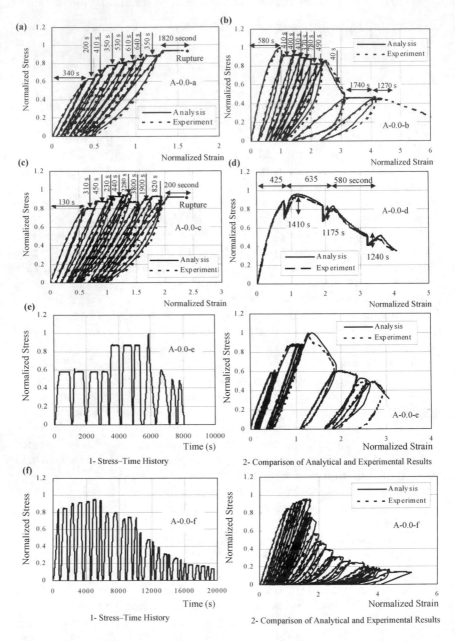

*Figure 8.11* Generic stress–strain paths and validation for unconfined concrete.

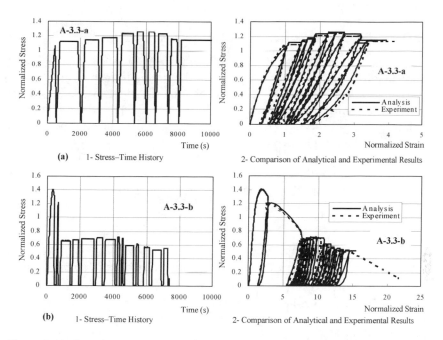

*Figure 8.12* Generic stress–strain paths and validation for normally confined concrete (3.3 MPa).

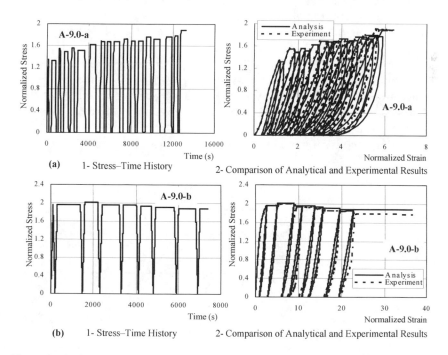

*Figure 8.13* Generic stress–strain paths and validation for largely confined concrete. (9.0 MPa).

## 8.2 Time-dependent softening of RC beams in flexure

In this section, the authors apply a generic time-dependent constitutive model formulated in Section 8.1 to collapse analysis, aiming to investigate the rate effects on post-peak structural mechanics and to examine the versatility of the meso-scale modeling for compression of structural concrete. Experiments on displacement-controlled RC beams are presented, and the direct time-integral analysis of collapsing structural concrete is discussed as a strategic way of analyzing the post-peak process under extreme loads. This is a preliminary study for completing the most macroscopic modeling of structural concrete linked with microscopic modeling (Chapter 9 and Chapter 10).

### 8.2.1 Experiment

The purpose of the experiment is to investigate the time dependency of post-peak softened RC members with compressive localization. To accomplish it, the authors used over-reinforced beams in flexure. The experiment was carried out under controlled loading with varied rates to confirm behavioral time dependency as well as the confinement effects on post-peak softening in specific terms. The concrete mixture, curing conditions and other details were reported by El-Kashif and Maekawa (2004b). The beam specimens and testing conditions are summarized in Table 8.1 and Figure 8.14. In order to avoid shear failure, the shear mode capacity was designed to be 1.6 times greater than the flexural mode capacity. A controlled displacement was induced for all specimens. The beams were supported by two sliding-rollers that permitted free rotation and translation in the longitudinal direction. A pin-connected spreader beam was used for distributing force on the two loading points, and the same force was applied under pre- and post-peak conditions. Two thick steel plates ($25 \times 100 \times 250$ mm) were attached to the surface of the concrete to distribute the bearing stress of rollers so as to avoid the splitting of tension cracking above the supports and loading points.

*Table 8.1* Beam specimen details

| Beam notation | Loading condition (target) | Dimension B-H-L (cm) | (Stirrups) L1 | | | (Stirrups) L2 | | |
|---|---|---|---|---|---|---|---|---|
| | | | D (mm) | S (cm) | L1 (cm) | D (mm) | S (cm) | L2 (cm) |
| Beam 1 | 1.00 mm/min. | 15-30-270 | 10 | 5 | 95 | - | - | 80.0 |
| Beam 2 | 0.15 mm/min. | 15-30-270 | 10 | 5 | 95 | - | - | 80.0 |
| Beam 3 | 1.00 mm/min. | 15-30-270 | 10 | 5 | 95 | 10 | 10 | 80.0 |
| Beam 4 | 0.15 mm/min. | 15-30-270 | 10 | 5 | 95 | 10 | 10 | 80.0 |

* Yield strength of longitudinal reinforcement (D22) and stirrups (D10): 625 MPa and 350 MPa, respectively.

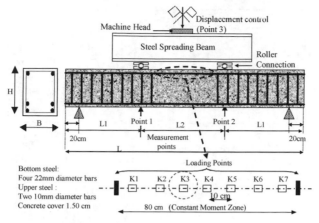

*Figure 8.14* RC beam specimens.

Strain transducers 30 mm long were embedded in concrete over the constant moment zone to capture the localized region of compressive deformation. Gauges were placed 3.5 cm from the top extreme fiber at intervals of 10 cm, as shown in Figure 8.14. As all the beams were expected to fail in flexural compression, a displacement control device featuring closed feedback was installed at Point 3 for stable equilibrium in the post-peak region, as shown in Figure 8.14.

The experimental results are presented in the form of load-displacement diagrams and the strains of concrete at the compressive fiber. The main points of discussion are the loading rates and degree of confinement supplied by web reinforcement (Table 8.1) in the constant moment zone (*L2* in Figure 8.14). If the loading rate were so high as to provoke stress waves, the experimental verification of the time-dependent plasticity and fracturing model would hardly be completed. Thus, experimental rates of loading were determined for verification so that time-dependent plasticity and fracturing would substantially develop in structural concrete during loading.

## 8.2.2 Time-dependent displacement

Figure 8.15 shows the average shear force–displacement relationships for Beams 1 and 2. The displacements shown here are the average readings taken from Points 1 and 2 (Figure 8.14). Flexural cracks gradually propagated first and the mode of failure was flexural compression without yield of reinforcement. Then, the compressive mechanics of concrete became substantial. The low rate loading of 0.15 mm/min. at Point 3 slightly decreased the peak capacity and stiffness under conditions of higher stress. The high rate loading of 1 mm/min. elevated the peak capacity a little, but a more rapid reduction in residual force was observed after the peak. Before and after the

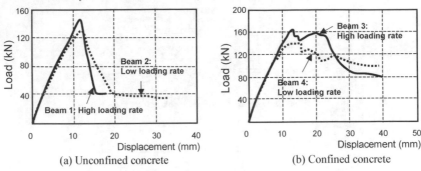

*Figure 8.15* Shear force–average displacement relations.

peak capacity, the same rates of stiffness and displacement were confirmed at individual loading points of Beam 1. In the case of Beam 2, the symmetry of deformation was slightly broken after the peak.

On the whole, the time-dependent mechanics of RC beams is similar to those of concrete material. A higher strain rate applied to the plain concrete mass enhanced its concrete strength, showing a steep descending branch of a softened stress–strain curve. If the mode of failure had been flexural tension associated with yielding of reinforcement, the time-dependent characteristics would have been less pronounced because compressive deformational characteristics have almost nothing to do with flexural capacity.

Beams 3 and 4 represent cases where the concrete was confined in the constant moment zone at the span center. The effects of loading rates on confined beams are shown in Figure 8.15. It is clear that confinement effectively improved ductility. Furthermore, there was greater similarity in time dependency in the post-peak responses. Almost the same rates of displacement developed at each loading point on the whole program of loading.

Lateral confinement by web reinforcement played an important role in improving the post-peak ductility in addition to loading rates, as outlined in Figure 8.15. The mechanics of confined concrete have widely been investigated for high loading rate cases. General experimental observation revealed that the load-carrying capacity was improved at both high and low loading rates. After the spalling of small concrete covers, the lateral ties continued to restrain the core concrete from further spalling. Cracks were found in the longitudinal direction along the compression steel, which was followed by large transverse displacement of reinforcing bars in the buckling mode in both Beams 1 and 2. Here, the buckling length was more or less half the span length. On the other hand, the buckling length became short in confined beams. The enhancement of ductility in confined beams is thought to be attributable to the ductile characteristics of confined concrete and post-buckling steel stress, which depends on the spacing of stirrups. However, it can be stated that the time dependency of the confined beams was a result of the characteristics of the concrete, since steel can be assumed to be time independent and the experimental buckling length was almost constant regardless of the loading rate.

### 8.2.3 Damage localization

Alca *et al.* (1997), Jansen *et al.* (1995), Jansen and Shah (1997) and Weiss *et al.* (2001) showed that the portions adjacent to localized damage regions change to unloaded states after the peak. A similar deformational field can be observed in over-reinforced beams under flexure. Figure 8.16 shows the profile of strains measured with embedded gauges (gauge length = 3 cm) and an overview of corresponding damage areas. Before reaching the peak capacity, the strain profile over the constant moment span was almost uniform.

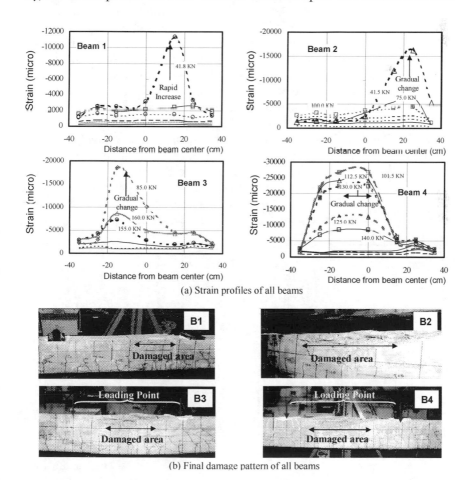

(a) Strain profiles of all beams

(b) Final damage pattern of all beams

*Figure 8.16* Strain distributions along constant moment zone at 3.5 cm from top fiber and overview of damaged area.

Just after peak capacity, predominant strain localization developed. For unconfined Beams 1 and 2, the size of the strain localization areas was approximately 20 cm (Lettsrisakulrat *et al.* 2000). The localized size of Beam 2 subjected to low loading rates appeared to be slightly larger.

With regard to the confined RC beams, a localized zone appeared in the concrete, too, but its size was definitely larger than in the unconfined beams, and when a low rate of loading was applied, weak localization was observed. It can be concluded that the size of compressive localization is not constant but varies in accordance with confinement as well as the rate of loading. The maximum size of coarse aggregates, which may also affect the compressive localization size, was kept unchanged during the experiment.

Figure 8.17 shows the strain history of each location. In Beam 1, a drastic increase in strain was observed at point *K6*, while the rest of the points showed decreases in strain. For Beam 2, two locations denoted as *K5* and *K6* exhibited an increase in deformation with no setback.

Localization under a low rate of loading tended to weaken. It is certain that the strain rate was not constant but suddenly increased immediately after the concrete softened. Although confinement may have weakened the localization, a rapid increase in strain rate was also observed for the confined beams, which is a trend similar to that observed in unconfined members.

Figure 8.18 shows the relationship between the applied load and the local strain. Compared with the unconfined beams, greater residual force was maintained even though larger strains developed at the span center, which was laterally confined by web reinforcement. Some gauges indicated progressive strain after the peak, even where no lateral confinement was provided (Beam 1). Nonetheless, when the rate of loading was high, only one gauge

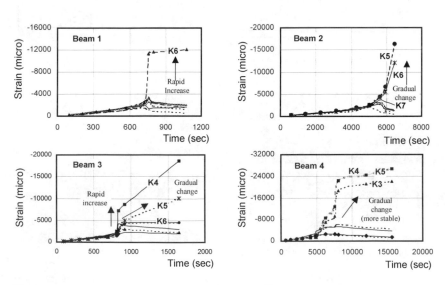

*Figure 8.17* Strain hysteresis at 3.5 cm from top fiber.

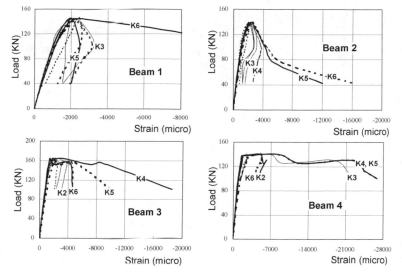

*Figure 8.18* Load–local strain relationship at 3.5 cm from top fiber.

remained to demonstrate the evolution of strain. If under a low loading rate, a large magnitude of stress were continuously applied, the two gauges would finally have shown localization even though the residual force decreased to the stability of softening.

## 8.3 Post-peak collapse analysis

### 8.3.1 Scheme of computation

To analyze the tested beams, the authors adopted the main frame of two-dimensional non-linear finite element analysis code *WCOMD* (Okamura and Maekawa 1991). The multi-directional fixed crack model (Maekawa *et al.* 2003a) for in-plane RC elements was the core technical aspect here. This constitutive modeling consisted of (1) concrete tension stiffening normal to cracks, (2) shear transfer based on a contact density model along cracks, and (3) concrete compression parallel to cracks. These three sub-models had been developed for the computation of ultimate limit states, and time dependency was not explicitly included in the formulation. Thus, the authors replaced the original elasto-plastic and fracturing concrete model with a newly proposed time-dependent constitutive model in Section 8.1.

Since the non-linear behavior of RC beams in flexure is hardly affected by crack shear transfer, time dependence can be ignored. Thus, the original contact density model (Okamura and Maekawa 1991) was used without modifying the time-dependency aspect. Post-cracking tension stiffening models are known to be time-dependent (Section 8.6) and the deflection of RC beams is chiefly affected by post-cracking tension stiffness modeling.

However, it is recognized that the time dependency of tension stiffness is a factor of long-term RC creep deformation but negligible in the short-term collapse analysis. Thus, the authors used the original tension stiffness modeling for collapse simulation of the beams tested. Code *WCOMD* operates a reinforcing bar model applicable to the pre- and post-buckling situations (Dhakal and Maekawa 2002; Maekawa *et al.* 2003a). This model assumes a buckling length that is associated with lateral confinement by steel, and computes the average post-buckling stress–strain relationship according to the size of finite elements with consistent energy (Maekawa and El-Kashif 2004). Time dependency was not taken into account for the reinforcing steel.

### 8.3.2 *Softened structural concrete model in compression*

Any constitutive model explains a space-averaged stress–strain relationship on the referential volume (REV) on which field averaging is conducted. The size of REV of the base model described in Section 8.1 is set at 20 cm (Lettsrisakulrat *et al.* 2000). If this model is applied to finite elements whose size coincides with that of REV, the material model and structural analysis represent the consistency of fracture energy in compression.

However, if the original constitutive model is applied to larger size finite elements, as shown in Figure 8.21 for example, physical inconsistency may be observed in the post-peak softening states. In reality, the uniformity assumption for the strain field is violated inside the finite element under a uniform stress field. The compressive localized volume develops within a larger finite element while the rest of the volume turns to the unloaded states. When the space-averaging procedure is applied for the size of a large finite element, the average stress–strain relationship on the finite element subjected to uniform stress becomes different from the specified-REV-based relationship. In general, the average stress–strain relationship defined on the large finite element shows a sharp decrease in residual stress, as shown in Figure 8.19. When using a smaller finite element, we have the opposite drift of the

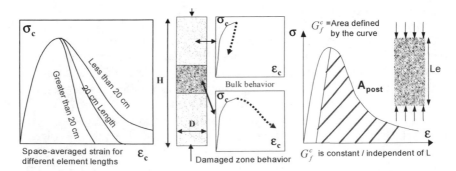

*Figure 8.19* Average stress–strain relationship associated with size of elements and compressive localization.

curve. In this case, some mechanics-based adaptation is required to achieve mathematical consistency.

Such adaptation has been conducted in post-cracking tension based on fracture mechanics. The authors followed logical procedures similar to those used for the post-peak analysis of tension and compression (An *et al.* 1997; Jansen *et al.* 1995; Jansen and Shah 1997; Nakamura and Higai 1999) and applied them to the fracture energy-based adaptation of the post-peak stress–strain relationship. Theoretically, this relationship can be computed in terms of the size of the finite element by solving the constitutive relationships of the localized (loading state) and non-localized (unloading state) volume with the assumption of stress uniformity. As implicit non-linear formulae are produced, the authors present an equivalent explicit-type computational formula in consideration of an easy and stable process of iterative computation. For the original model corresponding to REV = 20 cm, standard damage potential $F_k$ is proposed in Equation 8.6. Here, this fracturing-related function is generalized to arbitrary REV with regard to the control index denoted by $C$ as

$$\varepsilon_{ec} = \varepsilon_e^{\Omega} \quad if \; \varepsilon_e \geq \Re; \quad \varepsilon_{ec} = \varepsilon_e \quad if \; \varepsilon_e < \Re$$

$$\Re = 2.00 - 1.15 \gamma^{0.3}$$

$$\Omega = \left\{0.75 + 0.87 \exp(-\varepsilon_e / 0.8 \Re)\right\}^C, \gamma = \frac{\sigma_c}{\sigma_c + 4\sigma_l}$$

(8.14)

where $\sigma_c$ is the uniaxial compressive strength, and $\sigma_l$ is the lateral confining pressure. The stress–strain relationships given by Equation 8.6 and Equation 8.14 are shown in Figure 8.20 according to the control index. First, systematically arranged parametric studies were conducted with varied size of elements, loading rates and magnitude of confinement, and the resultant average stress–strain curves were obtained beforehand by using the original time-dependent constitutive equations. It was confirmed that the set of adapted curves shown in Figure 8.20 was almost identical to the exact curve, which had been derived from the implicit iterative computation, as stated before. This algebraic process of adaptation was conducted only to the part

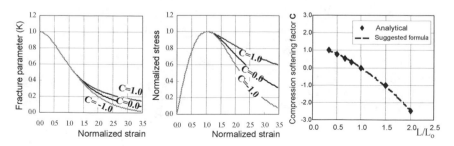

*Figure 8.20* Control parameter of compressive localization.

of fracture development, because the fracture parameter represented the internal continuum damage.

In order to specify parameter C, the authors assumed the constant specific fracture energy in compression (Lettsrisakulrat *et al.* 2000). To satisfy this, we have

$$L_O \cdot G_f^c = L \cdot \int \sigma d\varepsilon$$

$$L_O \int E_O K(C = 0).(\varepsilon - \varepsilon_p) d\varepsilon = L \int E_O K(C).(\varepsilon - \varepsilon_p) d\varepsilon \tag{8.15}$$

where $L_o$ is the size of referential volume (REV) as 20 cm, $L$ is the characteristic length of the finite element, and $G_f$ is the specific compressive fracture parameter. For the integral with respect to strain above, the released energy in the ascending portion of the stress–strain curve was extracted as a nonlocalized energy (see Figure 8.19). The numerical solution of Equation 8.15 with regard to $C$ is shown in Figure 8.20, and finally, the following equation was yielded:

$$C = 1.25 - 0.62 \left[ \frac{L}{L_O} \right]^2 - 0.63 \left[ \frac{L}{L_0} \right] \tag{8.16}$$

For example, Figure 8.21 shows the adapted average stress–strain curves for the different lengths of element. In the case where $L/L_o = 1$, the ascending portion coincided with the original modeling. In the descending branch, the strain rate when softened compressive stress became almost zero was about half that in the standardized case where the descending branch was simplified down to a linear line. Conceptually this is the same as the modeling by Cervenka, based on Van Mier's experimental facts (Van Mier 1986).

Confining the stress fields generally leads to stress deviation owing to the structural geometry and irregularity of the stress field, which may reduce the effectiveness of confinement. Pallewatta *et al.* (1995) proposed a confinement

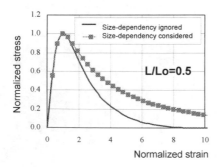
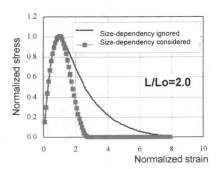

*Figure 8.21* Size-consistent stress–strain relationship.

effectiveness index, which can be computed through full three-dimensional non-linear analysis to evaluate the mean stress–strain relationship of the confined concrete core (Maekawa *et al.* 2003a). Based on these, the equivalent confining pressure applied to the constant moment zone was evaluated to be 2.2 MPa. The evaluated confining pressure was used for the internal parameter (γ of Equation 8.14) of the already verified time-dependent elasto-plastic and damaging model (El-Kashif and Maekawa 2004a).

### 8.3.3 Loading rate effects on softened structural responses

The computed and experimental results for unconfined members are compared, as shown in Figure 8.22. The simulation of time dependency successfully captures structural behaviors for both low and high loading rates, except for larger displacements of more than 20 mm. Here, compressive reinforcement was laterally deformed in the buckling mode and the residual force after the development of compressive localization relied on the modeling of reinforcement under extreme deformation. According to the computed trajectory of longitudinal compressive strain, slightly different strain distributions for different loading rates are seen. For a low loading rate, the localized strain area is slightly expanded, as shown in Figure 8.22, and the post-peak stress release becomes less pronounced than in the case of a high strain rate. These behavioral characteristics coincide with the experimental facts. A sensitivity analysis was performed by intentionally setting the time-dependent terms of plastic and damage evolutions to zero. For both rates of loading, time-independent analysis yields almost the same results as shown in Figure 8.22.

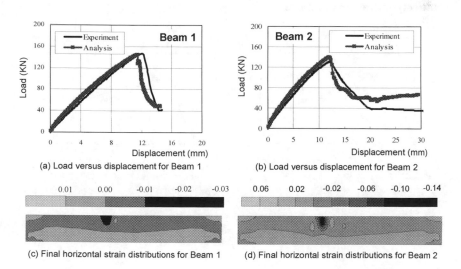

(a) Load versus displacement for Beam 1

(b) Load versus displacement for Beam 2

(c) Final horizontal strain distributions for Beam 1

(d) Final horizontal strain distributions for Beam 2

*Figure 8.22* Computed post-peak displacement and deformation of unconfined RC beams (time-dependency considered).

The simulation results for confined beams are compared with the experimental results as shown in Figure 8.23. In the analysis, the effects of confinement on concrete in compression are well simulated in terms of ductility and stress decay in the post-peak region. It must be noted that the computed size of the compressive localized zone was larger than that for unconfined members in both the high and low rates of loading. These results agree with observed facts. In the experiments and analysis alike, the high loading rate creates an apparent plastic plateau from 10 to 25 mm as well as a slightly higher flexural capacity followed by a sharp drop in residual force. For the case of low loading rate, on the contrary, no clear plastic plateau is observed and the weak softening of structural post-peak responses followed. Time-dependent analysis may well simulate these behavioral characteristics. In analytical terms, the low rate of loading brought about an expanded localized zone where damage continued to evolve, which agrees with reality.

To investigate the effect of time-dependent evolution, sensitivity analysis was conducted to cut off the term of time dependency, as shown in Figure 8.24. Compared with the low rate of loading, the time dependency of the post-peak response was clearly identified, although the pre-peak behavior was hardly affected. On the whole, the softened rate of residual force becomes larger when time dependency is ignored. In the case of a high loading rate, there are certain differences from the time-independent analysis. The release rate of residual force is almost identical but the apparent plastic plateau is different. The intensity of localization is rather small. It can be said that considering the time dependency of material non-linearity has a more significant impact on the confined concrete analysis. To further explore the effect of time on localization phenomena, local straining is discussed quantitatively in the next section.

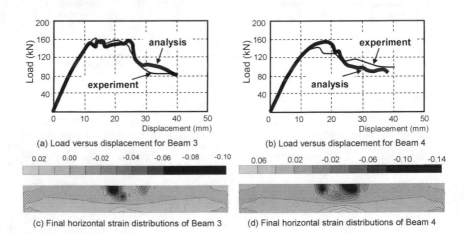

(a) Load versus displacement for Beam 3

(b) Load versus displacement for Beam 4

(c) Final horizontal strain distributions of Beam 3

(d) Final horizontal strain distributions of Beam 4

*Figure 8.23* Computed post-peak displacement and deformation of confined RC beams (time-dependency considered).

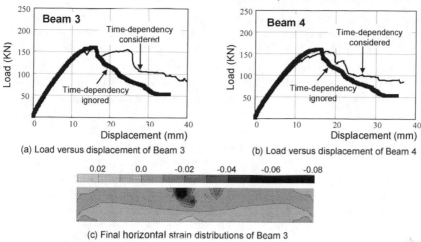

(a) Load versus displacement of Beam 3      (b) Load versus displacement of Beam 4

(c) Final horizontal strain distributions of Beam 3

*Figure 8.24* Computed post-peak displacement and deformation of confined RC beams (time-dependency ignored).

## 8.3.4 Loading rate effect on localized deformation

Figure 8.25 shows the relationship between the strains and elapsed time for each finite element. The strain values here are space-averaged over individual finite elements placed on a span of constant bending moment. Focus here is on the normal strain component along the member axis. In all cases, one

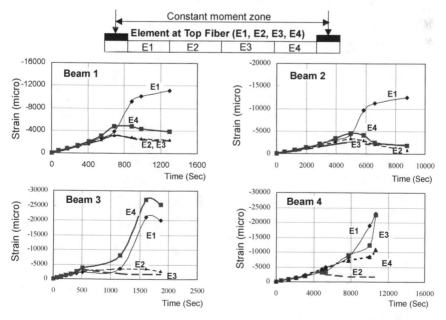

*Figure 8.25* Computed localized average strains in finite elements (20 cm).

element (20 cm) bears evidence of a high deformational rate and, broadly speaking, localized strain shows similar values to those in the experiments represented in Figure 8.17. In the case of unconfined beams, other elements changed to the unloading state where no instantaneous evolution of plasticity and damaging took place, regardless of the rate of loading. On the other hand, two or three elements (40–60 cm) display a rapid increase in strain. This means that the strain localization zones expand where lateral confinement is provided, which is consistent with the experimental observation. Figure 8.26 shows the relation of computed force and local strains in the finite elements placed over the span center of the beams, which can be compared with the experimental results shown in Figure 8.18. The relation of softening force and the corresponding local strain is similar to the experiment. The computed damage zones of confined beams are widely extended, unlike the those of unconfined beams. In the post-peak-capacity process, some elements are found to change to the unloaded condition after showing higher deformation than the peak strain borne by the compressive strength.

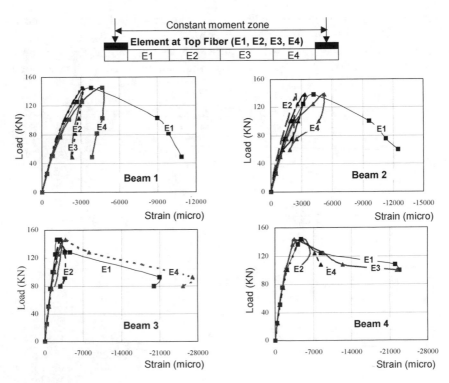

*Figure* 8.26 Relationships between computed force and local strain in each element (20 cm).

### 8.3.5 Averaged stress and strain relation of concrete in structures

Since time dependency was considered in the general constitutive law used in the analysis here, the resultant stress–strain relationship developed inside the structures was not unique but dependent on time and location. Given this, the authors extracted the actualized stress–strain paths of structural concrete at the span center along the member axis, as shown in Figure 8.27, which illustrates the case of unconfined beams. For higher loading rates, the stress–strain relationship developed at the strain-localized element is almost the same as that of the time-independent modeling. Other finite elements indicate unloading paths. For high loading rates of unconfined concrete, it is possible to apply a time-independent, unique stress–strain relationship in compression and perform reasonable post-peak analysis.

Under the low rate of loading, however, the resultant stress–strain relationship is not unique, as shown in Figure 8.27. This is caused by the different strain rates produced at each location, which accompanied strain localization. Here, careful attention should be given to the post-peak process. Although the external loading rates are very different, the slope of the descending branch of *E1* is almost the same as that of *E1* in Figure 8.27. The stiffness of the unloading condition is also similar to that of the high loading rate, regardless of the loading rate. It is the local strain rate that is at issue here. Immediately after the initiation of localization, the strain rate under the loading and unloading conditions suddenly increases despite the small rate of external displacement. The incremental stress–strain relationship becomes equivalent to that of the time-independent modeling. It can be concluded that without the time-dependent modeling, the post-peak structural behavior can consequently be simulated due to the fact that strain localization inevitably leads to high strain rates and removes the time-dependent plasticity and damage accumulated in the past mechanical history.

The resultant stress–strain relationships of confined RC beams are shown in Figure 8.28. It can be seen that the high loading rate hardly produced

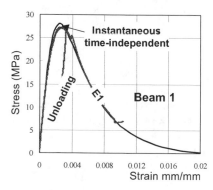
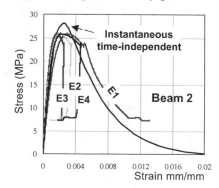

*Figure 8.27* Stress–strain curves for unconfined beams along constant moment span at the top fiber.

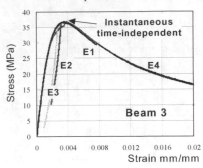
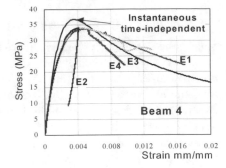

*Figure 8.28*  Stress-strain curves for confined beams along constant moment span at the top fiber.

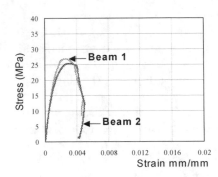
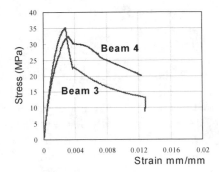

*Figure 8.29*  Extracted average stress–strain relationship over full center span of the beams.

time-dependent non-linearity at any location. In contrast, the stress–strain curves developed along the axis of the member varied in accordance with locations under the low loading rate. The softening region of the curves deviated from the instantaneous envelope, and the slope of descending parts was not parallel to the instantaneous slope, in contrast to the situation with the unconfined beams. The time-dependent effect in past hysteresis did not disappear but remained in the post-peak regions. Thus, the time-dependent constitutive model is required to enhance the accuracy of post-peak analysis.

As discussed before, the time dependency of material mechanics affects the local stress–strain relationship of structural concrete as well as the volume of strain-localized zone. These two factors have much to do with the structural post-peak behaviors. The average stress–strain relationships of concrete between the two loading supports, which are the result of two factors affected by the loading rate, are shown in Figure 8.29. The averaged stress–strain relationship at the top fiber was not pronouncedly time-dependent in the case of unconfined beams (1, 2), while it was significantly time-dependent in the case of confined beams (3, 4). In view of RC structural mechanics, the time dependency in the post-peak region carries greater weight in the practice of collapse analysis.

## 8.4 Cyclic cumulative damaging and modeling in compression

As far as material properties are concerned, a cyclic deterioration of concrete is clearly evident in compression, and low-cycle (N<100) fatigue modeling has been empirically presented (Gao and Hsu 1998; Hsu 1981; Hwan 1991), mainly for seismic analyses. It is also generally known that RC members exhibit stable restoring force characteristics in flexure before their capacity (Mutsuyoshi and Machida 1985) but that the structural deterioration caused by repeated loads becomes predominant after their peak capacities. Here, it must be noted again that the post-peak stress–strain relation is to a great extent time-dependent. The post-peak cyclic responses and stress decay are also dependent on the strain rate as well as the number of repetitions in nature. When the strain rate or frequency of repeated stress is changed, the apparent cyclic responses of material vary accordingly.

For extracting the authentic damage progress solely caused by repeated actions, systematically arranged experiments under different frequency are required, or a reasonably accurate time-dependent constitutive model is indispensable. In fact, time-dependent non-linearity also grows during cyclic loading under higher strains. This section's aim is to subtract the time-dependent evolution of non-linearity from the overall cyclic stress–strain relations by using a time-path dependent elasto-plastic and damaging model (Section 8.1), and the main focus is on post-peak behaviors.

Repeated stress creates additional plastic strain and a gradual decay of unloading stiffness in the pre-peak conditions, as shown in Figure 8.30. In this experiment, it is not necessary to keep the stress or strain rates constant for extracting the evolution rates of non-linearity but the precise measurement of induced stress and strain in the time domain are quite critical for model formation and verification before and after peak capacity. It should be kept in mind that both plasticity and unloading and reloading stiffness degradation (defined as "damaging") go forward even under constantly sustained stress states. Thus, the apparent non-linearity results from both time-dependent evolution and iteration of straining.

The computed strain based on the time-dependent constitutive model in Section 8.1, which does not include the effect of stress repetition, is shown in Figure 8.30. In this case, about 75% of the strain progress is attributed to time-dependent non-linearity and the rest, or 25%, is rooted in repetition of stresses. When the maximum total strain is kept unchanged under cyclic loads in the post-peak region, stress decay is experienced with advanced plasticity (Figure 8.30). A similar tendency is seen in the relaxation path where the total strain is kept constant. Thus, subtraction of time dependency is necessary to verify the effect of stress or strain repetition. Figure 8.30 shows the

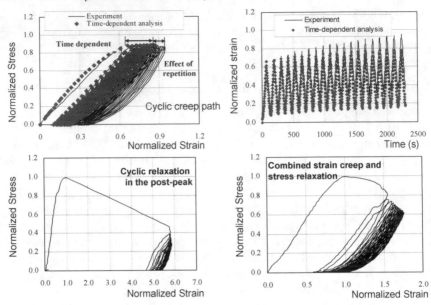

*Figure 8.30* Time-dependent cyclic stress–strain relation of concrete in compression (normalized by uniaxial compressive strength and its strain).

combined path of cyclic creep and relaxation. The evolution of plasticity and damaging can be seen in this path, too.

In this section, a time-dependent elasto-plastic and damaging model (Section 8.1) is implemented to assess pure time-dependent deformation so that the unmixed effect of stress or strain iteration can be identified. As it is crucial to verify the knowledge from structural post-peak responses, post-peak behaviors of in-plane shear walls subjected to different loading rates are used for multiple checking of the modeling. And it is aimed to import the progressive damaging caused purely by repetition of strain paths. The constitutive model was experimentally verified by using plain concrete specimens and over-reinforced concrete beams in flexure in Section 8.3, but the main point was directed to short-term (1–1,000 s) behaviors. Longer-term (about a year) non-linearity under higher stresses is also verified in this section.

Figure 8.31 shows the long RC column subjected to eccentric compression (Claeson and Gylltoft 2000). Reinforcement was made up of four deformed 16-mm bars and 8-mm stirrups at a distance of 13 cm with 656.0 and 466.0 MPa, respectively. The clear concrete cover was 15 mm and additional reinforcement was provided to prevent failure in the anchorage zones. Because of eccentricity of loading, failure is accelerated due to additional moment caused by the additional displacement in the middle span of the column. For transient non-linear analysis, the time-dependent constitutive law in Equations 8.1–8.4 and Equation 8.12 was imported in the fiber finite element (Maekawa *et al.* 2003a). Figure 8.31 shows the comparison of the

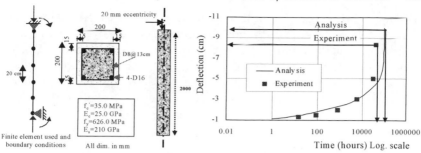

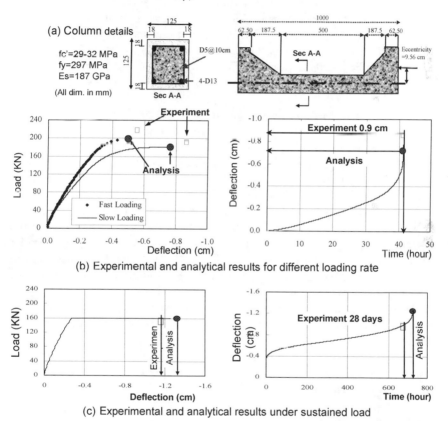

*Figure 8.31* Delayed creep failure of RC column in eccentric compression (large geo-metrical non-linearity).

*Figure 8.32* Delayed creep failure of RC column in eccentric compression (no geo-metrical non-linearity).

experimental results and analytical ones. A fair agreement is seen for the short column represented by the monotonic loading.

Figure 8.32 gives the short column subjected to eccentric compression by Viest *et al.* (1956) and the computed transient process of column deflection and applied loads is presented. The experimental data are reported only at the states of collapse. Geometrical non-linearity is comparatively small enough to directly check the validity of the time-dependent material model. The constant rates of load exhibit different load-displacement curves, which are well simulated. Under the sustained load, delayed creep failure was recognized and its elapsed time to failure is reasonably predicted.

### 8.4.1 Cumulative damaging by cyclic strain paths

Repeated mechanical action is known to introduce the extension of crack ligament into solid continua. The Paris law for fatigue life prediction is formulated based upon the fact that the incremental crack propagation per unit cycle of action is proportional to the amplitude of the stress intensity factor which represents the local elastic stress state close to the tip of a single crack. For the concrete continuum, the assembly of distributed microcracks has to be treated before peak capacity, and the localized macro-scale crack is the target in the post-peak region within the referential control volume.

Formulation of the cumulative damage by cyclic mechanical action is discussed here in terms of fracture parameter ($K$ in Equation 8.3 and Equation 8.4). In fact, as shown in Figure 8.30, degradation of unloading stiffness is seen during the repeated stress paths. It means that the fracture parameter denoted by $K$ in Equation 8.3 and Equation 8.4 changes by stress repetition as well as sustained load in continuum. Here, it is necessary to formulate the cyclic effect independently from the time-dependent evolution.

We also pay attention to the effect of stress or strain repetition on the development of plasticity. Apparently, progressive plastic strain is seen under cyclic loading, but it should be noted that some of the evident plastic evolution under cyclic loads can be caused by the cumulative damage resulting from stress repetition. Thus, it is not easy to extract only the effect of cyclic mechanical action on plasticity. The authors therefore first formulate cumulative damaging and afterwards verify the necessity of considering cyclic plasticity.

In order to include the effect of cyclic load into the main frame of the elasto-plastic and damaging model, the derivative of the fracture parameter in Equation 8.4 can be generalized as

$$\left(\frac{\partial K}{\partial \varepsilon_e}\right) = \lambda \quad \text{when } F_k < 0,$$

$$\left(\frac{\partial K}{\partial \varepsilon_e}\right) = -(\partial F_k / \partial \varepsilon_e)/(\partial F_k / \partial K) + \lambda \quad \text{when } F_k = 0$$

(8.17)

where $\lambda$ is defined as fracture degradation rate to consider the effect of repetition of elastic strain. Since the fracture degradation rate is specified by the elastic strain, cumulative plasticity which is unavoidable in the experiments can be unconnected for the formulation of fatigue fracturing.

The fracture degradation rate is thought to be small when internal damage of the concrete continuum is low. In other words, the fracture parameter may be associated with the entire extension of micro-cracks when the parameter is between unity and 0.5, a range where micro-defects can be assumed to be uniformly dispersed (Maekawa *et al.* 2003a). Thus, the fracture degradation rate will be larger according to the progress of damaging (1>K>0.5). The compressive damage localization occurs when the fracture parameter goes below 0.5 in the post-peak state. In this state, the effective volume where consequent damage can develop diminishes and the macroscopic fracture band is being built. Thus, the rate of cumulative damaging in the post-peak region is thought to decelerate as illustrated in Figure 8.33. As an extreme case, the progressive rate of damaging will be null when the remaining effective volume (indicated by $K$) which can bear stresses is completely lost ($K$ = 0). For considering these mechanics, the following formula is first proposed for full amplitude (schematically expressed in Figure 8.33) as

$$\lambda = K^3 \cdot (1 - K)^4 \cdot g,$$
$$g = 0.6 \text{ when } d\varepsilon_e < 0, \quad \text{otherwise}, g = 0 \tag{8.18}$$

where cumulative damaging is assumed to occur when relaxation of the internal stress intensity is performed. The full amplitude of repetition is represented by the elastic strain increment. The rate of cumulative damaging per unit of active volume that can substantially bear stress is also shown in Figure 8.33. The fracture degradation rate reaches its highest value around K = 0.4–0.5 where compression softening may start. This corresponds to the fact that the rate of total strain accelerates greatly just before fatigue failure, even under constant stress amplitude. This is similar to concrete creep failure, which can also be simulated by the elasto-plastic and damaging scheme in Section 8.1.

This assumption yields the elasto-plastic and damaging paths as shown in Figure 8.34 and Figure 8.35. Under the constant amplitude of stress, the total and plastic strains increase and the stiffness in each cycle gradually decreases.

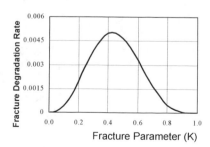 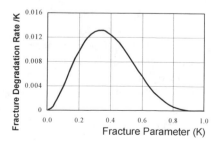

*Figure 8.33* Modeling of fracture degradation rate.

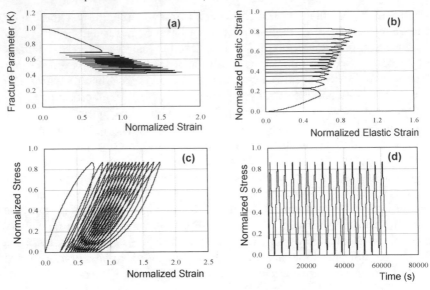

*Figure 8.34* Transient elasto-plastic and damaging path under cyclic constant amplitude of stress.

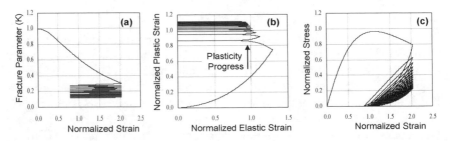

*Figure 8.35* Transient elasto-plastic and damaging path under cyclic constant amplitude of strain.

As the fracture parameter is steadily reduced as shown in Figure 8.34a, the elastic strain correspondingly increases even though the maximum stress remains unchanged (Figure 8.34b). Here, the plasticity is accelerated in terms of time and instantaneous elastic strain path as formulated in Equation 8.3 and Equation 8.4. Thus, the accelerated plastic evolution by cyclic paths can be computed just by considering the cumulative damaging of stress repetition with regard to $K$.

Figure 8.35c shows the transient course of internal non-linearity under a cyclic relaxation path on which the maximum total strain is kept constant. The damage progress leads to decay of both total stress and amplitude of elastic strain, as shown in Figure 8.35a. Even though the maximum elastic strain decreases, plastic evolution is computed to occur owing to the time-dependent term of Equation 8.3 and Equation 8.4 and as shown in Figure 8.35b. It is computationally confirmed that the cyclic plastic evolution can logically arise from the modeling of cyclic damaging. However, based on only the limited information discussed here, it is still not clear whether cyclic plasticity needs to be further implemented as a mechanics of concrete. This issue will be discussed through experimental verification in later sections.

### 8.4.2 Verification

In order to verify the proposed equation, experiments were selected to cover typical cases of structural analysis. As the post-peak behavior under cyclic loads represents the case of damaged structures close to collapse, it is essential to cover this area in practice. Figure 8.36a shows the analytical and experimental results of the specimen loaded at the age of 47 days with concrete compressive strength of 32.0 MPa. The comparison of the experimental and analytical results is also shown. A fair agreement between both results is noticeable. The computed result of the time-dependent non-linearity alone is also shown in Figure 8.36b for comparison. In fact, the derivatives of plasticity and damaging with respect to time were set equal to null in Equation 8.2. It can thus be said that cyclic repetition should be mechanically considered as an independent solid characteristic of concrete composite.

Figure 8.37a shows the analytical and experimental results of the concrete specimen loaded at the age of 54 days with uniaxial compressive strength of

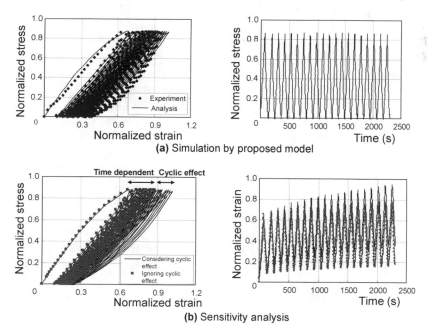

*Figure 8.36* Verification of modeling under constant amplitude of stress.

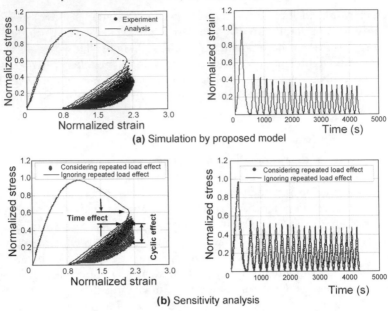

(a) Simulation by proposed model

(b) Sensitivity analysis

*Figure 8.37* Verification of modeling under constant amplitude of strain.

32.0 MPa. The post-peak cyclic relaxation path of stress decay is induced to a concrete cylinder with a height of 20 cm. The stiffness declines to the extent that a large number of cycles are applied. When this specimen reaches the final cycle, the increment of stiffness degradation seems to be very small compared to the beginning of cycle loading. The computational model can predict the residual strain as well as the stiffness degradation. Consequently, the total strain progress can be observed as shown in Figure 8.37a. The simulation of neglecting the cyclic effect but considering the time-dependent plastic and fracturing is shown in Figure 8.37b. The way of conducting sensitivity analysis is the same as the case of Figure 8.36. Broadly speaking, 30% of the total stress decay is attributed to the time-dependent evolution of non-linearity and the remaining 70% is associated with the effect of strain repetition.

Figure 8.38 shows the combined path of loading. When monotonic reloading is applied after large cyclic numbers, the stress–strain path is seen to come up to the envelope made by the pure monotonic loading from the beginning. This means that the cyclic cumulative evolution of fracture and plasticity may be renewed when the mechanical state exceeds the past maximum deformation and stresses to a large extent. The reasonable consistency of analytical and experimental results is recognizable on this sort of path-dependency. Thus, the authors deduced that the cumulative damaging of concrete under repeated action simulates the overall non-linearity that appears at the material level. However, the evolution function of compressive damaging can

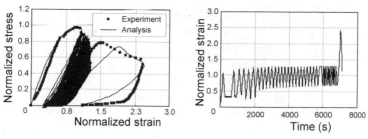

*Figure 8.38* Verification of modeling under varying amplitude of stress and strain.

be quantitatively improved in future, especially in the post-peak conditions where the rate of non-linear evolution is comparatively high.

Overall behavior of reinforced concrete members has been investigated by introducing the stiffness degradations factor (Matsushima 1969; Takeda *et al.* 1970) so as to consider the effect of cyclic loads on shear failure. In addition, a quantitative assessment of damage has been presented through the use of damage indicators (Izuno *et al.* 1993). These indictors are usually defined in terms of ductility or energy-related factors. Recent experimental studies (El-Bahy *et al.* 1999; Kono *et al.* 2002; Park and Ang 1985) on RC members have led to the empirical formulation that allows for concrete flexural resistance. However, the versatility of this sort of member-based approach is limited. In this chapter, pre- and post-peak computed responses of reinforced concrete under high cycles of external loads are investigated and experimental verification of the cyclic modeling is performed.

### 8.4.3 Application to high-cycle fatigue

As the proposed model is aimed mainly at simulating the post-peak failing process of structures, high-cycle fatigue loads under serviceable stress levels are not apparently considered. Under middle or lower stress levels, the effect of time dependency of deformational non-linearity on the computed fatigue life is comparatively small and the computed cycle numbers are mostly dependent on the model of fracturing degradation factor, as in in Equation 8.18. A parametric study was conducted in terms of fracturing the degradation factor and the parameter g (= 0.6) in Equation 8.18 was empirically found to be convertible to the high-fatigue S-N curve of concrete as

$$ g = 0.6 \left( \frac{9\gamma^8}{1 + 10^{(30K-22)}} \right), \quad \gamma \equiv -\frac{\varepsilon - \varepsilon_{e,tp}}{\varepsilon_{e(max)}} \tag{8.19} $$

where $\gamma$ indicates the normalized amplitude corresponding to updated stress variation, and $\varepsilon_{e,tp}$ denotes the turning point of compressive elastic strain.

The modified function by Equation 8.19 for smaller stress states is shown in Figure 8.39a. The full cyclic stress path was outlined by large numbers of discrete time steps. Failure was identified by the drastic increase of strain rate and following the collapse of static equilibrium. The minimum stress is programmed as zero (single-side amplitude) and the computational rate of fatigue stress was 1.0 Hz. The constant amplitude was applied to the concrete finite element until the failure. The design values of the S-N diagram are also shown together. The computed S-N diagram is close to the design specification model derived from many fatigue tests. For a sensitivity check, the slow rate of loading of 0.01 Hz is conducted as well. The fatigue strength is generally reduced by 20–30% of the slow loading when more time-dependent fracturing and plasticity may evolve (Award and Hilsdorf 1974; Raithby and Galloway 1974). Under small stresses, the loading rate effect becomes negligible, because the time dependency is of a much lesser magnitude.

For verification of the fatigue life with loading rates, the experiment by Award and Hilsdorf (1974) was used. The maximum stress is 90% of the uniaxial compressive strength and three levels of amplitude are applied under different stress rates. The degraded fatigue cycles can be seen in Figure 8.39b under the slower rate of stress and the proposed constitutive model basically matches the facts. In the case of small amplitude (= $0.1fc'$), it should be noted that time-dependent plasticity and damaging grow to be large and a small stress deviation may result in a large difference of apparent life because the time-averaged stress is high. In the case of middle amplitude (= $0.5fc'$), the non-linear creep generally reduces during the stress repetition. With consideration of these behavioral characters and the reproducibility of experiments, this non-linearity is thought to be fairly simulated.

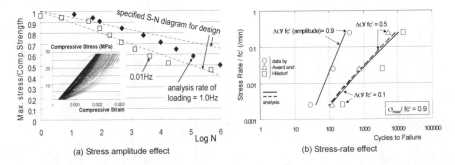

(a) Stress amplitude effect          (b) Stress-rate effect

*Figure 8.39* Computed fatigue life in terms of stress amplitude and rate of stresses.

## 8.5 Short-term cyclic damaging of RC structures

### 8.5.1 RC columns under cyclic loading for pre- and post-peak regions

Experimental results (Kawashima 1992) of laterally confined RC columns are adopted for verification. Geometrical and mechanical properties of specimens A and B are shown in Figure 8.40a and summarized in Table 8.2. Two columns with different amounts of reinforcement have a hollow circular cross-section with an outer diameter of 80 cm and an inner diameter of 51.8 cm. The columns were subjected to cyclic lateral displacement of a constant rate of 0.2 cm/min at a height of 247.5 cm. It was predicted that under large deformation and drift angle, spalling of the cover concrete might occur under cyclic deterioration. Two columns with different steel arrangements were cast continuously with rigid footings. The experimental results are shown in Figure 8.40b.

The proposed time-dependent cyclic constitutive model of concrete in compression was imported into three-dimensional fiber analysis (Maekawa *et al.* 2003a). The pre- and post-buckling model of reinforcing bars developed by Dhakal and Maekawa (2002) is applied in the program to simulate the large deformation that occurs beyond peak capacity. The size of the finite element was selected to be the same as the size of the referential volume of

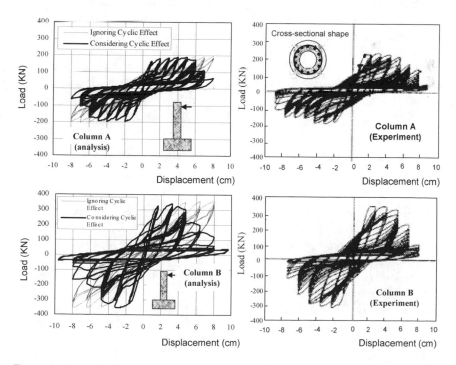

*Figure 8.40* Cyclic load effects on RC columns in flexure (a, b: upper, c, d: lower).

*Table 8.2* Dimensioning and detailing

|  | *Column A* | *Column B* |
|---|---|---|
| Diameter (cm) | 80 (outer) | 80 (outer) |
| Main reinforcement | 30-D10 | 30-D13 |
| Cover thickness | 5 cm | 5 cm |
| Comp. strength | 35.4 MPa | 34.9 MPa |
| Tensile strength | 1.50 MPa | 1.50 MPa |
| Young's modulus | 206 GPa | 206 GPa |
| Yield strength | 635.0 MPa | 635.0 MPa |

20cm, on which the averaged stress–strain relation of concrete is formulated. Thus, adaptation of the constitutive model in terms of compressive fracture energy is not necessary (Lettsrisakulrat *et al.* 2000). Figure 8.40c shows the computed results.

As the buckling could not be visually observed in reality, the strength deterioration can be attributed mainly to the time-dependent cyclic damaging and spalling of cover concrete. As can be seen in Figure 8.40, the analysis could successfully capture the strength degradation, load and deformation at the peak and post-peak responses, respectively. A softening branch followed by cyclic degradation can be seen. The stress decay under the constant amplitude of enforced displacement is significant in the post-peak state and well simulated by the cyclic modeling. However, if the cumulative damage by cyclic action is ignored in the analysis ($\lambda = 0$ in Equation 8.17) as shown in Figure 8.40, the stress degradation by stress iteration becomes much milder and greatly deviates from reality. Thus post-peak degradation cannot be explained only by the time-dependent progress of non-linearity in this case.

### 8.5.2 RC shear wall under cyclic loads with different rates of loading

Experimental verification of time-dependent cyclic modeling is directed to in-plane problems. Esaki and Ono (2001) investigated the rate effect on the restoring force characteristics of RC shear walls subjected to highly reversed cyclic actions and concluded that time dependency is hardly identical in comparisons with the reproducibility of experiments. This experimental fact may be satisfactorily close to the capacity of shear wall structures and the same conclusion is applicable to other RC structures in view of a seismic response (Mutsuyoshi and Machida 1985).

However, this issue is questionable in the transient states of collapse after the peak. Matsuoka *et al.* (2001) intensively conducted cyclic loading experiments of shear walls under different rates of displacement covering the post-peak regions, and they reported that the restoring force characteristics are influenced practically in accordance with the rate of loading. As the softening of RC in-plane panels is directly associated with compression softening of concrete along cracking, the post-peak cyclic behaviors of time dependency are appropriate for verification. Thus the authors imported the proposed time-dependent cyclic constitutive model of concrete in compression into the non-linear finite element code (Maekawa *et al.* 2003a, Okamura and Maekawa 1991).

The in-plane RC constitutive law consists of sub-models, as shown in Figure 8.41. The uniaxial stress–strain relation of concrete is applied to the stress-carrying mechanism along cracks. The original multi-directional fixed-crack modeling (Maekawa *et al.* 2003a) employs the elasto-plastic and fracturing laws of concrete in compression. Since the newly enhanced computational model concerning time dependency and cumulative damage of repeated action is built up on the same scheme of non-linear mechanics as the original, simple replacement of the sub-routine was required and sufficient. In this formulation, the tension stiffness and shear transfer of cracked concrete are unchanged and remain time-independent. In fact, the time dependency of tension stiffness was reported to be negligible in short-term loadings for ultimate limit state computation. As isotropically arranged RC panels subjected to simple in-plane shear are not affected by the shear transfer along cracks, shear transfer can be left out of discussion for popular RC shear wall

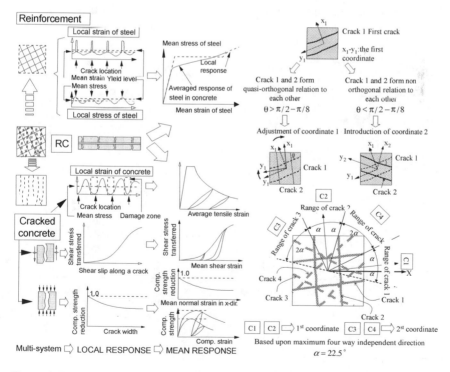

*Figure 8.41* RC in-plane constitutive modeling and multi-directional cracking.

designs under low-cycle fatigue. The post-buckling model of reinforcement is not necessary for this analysis because the absolute deformation of in-plane walls is limited. For reversed cyclic loads, multi-directional cracking is created within the individual finite element, as shown in Figure 8.41. In this case, crack-to-crack interaction with respect to shear slip along each crack plane is strictly taken into account within the scheme of the active crack method proposed by Okamura and Maekawa (1991).

Figure 8.42 shows the shear walls experimented on by Matsuoka *et al.* (2001) and used for verification of the modeling. The details of shear walls are summarized in Table 8.3. Finite element discretization and the loading point are indicated together. Computed and experimental shear forces versus drift angle relations are shown in Figure 8.43 under monotonic loading. In general, the computed results match the experiments in both pre- and post-peak states under three levels of deformational rates (0.01, 0.10 and 1.00 cm/sec). The capacity of the shear walls is slightly influenced by the loading rate. The high-speed loading gives rise to a greater capacity in reality. However, the analysis shows almost the same capacity and similar restoring force characteristics. As the yield of reinforcement is one of the limit states of this shear wall and the compressive strength of concrete plays a minor role in the capacity, the experimental capacity gain is thought to be caused by the strain rate effect on yield strength of reinforcement (Cowell 1965; Malvar 1998). In Figure 8.43d, the simulation results are shown with 10% increased yield

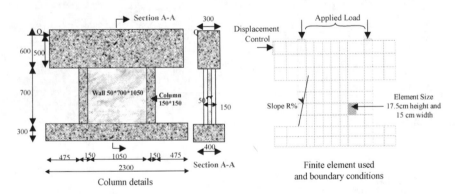

*Figure 8.42* Detailing of shear wall and finite element discretization.

*Table 8.3* Experiment details for shear walls

| Wall | Load pattern | Reinforcement ratio: side column | R/F: wall | Compressive strength |
|------|-------------|----------------------------------|-----------|---------------------|
| 1 | Monotonic | 3.40% in longitudinal dir. $fy$ = 334 MPa | 0.50% in both dir. $fy$ = 178 MPa | 26.5 MPa |
| 2 | Cycle 1 | 0.33% in horizontal dir. $fy$ = 178 MPa | Diameter 4 mm | 27.7 MPa |

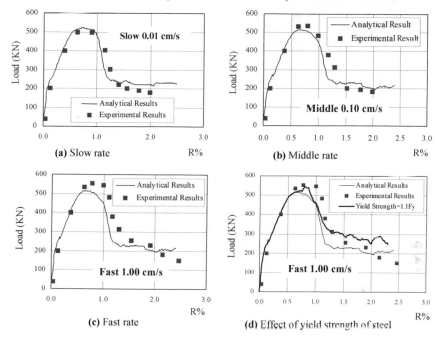

*Figure 8.43* Computational and experimental relations of monotonic shear force versus deformation.

strength of reinforcement for sensitivity analysis. The strength gain under the fast-loading rate can be explained by the increase in yield strength of reinforcement. However, the compression modeling of concrete is important in the post-peak regions. Thus, in both experiment and analysis, the rate effect on the post-peak ductility is not clearly identical under monotonic loads, as discussed in the past.

For shear walls of the same dimension and detailing, the reversed cyclic shear was applied under different rates of loading (slow = 0.01, middle = 0.10, fast = 1.00 cm/sec), as shown in Figure 8.44. Type B presents the path of large numbers of repetition. Consequent pre- and post-peak responses are

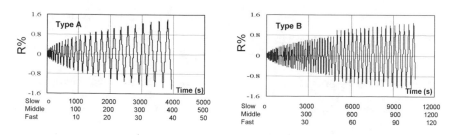

*Figure 8.44* Applied shear deformation under different rates of loading.

shown in Figure 8.45. The single-sided results are exhibited as the symmetry of load-displacement curves is shaped. For comparison, the simulation results of ignoring the cumulative damage by cyclic paths (only time-dependent deformation is considered) are presented together. The coupling of the effect of strain cycles with time dependency improves the accuracy of the analytical post-peak responses. For simulating the response until capacity is reached, it seems that cyclic effect is negligible.

To check the cyclic effect in detail, high-cycle loading was investigated (Figure 8.46). The total number of repetitions is larger than in the cases of Figure 8.45 and the decay of restoring forces after peak capacity is greater also. The larger deviation of residual shear force from the monotonic loading cases can be seen in both experiment and analysis. It can be concluded that the cyclic damaging model of concrete is significant, especially in the post-peak states accompanying the localization of compressive deformation rather than as the effect of loading rate. Thus, the coupling of the time effect with cyclic cumulative damage gives substantial insight into the effect of repetition, which is of greater practical importance for actual seismic analyses of wall-type structures than is the time effect.

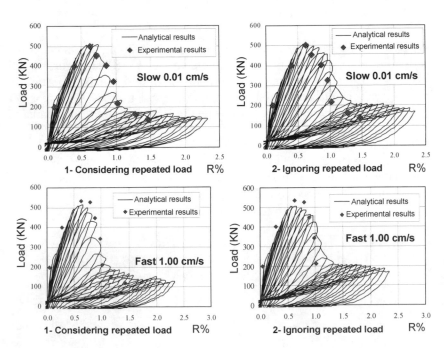

*Figure 8.45* Computational and experimental relations of cyclic shear force versus deformation (Type A loading).

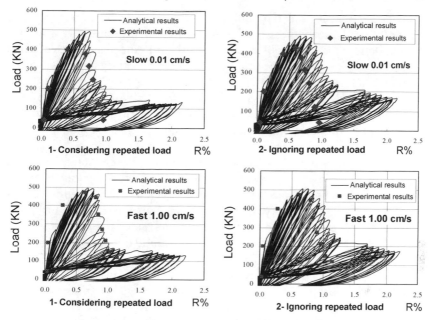

*Figure 8.16* Computational and experimental relations of cyclic shear force versus deformation (Type B loading of large numbers of cycles).

## 8.6 Long-term constitutive model of cracked concrete

Among the many industrial materials in common use, concrete is known as one that exhibits particularly large creep deformation, even under daily working stresses. As a consequence, great efforts have been expended on elucidating the time-dependent characteristics of cementitious composites (e.g., American Concrete Institute Committee 209, 2005). The creep behavior of structural concrete has also been investigated, mainly in relation to the design and management of pre-stressed concrete (PC). The accumulated knowledge addresses transient cable tension and its practical control particularly. In the case of slender reinforced concrete (RC) columns and thin space-shells, creep deflection results in an additional bending moment that may cause delayed buckling under axial compression. These problems generally relate to the pre-crack states, and crack-induced non-linearities are not associated with structural creep problems.

In contrast, some need has been recognized for the understanding of post-cracking creep in structural concrete in connection with both the enhanced serviceability limit state of new structures and the maintenance of existing structures with inherent defects. Current design codes generally specify a conservative value of concrete stress limit under the service load so as to avoid large inelastic creep. This kind of provision can occasionally lead to excessive thickness in the design of underground RC structures. More rational designs will depend on our ability to specify methods for long-term behavioral prediction. Excessive time-dependent deflections have also been noted in slender members when concrete with large shrinkage is used. These experiences indicate some need for improved serviceability prediction methods after cracking.

Nagataki and Sato (1983), Sato *et al.* (1987, 1995) and Gilbert *et al.* (1999, 2004, 2005) investigated the long-term deflection of RC beams/slabs and proposed computational models in consideration of drying shrinkage. Bond creep and cracked-concrete mechanics under sustained loads have also been intensely investigated with respect to concrete volume changes. Building on these original investigations, the authors aim in this section to further enhance the modeling of post-cracking creep, which can also be linked with moisture migration analysis (Chapter 3). Space-averaged tension stiffness is implemented as a cracked-concrete creep model, since the space-averaging procedure has been extended to non-linear static and dynamic analysis in the past decade. A careful experimental verification is performed so as to offer reasonable simplicity and applicability to structural analysis.

Recent advances in computational mechanics enable engineers to implement direct integration of path-dependent constitutive models by tracing microscopic material states under all cyclic paths of stresses and strains. In the past decade, this scheme has successfully been used to simulate the seismic response of structures, while it has been specified in part as a tool for assessing structural performance during design. In this chapter, the direct path-integral scheme is employed as a more versatile means of creep analysis after cracking. The authors began their research on RC creep based upon the multi-directional fixed-crack model (Maekawa *et al.* 2003a), which has been in practical application for seismic and fatigue analyses. According to this scheme, a two-dimensional space-averaged constitutive model is formed by combining one-dimensional stress–strain relations for cracked concrete in tension, compression and shear. The method of obtaining stress–strain field compatibility by assuming a one-dimensional compressive stress flow originates from the modified compression field theory of Collins and Vecchio (1982). In this study, the concurrence of principal axes of stress–strain is not necessarily assumed, but the shear transfer across crack planes is explicitly taken into account based on non-orthogonal fixed-crack interactions (Maekawa *et al.* 2003a).

Each constituent model is strain-path dependent. The time-dependent concrete compression model has been verified at high-strain ranges for creep failure analysis in the previous section. As with the compression model, the tension stiffening–softening and shear transfer models must be enhanced to reflect their time-dependency. In this section, simple formulae for pre- and post-cracking tension models are adopted with respect to incremental time and strains, and these are installed together with the compressive model as discussed in Section 8.5. As flexure is the main mode of long-term deflection, time-dependent shear transfer (Frenaij *et al.* 1988) is not considered in this study.

### 8.6.1 Tension stiffness: bond creep and delayed cracking

From a phenomenological viewpoint, the tensile creep of plain concrete is related to micro-crack propagation (Cornelissen and Reinhardt 1984; Reinhardt *et al.* 1986; Subramaniam *et al.* 2002). The apparent reduction in tension stiffness of an RC element after the first cracking can be attributed to bond creep and subsequent cracking in a new section. In order to represent these phenomena, the space-averaging scheme is used in the formulation. Overall stiffness and average strain are adopted as the main state variables rather than the running ligament of a single crack. Tension non-linearity before and after cracking is assumed to be governed by the fracture damage rooted in cracks and is simply formulated as

$$\sigma = E_o K_T \cdot \varepsilon$$
$$\varepsilon = \varepsilon_T - \varepsilon_{sh}$$

(8.20)

where the stress and strain tensors ($\varepsilon$, $\sigma$; positive in tension) are normalized with respect to the positive-defined uniaxial compressive strength and the corresponding peak strain, $E_o$ (= 2.0) is the non-dimensional initial stiffness of the concrete solid, and ($\varepsilon_T$, $\varepsilon_{sh}$) denote the total and free shrinkage strains, respectively. The tensile fracture parameter $K_T$ (initially equal to unity and falling with damage; positive definite) is a scalar standing for path-dependent instantaneous fracturing, time-dependent tension creep and accumulated fatigue damage. Thus, the total incremental form is yielded as

$$dK_T = F dt + G d\varepsilon + H d\varepsilon$$

(8.21)

The derivative denoted by $H$ indicates the instantaneous evolution of a tension fracture, as represented by the tension stiffness formula for RC and the tension softening one for plain concrete (Maekawa *et al.* 2003a) as

$$H = -(1+\alpha)\left(\frac{f_t}{E_o}\right)\varepsilon_{cr}^\alpha \cdot \varepsilon_{max}^{-(\alpha+2)},$$
*when* $d\varepsilon > 0$ *and* $\varepsilon = \varepsilon_{max}$
$$H = 0, \quad \text{when } d\varepsilon \leq 0 \text{ or } \varepsilon < \varepsilon_{max}$$

(8.22)

where $\varepsilon_{cr}$ is the crack strain defined equal to $2f_t/E_o$ and $\varepsilon_{max}$ is the updated maximum value of tensile strain $\varepsilon$ in the past history. Derivative $H$ characterizes the envelope of the tension stress–strain curve after cracking. This formula covers the tension stiffness of RC associated with bonding ($\alpha = 0.4$). If the element is allocated to plain concrete with a crack localized inside, this formulation also describes the tension softening or bridging stress across the fracture process zone of the concrete. In this case, the value of $\alpha$ depends on the fracture energy in tension and the sizes of the finite elements (Maekawa *et al.* 2003a). The derivative denoted by $G$ represents the effect of stress–strain repetition, and is negligible in this chapter. This is the predominant factor for high-cycle fatigue and will be discussed in Chapter 10.

The derivative of $F$ indicates the rate of fracturing. For simplicity, post-cracking non-linearity is represented by delayed cracking and local bond creep using the formulation proposed by Hisasue and Maekawa (2005) as

$$F = -10^{-5} \cdot S^3 \cdot (K_T - 0.5)^2, \quad when\ \varepsilon_{max} < \varepsilon_{cr}\ pre-cracking$$

$$F = -10^{-6} \cdot S^6, \quad when\ \varepsilon_{max} \geq \varepsilon_{cr} \qquad\qquad (8.23)$$

$$\qquad\qquad\qquad\qquad post-cracking$$

$$S \equiv \frac{E_o K \varepsilon}{f_t}$$

where $f_t$ is the uniaxial tensile strength. The damage rate $(dK_T/dt)$ is assumed to be dependent on the stress magnitude denoted by $S$ and the updated degraded stiffness as $K_T$.

This formulation derives from uniaxial tension creep experiments on RC members with cracks. It corresponds to wet and dry conditions under exposure to a normal climate (20°C and 60% RH, on average), as shown in Figure 8.47. To identify the computational model, long specimens of approximately 3 m were used to obtain stable readings of mean tensile strain with displacement transducers. In order to avoid self-weight, the specimens were suspended on distributed springs. The mix proportion and mechanical properties at the beginning of loading are shown in Figure 8.47. Self-equilibrated stress is induced by drying and the cracking stress is apparently degraded if the initial internal stress is ignored in the analysis. The measured mean strain is a macroscopic representation of concrete creep, delayed cracking, local bond creep and volume changes dependent on moisture content.

Generally, the intrinsic free shrinkage of an infinitely small volume cannot be measured in experiments. Using concrete specimens of finite volume, only the volumetric average of local free shrinkage can be obtained. This can be attributed to the non-uniform profile of moisture in concrete. Thus the average member shrinkage is usually estimated in structural analysis according to the surface-to-volume ratio or the member thickness. Furthermore, it is necessary to make an assumption of apparent tensile strength when considering early cracking caused by self-equilibrated tension at the member surface.

However, the latest thermo-hydro physics may point to capillary tension and disjoining pressure as the driving force for shrinkage (Maekawa *et al.* 1999). This enables us to predict the local intrinsic shrinkage strain, denoted by $\varepsilon_{sh}$ in Equation 8.20. In this chapter, the direct use of local intrinsic shrinkage is proposed for structural analysis, as shown in Figure 8.48, and the efficiency of this approach will be discussed later. Solidification modeling is used to compute the micro-skeleton deformation of cement hydrate (Chapter

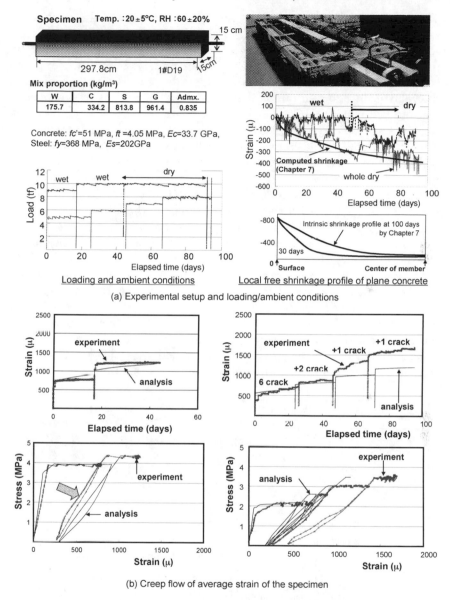

(a) Experimental setup and loading/ambient conditions

(b) Creep flow of average strain of the specimen

*Figure 8.47* Computed and experimental post-cracking tensile creep of RC.

7), which is linked with the thermo-hydro dynamic modeling as stated above. This system should cover drying shrinkage, autogenous shrinkage and their combinations.

Figure 8.47 shows the experimental and analytical mean strain evolution under sustained stress. In the analysis, the computed free shrinkage strain profile was used under assumed constant temperature and relative humidity.

The analysis domain was divided into two-dimensional finite meshes, as shown in Figure 8.48. The averaged shrinkage strain under no external stresses generally matches the experimental data, although daily fluctuations in temperature and relative humidity are present. Generally, the simple computational model gives results that reflect the actual situation. However, the analytically assumed rate of deformation just after the stress increment tends to be smaller than in reality. On the contrary, the creep rate becomes excessive when the stress level is close to the yield strength of reinforcement. Thus, the overall applicability of the model to structural analyses needs to be further examined based on verification of the structural level, as discussed in the following section.

If the free shrinkage strain of concrete is measured, it is recommended to use it in a direct manner for accuracy. In this case, the measured shrinkage is generally a sectional averaged value, and the non-uniform profile of local intrinsic shrinkage cannot be strictly considered, especially at the beginning of exposure to dry air.

Although the tension stiffness approach may macroscopically incorporate the local bond creep with delayed cracking of concrete, the crack spacing and width cannot be directly acquired. For these crack details, Nagataki and Sato (1983) and Sato *et al.* (1987) presented a discrete crack approach to RC flexural analysis with creep modeling of local bond. This method combined with the in-plane hypothesis is highly practical for slender members in flexure. The authors further attempt to build this time-dependency of local

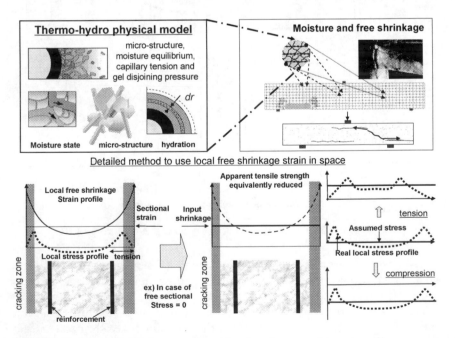

*Figure 8.48* Effect of shrinkage strain on local and averaged stress profile.

bond for more wide-ranging stress paths and strain fields by means of multi-directional crack modeling.

### 8.6.2 Compression creep model of cracked concrete

The authors make use of the elasto-plastic and fracturing model presented in the previous section. Time-dependent plasticity and fracturing are significant under higher stresses. Creep rupture under compression is numerically simulated as the combination of fracturing and plastic evolution accompanying increasing elasticity. These constitutive equations for compression were verified at the material and structural member levels under short-term high-stress states close to failure.

For long-term analysis, however, slow-rate creep associated with ambient states in micro-pores becomes predominant at the middle or lower stress level. Here, both creep components are simply added to cover the whole range of stresses applied to structural concrete. The authors added a simple linear creep rate represented by the linear viscous plasticity as

$$\kappa = -\frac{1}{C_v}\left(\varepsilon_p - C_{\lim}\varepsilon_e\right), \quad \kappa < 0$$

$$\frac{\partial \varepsilon_p}{\partial t} = \phi\left(\frac{\partial \varepsilon_p}{\partial t}\right)_b + \kappa, \quad \left(\frac{\partial \varepsilon_p}{\partial t}\right)_b = 0.034\left(\exp\left(\frac{\varepsilon_e}{4}\right) - 1\right)$$

(8.24)

where $C_v$ and $C_{\lim}$ correspond to the creep intrinsic time and the creep coefficient at the infinite time. These values can be identified through direct creep tests on the concrete of interest or through multi-scale simulation of concrete creep described in Chapter 7. If creep data are not available, $C_{\lim} = 3.0$ and $C_v = 40$ (day) are recommended as default values under the constant temperature and humidity of 20°C and 60%, respectively. The component denoted by $\kappa$ becomes comparatively greater than the non-linear term represented by $\phi$ in Equation 8.24.

## 8.7 Long-term time-dependent structural responses

### 8.7.1 Profile of intrinsic shrinkage

First, verification of the proposed model is carried out against the thin slab and beam experiments reported by Chong *et al.* (2004). A singly reinforced concrete slab was subjected to a sustained bending moment and moisture loss. Detailed ambient conditions around the specimens were not reported for these experiments, but creep test results for the concrete are available. The mean shrinkage strain was also measured for load-free reference specimens having similar dimensions to the slab and the beam.

Here, three types of analysis were carried out. First, the most detailed analysis considers the local profile of intrinsic shrinkage strain. In the experimental series, the sectional averaged shrinkage strain was directly measured, but the local profile of free-stress shrinkage strain cannot reasonably be obtained from experiments. The local profile of intrinsic free shrinkage was computed so that the sectional averaged value of each shrinkage profile coincides with the experimentally measured value (see Figure 8.49). The second analysis simulates the time-history of the beam deflection by assuming a uniform sectional averaged strain. In this case, the overall stresses of the constituent materials are computed properly, although the accuracy of local stress prediction is reduced. The third analysis is with the equivalently reduced tensile strength of concrete accompanying no drying shrinkage.

The tested creep coefficient at 200 days was about 1.4, which is nearly half the default value of $C_{lim}$. The authors then ran the computation with $C_{lim}$ = 1.5 and $C_v$ = 40 (days). Mindlin shell finite element was applied for space discretization. The slab thickness was divided into seven layers in which the

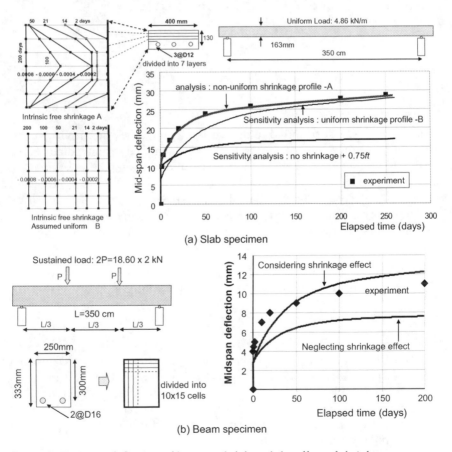

(a) Slab specimen

(b) Beam specimen

*Figure 8.49* Creep deflection of beam and slab and the effect of shrinkage.

intrinsic free shrinkage was set forth as shown in Figure 8.49. The one-way slab was divided into seven shell elements with the same dimensions.

It is clear that this detailed computation coincides mostly with reality. The second analysis with the uniform profile of local shrinkage strain also predicts the overall deflection flow except for the initial stage of loading. As the local tensile stress is underestimated, the simulation predicts delayed cracking. Simplified analysis with equivalently reduced tensile strength and no shrinkage captures the short-term response just after the loading, but the long-term deflection is underestimated. Thus, the effect of shrinkage on long-term deflection cannot be obtained by this simplified method. So it is the third method that is solely effective as a means of short-term failure analysis.

In the same manner as for the most detailed analysis above, the flow rate of beam deflection was simulated with 9 fiber-beam elements whose sections were divided into 10 × 15 cells as shown in Figure 8.49. In both the beam and the slab in question, the effects of creep and the time-dependent characteristics of the concrete on the flow of deflection are of similar magnitude.

### 8.7.2 Creep curvature under constant flexure

Experimental verification is extended to the creep curvature of RC members subjected to different levels of flexure. Mozer *et al.* (1970) reported on creep bending experiments as shown in Figure 8.50. The singly reinforced concrete sections were cast with deformed bars. Since the concrete creep reported in the paper was similar to that predicted by the model in Section 8.1.1, no modification of the long-term plasticity coefficients was made. As sectional stiffness is non-symmetric in mechanics, uniform shrinkage may cause some warping. For accuracy, the shrinkage strain was directly considered in the analysis of creep deflection. According to the shrinkage strain reading of the reference beam without loading, values of 400 μ at 91 days, 250 μ at 30 days and 120 μ for seven days were assumed and the same method as described in Section 8.7.1 was applied for the profile of intrinsic shrinkage. The space discretization applied is similar to the one as shown in Figure 8.49b.

In case of the low bending moment ($0.38M_y$; $M_y$ = yield moment) and medium bending moment ($0.65M_y$), the rate of change of deflection is small and similar to the compression creep rate of concrete. In fact, a progressive strain increment of flexural compression was observed but a much smaller increment was seen on the tension side in both analysis and experiment. However, in the case of sustained yield moment ($M_y$), creep strain in both tension and compression arises. As well, the tension creep after cracking is also substantial. A reasonable matching of analysis and experiments is shown in Figure 8.50.

To investigate the effect of shrinkage, sensitivity analysis was carried out under the assumption that there was no volumetric shrinkage at all. For a sustained moment of 0.38 $M_y$, a 50% reduced creep deflection is shown in Figure 8.50 and the compression fiber strain becomes smaller due to

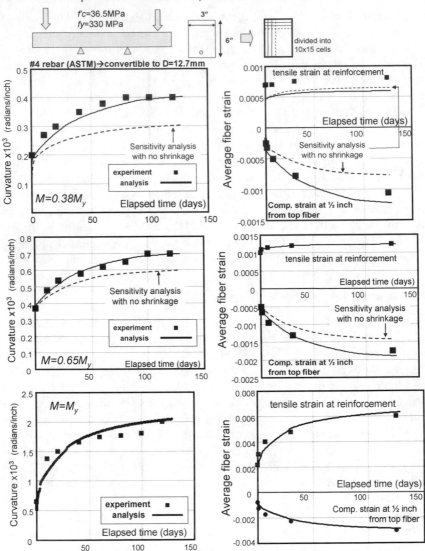

*Figure 8.50* Creep curvatures under constant flexural moments.

shrinkage, because the concrete shrinkage is immersed in growing crack widths. The tension fiber strain slightly increases. In this case, concrete shrinkage is the main driver of curvature, resulting in progressive deflection. For a sustained moment of 0.65 $M_y$, similar behavior is seen but the effect of shrinkage is less.

When the applied load is close to the yield moment, the shrinkage effect finally becomes negligible. In particular, the fiber strain of the cracked concrete in tension greatly evolves, as shown in Figure 8.50, although shrinkage has no substantial influence. For further sensitivity analysis, tension creep after cracking was also taken out of the analysis. In this case, there was no change in deflection. The tensile deformation after yielding of the steel is provoked by the reduced flexural compression accompanying non-linear creep.

### 8.7.3 Creep deflection subjected to uniform sustained loads

The next discussion concerns the long-term deflection of RC beams as investigated by Washa and Fluck (1952). The symmetrical member sections shown in Figure 8.51 were studied for verification in order to avoid warp-mode deflection caused by shrinkage. The specimens were loaded with a dead weight as well as additional concrete blocks or bricks for 2.5 years and the mid-span deflections and reinforcement strains were measured.

Normal concrete made with Type I Portland cement was used. After molding, five days of wet curing was carried out with wet canvas and burlap until the side forms were removed. After form stripping, specimens were exposed to dry air and the loading tests began two weeks later. During the 2.5-year tests, loaded specimens were kept at interior ambient conditions of approximately 70–85°F temperature and 20–80% RH. Although the drying shrinkage and the basic creep characteristics of the concrete used are not well known, the immediate deflection of each beam was reported. The apparent tensile strength of structural concrete, which is generally 15–50%

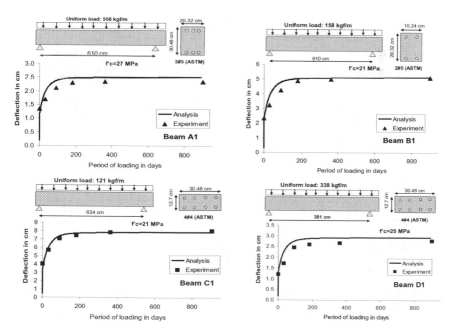

*Figure 8.51* Time-dependent deflection of RC beams subjected to uniformly distributed loads.

lower than the specimen strength due to the non-uniform profile of intrinsic drying shrinkage as discussed in Section 8.7.1, can be inversely calculated to yield the same cracking load as 2.0 MPa. In these cases, the tensile specimen strength estimated from the compressive strength is 2.4 MPa. This identifies the reduction as being about 20%. This seems like a reasonable value according to past experience. As the bending mode is predominant, 10 fiber-beam elements were used with 10 × 15 cells for the section, as shown in Figure 8.49b.

Figure 8.51 shows the creep flow deflection after loading for more than two years. The greater part of creep took place within the first 200 days in both experiment and analysis. The computation based on standard values of long-term plasticity rate is not very far from reality. The strain evolution for the compression reinforcement was reported to be between two and four times that of the tensile reinforcement.

### 8.7.4 Relaxation of concentrated load under sustained deflection

Force settlement under sustained displacement was investigated by Chali *et al.* (1969), along with creep deflection under constant external force. In view of the relaxation of concrete creep in tension and compression, the measured settlement history obtained in this work is useful for verification. Normal concrete of W/C = 62% by weight and Type I cement was used. Deformed bars were used for the longitudinal reinforcement. A symmetric arrangement of reinforcement was selected to minimize the warping of specimens as a result of shrinkage. Specimens were kept indoors at 18–22°C and 50–65% RH.

A concentrated load was applied to the beam at 13 days of age and sustained for 240 days, as shown in Figure 8.52. Since the properties of drying shrinkage and creep were not reported, standard values of creep modeling as given in Section 8.6.1 were used and the apparent cracking strength was estimated to be 3.0 MPa. This is almost the same as the estimated tensile strength. A fiber model for discretization was used in numerical analysis. As discussed in Section 8.7.3, the computed time-history of deflection matches the experimental results.

Next, a step-by-step relaxation path was followed as shown in Figure 8.52. Forced displacement was applied at the mid-span for 30 minutes and the reaction was measured with the displacement clamped. Subsequently, the settlement was measured at different displacements under cycling. Computation of the relaxing force is associated with both instantaneous stiffness and the transient flow of deformation. The computed results generally agree with the experimental values reported by Chali *et al.* (1969). As this settlement has some influence on designs for statically indeterminate continuous members, British standards specify a creep deflection coefficient of 2.0–0.8, according to the arrangement of the compressive reinforcement.

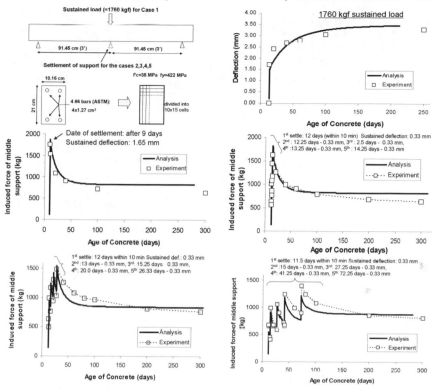

*Figure 8.52* Load relaxation of RC beams under sustained deflection.

## 8.7.5 Creep deflection of RC slabs subjected to sustained out-of-plane action

Verification for the case of complex three-dimensional structural behavior was carried out against experiments with continuous multi-span RC slabs subjected to sustained uniform loads by Guo and Gilbert (2002). As the primary mode of deformation is biaxial flexure, RC Mindlin shell elements were used for discretization of the analysis domain, as shown in Figure 8.53. The legs supporting the flat slab were idealized by fiber beam-column elements based on the in-plane hypothesis. Since the creep coefficient at each age and the free-drying shrinkage strain of the slab members were available, these values are directly used as input data in this analysis. The default values of long-term plastic flow rate coefficients (Section 8.6.2) were used and the profile of drying shrinkage strain was simply assumed to be uniform because the slabs were thin, at 9–10 cm thickness.

In the case of specimen *S1*, the computed deflection at the first stage of loading agrees fairly well with the experiment. Deflection readings at the symmetric points (#4-6-11-13, #1-2-15-16, #5-12) were reported to be almost equal within a small margin. As the multi-leg slab is statically indeterminate,

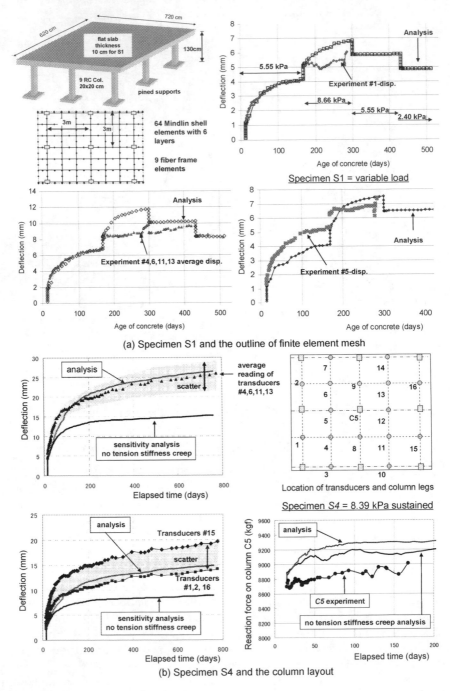

(a) Specimen S1 and the outline of finite element mesh

(b) Specimen S4 and the column layout

*Figure 8.53* Creep deflection of slab subjected to multi-axial bending.

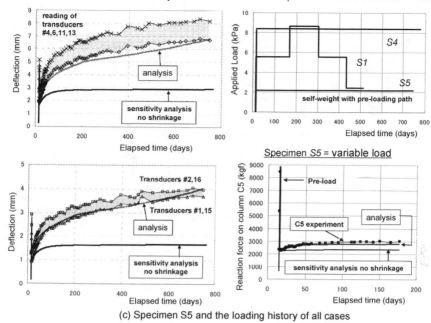

(c) Specimen S5 and the loading history of all cases

*Figure 8.53 (continued)* Creep deflection of slab subjected to multi-axial bending.

a slight deviation from mechanical symmetry may cause further successive deviation. At the following stages of loading, some deviations in deflection readings at the symmetric points were reported, although they were controlled within an acceptable range in practice.

In the case of specimen S4, under a highly sustained uniform load, computed deflections are slight overestimates, but the general tendency is reasonably reproduced. Since the deviation in deflection readings at the symmetric points was comparatively larger than with specimen S1, the range of scatter is also shown around the mean value in Figure 8.53.

In order to verify the modeling under a rather small load, case S5 was selected for experimental verification. The sustained load is the dead weight only, but the initial cracking was induced by the pre-loading. Here, the tension non-linearity after cracking is active even though the load level is small. The initially cracked member subjected to dead weight only shows the large creep deflection and gradual increase in the reaction at the center column C5, as shown in Figure 8.53, and quantitative assessment seems reasonable. The shrinkage of concrete is seen to greatly influence the self-weight deflection.

As tensile modeling is a substantial part of the overall model in these cases, further sensitivity analysis was carried out for the concrete tensile strength. The results are responsive to computed deflections as shown in Figure 8.54. If the applied load is rather high (S4), the sensitivity to tensile strength becomes comparatively less. In other words, specimen S5 is appropriate for verification of post-cracking tension stiffness modeling. To understand how

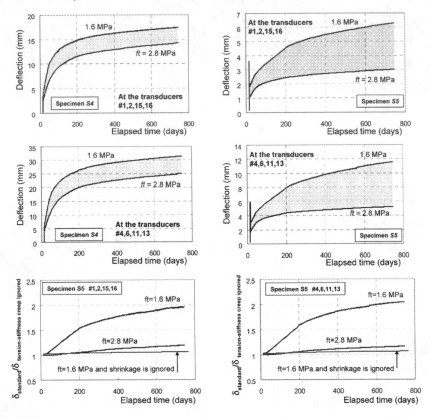

*Figure 8.54*  Sensitivity analysis for tension modeling.

influential the post-cracking tension creep model is, the computed deflections with and without post-cracking tension creep according to Equation 8.23 are compared for specimen S5, as shown in Figure 8.54. When the smaller tensile strength is assumed, the effect of post-cracking tension creep becomes predominant on the structural level. However, if concrete shrinkage is ignored, the time-dependence of tension stiffness plays a minor role. It can be said in general that the post-cracking tension creep model plays a primary role in long-term structural analysis when the tensile strength and/or reinforcement ratio is small, and the volumetric shrinkage is coupled.

### 8.7.6 Creep deflection of PRC beams subjected to sustained loads

Verification for the case of pre-stressed concrete was conducted against experiments with continuous PRC beams under sustained concentric loads by Tsuda *et al.* (1995). Pre-stressed forces with different magnitudes (131, 212 and 313 kN) were introduced to three RC continuous beams, as shown in Figure 8.55, and the average deflections at the loading points are compared

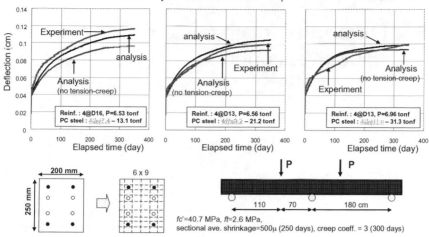

*Figure 8.55* Creep deflection of PRC beams with cracking.

to analysis results with and without post-cracking tension creep. The measured sectional averaged shrinkage strain of the reference specimen was directly used in the analysis. The long-term plasticity rate was validated by the reported creep test result of concrete as $C_{lim} = 2.0$ and $C_v = 60$ (day) in Equation 8.24.

For the case of high pre-stressing, cracking is almost blocked out and the effect of post-cracking tension creep is hardly seen. Here, the structural creep is governed by the concrete time-dependency. The unsmooth flow rate of experimental deflection might be attributed to temperature change over the seasons, which is not taken into account in Equation 8.24. In contrast, the time-dependent tension stiffness after cracking is confirmed to be effective for behavioral simulation of lightly pre-stressed members, as shown in Figure 8.55.

# 9 Durability mechanics of damaged structural concrete

This chapter aims to numerically simulate corrosion-induced cracking, its propagation over sections of members and the penetration of corrosive gel product into crack gaps. A coupled steel core and surrounding corrosion product are mechanically represented by a fictitious growing composite, with which the corrosive cracking initiation and subsequent propagation are simulated by non-linear crack analysis. The injection of corrosive gels into evolving cracks is substantiated in cases where corrosive cracks stably propagate, such as large covers and/or comparatively small diameters of steel, and the coupled system of gel formation, migration and crack propagation is newly presented. The simulation scheme was verified through reinforced concrete (RC) sections subjected to accelerated corrosion by electric current with regard to crack patterns and critical corrosion rates when cracks reach the outer surface of members.

A multi-mechanical model that explicitly takes into account corrosion cracks in structural safety performance is presented to deal with materialized corrosive substances around steel bars and equilibrated damage in structural concrete. The multi-mechanics of corrosive product in Chapter 6 and cracked concrete in Chapter 8 are integrated with non-linear, time-dependent multi-directional fixed-crack modeling so that corrosion cracks can be simulated in a unified manner. Structural analysis of corroded RC beams is carried out for experimental verification of the multi-mechanical model in terms of shear capacity and ductility. RC beams, which primarily fail in shear or flexure, are discussed and special attention is addressed to the conversion of failure modes and absolute capacity. Consideration of inherent cracking on corroded RC members is proven to be crucial for structural performance assessment, and the anchorage failure of longitudinal reinforcement is found to cause considerable decay of member capacity.

The mechanical effect of fractured web reinforcement on structural safety was experimentally investigated by intentionally avoiding hooks or anchorage devices at the extreme ends of stirrups, which were replicas of web steel damaged by corrosion or alkali-aggregate reaction of concrete. Significant reduction in shear capacity was experimentally found without yielding of web reinforcement when the web steel anchorage was incomplete. A marked difference in failure crack patterns was also observed compared with the sound RC specimen. Longitudinal cracks were ultimately formed along the main reinforcement where the unprocessed edges of the stirrups lie. Non-linear finite element analysis was employed to investigate and simulate failure processes and static capacity. The bond deterioration zone of the web reinforcement was computationally assumed to be ten times the diameter of steel bars from the cut-off trimming of steel bars. This simple assumption in non-linear computation was verified to be acceptable for performance assessment of damaged reinforced concrete with fractured stirrups.

## 9.1 Multi-mechanics of corrosive substances and concrete (micro-level)

Steel corrosion in reinforced concrete structures has been recognized as one of the major threats for buildings and infrastructure in service. The safety and serviceability of corroded RC structures has been widely addressed by practicing engineers. In some durability designs, the initiation of corrosion is chosen as a limit state of durability performance with regard to carbonation and chloride penetration. This is due to the current difficulty of predicting the lifetime of mechanical damage of the concrete cover after corrosion initiation. Since the 1980s, theoretical proposals and experimental investigations on steel corrosion in concrete have been intensely carried out and reported in the literature (e.g., Bazant 1979; Browne 1980; Tuutti 1982; Ravindrarajah and Ong 1987). Here, not only experimental investigation but also a reliable numerical approach for predicting critical corrosion mass loss for concrete cover cracking is indispensable as a tool for rational durability assessment.

Morikawa *et al.* (1988) and Tsunomoto *et al.* (1990) conducted the analysis of critical weight loss for cover cracking due to reinforcement corrosion. In their analyses, the target specimens have smaller covers compared to the reinforcing bar diameters (C/D ratio is 1.2 to 2.8). The analytical result of weight loss for cracking was close to the experimental one at the high corrosion rate. However, in the case of long-term corrosion, the analysis underestimated the critical weight loss compared to reality. Molina *et al.* (1993) proposed the numerical model and verified it against their own experimental results. In their analysis, not only the critical mass loss but also the numerically estimated crack width was computed, and the analytical results showed overestimation of corrosion weight loss of steel. The analysis of corroded RC specimens having a relatively large C/D ratio was conducted by Tsutsumi *et al.* (1998). Their model was verified by RC specimens having C/D of 2.31, 3.85 and 5.26. It was found that the weight loss was much underestimated, especially in the case of large C/D ratios. Pantazopoulou and Papoulia (2001) investigated simple analytical modeling for cover cracking. The verification of the model was made using many past experimental results. While a qualitative tendency was seen, it was quantitatively found that the computed weight loss was greatly underestimated compared to the experimental results. In their conclusions, an important comment was made on a more reliable approach and the need for experimental evidence on the penetration of corrosion products into cracks. In fact, this point raised by Pantazopoulou (2001) is the main target of this section.

The non-linear analysis with multi-mechanics of corrosive substances and structural concrete has been proposed for simulation of post-corrosion cracked members (Toongoenthong and Maekawa 2004b, 2004d). Here, a core of non-corroded steel and its surrounding rust substances are mechanically represented by a comparable growing material, and the corrosion degree is dynamically associated with the numerical simulator based on thermodynamic multi-chemo physics (Maekawa *et al.* 1999), or non-destructive testing and inspection data in reality.

This approach has been verified chiefly in terms of structural safety assessment of RC members after the formation of section-penetrating cracks caused by the corrosive volumetric expansion of steel (Toongoenthong and Maekawa 2004a). However, the approach has not been discussed or verified with regard to predicted duration from the initiation of corrosion to spalling of the cover concrete (where the internal corrosion crack reaches the surface of members). In this chapter, the concept of the multi-mechanical approach with two phases to the corroded reinforcement system and concrete will be extended to the limit state evaluation of cover spalling. The authors employ the concept of multi-mechanics composed of corrosion product, remaining non-corroded steel and surrounding structural concrete. Examined as stated above is its capability for predicting the limit state of section damage, which is associated with spall-off of cover concrete for a wide range of cover dimensions relative to reinforcing bar diameters. The proposed computational approach is also extended to the non-linear post-cracking mechanics of the concrete cover in RC sections. The effect of creep in concrete is taken into account in the analysis as well, via the reduction of effective concrete stiffness.

The migration of corrosive gel products into stable crack gaps is quantitatively highlighted in the coupled computational system of chemophysics and mechanics with substance formation. In this section, the authors will concentrate on this gel product migration into crack gaps as the original focus of this book.

For simulation of corrosion occurring in RC sections, the authors assume a two-phase mechanical system composed of corrosion product with non-corroded steel as one integrated phase and nearby concrete as another. After the initiation of corrosion, the volumetric expansion of rust is constrained by the surrounding concrete. Accordingly, compressive stress is introduced to the corroded reinforcement system, while tensile stress develops in the surrounding area. Internal equilibrium and deformational compatibility may lead to pre- and post-cracking fields of stress and strain. Here, the combined corrosion product and non-corroded steel are modeled together as a growing compatible composite. For surrounding structural concrete, it was represented by smeared crack elements having coherent tension softening after cracking with a finite element size and fracture energy of concrete (Okamura and Maekawa 1991; Maekawa *et al.* 2003b).

Figure 9.1 shows a schematic diagram for the expanded geometry of corroded steel and concrete. The uniform corrosion accumulation around the reinforcement is simply assumed in this computation. Although actual corrosion in RC structures may not be uniformly distributed due to interfacial zones and bleeding beneath the bars, the hypothesis of this uniformity is thought to represent the critical index for corrosion estimation of cover cracking. When considering the possibility of the migration of the corrosion product into crack spaces or adjacent concrete voids, it is not only volume loss of corroded steel (denoted $V_{loss}$) that causes the build-up of stress in the nearby concrete. Some amount of corrosion product may possibly occupy interior spaces bordered by a pair of cracked planes or bleeding gaps beneath steel bars before cracks reach the concrete surface. Let $V_{effective}$ denote the effective volume loss per unit length from the parent steel that really causes the stress build-up, and let $Q_{cr}$ denote the rest of the corrosion product that may be injected inside concrete cracks and micro-voids. Using the analogy of *plastic theory*, the effective volume loss that causes stresses and cracking to cover concrete can be expressed as

$$V_{effective} = V_{loss} - Q_{cr} \tag{9.1}$$

In order to estimate $V_{loss}$, we need some non-destructive testing or other means to identify this value for existing structures. For durability design, this total amount of volume loss, $V_{loss}$, can be obtained from accumulated mass loss per unit of surface area of steel (kg/m²), which is obtained from the simulation (Chapter 6; Maekawa *et al.* 1999, 2003a).

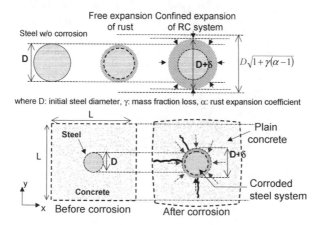

*Figure 9.1* Schematic representations for expansion of corroded steel in RC.

If there is no restraint to the corrosion product from the surrounding concrete, all effective corrosion substances can expand freely with no confining stress. First, the corrosive volume loss from the mother steel is calculated by

$$V_{effective} = \gamma \cdot A_{steel} = \gamma \cdot \pi D^2 / 4 \qquad (9.2)$$

where $A_{steel}$ is the original steel area, $D$ is the original steel diameter (m), and $\gamma$ is the effective volume fraction loss (non-dimensional effective mass loss) of steel per unit length ranging from 0 to 1 (zero means no corrosion, while unity means complete loss of the reinforcement section).

Consequently, the mother steel exhibits a smaller diameter due to corrosion loss. The remaining diameter of uncorroded steel (denoted by $D_{corroded\text{-}steel}$) yields

$$D_{corroded-steel} = D\sqrt{1-\gamma} \qquad (9.3)$$

As the volume loss results in expansive corroded product, we have the final volume of corroded product ($V_{corr}$) as

$$V_{corr} = \alpha \cdot \gamma \cdot \pi \cdot D^2 / 4 \qquad (9.4)$$

where $\alpha$ is defined as the coefficient of expansion of corroded substance (assumed $\alpha = 2$). Morikawa *et al.* (1988) conducted the X-ray diffraction of corrosion product in an acceleration test and found that $Fe_3O_4$ is the main component of the corrosion product. Moreover Liu and Weyers (1998) provided information on the relative volumes of iron and its corrosion products in many forms. Among various iron oxides, $Fe_3O_4$ has volume expansion of approximately two with regard to $\alpha$. Consequently, $\alpha = 2$ is used as the first assumption. This coefficient will also be verified through the benchmark experiments of structural concrete in the following sections.

The corroded product, which effectively creates internal stress, consequently forms around the mother steel. The mean diameter of the corroded reinforcement system, which consists of mother steel and its corrosive product, is obtained in the case of stress-free and geometrically uniform expansion as

$$D_{corrodedlayer} = D\sqrt{1+\gamma(\alpha-1)} \qquad (9.5)$$

Then, the reinforcement free-expansion strain ($\varepsilon_{s,\,free}$) of isotropy yields

$$\varepsilon_{s,free} = \sqrt{1+\gamma(\alpha-1)} - 1 \qquad (9.6)$$

The value stated above is the mean strain of the corroded system with respect to the initial diameter of reinforcing bars, and it is a function of the specific level of corrosion '$\gamma$'. The stress induced by restraint of the surrounding concrete is computed by multiplying the total strain of the corroded system with its average stiffness, representing multi-mechanical properties. The average stiffness of the corroded system, which is denoted by $E_{s,eq}$, is computed based upon the steel and corroded layers. As the corroded system is assumed to have uniform characteristics, the average stiffness of the corroded system can be expressed based upon the volume fraction as

$$E_{s,eq} = \frac{1 + \gamma(\alpha - 1)}{\left(\frac{1 - \gamma}{E_s}\right) + (\gamma\alpha/G)} \tag{9.7}$$

where $E_s$ is the stiffness of steel (~200 GPa), and $G$ is the stiffness of corroded substance (first assumed as 7 GPa). Note that the linear constitutive law of the corroded substance is assumed in this formulation. Moreover, a wide range of corrosion substance stiffness has been proposed in the literature. A stiffness value as low as 0.02 GPa was suggested by Yoshioka and Yonezawa (1982).

Molina *et al.* (1993) used the value of 2–4 GPa in their modeling and concluded insignificant influence on computed critical mass loss. Lundgren (2002) proposed the stress level-dependent stiffness of corrosion substance, which can be up to 14 GPa under high confinement. So, the assumed value of $G$ is also the target of experimental verification in the following section.

By employing the computed stress-free expansion strain and equivalent stiffness of the corroded reinforcement system, self-equilibrated stress is computed by the following equivalent elasticity of the steel–rust composite elements as

$$\sigma_{ij} = D_{ijkl}(E_{s,eq}) \cdot \left(\varepsilon_{kl} - \delta_{kl} \cdot \varepsilon_{s,free}(\gamma)\right) \tag{9.8}$$

where $D_{ijkl}$ is the constitutive model function for corroded steel continuum, and $\delta_{kl}$ is Kroneker's delta, defined on X–Y coordinates including the RC cross-section concerned. Here, the isotropic linear matrix of $D_{ijkl}$ was assumed with Poisson's ratio of 0.3 for the analyses of RC sections.

Morikawa *et al.* (1988) pointed out the substantial effect of concrete creep on corrosive cracking initiation. Consequently, in the two-dimensional crack analysis, the effect of concrete creep is simply taken into account by factoring the stiffness of concrete with the creep coefficient. Comparisons with non-creep analysis will also be made.

## 9.2 Critical corrosion for cover cracking

### 9.2.1 Equivalent growing material of rust–steel composite

The galvanostatic corrosion experiments of 13 specimens were selected for experimental verification from the existing literature, as listed in Table 9.1. Expansion of corrosion product around corroded steel is thought to exhibit

Table 9.1 Experimentally observed critical mass loss of steel and FEM prediction for concrete cover cracking

| Reference | Min. covers (specimen length) in mm. | Steel diameter (mm) | C/D ratio | Experimental critical mass loss (%) | Computed critical mass loss (%) | (Exp./FEM) | Buffer capacity, $Q_{cr}$ (%) | Creep effect (long period corrosion) |
|---|---|---|---|---|---|---|---|---|
| Andrade et al. | 20×20 (380) | 16 | 1.25 | 0.37 | 0.30 | 1.23 | - | X |
| | 20×30 (380) | 16 | 1.25 | 0.448 | 0.42 | 1.07 | - | X |
| | 30×30 (380) | 16 | 1.875 | 0.45 | 0.44 | 1.02 | - | X |
| Al-Sulaimani et al. | 70×70 (150) | 10 | 7.00 | 4.50 | 4.45 | 1.01 | 3.62 | O |
| | 65×65 (150) | 20 | 3.25 | 1.80 | 1.34 | 1.34 | 0.88 | O |
| | 68×68 (150) | 14 | 4.86 | 2.89 | 2.18 | 1.33 | 1.35 | O |
| Cabrera and Ghoddoussi | 69×69 (150) | 12 | 5.75 | 2.25 | 2.16 | 1.04 | 1.34 | O |
| Morikawa et al. | 30×30 (200) | 25 | 1.2 | 0.08 | 0.10 | 0.80 | - | X |
| | 30×30 (200) | 25 | 1.2 | 0.20 | 0.18 | 1.11 | - | O |
| | 50×50 (200) | 25 | 2 | 0.12 | 0.158 | 0.76 | - | X |
| | 50×50 (200) | 25 | 2 | 0.30 | 0.28 | 1.07 | - | O |
| | 70×70 (200) | 25 | 2.8 | 0.15 | 0.21 | 0.71 | - | X |
| | 70×70 (200) | 25 | 2.8 | 0.30 | 0.29 | 1.03 | - | O |

uniformity under the accelerated corrosion with galvanostatic process. Thus, this electrically accelerated condition, which actually differs from the mode of real-time scale corrosion, is preferable for experimental verification of the numerical modeling. Although there is a wide variety of specimen dimensions and reinforcement arrangements, the experimental common objective is to investigate the critical mass loss of steel during cover cracking and spalling. The experimentally measured corrosive mass loss is directly used for computing the equivalent stiffness of the corroded reinforcement system and stress-free expansion strain, as explained in the previous section. By applying Equation 9.8 for finite elements to represent steel bars, varying stiffness and free expansion, which are derived from the measured corrosion mass loss, are incorporated into two-dimensional finite element mesh (FEM) analysis. The experimentally measured mass loss of steel at failure, computed critical mass loss and specimen details are also given in Table 9.1. Table 9.2 summarizes the concrete material properties used in FEM analysis, including experimentally obtained compressive strength of concrete, estimated tensile strength and tensile fracture energy of concrete.

An example of the FEM used in the analysis is shown in Figure 9.2. The mesh is composed of three types of elements, i.e., the expansive material element, plain concrete and interface elements. The elastic element with the equivalent expansive material model is used to represent the corroded reinforcement system, as discussed in Section 9.1. The concrete domain is

*Table 9.2* Material properties for FEM simulation analysis

| References | Concrete compressive strength from experiment | Concrete tensile strength used in FEM (*) | Calculated fracture energy (**) |
|---|---|---|---|
| Andrade *et al.* | 30 MPa | 2.58 MPa | 0.066 Nmm/mm² |
| Al-Sulaimani *et al.* | 30 MPa | 2.58 MPa | 0.066 Nmm/mm² |
| Cabrera and Ghoddoussi | 56 MPa | 3.91 MPa | 0.101 Nmm/mm² |
| Morikawa *et al.* | 35.6 MPa | 2.89 MPa | 0.074 Nmm/mm² |
| Tachibana *et al.* | 35.6 MPa | 2.89 MPa | 0.074 Nmm/mm² |
| Cabrera | 35 MPa | 2.86 MPa | 0.073 Nmm/mm² |
| Toongoenthong and Maekawa | 34 MPa | 2.80 MPa | 0.072 Nmm/mm² |
| Matsuo *et al.* | 47.8 MPa | 3.52 MPa | 0.091 Nmm/mm² |

(*) $f_t = 0.58(f_c')^{2/3}$
where  $f_t$ = concrete tensile strength (kg/cm²)
    $f_c'$ = concrete compressive strength (kg/cm²)
(**) $G_f = \alpha(f_{cm}/f_{cm0})^{0.7}$
where  $G_f$ = fracture energy
    $\alpha$ = a coefficient based on maximum aggregate size (N·mm/mm²)
    $f_{cm}$ = cylindrical compressive strength (kg/cm²)
    $f_{cm0} = 100$

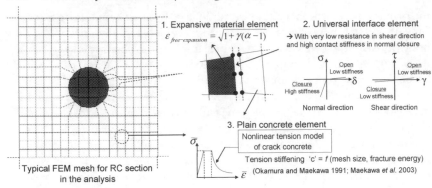

Typical FEM mesh for RC section in the analysis

1. Expansive material element
$$\varepsilon_{free-expansion} = \sqrt{1+\gamma(\alpha-1)}$$

2. Universal interface element
→ With very low resistance in shear direction and high contact stiffness in normal closure

3. Plain concrete element
Nonlinear tension model of crack concrete

Tension stiffening 'c' = f (mesh size, fracture energy)
(Okamura and Maekawa 1991; Maekawa et al. 2003)

*Figure 9.2* FEM mesh and elements used in analysis of corrosive crack propagation.

modeled with the smeared-crack plain concrete elements having tension softening that is consistent with the finite element size and the fracture energy of concrete (Okamura and Maekawa 1991; Maekawa *et al.* 2003a). The interfacial zone between corroded steel elements and contacting concrete is modeled by the joint interface elements to represent the stress release at separation and stiffness recovery at re-contact. For the normal direction, a large stiffness in the closure mode and negligibly small stiffness in the opening mode are assumed. For the shear direction, the interface element has zero stiffness both in the closure and opening modes. Here, the authors conducted trial analysis to clarify the importance of interfacial zone modeling.

As shown in Figure 9.3a, when the steel surface and concrete were unrealistically assumed to be perfectly connected without joint elements, overestimation of the cracking damage around reinforcing bars was observed, which is obviously different from reality as observed in the experiment. On the other hand, a reasonable cracking damage condition can be obtained by more realistic modeling with the interfacial zone, as shown in Figure 9.3b. To avoid the freedom of rigid body motions of the entire analysis domain, the central nodes of reinforcing bars were numerically fixed by displacement.

First, let us concentrate on the simulation of small cover cases of RC specimens. For the specimens with relatively small cover to the bar diameter, the occurrence of cover cracking tends to be very rapid, hence providing no time for corrosion gel penetration into cracks or concrete voids. Thus, $Q_{cr} \approx 0$, due to an insignificant amount of corrosion product injected into cracks

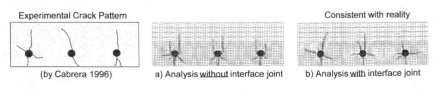

Experimental Crack Pattern

(by Cabrera 1996)          a) Analysis without interface joint          b) Analysis with interface joint

Consistent with reality

*Figure 9.3* Importance of proper interface zone modeling in FEM.

and, consequently, we have $V_{effective} = V_{loss}$ where $Q_{cr} = 0$ in Equation 9.1. As a result, the computed effective critical mass loss for small C/D cases corresponding to experimentally observed cover cracking reasonably matches the experimental measurement as shown in Table 9.1. This successful simulation for small cover cases confirmed the numerical approach concept similarly to past research, where the amount of corrosion penetration was so small as to be numerically negligible. Figure 9.4a shows the numerical crack pattern of the fair simulation. The experimental and computed values of critical mass loss are close to each other without considering the amount of corrosion gel penetration. Here, the creep effect of concrete is taken into account. Especially, it is somehow influential for the cases where crack initiation takes a long time, for example when there is a large cover depth relative to the bar diameter (Al-Sulaimani *et al.* 1990; Cabrera and Ghoddoussi 1992), or in the case of a long-term accelerated corrosion test (Morikawa *et al.* 1988).

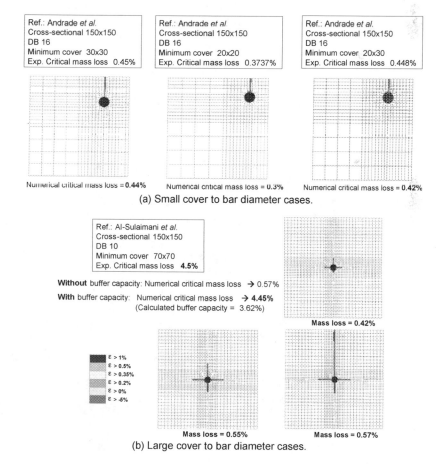

*Figure 9.4* Numerical crack propagation by FEM with and without buffer capacity of crack gaps.

For the sake of simplicity, concrete mean elasticity was uniformly reduced using a creep coefficient of 2.0. As Morikawa *et al.* (1988) pointed out, the creep of concrete results in higher weight loss associated with cover concrete cracking in experiments when a slow corrosion rate develops. By analysing the specimens of Morikawa *et al.* (1988), this creep effect is also captured in this simulation as discussed later (Figure 9.5a).

However, in the case of a large cover specimen (large C/D ratio), the only computed effective critical mass loss gives the underestimated result of critical total mass loss in general. Some finite element discretized meshes of large cover specimens are shown in Figure 9.4b. In these cases, finer sized finite elements are appropriate for discussing the stability of crack propagation in detail. A relatively smaller computed mass loss of 0.57% is obtained compared with the experimental measurement of critical total mass loss of 4.5%. The large discrepancy between the predictive and experimental values is thought to be caused by the migration of corrosive substance into crack gaps (buffer effect), as discussed in the following section.

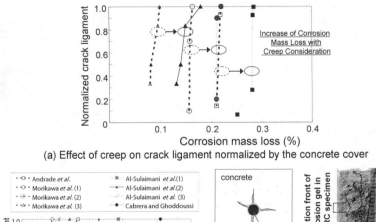

(a) Effect of creep on crack ligament normalized by the concrete cover

(b) Effect of crack gap volume.

*Figure 9.5* Corrosive crack propagation with effective corrosion degree.

### 9.2.2 Penetration of corrosive product into cracks

Penetration of corrosive product into crack gaps of concrete is thought to occur because the corroded substance is a form of gel with some liquid characteristics. If this penetration is substantial, internal pressure around the corroded steel may be reduced and an internal gap formed by a pair of crack planes may work as a buffer or a reservoir. In the case of a large cover thickness relative to the bar diameters, for example, crack propagation is slow and stable enough to migrate into crack gaps and these spaces can be large (the assumption of $Q_{cr} \approx 0$ is no longer valid). This buffer capacity is defined as the volume surrounded by crack planes, which can be occupied by corrosion product at most. Hence, the total critical corrosion mass loss should include this buffer capacity. The created spaces surrounded by crack planes can be simply computed by integrating the computed crack width with the increment of crack ligament. The crack width is obtained by the relative displacements of nodes crossing a strain localization zone as a crack band. In this case, the volume change of the concrete continuum is negligible compared with the spaces newly created by cracking. Thus, we can simply estimate the total buffer capacity defined by $Q_{cr}$ in Equation 9.1 as

$$Q_{cr} \approx \oint_{domain} (\bar{u} \cdot \bar{n}) ds \qquad (9.9)$$

where $u$ and $n$ denote the displacement vector and the unit directional one normal to the boundary surface of analysis domain. The simultaneous coupling of boundary integral Equation 9.9 with the finite element field ones is the first original challenge on the way to computing corrosion crack propagation associated with rust gel injection into the analysis domain.

Figure 9.5 shows the computed crack ligament in the analysis with the effective corrosion mass loss. It can be seen that a relatively large cover thickness to bar diameter specimen of Al-Sulaimani *et al.* (1990) and Cabrera and Ghoddoussi (1992) exhibits stable crack growth before cracking reaches the surface of the cover concrete. It takes time until complete crack growth to the surface occurs. This means that the rust gel may possibly migrate into cracks if given sufficient time. As a result, the corrosion product may occupy the cracking spaces around the corroded reinforcement before unstable crack growth driven by a small fraction of consecutive corrosion. Here, this capacity of corrosion product accumulated in the crack gaps is denoted as buffer capacity and its computation by two-dimensional analysis is shown in Table 9.1. By incorporating the buffer capacity by Equation 9.9 with the effective one into the total computed mass loss (see Equation 9.1), the analysis can give a fair estimation of critical mass loss at cover cracking that compares well with the experimental value as shown in Figure 9.4.

In order to confirm the existence of the phenomenon of corrosive gel penetration into crack gaps, the authors also conducted a simple experiment on a corroded RC specimen with a relatively large cover. Figure 9.6a shows the specimen and experimental set-up for the accelerated corrosion process. In the experiment, to verify the buffer phenomenon, the cathodic accelerated corrosion process was terminated before the corrosive crack plane could reach the surface of the concrete block. After that, the concrete was carefully chipped off to reveal the corrosive gel penetration near the corroded reinforcement. The specimen was a cube measuring 20 cm on each side. A single reinforcement 16 mm in diameter was embedded at the middle of the specimen (C/D ratio = 5.75). The migration of corrosion product away from the parent steel was experimentally confirmed as shown in Figure 9.6b and Figure 9.6c, which gives the propagation front of corrosive gel inside the RC specimen. The experimental mass loss of corroded steel was measured as 0.77%.

Figure 9.6d shows the corrosive crack propagation by FEM based on the experimental measured mass loss as input, taking into consideration the buffer capacity. The tested compressive strength of concrete was 35.1 MPa. In this case, computationally obtained effective mass loss was 0.65%. Thus, the mass loss of 0.12 (= 0.77–0.65%) stored in the crack gaps cannot be ignored compared with the total mass loss of steel. The propagation front of cracking is similar to the experimental one. The ratio of buffer to the effective mass loss is not constant, instead varying over a wide range according to the geometry of the sections and the crack ligament length.

On the other hand, instantaneous crack propagation to the concrete cover is numerically observed in the case of relatively small cover depth to bar diameter specimens of Andrade *et al.* (1993) and Morikawa *et al.* (1988), as shown in Figure 9.5a. Here, the corrosion product is hardly accumulated inside the crack spaces due to the immediate propagation of corrosive cracks

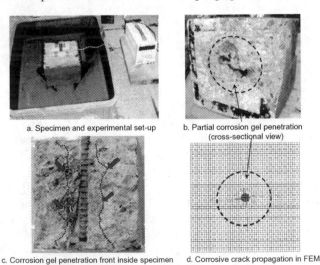

a. Specimen and experimental set-up

b. Partial corrosion gel penetration (cross-sectional view)

c. Corrosion gel penetration front inside specimen

d. Corrosive crack propagation in FEM

*Figure 9.6* Observed corrosive gel penetration into crack gaps before failure of cover concrete in experiment and analysis.

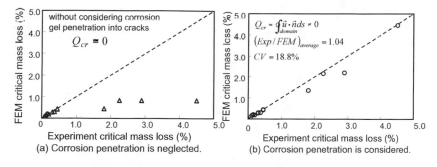

*Figure 9.7* Comparison of critical mass loss: experiment and FEM computation.

to the surface and insufficient time for gel migration. For these cases (small C/D ratio), computation can give fair agreement of critical mass loss only by considering the effective mass loss around the mother steel. In this case, the creep effect (linear creep coefficient: 2.0) is seen to be substantial with regard to the corrosion rate and the cracking time. Figure 9.7a and Figure 9.7b summarize the comparison of the experimental and the computed total critical mass losses with and without consideration of the amount of corrosion gel penetration by non-linear finite element analysis. Consideration of the buffer effect is indispensable for mechanical damage simulation of cover concrete. As the capacity of $Q_{cr}$ is so large compared with the total corrosion mass, the effect of bleeding and interfacial zones on the lifetime of cracking may tend to be less for thick concrete covers.

### 9.2.3 Corrosion cracking around main reinforcement

In Section 9.2.2, the numerical critical mass loss at cover cracking was verified with the experimental results of RC specimens with embedded single bars. In this section, the results of further analysis performed on RC sections with multiple arranged reinforcing steels are discussed. This section also aims to verify simulation with regard to the corrosive crack propagation pattern when RC sections contain more than single steel. Here, the experimental results for RC beams obtained by Tachibana *et al.* (1990) are used for verification of corrosive crack propagation. The targeted RC beam section is 150 × 200 mm, with two D16-mm reinforcing bars. The beam length is 2 m. The minimum cover at the bottom is 30 mm. The compressive strength of concrete is reported as 35.6 MPa. The RC beam was corroded with a galvanostatic corrosion system employing the electric current connected to the reinforcement. The experimental growth of cracks after a 15-day current application period is shown in Figure 9.8.

In order to verify the applicability of the corrosive cracking propagation analysis, finite element mesh of the same dimensions is used and input with the experimental material properties of concrete and reinforcement. The

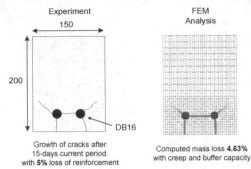

*Figure 9.8* Growth of corrosive cracking in experiment and FEM analysis.

reinforcement is represented by the evolving equivalent elastic elements with isotropic expansion in each time step (Section 9.1). The computation terminates when the numerical crack propagation reaches the surface of the member as experimentally observed. Due to the relatively long corrosion period, the effect of creep is also considered in the analysis through the reduction of concrete stiffness by a creep coefficient of 2.0. Figure 9.8 shows the comparison of the experimental corrosive crack patterns with the numerical ones. The computed mass loss of reinforcement was also given by considering the possible buffer capacity as $Q_{cr}$, which reflects the possibility that corrosion product can occupy corrosive cracking spaces. A slightly smaller analytical value of estimated reinforcement mass loss might be attributed to the fact that some of the corrosion product may be discharged outside just before or after cover cracking. This leakage of rust gel is not taken into account in the numerical simulation until cracking reaches the surface of concrete members.

Next, RC sections with multiple bars are discussed by employing the detailed characters on corrosive cracking by Cabrera (1996). Three selected specimens in total have the same cross-sectional dimension of 100 × 300 mm, with a length of 200 mm along steel bars. Each specimen has three steel bars embedded inside, as shown in Figure 9.9. The three different sizes of reinforcement diameters are the points of discussion, i.e., DB12, DB16 and DB20. The three specimens were corroded by the galvanostatic corrosion system with electric current. At the end of a 28-day current application period, the corrosive cracking patterns were recorded along with measurement of the

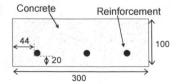

| Specimen | Steel Diameter (mm) | Experimental Mass Loss | Max. surface crack width |
|----------|---------------------|------------------------|--------------------------|
| A-12 | 12 | 17.73% | 0.38mm |
| A-16 | 16 | 12.8% | 0.56mm |
| A-20 | 20 | 6.76% | 0.56mm |

*Figure 9.9* RC specimens with multiple reinforcements by Cabrera (1996).

reinforcement weight loss and maximum surface crack width, as shown in Figure 9.10.

The numerical analysis of crack propagation was carried out for all experimental cases with consideration of creep. The computation is terminated when the computed numerical crack width reaches the experimentally observed value, under the assumption that the same corrosion damage level is thus reached. The reinforcement mass loss with corrosion product injected inside the surrounding crack spaces is estimated to correspond to the observed crack propagation. The comparison of the final crack pattern is shown in Figure 9.10, along with the estimated numerical reinforcement mass loss. Here, the numerical estimated mass loss is slightly underestimated, which might be explained by the possible leakage of corrosion product from inside the RC specimen just before or after cracking during the long-term corrosion process. The present numerical computation does not include this leakage of corrosion gel, as stated before.

Figure 9.11 shows the sequential process of corrosive crack evolution for case A-16 at different corrosion levels. The initial crack was observed to form at the bottom cover side and then propagate further to the main body of the specimen. At the beginning stage of crack propagation, the computational crack pattern is almost symmetric. However, this symmetry is then lost, similarly to the experiment according to the evolution of corrosion. In the analysis, the predictor–corrector method was applied to search for the unique

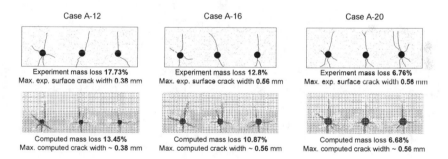

*Figure 9.10* Experimental (Cabrera 1996) and numerical corrosive crack patterns with computed mass loss.

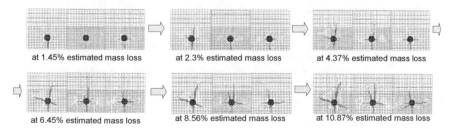

*Figure 9.11* Crack evolutions with increasing mass loss (case A-16).

stable equilibrium when cracking proceeds. Thus, the higher energy solution of perfect symmetry of crack damage was not computationally selected but the convergence of solution reached the asymmetric solution of the lowest strain energy (Maekawa *et al.* 2003a).

The corroded RC beams tested by Toongoenthong and Maekawa (2004a) were also analysed for verification. All three sections of the RC beams were corroded by the accelerated galvanostatic method. After a period of several weeks of corrosion, the damage in terms of corrosive crack width was measured with focus on the mass loss of corroded steel after the structural loading test. Table 9.3 shows the recorded corrosive crack width information and the measured mass loss from the experiment for all three beams. The RC sections were designed to be a single layer arrangement of reinforcement with relatively small cover depth and clearance between bars, as shown in Figure 9.12. With this dense, single-layer reinforcement arrangement, the corrosive horizontal crack then quickly propagated and resulted in a localized horizontal crack plane being observed in both the experimental and the numerical crack patterns, as shown in Figure 9.12. The buffer capacity at the crack propagation to the surface of structural concrete is thought insignificant, since sudden propagation is numerically predicted within a short time. The numerically computed crack width for all three cases is given

*Table 9.3* Experimentally obtained corrosion and computed corrosion crack width

|  | Compression mid-span | Tension mid-span | Tension near support |
|---|---|---|---|
| Experimental crack width (mm) | 0*)~0.3~0.5 mm | 0.2~0.7 mm | 0.3~0.5 mm |
| Experimental mass loss (%) | 1.89% | 8.76% | 6.7% |
| Computed crack width (mm) | 0.1 mm | 0.41 mm | 0.31 mm |

\*   Visible cracking (0.3~0.5 mm) in one side but cracking of the other side was hardly detected

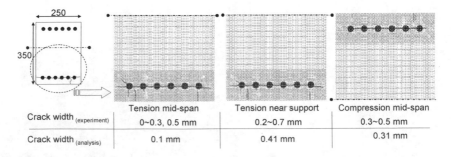

*Figure 9.12* Experimental and numerically created horizontal corrosive crack plane of RC beams.

in Table 9.3 with the assumed creep coefficient of 2.0 to take into account the long-term corrosion period. The computed crack width fairly agrees with the experimental observation.

Finally, verification of the analytical approach was attempted by using the experimental data of Matsuo *et al.* (2002) on corroded RC member experiments. The three RC beams with sectional dimension of 20 × 40 cm, with two 22-mm reinforcing bars, were used. The length of all beams was 3.0 m. The beams were corroded with the chemo-electrical process of galvanostatic by using electric current. Different corrosion periods of 18, 36 and 72 hours were used for the three beams. The location of corrosive cracking was recorded for all beams and is shown in Figure 9.13. Most of the corroded beams show corrosive cracking running along two longitudinal bars on the bottom faces. The information of mass loss of the corroded bars was also reported along with the maximum crack width, as shown in Table 9.4. Analysis of the RC section was then carried out and forced to stop when the numerical surface crack width reached the experimentally observed one. After this, the numerical mass loss was estimated with consideration of the possible migration of corrosive product inside the corrosive crack spaces as listed in Table 9.4.

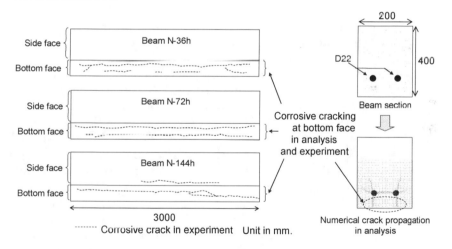

*Figure 9.13* Experimental and numerical corrosive cracking damage for corroded beams.

*Table 9.4* Experimental and numerical crack width and corroded mass loss

| Specimen No. | N-36h | N-72h | N-144h |
|---|---|---|---|
| Experimental max. crack width (mm) | 0.3 mm | 0.4 mm | 1.4 mm |
| Experimental mass loss (%) | 2.2% | 6.9% | 17.7% |
| Computed crack width (mm) | 0.3 mm | 0.4 mm | 1.4 mm |
| Computed max. mass loss (%) | 6.2% | 8.7% | 30% |

In the experiment, maximum crack width was reported over the domain of specimens because of the scatter of crack widths place by place, as shown in Figure 9.13. However, in the analysis of RC beam sections with unit breadth, non-uniformity of crack width along the beam span was not considered. Thus, strictly speaking, experimental and analytical values cannot be directly compared, but the analytical crack width is thought to be a mean value over the longitudinal span under the assumption that no leakage of corrosion product outside occurs. Thus, it would be reasonable to have relatively larger values of computed mass loss corresponding to the mean crack widths of analysis. In fact, this length of reinforcement (3,000 mm) is larger than the other cases discussed in the previous sections (150–380 mm for the data in Figure 9.7, 2,000 mm in Figure 9.8 and 200 mm in Figure 9.10). It seems that in the cases of N-36h and N-144h, uniformity of surface crack width over the bottom surface is comparatively less than in the case of N-72h. The analytically obtained mass loss is closer to reality for apparently uniform crack width development of N-72h.

## 9.3 Multi-mechanics of corrosive substances and structural concrete (meso-level)

Over the life-cycle of any RC structure, environmental actions are more or less unavoidable. Although alkalinity of concrete can effectively prevent reinforcement corrosion, it has been recognized as one of the primary sources of durability problems. Questions on safety and remaining serviceability of such corroded reinforced concrete are raised as an engineering issue of critical importance (Okada *et al.* 1988; Takewaka and Matsumoto 1984; Enright and Frangopol 1998; Li 2003). Here, in order to quantitatively assess the service life, ambient conditions causing steel corrosion need to be evaluated with a reliable approach in both material and structural stages.

At present, non-linear mechanics-based simulation is successful for initially non-damaged RC structures in which inherent defects are not necessarily considered in general. In order to effectively utilize a computational approach to performance assessment, possible defects or damages should be properly included in the analysis as initial conditions. Possible damages are attributed to loss of cross-sectional areas of reinforcing bars and volume expansion of corroded substances (Lee *et al.* 1998; Yoon *et al.* 2000; Coronelli and Gambarova 2004). Once the corrosion starts, the risk of splitting cracks that directly affect the macroscopic bond is also raised. Without considering these possible damages, reliable evaluation of corroded RC structures will not be achieved. In the use of simulation taking into account inherent damages, structural performance assessment throughout the life of an RC structure and appropriate maintenance planning can be achieved, as schematically shown in Figure 9.14, where damage brought on by ambient actions is an input to the non-linear behavioral simulation of structural concrete (Okamura and Maekawa 1991; Maekawa *et al.* 2003). The damage information can be

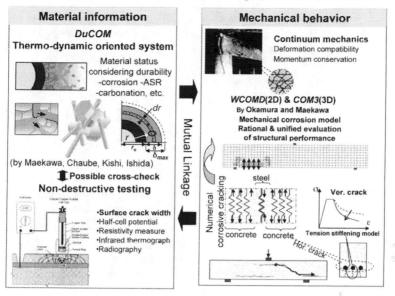

*Figure 9.14* Rational structural performance assessment of corroded RC.

obtained either by means of simulations based on thermodynamic multi-chemo-physics (Maekawa *et al.* 1999, 2003a) or non-destructive testing and inspection of an existing structure.

What is required here is an analytical approach that can take into account the mechanical damages caused by corrosion cracks (Yoon *et al.* 2000; Coronelli and Gambarova 2004). Here, a multi-mechanical model to deal with formed corrosion substances and cracking is proposed so that corrosion cracking can successfully be treated in a unified manner by integrating it into non-linear finite element analysis. As the crack-to-crack interaction becomes significant, the authors focus mainly on this aspect and employ multi-directional, non-orthogonal fixed-crack modeling (Okamura and Maekawa 1991; Maekawa *et al.* 2003b), because self-equilibrated stress cracking caused by corrosion and/or volume change of concrete does not necessarily intersect normally with cracks induced by external forces, unlike the case of seismic dynamic analysis under one-way reversed cyclic loads.

In order to consider cracking caused by an expansive corrosive product or silica-gel from an alkali-silicate reaction, an idealized material model having the mechanical properties of formed substances is placed inside an integrated RC element to reproduce the multi-mechanics of steel, corroded substances and concrete. Due to the group effect of reinforcement, the resultant force induced by expansive material around steel bars is assumed to be equivalent to the one-dimensional stress field for critical corrosion cracks, which penetrate the whole sections. Consequently, structural safety related to corrosion cracking can be computed under external forces.

The authors assume a two-phase problem composed of corrosion product with non-corroded steel as one integrated phase and surrounding concrete as another (see Figure 9.14 and Figure 9.15). When corrosion takes place, the expansion of rust is constrained by the surrounding concrete. Accordingly, compressive stress is introduced to the corroded reinforcement system, while tensile stress is induced to the surrounding concrete (Lundgren 2002). Internal equilibrium and deformational compatibility may lead to pre- and post-cracking deformation and stresses. In a pre-crack condition, induced tensile stress to concrete may cause cracking if its value exceeds the tensile strength. In a post-cracking state of corrosion, stress transfer across corrosion crack planes may be derived from this multi-mechanical modeling. Figure 9.15 shows a schematic diagram for the expanded geometry of corroded steel and concrete. Here, the averaged free shrinkage strain denoted by $\varepsilon_{s,\,free}$ and the equivalent stiffness $E_{s,eq}$ can be computed as formulated in Section 9.1.

Let us first consider the post-cracking state of corrosion in a finite element including reinforcement. In fact, the post-cracking model is indispensable to verify structural performance under loads as well as prediction of crack initiation. The authors direct their attention primarily to the post-crack and multi-mechanics aspects rather than to crack initiation. The crack plane is assumed in the transverse direction of the Y-axis, as shown in Figure 9.14. Here, the normal stress is transferred just through the multi-material system of corroded and non-corroded steel, and shear transfer is carried chiefly along the concrete crack by interlocking. Under the condition of crack separation, incremental normal strain of the control volume can be described as the sum of the concrete and multi-mechanical steel. The stress developing in the corroded steel is nearly inversely proportional to the reinforcement ratio, and

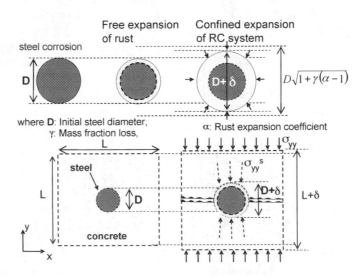

*Figure 9.15* Schematic representations for expansion of corroded steel in RC.

the strain field of the multi-mechanical system of corroded steel is a part of the total strain expressed as

$$d\varepsilon_{yy} = (1-\psi)\varepsilon_{c,yy} + \psi\varepsilon_s \cong (1-\psi)\frac{d\sigma_{yy}}{\psi E_c} + \psi\frac{d\sigma_{yy}}{\psi E_{s,eq}} \qquad (9.10)$$

where $\psi$ denotes the ratio of local corroded steel strain over the mean strain in the y-direction. When reinforcement is assumed to be uniformly suspended in the control volume or a finite element as shown in Figure 9.14, we have

$$\psi \cong p^{1/2} \qquad (9.11)$$

where $p$ is defined as the reinforcement ratio by the control volume or a finite element. Then, the element-based stiffness of coupled cracked concrete and multi-mechanical corroded steel yields

$$d\sigma_{yy} = \left(1-\psi+\psi\frac{E_c}{E_{s,eq}}\right)^{-1}\psi E_c d\varepsilon_{yy} \cong \psi E_c d\varepsilon_{yy} \qquad (9.12)$$

Here, let us consider the free expansion as the control volume after cracking. The transverse stress is equal to zero, the continuum strain of cracked concrete can be assumed to be nil, and the corroded steel strain coincides with $\varepsilon_{s,free}$. In considering the volume ratio of corroded steel over the control volume or the finite element, we have $p^{1/2}\varepsilon_{s,free}$ as the normal expansion free strain of the multi-mechanical system, which consists of cracked concrete, corroded product layers and the non-corroded mother steel. In considering the stress-free strain of the element and the derivative of total stress–strain expressed by Equation 9.6, we have

$$\sigma_{yy}{}^{s} = \psi E_c \cdot \left(\varepsilon_{yy} - \psi \cdot \varepsilon_{s,free}\right) \cdots \quad 0 \le \varepsilon_{yy} < \psi \cdot \varepsilon_{s,free}$$

$$\sigma_{yy}{}^{s} = -\varepsilon_{s,free} \cdot \psi^2 \cdot E_c \quad \cdots\cdots \quad \varepsilon_{yy} < 0 \qquad (9.13)$$

$$\sigma_{yy}{}^{s} = 0 \quad \cdots\cdots\cdots\cdots \quad \varepsilon_{yy} > \psi \cdot \varepsilon_{s,free}$$

where the total transverse stress becomes null when the total strain exceeds $\psi\varepsilon_{s,free}$. When the transverse strain turns negative, cracks will close and start to transfer compressive normal stress. In this case, cracked concrete primarily bears the stress transfer, and this stress can be computed based on the smeared-crack constitutive modeling. Then, the additional stress transferred by the corroded steel in Equation 9.13 can be overlaid on the original multi-directional, fixed-crack approach.

As far as shear transfer across a crack is concerned, corroded steel may play a negligible role compared with that of concrete rough-crack interlock. Normal stress transfer in the longitudinal direction is carried by the reinforcement and can be modeled as previously done. Thus, the transverse component of reinforcement stress is added to the conventional framework of multi-directional, fixed-crack modeling of RC elements (Maekawa *et al.* 2003b). This constitutes the newly proposed approach in the spatial averaging scheme of smeared cracks. In this approach, creep of the concrete continuum and the non-uniformity of concrete stress field, which are important for the pre-crack state, need not explicitly be taken into account, since crack kinematics may absorb creep deformation, thus making continuum deformation of concrete insignificant in relation to the overall deformation. This modeling will be verified in the post-cracked RC structural behaviors as discussed in the following chapter.

In this chapter, the multi-mechanical model for the corroded steel system is integrated into non-linear finite element analysis (Okamura and Maekawa 1991; Maekawa *et al.* 2003b). Verification of multi-mechanics is aimed for through the analysis of RC members subjected to combined corrosion and external loads. Numerical simulation of corroded RC members is conducted for flexural and shear modes of failure.

When reinforcement corrosion takes place inside a member, a loss of effective reinforcement area is observed. When corrosion advances further, the tensile stress inside the surrounding concrete is raised due to self-equilibrated force to restrain the expansion of corrosive product. After a critical stress, cracking occurs along main reinforcing bars and macroscopic bond deterioration is initiated since the layer of steel and the concrete beam web are mechanically divided. Here, the loss of corroded steel sectional area is simply represented as a reduced reinforcement ratio in numerical analysis. The bond loss (Auyeung *et al.* 2000; Coronelli 2002; Stanish *et al.* 1999) associated with longitudinal cracking along the main reinforcement is represented by a degradation in tension stiffening, which represents the ability to carry a certain amount of tension transferred from the reinforcement to concrete between cracks.

In this chapter, macroscopic bond deterioration is mechanically considered through the section-penetrating corrosion cracking as previously stated, and the microscopic bond deterioration around the main reinforcement is represented by an altered tension stiffening parameter (Okamura and Maekawa 1991; Maekawa *et al.* 2003b, 2007). Finally, the effect of cracking damage on the resultant member capacity is considered by employing the multi-mechanical model introduced in the non-linear analysis as

$$\sigma_{ij} = \sigma_{ij}{}^{c}(\varepsilon_{kl}) + \sigma_{ij}{}^{s}(\varepsilon_{kl}),$$

$$\sigma_{ij}{}^{s}(\varepsilon_{kl}) = \left\{ \begin{array}{c} \sigma_{xx}{}^{s}(\varepsilon_{xx}) \\ \sigma_{yy}{}^{s}(\varepsilon_{yy}, \gamma) \\ \sigma_{xy}{}^{s} = 0 \end{array} \right\} \tag{9.14}$$

where $\sigma_{ij}{}^{c}$ is the concrete stress with cracking and is formulated by multi-directional fixed-crack modeling (Okamura and Maekawa 1991, Maekawa *et al.* 2003b), and $\sigma_{ij}{}^s$ is the mean stress component carried by the reinforcement steel. When the longitudinal direction is specified as the x-axis, $\sigma_{xx}{}^s$ denotes the longitudinal steel stress and is computed in terms of the mean normal strain of the element along the reinforcement. Stress component $\sigma_{yy}{}^s$ is the multi-mechanical transverse stress in terms of corrosion degree, and the transverse strain as given by Equation 9.13 and illustrated in Figure 9.16.

Figure 9.17 summarizes the analysis approach for structural performance evaluation of corroded members. Here, careful consideration should be given to the corrosion crack plane. A longitudinal crack plane over the thickness of members is crucial, since crack-to-crack interaction is mobilized in the structural analysis.

On the other hand, a vertical splitting crack along reinforcement has less influence on crack-to-crack interaction, but the free elongation of reinforcement is hardly restricted due to the local bond deterioration caused by corrosion. Thus, bond deterioration due to splitting corrosion cracks is considered in terms of tension stiffness (Okamura and Maekawa 1991, Maekawa *et al.* 2003b, 2007).

It is known that the tension stiffness characteristic or bond performance depends on the magnitude of corrosion (Auyeung *et al.* 2000; Coronelli 2002; Stanish *et al.* 1999). In fact, local bond is enhanced under light corrosion due to corrosion-induced pressure, but it undoubtedly becomes degraded after bond-splitting cracking. As this section focuses on the ultimate limit state that accompanies bond cracking, tension-softening characteristics corresponding to plain concrete are solely applied to the bond deteriorated

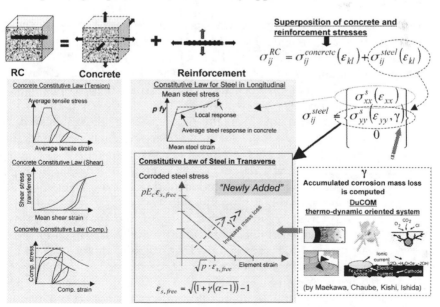

*Figure 9.16* Assumed non-linear stress–strain relationship of corrosion product.

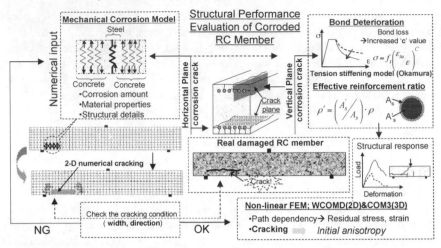

*Figure 9.17* Analysis approach for structural performance evaluation of corroded RC.

finite elements. This means that the tensile stress transfer by bond is simply ignored. This assumption will be evaluated in the following sections of experimental verification.

## 9.4  Structural performance of damaged RC members: corrosion cracking

The focus here is on RC beams, which are designed to have greater flexural capacity than shear strength. Thus the shear resistance mechanism is significant and subjected to longitudinal cracking by corrosion. Indeed, it was confirmed by Pimanmas and Maekawa (2001a, 2001d) that there exists a strong interaction of non-orthogonal cracks in the shear behaviors of RC members. The shear capacity of corroded RC beams with longitudinal deep cracking in space was also experimentally investigated and the significance of the location of these partial local damages was clarified (Toongoenthong and Maekawa 2004a). The effect of local longitudinal cracking on shear capacity of RC damaged members can be adverse, insignificant or even favorable, depending upon the specific damage location in space and member structural detail (Toongoenthong and Maekawa 2004a).

Consequently, non-linear analyses that can take into account corrosion cracking may prove more useful. Following are the authors' analyses of damaged RC beams with partial corrosion and artificial defects for magnifying the mechanical effect of corrosion cracks. In order to verify the applicability of the developed non-linear FEM, chiefly with regard to crack-to-crack interaction, RC beams with different patterns of local cracking and defects are analyzed by employing the multi-mechanical corrosion model.

### 9.4.1 Corroded RC beams in shear

#### 9.4.1.1 Analysis with local damage induced inside the shear span

An experiment on RC beams with longitudinal cracking limited to a small area was conducted (Toongoenthong and Maekawa 2004a), i.e., compression fiber-damaged beam (Case 1), tension fiber-damaged beam (Case 2), corroded cracking confined to a beam support inside a shear span (Case 3) and a combination thereof (Case 4). Figure 9.18 shows the specimen dimensions and reinforcement arrangement together with the FEM mesh used in the analysis. The analytical result of the reference with no corrosion is shown in Figure 9.19. The computed shear capacity is 162 kN, which favorably conforms to the experimental value.

An RC beam with corrosion cracking around compression fibers (Case 1) is shown in Figure 9.20. The damaged beam with accelerated corrosion around the compression fiber shows evidence of a crack width of approximately 0.3–0.5 mm on one side and zero on the other, and the mass loss of

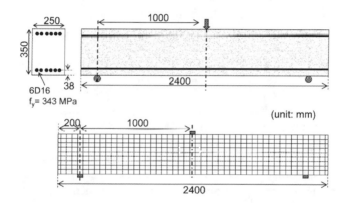

*Figure 9.18* Specimen dimensions, reinforcement arrangement and FEM mesh for the analysis.

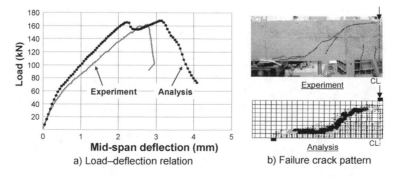

a) Load–deflection relation  b) Failure crack pattern

*Figure 9.19* Experiment and analysis results of referential non-damaged beam.

the corroded reinforcement of 1.89% as tabulated in Table 9.5. Reduction in shear capacity is approximately 8% compared to the non-damaged case. Non-linear analysis was performed with the remaining reinforcement ratio after corrosion. The local bond deterioration by corrosion cracks was considered by applying tension stiffening equivalent to plain concrete (Okamura and Maekawa 1991; Maekawa *et al.* 2003b). To take into account the pre-existing corrosion cracking, numerical cracking was introduced into the target body in the first step followed by subsequent shear loading. Experimentally obtained mass loss of the reinforcing bars was directly input into the multi-mechanical model and numerical corrosion cracks were computationally reproduced as shown in Figure 9.20. The computed corrosion crack width was 0.155 mm and the experimental crack width, taken as the average of both web sides, was 0.15–0.25 mm (see Table 9.5).

Figure 9.20 shows the load–deflection responses of the compression fiber-damaged and non-damaged reference beams. A small reduction in shear capacity and stiffness similar to the experimental one were computed. The crack pattern of the first step at the production of corrosion cracks is shown along with the failure crack pattern in Figure 9.20. The merging of diagonal shear and corrosion cracks around the compression fiber was observed in the analysis to be similar to that occurring in the experimental observation.

Next, the damaged beam with pre-cracking around tension fibers (Case 2) was analyzed. The experiment shows the crack width to be 0.2–0.7mm after accelerated corrosion. The percentage of mass loss of the corroded reinforcement was reported as 8.76% and the computed crack width was 0.714 mm, which is a little larger than in reality. A 13.6% decrease in shear capacity was reported compared with the non-damaged beam. The analysis procedure was the same as in the case of the compression-damaged one. Figure 9.21 shows the analytical result and the small reduction in shear capacity, which favorably conforms to the experimental facts. The induced crack pattern and failure mode are shown along with the experimental ones. The crack-to-crack interaction in these cases is small enough to influence the safety performance, although some geometrical merging of corrosion and diagonal shear cracks occurs.

*Table 9.5* Crack width and mass loss of steel from the accelerated corrosion method

| | Compression mid-span Case 1 | Tension mid-span Case 2 | Tension near support Case 3 |
|---|---|---|---|
| Crack width (mm) | 0*)~0.3~0.5 | 0.2~0.7 | 0.3~0.5 |
| Mass loss (%) | 1.89 | 8.76 | 6.7 |
| Computed crack width (mm) | 0.155 | 0.714 | 0.551 |

\* Visible cracking (0.3-0.5mm) in one side but cracking of the other side was hardly detected.

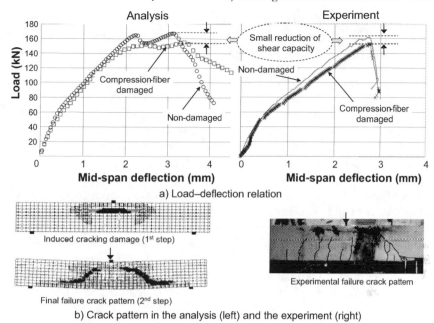

a) Load–deflection relation

b) Crack pattern in the analysis (left) and the experiment (right)

*Figure 9.20* Analysis and experimental results of compression damaged beam.

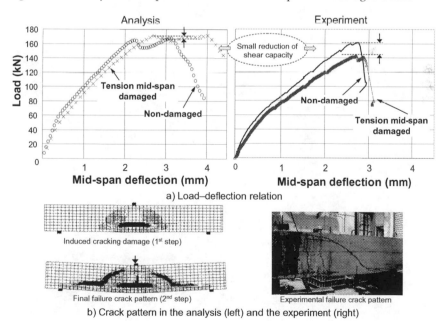

a) Load–deflection relation

b) Crack pattern in the analysis (left) and the experiment (right)

*Figure 9.21* Analysis and experimental results of tension-fiber damaged beam.

An RC beam with corrosion cracking near the beam support inside the shear span (Case 3) was targeted. The experimental shear capacity was 154.5 kN. The corrosion crack width was observed to be approximately 0.3–0.5 mm on both webs and the computed crack width was 0.551 mm, as shown in Table 9.5. The multi-mechanical model yields a slightly larger crack width through corrosion in simulation. This is attributable to the seepage of corrosive gel, which is not explicitly considered in modeling. The major corrosion cracks were found lying on the bottom face, as shown in Figure 9.22. This Figure also shows the computed load-deflection and crack patterns. The computed shear capacity is 148 kN, or a reduction of approximately 11.3% compared with the reference case. The merging of diagonal shear with existing corrosion cracks around the support can be observed in both analysis and experiment. Here, reduction in shear capacity of just 4.6% was seen in the experiment. As mentioned, the major actual corrosion cracks were found on the bottom face of this beam. As a sensitivity analysis, only the tension-softening equivalent to plain concrete and a reduced reinforcement ratio were assumed without cracking damage, as shown in Figure 9.22. The trial analysis result is closer to the experimental one, but no shear stiffness degradation was found until failure, unlike in reality. The experimental facts that were observed lie somewhere between these two extreme cases. This implies that the microscopic bond deterioration by corrosion is not absolutely critical.

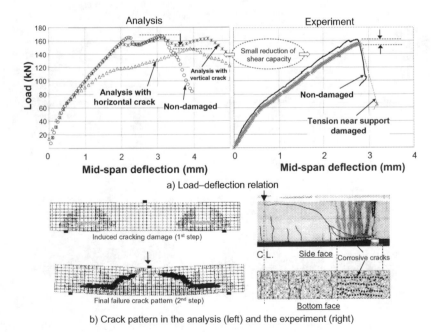

a) Load–deflection relation

b) Crack pattern in the analysis (left) and the experiment (right)

*Figure 9.22* Analysis and experimental results of tension-near-support damaged beam.

### 9.4.1.2 Analysis with local damage extended to anchorage zone

The experimental verification regarding crack-to-crack interaction was further carried out by use of artificial damage devices extended to the anchorage zone. The main aim was to ensure the ability of simulating anchorage performance by the multi-mechanical and multi-directional fixed-crack approach before the discussion in the following sections. Figure 9.23 shows the detail of the specimen, including an artificial crack-like defect (Pimanmas and Maekawa 2001b, 2001c) tested in the experiment. Instead of corrosive rough cracking, an artificial inherent crack-like defect produced by a flat paper plate was introduced during reinforcement fabrication. As this defect has a smooth surface with negligible shear transfer and a sharp interaction with diagonal shear cracking, the anchorage performance is thought to be much deteriorated.

The specimen has a shear span-to-depth ratio of 3.2 and a tensile reinforcement ratio of 1.29%. The damaged anchorage length at the beam end is 0.2 m and the compressive strength of the concrete is 34 MPa. Yield strength of reinforcement is 343 MPa. The experimental result shows a small shear capacity of 60 kN compared with the reference case of 150 kN accompanying the sliding along the defect outside the shear span. In the analysis, the pre-cracking damage was first introduced by using multi-mechanical modeling, as shown in Figure 9.24. Here, the shear transfer of this fictitious cracking is assumed to be zero to represent the flat plane geometry. Monotonic loading was subsequently applied at mid-span. Figure 9.24 shows the analytically obtained load-deflection relation and crack patterns compared to the experiment. The analysis gives the maximum resultant shear capacity of 59 kN for the beam with embedded damage, and the drastic reduction in member shear capacity matches reality. The crack pattern exhibits merging of the initial defect and external force-induced cracking, as the experiment did. The computed post-peak residual force is sustained to a greater extent than in reality. This difference may be attributed to the small deviation of the direction of diagonal crack propagation from the tip of the artificial defect, since a rather large finite element was applied even in the crack tip area. According to this verification, it can be said that the analysis is applicable to capacity estimation at least up to peak strength.

The short-term creep deflection over one day is computed under a sustained load (25 kN). This load level is about 20% of the maximum loading

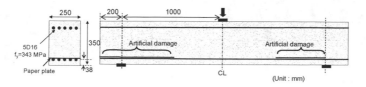

*Figure 9.23* Specimen detail in damaged beam with cracking until anchorage.

capacity for the sound structural member (normal serviceability) but it is more than 50% in the case of the pre-damaged one. Figure 9.24 also shows the time-dependent deflection at the center span together with short-term failure analysis. For the case without pre-damage, the apparent creep coefficient by deflection at 100 days is about 1.0, which is similar to the beams described in Section 8.7.3 and Section 8.7.4. In contrast, the pre-damaged RC beam exhibits much greater creep deflection. It can be said that the time-dependent properties of materials and structures should be considered in performance simulations of members damaged by steel corrosion, alkali aggregate reaction (ASR) and volumetric shrinkage. The overall behavior is captured by the multi-scale computational model (Chapter 8). The incremental deformation of the damaged member is primarily caused by non-linear tension creep, and its mode is not flexure but shear. It can be said that the proposed model may function for mechanistic simulation of flexure, shear and their combination. Further investigation is needed to enhance applicability quantitatively for practice.

### 9.4.1.3 Analysis with local damage in a whole shear span

Here, discussion and verification will be extended to defects over the whole shear span of beams. There are two key issues. One is the loss of macroscopic bond. The web concrete is mechanically detached from the main reinforcement layer by corrosion cracking over the thickness. Ikeda and Uji (1980) experimentally demonstrated that the shear capacity of RC beams is elevated when the bond between tension main reinforcement and concrete disappears due to retarded formation of diagonal shear crack and an arch

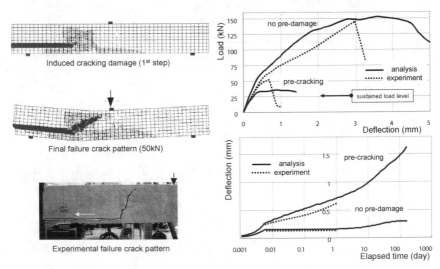

*Figure 9.24* Analysis result of beam with inherent crack-like defect until the anchorage zone.

action. Another issue is the increased risk of failure at the beam anchorage, which also leads to degradation in capacity. Thus, two possible factors co-exist with this problem where the multi-mechanical model is to be verified. Figure 9.25 shows the details of the specimen, reinforcement arrangement and location of longitudinal cracks in the experiment (Toongoenthong and Maekawa 2004a).

In this case, a premature failure around the anchorage zone was experi-mentally observed, unlike the cases previously discussed. The resultant shear capacity is 146 kN, which is slightly lower than the capacity of the non-damaged beam (167 kN) as shown in Figure 9.26. The anchorage capacity of the beam details might not be enough to sustain the arch action capacity. Non-linear finite element analysis was performed to verify this phenomenon with the multi-mechanical model in a similar way to the previous cases. The computed load-deflection response was compared with the experimental results, as shown in Figure 9.26. The analysis showed fair agreement with the experimental results. The premature failure around the anchorage zone was well captured in the analysis, too.

As previously stated for the longitudinal cracking in the whole shear span, the shear strength can be preserved when sufficient anchorage of longitudi-nal reinforcement is provided. In order to confirm this fact, significant web

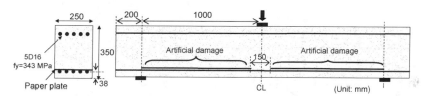

*Figure 9.25* Specimen dimension and reinforcement arrangement with whole span damage.

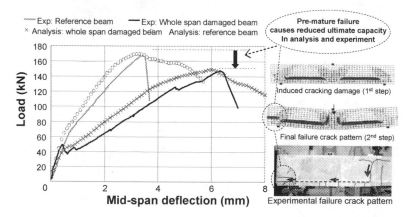

*Figure 9.26* Analysis result of the whole span damaged beam with insufficient anchorage.

reinforcement of 3% was numerically placed to prevent premature failure of the anchorage zone. Similarly, induced pre-cracking damage in the whole shear span was performed and followed by monotonic loading until failure. Figure 9.27 shows the expected enhanced shear capacity when an arch action effectively develops with adequate anchorage capacity. The computed crack pattern is also shown in Figure 9.27. In this pattern, no premature failure is seen at the beam end. Here, it is obvious that anchorage performance is crucial, and special attention is required when a complete shear span is subjected to cracking where higher anchorage capacity is sustained, to satisfactorily form an arch action.

Verification was conducted on the RC beam with longitudinal corrosion cracks over the whole shear span, as reported by Satoh *et al.* (2003). Figure 9.28 shows the specimen details and reinforcement arrangement. These RC specimens are composed of three beams. One is the reference, non-damaged beam. The other two were subjected to corrosion by submerging each specimen into a sodium pond for accelerated corrosion, resulting in uniformly distributed reinforcement corrosion along the entirety of the longitudinal steel. The two corroded beams were reported to have different levels of corrosion along the main reinforcement. The lightly corroded beam had a mass loss of 2.1% and the heavily corroded beam had a mass loss of 3.3%.

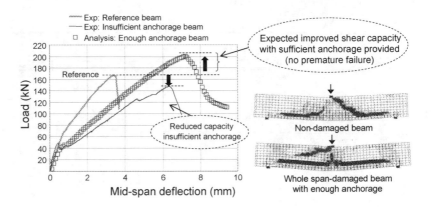

*Figure 9.27* Analysis result of the whole span damaged beam with sufficient anchorage.

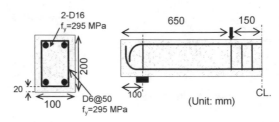

*Figure 9.28* Specimen detail and reinforcement arrangement by Satoh *et al.* (2003).

In this experiment, the main reinforcing bars were bent up 90 degrees inside the anchorage zone, as shown in Figure 9.28. Thus, a satisfactory anchorage capacity is expected despite the existence of corrosion.

Figure 9.29 shows the experimental load-displacement relation and crack pattern of all RC specimens. Both corroded beams show a favorable shear capacity compared with the non-damaged beam. As illustrated also in Figure 9.29, there are clear differences in failure crack patterns between the two corroded damaged beams and the non-damaged reference beam. Although the diagonal shear failure was observed in the non-damaged beam, no localized diagonal shear crack was formed in the case of the lightly corroded beam.

Similar to the previous cases of analysis, the target specimens were analyzed by introducing numerical pre-cracking at the location similar to that of the experiment in the first step of analysis. Then the monotonic load was applied until failure. Figure 9.30 and Figure 9.31 show the load-displacement relations and failure crack patterns obtained for the cases of the lightly and heavily corroded beams, respectively. It can be seen that the analytical results of both cases are in good agreement with the experimental facts in both the load-displacement curves and failure crack patterns. Furthermore, analysis of both corroded beams without pre-corrosive cracking was also performed

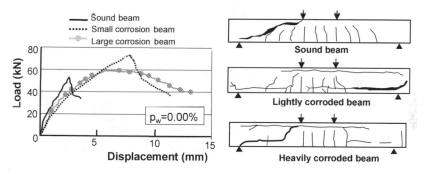

*Figure 9.29* Behaviors of corroded beams by Satoh *et al.* (2003).

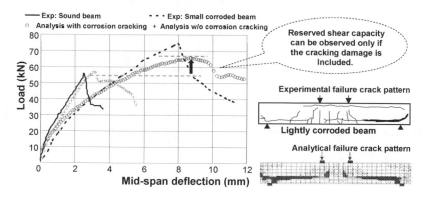

*Figure 9.30* Simulated shear capacity and cracking of small corrosion beam.

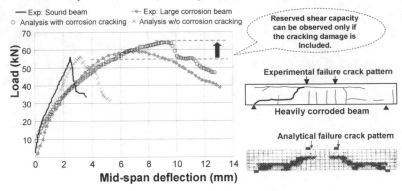

*Figure 9.31* Simulated shear capacity and cracking of large corrosion beam.

to ensure the significance of the analytical approach by considering only the reduced reinforcement ratio and tension stiffening. As illustrated in Figure 9.30 and Figure 9.31, the analysis result without considering pre-cracking is clearly far off the experimental fact.

### 9.4.2 Corroded RC beams in flexure

It is numerically confirmed that longitudinal corrosion cracking can be insignificant, adverse or even favorable on resultant shear capacity and that it depends upon the damage location and structural details. It is also known that flexural capacity is closely associated with the remaining sectional areas of steel (Rodriguez *et al.* 1997; Lee *et al.* 1998; Mangat and Elgarf 1999). In other words, corrosion cracking has little influence on flexural capacity.

However, it should be noted that there is also an alternative view of ductility and post-peak behaviors. Past studies (Tachibana *et al.* 1990; Coronelli and Gambarova 2004) have shown that the transition of the failure mode from steel yielding to bond-shear appears in the beam subjected to corrosion. Figure 9.32 shows the specimen details and reinforcement arrangement. The concrete compressive strength is 35.6 MPa and the yield strength of reinforcement is 353 MPa with a tensile reinforcement ratio of 1.3%. The shear span-to-depth ratio is 4.2. The observed corrosion cracking is also shown in Figure 9.32. Corrosion cracks developed up to the bottom surface as well as the side face, with a maximum crack width of 0.75 mm (Tachibana *et al.* 1990).

The corrosion-induced pre-cracking was numerically introduced first, followed by conventional loading. Figure 9.33 shows the analysis result with different magnitudes of numerical cracking damage. The transition of the failure mode from flexure to bond-shear has also been successfully captured, as shown in Figure 9.33, with crack patterns and the post-peak degradation in load-displacement relation, which differs from the typical ductile behavior of the non-damaged beam.

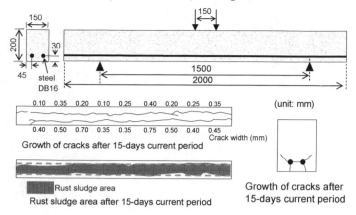

*Figure 9.32* Experimentally observed corrosion cracking by Tachibana *et al.* (1990).

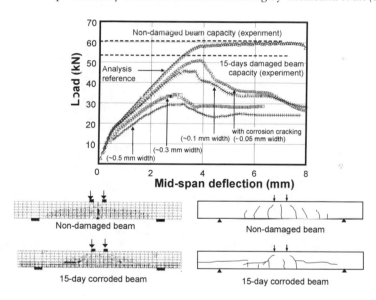

*Figure 9.33* Numerical simulation and experimental shear failure with corrosion-induced pre-cracking by multi-mechanical model (standard analysis).

In order to refocus on the importance of pre-cracks penetrating the whole section, pre-corrosive cracking was intentionally excluded in the sensitivity analysis. Here, reduced reinforcement ratio is defined according to the lost sectional area, and bond deterioration is represented by reduced tension stiffness equivalent to plain concrete (no local bond assumed) in the analysis. Figure 9.34 shows the analytical result and the failure crack pattern. The flexural failure mode accompanying the yield of the main reinforcement was obtained without considering pre-corrosive cracking. The failure crack patterns in the fictitious analysis and experiment do not match. Thus, it is

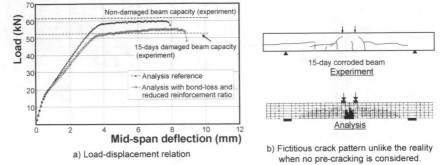

a) Load-displacement relation

b) Fictitious crack pattern unlike the reality when no pre-cracking is considered.

*Figure 9.34* Sensitivity analysis of 15-day corroded beam without pre-cracking.

clearly seen that the consideration of corrosion cracking is substantial in the structural performance assessment of corroded RC members and that the macroscopic bond deterioration associated with corrosive penetrating cracks dominates the overall feature of ultimate limit states.

### 9.4.3  Seismic performance of corroded RC box culvert

It is not easy to inspect underground culvert structures because of accessibility issues in general. Furthermore, strengthening or repair is also challenging if necessary, because internal spaces are frequently occupied by water, heavy traffic and so on. Thus, structural performance with corrosion damage has to be rigorously examined for necessary and sufficient maintenance. Figure 9.35 shows a part of the experimental program conducted by the Central Research Institute of the Electric Power Industry (Matsuo *et al.* 2002) and the Japan Society of Civil Engineers Committee (Matsuo *et al.* 2007). An electric charge was applied to produce corrosion and associated cracking along the main reinforcement. Afterwards, reversed cyclic forces were applied to the damaged structures in consideration of the deformational mode of underground culverts in an earthquake.

As the structure is statically indeterminate, local damage and corresponding stiffness degradation cause stress redistribution from the damaged area to more sound portions of comparatively higher stiffness. Due to this higher redundancy, the ultimate strength of damaged culverts is not so affected, but ductility is degraded under cyclic actions as shown in Figure 9.35. Figure 9.36 shows the computational simulation of damaged RC culverts. The corrosion mass loss can be assumed so that the computed crack width matches reality and, subsequently, the seismic ductility needs to be examined for a risk assessment of corrosion damage. The failure mode and the restoring force characteristics can be fairly reproduced.

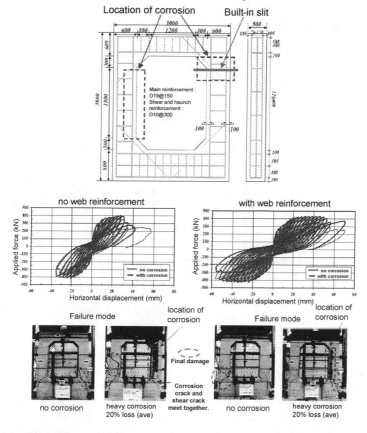

*Figure 9.35* Corroded RC box culverts and corresponding seismic ductility and strength.

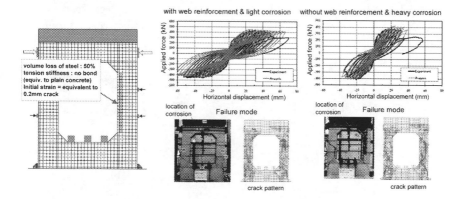

*Figure 9.36* Multi-scale simulation of corroded RC culverts.

## 9.5 Structural performance of damaged RC members: anchorage of stirrups

Practitioners in charge of infrastructure maintenance and management are currently facing serious questions on the remaining functionality of reinforced concrete damaged by weathering. A computational approach to assessing RC structural safety performance over service life has been proposed (Maekawa *et al.* 1999; Maekawa *et al.* 2003a), and pre-existing cracks in structural concrete have been discussed from the perspective of non-linear mechanics. Inevitably, real RC structures are subjected to multiple forms of environmental risks, such as corrosion and alkali aggregate reaction of concrete, to a greater or lesser degree, and the resulting damage in reinforcement as well as concrete needs to be taken into consideration.

It is well known that chloride ingress and/or carbonation raise the risk of reinforcement corrosion, which has much to do with the alkalinity of the original concrete cover. Steel corrosion results in the loss of reinforcement sections and ultimately in the rupture of steel, especially at the bent portions. Moreover, in the case of the alkali aggregate reaction, the volumetric expansion of gel products around active aggregates may induce cracking in structural concrete as well as excessive extension of reinforcement. Such volumetric expansion is prevented or retarded through confinement of concrete by steel reinforcement.

It was reported for some bridge piers in Japan that the confinement of volumetric expansion caused by the alkali aggregate reaction may lead to complete rupture at bent portions of stirrup steels and main longitudinal reinforcement at the anchorage zone and welding points (Miyagawa 2003), as shown in Figure 9.37. This failure raises questions about the safety of such a damaged RC structure. The observed declines in anchorage capacity and their performance against external shear have been deemed not to satisfy the required structural performance and are being investigated urgently to obtain a reliable safety assessment method and repair in line with maintenance planning.

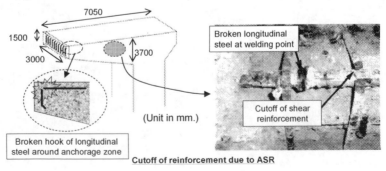

*Figure 9.37* Broken web reinforcement found in the real RC structures (Miyagawa 2003).

In this section, the authors investigate the influence of broken web reinforcement on overall member performances by conducting an experimental program on defective stirrups for RC beams under static loads. A numerical analysis method is also applied for safety performance assessment of steel-fractured RC members in shear and is experimentally verified for accuracy and applicability. As non-orthogonal cracking at intersections in concrete may occur under combined mechanical forces and environmental actions, multi-directional non-orthogonal fixed-crack modeling for concrete is applied. Here, non-damaged sound concrete is experimentally reproduced in order to focus on the effect of premature anchorage at the extreme ends of reinforcing bars. Thus, the coupling of broken reinforcement with damaged concrete will be subsequently investigated in future.

### 9.5.1 Experimental facts

#### 9.5.1.1 Specimen and materials

For evaluating the effect of broken stirrups in RC beams on their shear capacity and ductility (see Figure 9.37), an experimental program of RC beams with shear reinforcement, whose anchorage and bond were removed, was conducted. Figure 9.38 shows the specimen details and reinforcement arrangement. The shear span-to-depth ratio denoted by a/d was 3.2, with a tension reinforcement ratio of 1.46% and the anchorage length of the main tensile reinforcement was 0.2 m. The shear reinforcement ratio was 0.226%, with a spacing of 0.1 m as shown in Figure 9.38. The concrete cover of the stirrup was 22 mm. The tested compressive strength of the concrete after standard curing of 28 days was 35 MPa. The high-yield strength of both the tension and compression reinforcement was 645 MPa for producing a comparatively large flexural capacity to ensure that shear failure would occur. The yield strength of the shear reinforcement was 345 MPa. The loading was

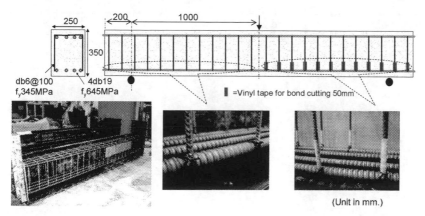

*Figure 9.38* Specimen set-up, dimension and reinforcement arrangement.

applied with combined flexure–shear forces. The loading mode was static at the middle of the beams. The rate of displacement at the loading point was controlled to be 0.2 mm/minute in all cases. The width and thickness of loading and support steel plates was 100 mm and 10 mm, respectively.

This set-up was designed as an analogy of corroded broken web reinforcement or ruptured stirrups at bent portions caused by the alkali aggregate reaction. The legs of shear stirrups near tensile reinforcement were unprocessed without curved hook fabrication before concrete placement. Premature anchorage of web steel is thought to cause degradation of reinforcement capability, and the loss of bond close to the tip of the cut-off steel may occur as a result. In the case of a smaller shear span-to-depth ratio, insufficient anchorage of stirrups would not change the load-carrying mechanism of the concrete to a considerable degree, because the so-called tied-arch mechanism is essentially formed regardless of the amount of web reinforcement. Thus, the authors did not use deep beams in the experimental program but selected instead a slender shape of a/d = 3.2 with a comparatively larger amount of stirrups.

The difference in surface condition at the legs of stirrups was additionally addressed for each side of the beam by wrapping the extreme ends of stirrup-legs with a 5-cm-long strip of vinyl tape (see Figure 9.38). The aim of the vinyl tape wrapping was to create smooth surfaces to eliminate bond close to the edge positions. The moment capacity of the non-damaged referential beam was computed by employing the material properties and detailed geometry of the cross-section. The shear capacity was predicted based on the Japan Society of Civil Engineers (JSCE) standard specification (Japan Society of Civil Engineers 2002; Okamura and Higai 1980). The sectional capacity of the non-damaged referential beam is shown in Table 9.6. A non-damaged

*Table 9.6* Calculated bending and shear capacity

| Properties | Values |
|---|---|
| Shear span: effective depth (a/d ratio) | 3.2 |
| Shear capacity of concrete (Vc) Okamura–Higai equation | 90.7 kN |
| Shear capacity of stirrups (Vs) | 60.5 kN |
| Yielding moment capacity (My) | 202.9 kN-m |
| Bending failure load (Py) | 405.8 kN |
| Shear failure load (Pv) | 302.4 kN |

Note: $V_c$ by JSCE 2002 (Okamura and Higai)
$V_c = \beta_d \cdot \beta_p \cdot f_{vcd} \cdot b_w \cdot d$
$\beta_d = \sqrt[4]{1/d} \leq 1.5$ (d: m)
$\beta_p = \sqrt[3]{100} \leq 1.5$
$f_{vcd} = 0.2\sqrt[3]{f_c'} \leq 0.72$ (N/mm²)
d: effective depth, p: longitudinal steel ratio
$f_c'$: design concrete compressive strength
$b_w$: web width

referential beam was also produced with standardized hooks of stirrups according to the provisional structural details in design codes.

It should be noted that the RC beam specimen was fabricated by providing 0.2 m anchorage length without a hook of main reinforcement around the anchorage zone, which obviously violated the general specified codes. This condition of anchorage was intended to represent the damage of the bar hook at the member's ends (leaving the risk of insufficient development length of anchorage main steel), as found in the ASR-damaged structure seen in Figure 9.37. In fact, the problem of fractured shear reinforcements and their effect on shear capacity is the main point of discussion. Hence, although an anchorage design code was not satisfied, the existing anchorage length proved to be sufficient to avoid anchorage failure in the referential non-damaged beam, as shown experimentally and numerically in later sections. The problem of the critical damage condition around the anchorage zone due to ASR or reinforcement corrosion is investigated and discussed in Section 9.4 and Section 9.6.

### 9.5.1.2 Loading test results

Figure 9.39a shows the load-deflection relation at the mid-span of the non-damaged beam. The load-deflection relationship indicates the change in stiffness approximately at the load level of 200 kN, which is close to the computed concrete shear capacity without web reinforcement. In reality, diagonal shear cracking was initiated in the web. After the first diagonal shear crack, somewhat dispersed diagonal cracks were formed owing to spatially

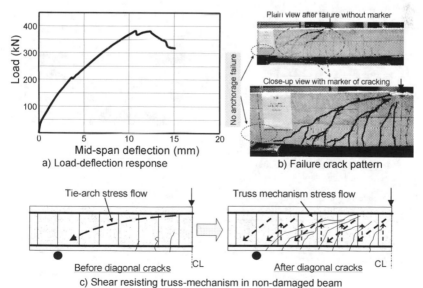

*Figure 9.39* Experimental result of non-damaged beam: reference case.

dispersed shear reinforcement. The final failure stage was then reached with the formation of an unstable shear failure path between the loading point and the beam supports. This resulted in a permanent decrease in the load-carrying mechanism.

The shear capacity at the peak was approximately 380 kN, which coincides fairly well with the shear strength computed by the JSCE standard specification (Japan Society of Civil Engineers 2002; Okamura and Higai 1980), taking into consideration that the computed shear capacity corresponds to the yield limit state of web steel. It is important to note that the shear force carried by concrete is approximately 53% of the total shear capacity. This means that the contribution of web reinforcement is not less than in the case of large-scale bridge infrastructures currently in service. The experimental crack pattern after failure is also shown in Figure 9.39b, which indicates dispersed diagonal shear cracks. When the beam experienced its peak capacity, the main reinforcement remained elastic. A decline in residual force was evidenced with a localized band of shear deformation in the web concrete after the peak.

After the reference case was reviewed, the RC beam with built-in damage in shear reinforcement was loaded. As mentioned earlier, this damaged beam was intentionally given imperfections by way of a straight cut-off of the stirrup legs close to the tensile reinforcement, as shown in Figure 9.38. Since the unprocessed edge of the shear reinforcement was made before concrete casting, the local interfacial condition between the stirrups and surrounding concrete was expected to be sufficiently maintained even for the broken stirrups. Figure 9.40a shows the experimental load-deflection curve

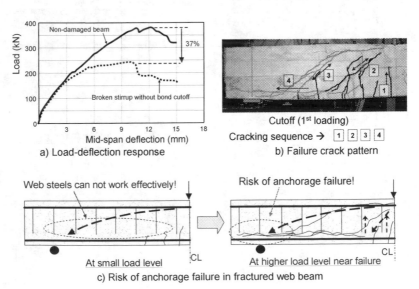

a) Load-deflection response

b) Failure crack pattern

c) Risk of anchorage failure in fractured web beam

*Figure 9.40* Experimental result of beam with broken stirrup without bond cut-off.

of the damaged beam exhibiting shear failure in the shear span containing the broken web reinforcement. A significant reduction in shear capacity can clearly be seen. The damaged beam containing cut-off shear reinforcement reached the ultimate capacity of 240 kN, which is a decrease in shear of approximately 37% compared with the non-damaged reference.

The different crack pattern of the insufficiently anchored stirrup was also closely observed at failure (see Figure 9.40b). Starting with a set of flexural cracks denoted by No. 1 in the Figure, we can observe propagation and varying crack angles at the higher loads that form the diagonal shear cracks around the web (crack set indicated by No. 2 in Figure 9.40b). Without the straight cut-off of the stirrup legs, further development of distributed diagonal shear cracks could be expected for the non-damaged referential beam.

However, because of the discontinuity of the shear reinforcement legs adjacent to the longitudinal steels, sufficient anchorage did not develop, resulting in localized diagonal shear cracks (cracks indicated by No. 3 in Figure 9.40b). As soon as the localized diagonal shear cracks are formed, high stress concentration at the straight edge of the stirrups close to the tensile reinforcement causes horizontal cracks penetrating along the tensile reinforcement (longitudinal cracks indicated by No. 4 in Figure 9.40b). It seems that the web steel may not resist crack propagation around its straight edge. This crack finally reached the anchorage zone of the main tensile reinforcement ahead of the beam support. At this moment, the stirrups could not effectively carry the tensile force and peak capacity was reached. Here, the truss mechanism is thought to decline and is replaced with the tied-arch system, as shown in Figure 9.40c. As a result, the resultant higher internal compression can flow to the anchorage zone for main reinforcement. Gradual decay of the residual force occurred in the post-peak region, as shown in Figure 9.40a. Here, the deterioration in efficiency of the shear reinforcements was experimentally seen with no yielding of fractured stirrups.

After the failure of the damaged beam, which nevertheless maintained the local bond, strengthening was performed by encasing the failed shear span with ductile non-metallic sheets (Igarashi 2002), in order to subsequently apply the reloading on the other live side of the beam. Here, imperfection of the steel–concrete bondage near the edge of the stirrups may exist. As mentioned previously, a 5-cm-long portion of the shear reinforcement legs from the extreme edge of the cut-off point (close to the tensile reinforcement) was wrapped with vinyl tape during the reinforcement fabrication, so as to achieve a smooth surface representing the absence of bond, as shown in Figure 9.38. The load–deflection relation of the beam with the broken shear reinforcement and erased bond is shown in Figure 9.41a. The shear capacity in this case was approximately 237 kN, which is nearly the same as the previous case where the monolithic steel–concrete bondage was retained, and represents a reduction in shear capacity of approximately 37% compared with the non-damaged reference.

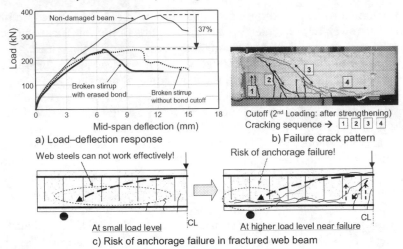

a) Load–deflection response

b) Failure crack pattern

c) Risk of anchorage failure in fractured web beam

*Figure 9.41* Experimental result of beam with broken stirrup with erased bond.

Nevertheless, a sharp drop in residual capacity after the peak was observed in this case of perfect unbonded stirrups at the extreme ends. This sudden drop in residual capacity is rooted in the complete loss of stress transfer between steel and concrete at the anchorage zone of web stirrups. Figure 9.41b also shows the failure crack pattern of the case where a partially unbonded stirrup was applied, which indicates a similar crack sequence and pattern as in the previous damaged beam. The less-dispersed diagonal shear cracks developed due to insufficient bond and anchorage. Longitudinal splitting cracks along the tensile reinforcement were similarly observed prior to and after peak capacity. They also invaded the anchorage zone of longitudinal reinforcement.

### 9.5.2 Bond deterioration zone

It was clarified and confirmed that shear reinforcement cannot work fully satisfactorily without hooks of web steel. In other words, full efficiency of the truss mechanism cannot be expected when there is insufficient anchorage of shear reinforcement. As observed in the experimental crack pattern of the previous section, the occurrence of localized longitudinal cracks along the tensile reinforcement was seen at the straight-edge location of the web reinforcement. In this section, the responses of RC members with fractured stirrups will be focused on for safety performance assessment, using a non-linear analysis. In order to investigate the failure process, a two-dimensional non-linear RC model with multi-directional fixed-crack modeling (Maekawa *et al.* 2003b) will be employed hereafter. Although the non-linearity of concrete compression softening and reinforcement post yielding can be considered in the model, their actual contribution to the ultimate capacity of the defective beam under experiment was very small in both analysis and experiment (no concrete compression softening and yielding of reinforcement).

Starting with the reference case of no damage, we have the analytical results, which match fairly well both with the experimental capacity and the cracking pattern. Figure 9.42a shows the load-deflection relation of the reference case with specified hooks. Distributed diagonal shear cracks can be simulated well, owing to the existence of web reinforcement as shown in Figure 9.42b. Here, the stirrup was numerically confirmed to have efficiency similar to the experimental result. The truss-like stress flow of structural concrete compression can be also validated over the shear span, as shown in Figure 9.42c.

Next, the damaged RC beam without hooks on stirrups was computationally investigated. Bond deterioration was assumed to take into account the insufficient anchorage capability. No axial tensile stiffness of steel was assumed in the area of bond deterioration length. The numerical sensitivity analysis by non-linear FEM was performed beforehand by varying the size of the bond deterioration zone around the cut-off location near tensile reinforcements, as shown in Figure 9.43b, to obtain the overview of the bond damage associated with the non-hooked stirrups. Consequently, it can be said from the numerical results that bond deterioration can be specified at the extreme end of steel bars of 10 times diameter (10D) by assuming null steel tensile stiffness.

This means that the web reinforcement does not function any more in this zone, even if cracking occurs within it. The schematic representation of the possible assumed bond deterioration zone is shown in Figure 9.43a, with expected variations in damage locations in reality. Figure 9.44a shows the

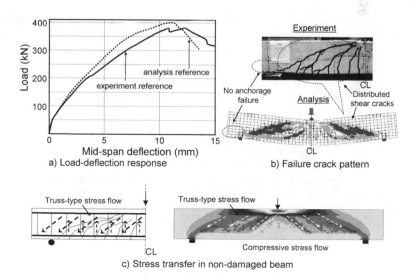

a) Load-deflection response

b) Failure crack pattern

c) Stress transfer in non-damaged beam

*Figure 9.42* Load displacement relation and crack pattern of non-damaged beam.

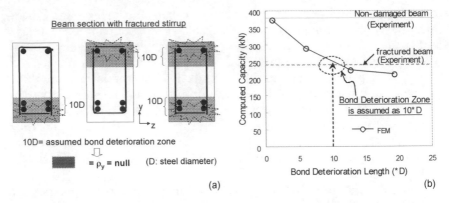

*Figure 9.43* Assumed bond deterioration zone for fractured web reinforcement.

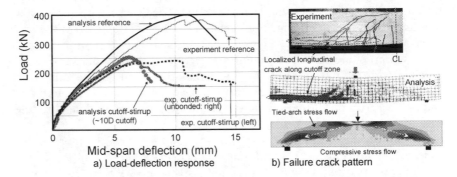

*Figure 9.44* Load–displacement relation and crack pattern for cut-off stirrup beam.

numerical results of the damaged beam with unhooked shear reinforcement by applying this rule of the 10D deterioration size of bond. It can be seen that a reasonable numerical result can be obtained based on this simple assumption. It is concluded from the analysis that the truss system by shear reinforcement is mechanically spoilt and that part of the functionality is lost due to the incomplete anchorage of stirrups. Here, a tied-arch action subsequently occurred due to the bond deterioration formed along the longitudinal reinforcement (Toongoenthong and Maekawa 2004c). This tied-arch stress flow of concrete can clearly be seen in Figure 9.44b. Accordingly, the overload of the anchorage zone for longitudinal reinforcement became increasingly significant and finally failure reached this anchorage beyond the beam support, as observed in both experiment and analysis (see Figure 9.44b).

The assumed bond deterioration zone of 10 times diameter is not merely an *a priori* hypothesis but is also rooted in the microscopic interaction mechanics of the concrete crack plane and intersecting reinforcement. To take into account discrete joint interface problems, such as the pull-out of anchored reinforcing bars from a footing, Mishima *et al.* (1992) enhanced Shima *et*

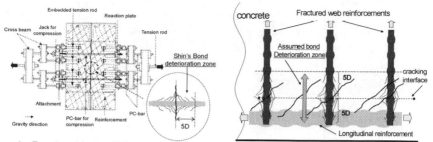

a)   Experimental setup   b) Assumed bond deterioration at bent portion of fractured web reinforcement

*Figure 9.45*  Assumed bond deterioration zone considering microscopic damage near crack interface.

*al.*'s original model (1987) by incorporating the bond deterioration zone with the local bond–slip–strain model. In this model of RC joint interface, bond deterioration was assumed at a distance of five times diameter (5D) from the concrete crack plane. The model was then experimentally verified by using RC plate specimens with a single crack plane, as shown in Figure 9.45a. Here, it should be noted that sound condition of surrounding concrete close to a crack interface was maintained.

However, in the case of fractured web reinforcement by ASR near the bent portions (equivalent to unhooked stirrups), the soundness of the concrete around the fractured steel may not always be guaranteed. Figure 9.45b shows the schematic image of the cracking damage near the ruptured stirrups. In consideration of this damage at the fractured stirrup leg, it can be said that the bond deterioration zone of 10 times diameter of web reinforcement is fairly reasonable from the microscopic viewpoint of bond deterioration.

### 9.5.3 Verification

In order to examine the applicability of the proposed modeling on bond deterioration of an unhooked web reinforcement, further analyses were carried out. Experimental results by Regan and Kennedy (2004) on shear strength of RC beams with defective stirrup anchorages were used for experimental verification. The experiment was composed of 13 RC beams arranged in four groups by varying reinforcement arrangement, effective depth and dimensions. Table 9.7 shows the material properties and details of defective stirrups (Regan and Kennedy 2004). Figures 9.46–9.49 show the specimen set-up and dimensions of all four groups. Defective stirrups were intentionally fabricated by straight reinforcement without hooks in both the top portion near the compressive bars and the bottom near the tensile bars, to simulate the broken web reinforcement. The exception was with Beam 3, where defective stirrups were an inverse U-shape with the straight edge of the main reinforcement.

A systematic analysis was conducted by applying the same approach

Table 9.7 Comparison of experimental and analysis result with assumed bond deterioration

| Group | Beam No. | d (mm) | Concrete f'c (MPa) | % Stirrup defective | Cover of main tensile bars | Experiment ultimate capacity (kN) | Analysis ultimate capacity (kN) | Capacity ratio (FEM/ Exp.) | Vc by JSCE code | Vs/Vc | Failure mode |
|---|---|---|---|---|---|---|---|---|---|---|---|
| A | 1 | 335 | 58.1 | - | Y | 314 | 320 | 1.02 | 78.1 | 0.89 | Bending |
| | 2 | | 56.6 | 75 | Y | 310 | 313 | 1.01 | 77.45 | 0.91 | Bending |
| | 3 | | 60.2 | 50 | Y | 314 | 315 | 1.00 | 79.05 | 0.89 | Bending |
| | 4 | | 52.8 | 50 | Y | 314 | 316 | 1.01 | 75.65 | 0.93 | Bending |
| B | 5 | 350 | 51.9 | - | Y | 250 | 265 | 1.06 | 77.45 | 0.46 | Shear |
| | 6 | | 43.8 | 75 | Y | 210 | 219 | 1.04 | 73.47 | 0.48 | Shear |
| C | 7 | 340 | 47.3 | - | Y | 414 | 424 | 1.02 | 81.94 | 0.97 | Shear |
| | 8 | | 53.3 | 67 | Y | 340 | 369 | 1.09 | 85.26 | 0.93 | Shear |
| | 9 | | 45 | 67 | N | 330 | 364 | 1.10 | 80.59 | 0.99 | Shear |
| | 10 | | 49.7 | 67 | N | 280 | 359 | 1.28 | 83.3 | 0.96 | Shear |
| D | 11 | 200 | 47.5 | - | Y | 252 | 258 | 1.02 | 54.75 | 0.91 | Shear |
| | 12 | | 47.5 | No stirrup | Y | 100 | 108 | 1.08 | 54.75 | 0.91 | Shear |
| | 13 | | 50.7 | 67 | N | 180 | 177 | 0.98 | 55.95 | 0.89 | Shear |

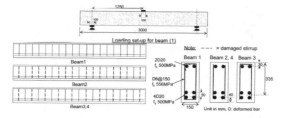

*Figure 9.46*  Specimen set-up and dimension for Group A: Beams (1), (2), (3) and (4).

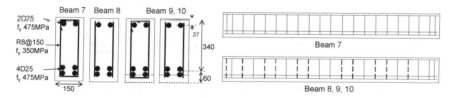

*Figure 9.47*  Specimen set-up and dimension for Group B: Beams (5) and (6).

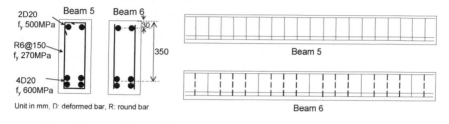

*Figure 9.48*  Specimen set-up and dimension for Group C: Beams (7), (8), (9) and (10).

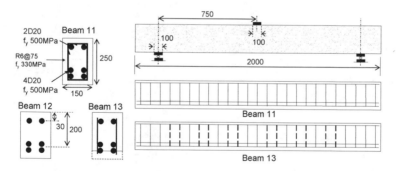

*Figure 9.49*  Specimen set-up and dimension for Group D: Beams (11), (12) and (13).

for the bond deterioration zone as for the previous analysis of the authors' experiment. The lack of concrete cover at the bottom of the beam was represented by assuming no tension stiffening (Okamura and Maekawa 1991) at the necked steel layers. Table 9.7 shows the analytical results along with the experimental ones. It can be seen that the analysis of the fractured

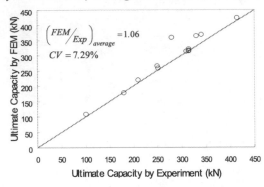

$$\left(\frac{FEM}{Exp}\right)_{average} = 1.06$$

$$CV = 7.29\%$$

*Figure 9.50* Comparison between experimental result and FEM-computed capacity.

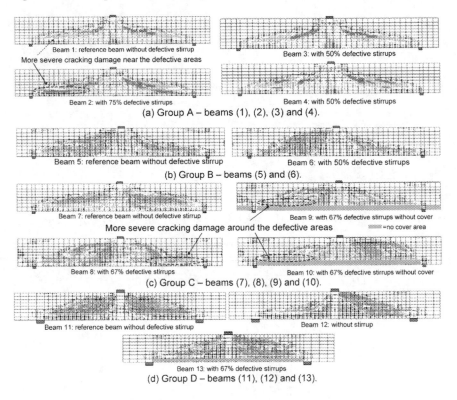

*Figure 9.51* Numerical crack patterns.

web stirrups yielded the precise ultimate capacity along with consistent failure modes. Figure 9.50 shows the comparison of the experimental and FEM computed capacities. Numerically obtained cracking patterns at failure are illustrated in Figure 9.51 for all four groups, with different reinforcement arrangement and detail of defective stirrups.

In Group A, lower flexural capacity was produced compared with the potential strength in shear. Thus, the ultimate capacities of all defective stirrup beams were mostly the same as the referential one, even though defective stirrups with insufficient anchorage were used. However, severe shear cracking damage was observed in Beam 2 containing the 75% defective web reinforcing bars, as shown in Figure 9.51a. The crack localization along the main reinforcement is characteristic, like that in the experiment, as shown in Figure 9.45. For both Beam 3 and Beam 4 containing 50% defective stirrup (around only the tension side for Beam 3 but around both tension and compression sides for Beam 4) there were similar failure crack patterns to the reference beam. The efficiency of the rest of the non-defective stirrups was experimentally and numerically confirmed.

In Group B, the shear force carried by web reinforcement was approximately 50% of the concrete contribution when structural details were satisfied, and the failure mode of the beams was originally in shear. Although the FEM computation revealed little difference in failure crack patterns between the defective beams and the reference beam as shown in Figure 9.51b, the deterioration of the shear capacity was observed in the cases of the defective stirrup compared with the reference shear failure. The numerical result considering the assumed bond deterioration at the extreme ends of stirrups was verified against the reasonably computed ultimate shear capacity.

Similarly, in Group C, the built-in damage of web reinforcement resulted in the degradation of shear capacity compared with the reference shear failure beam, too. In Group C, the amount of web reinforcement was comparatively larger than in Group A. The effect of concrete cover spalling was also included in Beam 9 (no concrete cover along the whole tensile steel) and Beam 10 (no concrete cover along the tensile steel between supports). Figure 9.51c shows the numerical failure crack patterns with severe cracking damage around the defective areas where bond deterioration was placed. The effect of the concrete cover spalling beneath the main tensile steels was verified to be small in both the experiment and the simulation as similar crack patterns were observed among Beams 8–10 in Figure 9.51c. The numerical computed shear capacities were reasonably predicted by employing the assumed bond deterioration zones with no axial stiffness of reinforcement.

Group D provided the data for beams whose shear capacity carried by concrete is relatively smaller compared to Group B and Group C. Unlike the reference distributed shear crack, Beam 11, the numerical crack reasonably captured localized diagonal shear crack for Beam 12 containing no stirrups, as shown in Figure 9.51d. For Beam 13 containing 67% defective stirrups with no concrete cover along the tensile steels between supports, a less distributed and more localized appearance of diagonal shear cracks was clearly seen, similar to that of Beam 12 without stirrups, as shown in Figure 9.51d. The effect of concrete cover was also small, similar to Beams 9 and 10 in Group C, due to the bond deterioration of fractured stirrups at hook portions near the tensile reinforcements.

In all combinations, computational assessment of shear capacity was highly accurate, as shown in Figure 9.50. Due to the remaining bond of stirrups placed around the anchorage zones of the main reinforcement, longitudinal anchorage (bond) failure was successfully prevented, unlike in the authors' experiments with short development lengths with ruptured stirrups. As the tied-arch mechanism was firmly reproduced, the load-carrying capacity of the concrete web was enhanced rather than the usual cases. On the other hand, the contribution of web reinforcement was moderated. As a whole, the total capacity is thought to have deteriorated accordingly.

As discussed in Section 9.5.1, shear capacity of slender beams is much influenced by the damaged anchorage of stirrups. However, if the amount of web steel is moderate or small, unlike the case discussed in Section 3.1, the cracking pattern and load-carrying mechanism of structural concrete are not expected to change much. Figure 9.52 shows the experimental results by Shioya *et al.* (2001) with and without anchorage of stirrups. In this case, stirrups (denoted as No. 1, without damage) were provided by welding steel plates ahead of the stirrup ends. On the other hand, no hook was devised in the case of No. 3 and the extreme edge of the stirrups was situated at the level of the main reinforcement layer. The shear span-to-depth ratio was similar to the case described in Section 9.5.2 but the shear fraction carried by the web steel was smaller compared with that of the concrete.

As shown in Figure 9.52, crack patterns are not affected by the anchorage

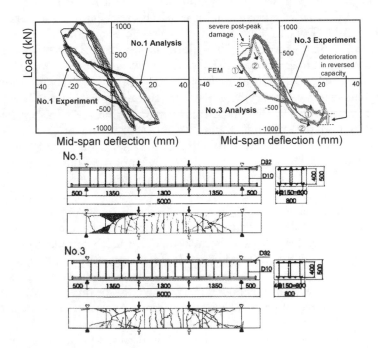

*Figure 9.52* Effect of stirrup rupture on slender RC beam with comparatively smaller web steel.

details and the reduction rate of shear capacity is comparatively smaller, as expected. The proposed computational approach can satisfactorily capture the capacity and reversed cyclic characteristics of the beams with and without hook anchorage. In this post-peak analysis, two loading points are forced to move together. This implies no rotational degree of freedom on the constant moment span and similar damage is forced to develop in both shear spans after the peak. If we allow the rotational kinematics of the post-peak range in analysis, the damage in the structural softened states becomes unsymmetrical. In fact, the displacements at two loading points are not known from the literature but the mid-span vertical displacement is available. According to roughly symmetrical experimental crack patterns as shown in Figure 9.52, the authors assumed almost the same displacements developed at both loading points in the experiment and applied this kinematics in the analysis.

The damaged beam experimentally and analytically exhibited a rather sharper drop in restoring forces in the softening stage. Two unloading–reloading paths were computed from different damage levels, as shown in Figure 9.52. The capacity in the reversed paths is affected by the post-peak maximum displacement in computation. Analytical assessment of damaged RC members may yield a reasonable estimation or even one that is rather on the safe side.

### 9.5.4 Size effect of damaged web reinforcement

In the previous section, the fracture of shear reinforcements inside RC beams was experimentally and numerically examined for its adverse effect on shear capacity. In the experiment conducted by the authors ,an important reduction in shear capacity of approximately 37% was observed in comparison with the non-damaged reference beam. In the experiment by Regan and Kennedy Reid (2004), a relatively smaller reduction in shear capacity, up to 29% in Group D, was observed. By considering the bond deterioration zone of 10 times the fractured web steel diameter, the resultant shear capacity of the damaged beam with fractured web reinforcement can be well estimated, as explained in Chapter 3.

Here, real-sized RC members will be discussed. The dimensions of real-scale structures may be several times larger than those of the laboratory specimens. Due to the limitations of fabrication using the discretized bar sizes available in industry, the diameters of reinforcing bars cannot be proportionally coordinated with the sizes of members. Thus, a detailed reinforcement arrangement such as that involving multiple layers of steel in real structures is usually required when the same reinforcement ratio is maintained. Figure 9.53 shows an example of an RC section in the authors' experiment and the assumed large-sized RC in the scale of actual members in service. Both sections have the same reinforcement ratio for the main and shear directions and the same material properties of steel and concrete, as shown in Table 9.6. With the same fractured web reinforcement legs at

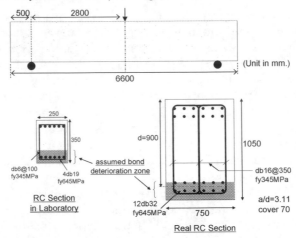

*Figure 9.53* Comparison of RC section in laboratory and real design.

the bent portion near tensile reinforcement, the extent of assumed bond deterioration of 10 times the web reinforcement diameter in the small RC section is located beyond the reinforcement layers, while in the large beam, the bond deterioration zone remains in the reinforcement layers.

Figure 9.54 shows the comparison in the degradation of shear capacity of the non-damaged reference beam and the fractured web reinforcement beam for the laboratory size and the large one of the real RC section. Less reduction in shear capacity of 16% compared with the non-damaged beam was observed in the case of the large-sized RC beam where multi-layers of reinforcement arrangement are demanded in practice. This results in the bond deterioration zone being located within the main steel layers. Although both sections have the same structural and material properties with the same shear span-to-effective depth ratio, the effect of fractured shear reinforcement on the shear capacity degradation does not vary linearly in dimension. Consequently, this size effect will be realized and considered carefully when

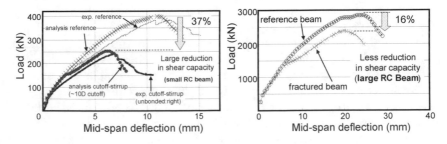

*Figure 9.54* Size effect in reinforcement of RC member with fractured shear reinforcement.

dealing with shear capacity assessment of the fractured web reinforcement beam in the real scale of RC members.

## 9.6 Structural performance of damaged RC members: anchorage of main reinforcement

### 9.6.1 Detailing around the anchorage zone

All beams as shown in Figure 9.55 are designed to fail in shear. These are mock-ups of deteriorated beams with lost anchorage and bond between main reinforcement and web concrete. The dimensions are 2,400 (length) × 350 (height) × 250 mm (breadth). Four D19 (fy = 645MPa) bars are arranged in tension and compression zones, and concrete strength is targeted to 34MPa. Design detail is described as follows: 1) Beam with no shear reinforcement (s0cr0); selected as the reference. The load capacity of this beam as calculated by the design formulae is 154kN in shear and 308kN in concentrated force. 2) Artificially cracked beam with no shear reinforcement (s0cr1); with no shear reinforcement. Artificial gap planes are inserted to kill the interaction of main reinforcement and web concrete at each end of the beam (as an extreme case). 3) Artificially cracked beam with shear reinforcement (s1cr1); there are shear reinforcements in this beam. D10 shear reinforcements whose strength is 340 MPa are set every 156 mm and artificial cracks are introduced in each end of this beam. If there is no artificial crack, load capacity of this beam is 352kN (in shear failure). 4) Artificially cracked beam with shear reinforcement (se1cr1); artificial cracks are introduced in each end of this beam. And only at the end of beam, there are D10 shear reinforcements set every 156 mm. The strength of shear reinforcement is 340 MPa. Shear reinforcement is expected to keep the main reinforcement and bottom up the load capacity. 5) Artificially cracked beam with bended main reinforcement (sb1cr1); artificial cracks are introduced in each end of this beam. There is no shear reinforcement in this beam, but the end of the main bar is bent up. By bending up, the divided parts may be unified and the load capacity will be kept.

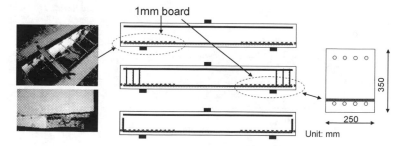

*Figure 9.55* Specimens with damage at anchorage zones of main reinforcement.

### 9.6.2 *Capacity and ductility*

Figure 9.56 shows the defect built-in RC members subjected to combined flexure and shear (Chijiwa 2006). This is a mock-up of corrosion cracks and is used for verification of the coupled mechanistic and thermodynamic analysis. It can be seen that the quasi-ambient induced damages in the anchorage zone are very severe for structural strength. Unlike the pre-induced damages inside the shear span, a distinct crack slip occurs and the tied-arch mechanism is hard to make appear if the damage is limited in the narrow anchorage zone. The damage concentrates between the tip of the quasi-ambient induced damages and the loading point, and the diagonal crack width becomes remarkably wide. Here, the enhanced contact density model (Bujadham and Maekawa 1992) can be used to describe plastic damaging shear transfer in wider range of crack widths and displacement paths. In considering the geometrical contact non-linearity at crack planes, more rational formulation is expected based upon the original contact density modeling to fairly simulate the behavior of the structural concrete with cracks. Larger crack

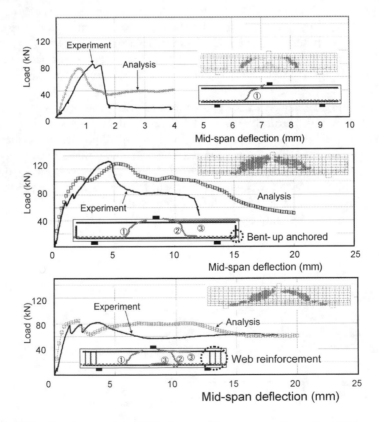

*Figure 9.56* Simulation of partially damaged beam along the main reinforcement.

widths with dramatically rotating axes of principal stresses, which are rare in non-damaged ones but frequent for structural concrete including corrosion cracking, must be considered for structures subjected to combined mechanical and environmental actions (Chijiwa 2006).

# 10 Fatigue life of structural concrete

In this chapter, path-dependent fatigue constitutive models for cracked concrete tension and shear transfer are proposed and directly integrated with respect to time and deformational paths actualized in structural concrete. The authors aim to verify the multi-scale modeling to be consistent with the fatigue life of materials and structural members under high repetition of forces. The mechanistic background of the extended truss model for fatigue design is also investigated. The coupling of fatigue loads with initial defects is simulated and its applicability is discussed. As well, analytical scrutiny is made of the shear fatigue behavior of reinforced concrete (RC) beams subjected to moving loads, and the mechanism for dramatically reduced fatigue life under moving loads is computationally discussed. A simplified relation for the prediction of fatigue life under moving loads is proposed for practical use. The effect of randomness in the position of loading is examined and its implication for the reliability of the current fatigue life assessment method is put forward.

The direct path-integral scheme is also applied to the life-cycle assessment of RC slabs as a tool for durability mechanics. The three-dimensional fatigue analysis successfully predicts the characteristic mode of failure under moving loads as well as in the case of fixed-point pulsation in punching shear. Importantly, the drastically shortened fatigue life under traveling wheel-type loads is mechanically demonstrated by implementing a multi-scale modeling of three-dimensional extent. A sensitivity study is carried out to clarify the influence of shear transfer decay and compression fatigue on RC slab performance.

A high-cycle fatigue constitutive model for concrete joint interfaces is proposed and the direct path-integral scheme for reinforced concrete and pre-stressed concrete structures with junction planes is presented as well. Both cyclic pullout and the associated dowel action of reinforcing bars are formulated at a crack/joint section. The proposed differential formula is verified by high-cycle fatigue experiments of dowel bars and pullout of reinforcement crossing a joint in structural concrete. The life-cycle analysis is applied to the assembly of pre-cast pre-stressed concrete members with RC joints, that is to say, the most macroscopic system in this book. A mechanics-based discussion is presented of the different fatigue life observed in pre-cast concrete (PC) slabs with localized discrete joints and in monolithically constructed RC, where dispersed cracking develops.

## 10.1 Two-dimensional fatigue model of cracked structural concrete

The truss analogy of reinforced concrete (RC) with web reinforcement in shear was first proposed by Ritter and Morch in the early twentieth century and has since been used in practice worldwide. Later, this load-carrying mechanistic analogy was upgraded to take into account the remaining force carried by cracked web concrete (denoted by $V_c$), and the simple summation law of $V_c$ with $V_s$ carried by web steel was generally recognized. This modified truss model became the basis of the subsequent compression field theory (Collins and Vecchio 1982) and strut-tie models (Schlaich and Weischede 1982; Marti 1985, etc). This macro-behavioral modeling of RC members is the reflection of a primary load-carrying mechanism.

This truss analogy was further extended to highly repeated shear (e.g., Hawkins 1974; Okamura *et al.* 1981; Ueda and Okamura 1982) and, at present, it serves some fatigue design codes by estimating stress amplitudes of web reinforcement. Figure 10.1 shows the typical relation of the repeated shear force and mean strain of web reinforcement. The following are characteristics of the phenomenon experimentally obtained by Ueda and Okamura (1982).

1   The increase in the mean strain of stirrups caused by fatigue loading is not highly dependent on the level of maximum shear force if the minimum shear is nearly zero. This means that the shear capacity for concrete decreases according to the number of cycles but the reduction rate is almost independent of the magnitude of shear forces.
2   The shear capacity remains about 60% after one million cycles and 70–75% after 10,000 cycles and this S-N relation of $V_c$ is not much dependent on the amount of web reinforcement.
3   Unloading–reloading curves have a unique focal point on the shear force versus stirrup strain diagram before yielding. The steel strain does not reach null even when the shear force is completely removed. This is the great difference from the flexural cracking and longitudinal reinforcement.
4   When the minimum shear force is large, the shear fatigue life is prolonged exponentially.

This behavioral knowledge is practical for assessing the stress amplitude of web steel with reasonable accuracy. However, its microscopic mechanism is still under discussion. Ueda *et al.* (1999) proposed a new finite element (FE) fatigue analysis method, too. Here, tailored constitutive models were applied in consideration of fatigue effects by reducing the stiffness and strength of

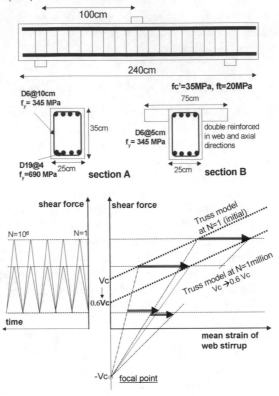

*Figure 10.1* Extended truss model for RC members in highly repeated shear.

material models. In this case, the following hypothesis is assumed: (a) that initial states of materials are non-damaged, and (b) that from the beginning to the end of loading, stress states do not vary much during the damaging process. This original study was successful in simulating highly cyclic responses, although the cyclic deterioration process of materials is not strictly traced. The authors understand that the above hypothesis holds for initially non-damaged cases and works well for fatigue design.

However, these hypotheses may not hold when environmental actions cause corrosion of steel, drying shrinkage and cracking. Stress amplitudes and paths may drastically change and proportional stress history can no longer be assumed. Here, it is necessary to follow each step of the damaging process, in a similar way to our following of non-linear seismic effects, as shown in Figure 10.2. This chapter aims to investigate the fatigue mechanism by tracing the exact transient process of gradual damaging under repeated shear by using a direct path-integral scheme. This means that the computerized RC members are created virtually, and high cyclic actions are reproduced by integrating the multi-scale constitutive modeling of materials with regard to damage accumulation. The analysis is extended to the fatigue life assessment

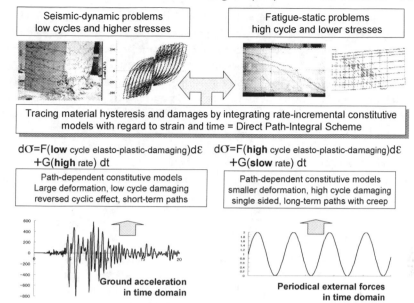

*Figure 10.2* Direct path-integral scheme for seismic and fatigue problems.

of initially damaged RC members as the coupled problem of fatigue and ambient actions. In this section, the fatigue life model of reinforcement is not discussed but the authors accept the S-N diagram as a well-established basis for steel reinforcement knowledge.

Recent advances in computational mechanics enable engineers to conduct the direct integral of path-dependent constitutive models by tracing microscopic material states under all cyclic paths of stresses and strains. In the past decade, this scheme has been successfully used to simulate the dynamic response of structures under seismic actions, and received some endorsement as a tool to assess structural performance in design. Here, the cyclic degradation is expressed by incremental plasticity and damaging. In this book, the direct path-integral scheme is intended as a more versatile means of fatigue analysis. Thus, the authors started their research from the multi-directional fixed-crack model (Maekawa *et al.* 2003b), which has been used for seismic analysis in practice.

According to this scheme, a two-dimensional space-averaged constitutive model is formed by combining one-dimensional stress–strain relations of cracked concrete in tension, compression and shear. Each constituent modeling is strain path dependent. Here it is important to note that all component behaviors are time dependent as well (Chapter 8; Maekawa *et al.* 2006b). Thus, the apparent effect of cyclic actions under high stress conditions is chiefly attributed to time-dependent plasticity and fracturing. All fatigue experiments more or less unavoidably include the time-dependent effect. For example, the apparent fatigue life of materials may vary according to stress frequency (Award and Hilsdorf 1974; Raithby and Galloway 1974). The authors extracted the accumulated fatigue damage of concrete solid by experimentally subtracting the component of time-dependent deformation and re-formulated a more generic constitutive model for compression in Chapter 8. This fatigue compression model is hereafter used in the context of multi-directional crack modeling (Section 8.5 and Figure 8.47). Similar to the compression model, the tension stiffening–softening and shear transfer models need to be enhanced to be able to consider the cumulative fatigue–time damaging for the direct path-integral scheme. In this book, simple formulae for tension and crack shear are also performed with respect to incremental time and strains, and melded together with the compressive fatigue model presented in Chapter 8.

### 10.1.1 High-cycle damaging in tension

As a phenomenon, the tensile fatigue of plain concrete is associated with crack propagation (Cornelissen and Reinhardt 1984; Reinhardt *et al.* 1986; Subramaniam *et al.* 2002) and the apparent reduction of tension stiffness in RC after the first cracking is attributed to bond fatigue and leads to subsequent new sectional cracking in tension members. In order to represent these microscopic phenomena in structural analyses, the space-averaging scheme is applied for formulation. Thus, in fracture mechanics overall stiffness and average strain are adopted as the main state variables rather than the running ligament of a single crack.

Tension non-linearity (positive $\varepsilon$ and $\sigma$) before and after cracking is assumed to be governed by the fracturing damage rooted in cracks, and no fatigue plasticity is simply assumed as discussed in Section 8.6. In this section, attention is directed to the accumulated damage term denoted by $G$ in Equation 8.21. This term is negligible for the creep analysis in Chapter 8, but it does play a substantial role for high-cycle fatigue analysis. Based on the experiment by Nakasu and Iwatate (1996), the cyclic fatigue damage in tension after cracking is idealized with regard to the derivative $G$ as

$$G d\varepsilon = K_T \left( \frac{\sigma_{tp}}{\sigma_{env}} \right)^{20} \cdot d\tilde{\varepsilon},$$

$$where, \sigma_{env} = f_t \left( \frac{\varepsilon_{cr}}{\varepsilon_{tp}} \right)^{\alpha} \tag{10.1}$$

$$d\tilde{\varepsilon} = 0, \quad when \quad d\varepsilon \geq 0$$

$$d\tilde{\varepsilon} = 9 \cdot \gamma^m \cdot \left( \frac{d\varepsilon}{\varepsilon_o} \right), \quad m = 8, \gamma \equiv -\frac{\varepsilon - \varepsilon_{tp}}{\varepsilon_{max}}, \quad when \quad d\varepsilon < 0$$

This formula is applicable to the non-cracked state with $m = 20$ and $\sigma_{tp}/\sigma_{env}$ = 1.0. The evolution of the fracture parameter results in increasing strain when the amplitude of stress is kept constant. The modeling formulated is applied to both smeared-crack tension stiffness and softening modeling, as discussed in the following sections. Figure 10.3 shows the computed S-N diagram for tensile fatigue strength derived from Equation 10.1 before cracking. The fatigue strength of one million cycles is approximately 60–70% of the static uniaxial strength. Broadly speaking, this is similar to the S-N curve for compression and the experimental reports also show this similarity of compression and tension (Tepfers 1979, 1982; Kodama and Ishikawa 1982; American Concrete Institute Committee 215 1982; RILEM Committee 36-RDL 1984). The computed strain response under high cyclic tension is also shown in Figure 10.3. Just before the fatigue rupture, the tensile maximum strain reaches approximately 1.2–2.2 times that of the first cycle, and this matches the experimental facts obtained by Saito and Imai (1983) and by Kaneko and Ohgishi (1991).

Figure 10.4 shows the comparison of analytical predictions derived from Equation 10.1 and experiments after cracking. The uniaxial tension fatigue experiment of full amplitude was conducted by Nakasu and Iwatate (1996) using an RC slender column with 2% reinforcement ratio. The averaged concrete stress transferred by bond was extracted as below. Generally, the average tension stress transferred through bonds is reduced 50% after one million cycles in both experiments and analysis. This is also quantitatively

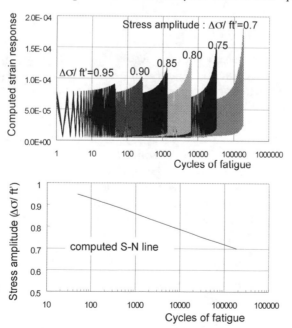

*Figure 10.3* Computed tensile fatigue life and strain responses.

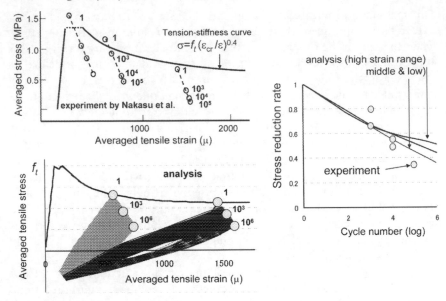

*Figure 10.4* Tension stiffness under uniaxial fatigue loading.

close to the assumption by Ueda *et al.* (1999) for their fatigue tension stiffness work.

### 10.1.2 Shear transfer fatigue

It is known that the shear transfer across crack planes degenerates under repeated slippage. Crack roughness and contact friction are thought to be degraded after high cycles. According to the contact density model (Li and Maekawa 1987; Figure 10.5) which is based on Walraven's innovative analysis and modeling of rough cracks (1981), the transferred shear can be formulated in terms of intrinsic shear slip or shear strain normalized by the crack opening or averaged normal strain (Maekawa *et al.* 2003b). Pruijssers *et al.* (1988) investigated fatigue damage evolution, and Gebreyouhannes (2006) carried out a fatigue experiment to determine shear transfer and associated dilatancy under medium loading levels. Here, the authors propose the stiffness reduction rate for extending the applicability of the original contact density modeling as

$$\tau = X \cdot \tau_{or}(\delta, \omega)$$

$$X = 1 - \frac{1}{10} \log_{10}\left(1 + \int |d(\delta/\omega)|\right), \quad \geq 0.1 \tag{10.2}$$

where $\tau_{or}$ is the transferred shear stress calculated by the original contact density model, $(\delta, \omega)$ are the crack shear slip and width, and $X$ is the fatigue

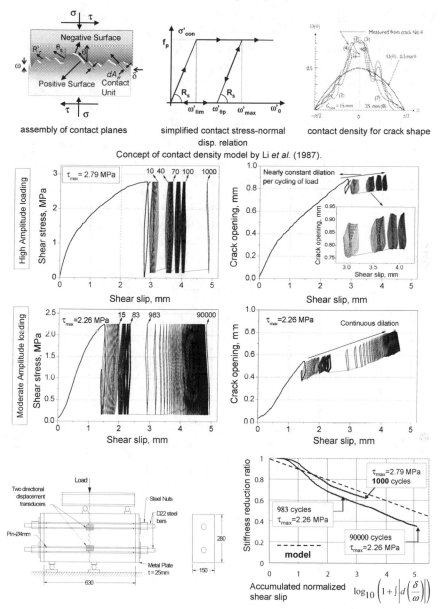

*Figure 10.5* Cyclic shear transfer stiffness and fatigue model.

modification factor to express the stiffness reduction with regard to accumulated intrinsic shear deformation. Provided that dispersed cracking is assumed in the finite element, $\delta/\omega$ can be replaced with the average shear-to-normal strain ratio as $\gamma/\varepsilon$ (Maekawa *et al.* 2003b).

The shear transfer fatigue is experimentally known to be predominant when the stress path is reversed cyclic. The above equation is thought to cover single-sided stress amplitude under dry conditions. Reversed cyclic paths may occur under moving loads, such as slabs, and do reduce the fatigue life of members dramatically. For RC beams in practice, however, the single-sided stress path is leading, unlike the case with columns.

### 10.1.3 Direct path-integral for fatigue simulation

Well-established non-linear dynamic analyses for RC structures can be extended to fatigue analyses in time-space domains by a direct path-integral of constitutive modeling. The essence is the material modeling, as explained in previous sections. As fatigue options of concrete compression (Section 8.4.3), tension (Section 10.1.1) and shear transfer (Section 10.1.2) can be added to the original modeling in terms of evolution derivatives. These extended material models also cover high non-linear behaviors. Consequently, the fatigue analysis system shares the same scheme as that of non-linear failure analysis under seismic loads (Section 8.5) and creep simulation (Section 8.6).

Repeated loads of high cycles can be directly input as the nodal forces in full time steps. However, it takes a great deal of computation time when a vast number of cycles are provided. In order to speed up the computation, a logarithmic time derivative can be applied for path-integrating the constitutive models. The time-dependent and damaging rates with respect to strain paths are simply integrated in the numerical scheme as

$$\Delta \varepsilon_p = \zeta \left( \frac{d\varepsilon_p}{dt} \right) \Delta t + \left( \frac{d\varepsilon_p}{d\varepsilon_e} \right) \Delta \varepsilon_e,$$

Compression: $\qquad\qquad\qquad\qquad\qquad\qquad\qquad\qquad$ (10.3a)

$$\Delta K = \zeta \left( \frac{dK}{dt} \right) \Delta t + \left[ -\left( \frac{\partial F_k}{\partial \varepsilon_e} \right) \left( \frac{\partial F_k}{\partial K} \right) + \zeta \cdot \lambda \right] \Delta \varepsilon_e$$

(see Equation 8.17)

Tension:  $\quad \Delta K = \zeta \cdot F \Delta t + \zeta \cdot G \Delta \varepsilon + H \Delta \varepsilon$ $\qquad\qquad$ (10.3b)

(see Equation 8.21)

Crack shear Transfer:  $\quad X = 1 - \dfrac{1}{10} \log_{10} \left( 1 + \sum \zeta \cdot \left| \Delta \left( \dfrac{\gamma}{\varepsilon} \right) \right| \right), \quad \geq 0.1$  (10.3c)

(see Equation 10.2)

where $\zeta$ is the integral acceleration factor. As material non-linearity in structures evolves significantly at the beginning of fatigue loadings, the value of $\zeta$ should be unity for accuracy, i.e., the direct exact paths of forces and corresponding internal stresses need to be tracked in real time. After many cycles, plastic and fracturing rates as well as fatigue damaging evolution tend to be much reduced and the increment of non-linearity via the cycle becomes exponentially small. At this stage, for computational efficiency, the value of

$\zeta$ can be larger for each load step. The single cycle of computational load is equivalent to $\zeta$–cycles. In Equation 10.3, only time and the strain path integral are magnified by $\zeta$. It should be noted that the terms of instantaneous non-linearity associated with evolution boundaries (plasticity and damaging) are not magnified by $\zeta$, though the path is strictly mapped out.

Figure 10.6 shows the computed strain and displacement histories of the compression specimen and an RC beam. These were obtained by simultaneously integrating Equation 10.3 with a different time-integral magnification. For a cycle of loading, 20 discrete steps were assigned. The maximum stress is 80% of the static strength and the computed fatigue life of the material is about 20,000 cycles. The maximum shear force is 160 kN for the beam fatigue simulation (section A in Figure 10.1). Two different time integrals are compared. One is the full-range time ($\zeta = 1.0$) and the other is the binary sequence ($\zeta = 1, 2, 5, 10, 20, 50, 100 \dots$) as shown in Figure 10.6. Both time intervals give almost the same results in terms of strain and displacement (less than 5%). The computation time is approximately 50 times shortened by the acceleration scheme and thus it becomes feasible to effect the computation. It takes no more than one hour to complete beam fatigue simulation for one million cycles.

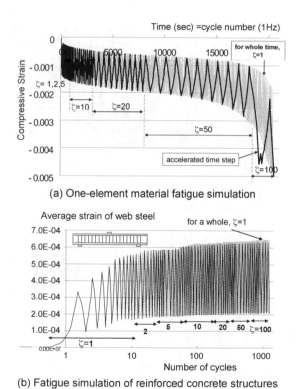

(a) One-element material fatigue simulation

(b) Fatigue simulation of reinforced concrete structures

*Figure 10.6* Magnified direct time and path integral.

Figure 10.7 shows the computed and experimental mean strains of the RC web with stirrups and applied shear. The design formula for shear capacity predicts the diagonal cracking shear of 90 kN and the web yielding of 150 kN. The computed shear capacity after yielding is 190 kN by push-over analysis. First, the authors checked the cyclic responses under rather high stresses in experiment. Partial yielding of web reinforcement occurred at the first cycle of loading and the number of repetitions was small (100 cycles). The mean tensile strain was recorded using displacement transducers of 300 mm gauge length crossing the main diagonal cracking. This verification is effective for comparatively higher stress states under plastic conditions.

## 10.2 Fatigue life simulation of RC beams and verification: fixed-point pulsation

In the extended truss model, the concept of a focal point on the unloading–reloading paths is most useful for indicating the limits of the design, as shown in Figure 10.1. The presence of the focal point on the diagram of shear force versus web stress (strain) implies that the residual strain proceeds according

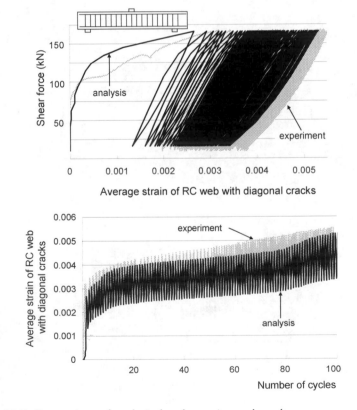

*Figure 10.7* Comparison of analytical and experimental results.

to load cycles, and the stiffness accompanying shear force and web steel strain declines. It is clear that the path-dependency of tension stiffness and the shear transfer will cause residual deformation of web reinforcement when the applied shear is removed. Thus, this chapter will discuss the applicability of computational fatigue analysis in line with the extended truss analogy, which has already been examined by systematic experiments (Ueda and Okamura 1982). Since the topic cannot be fully addressed by the experimental approach alone, a numerical sensitivity simulation and its comparison with reality may lead to greater understanding of the internal mechanism at play.

### 10.2.1 Full amplitudes

A simple supported RC beam was selected as the standard specimen for discussion. The shear span-to-depth ratio is 3.2, with a T-section flange as shown in Figure 10.1 in more detail. The nominal shear capacity of concrete is about 110 kN in analysis and almost the same as that calculated by Okamura and Higai's formulae (1980). The capacity with web reinforcement is 121 kN, estimated from the practical truss model. The flexural yield may occur at 420 kN of the shear force.

Figure 10.8 and Figure 10.9 show the analytical results on fully repeated shear versus the average strain relation of the stirrups up to 10,000 cycles. Broadly speaking, the growth of stirrup strains is not highly dependent on shear force amplitude when the maximum shear is not close to the diagonal cracking capacity denoted by $V_c$. This result is similar to reality, as pointed out by Okamura *et al.* (1981) and Ueda and Okamura (1982). Experimentally and analytically, this implies that the shear force carried by concrete may decrease irrespective of the maximum force levels as shown in Figure 10.9. This computed S-N relation of cracked web concrete shear

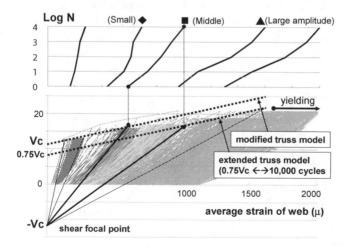

*Figure 10.8* Shear force versus mean strain of RC web zone with distributed stirrup.

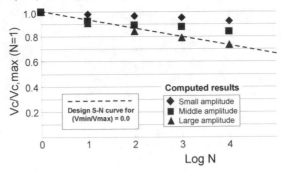

*Figure 10.9* Extracted S-N diagram of concrete in shear.

in the RC beam is not far from that proposed by Higai (1978) and Ueda and Okamura (1982), except in case of small amplitudes. The focal point of the unloading–reloading curves of the diagram is definitely identical in the analysis results as shown in Figure 10.8. The unloading lines converge on the unique point where reinforcement remains elastic. The dotted lines indicate the prediction made by the design formula of Ueda and Okamura (1982).

### 10.2.2 Middle and small amplitudes

The verification is performed with different minimum and maximum shear forces. In order to deal with several dimensions and detailings, the authors used in all the following examples a simple supported RC beam with a rectangular cross-section, as shown in Figure 10.1. The shear span-to-depth ratio is 3.2, which is the same as that discussed in Section 10.2.1. The nominal shear capacity of the concrete is about 90 kN and that of the web reinforcement is expected to be 60 kN. Flexural yield may occur at 210 kN of the shear force. In this case, half the reinforcement signaled in the previous section was used. Figure 10.10 shows the shear force and mean stirrup strain relation. The amplitude is unchanged at 50 kN, but the minimum shear force is variable. The ratio of the minimum shear to the maximum shear, denoted by $r$, ranges from 0.44 to 0.6. Based upon systematically arranged experiments, Higai (1978) proposed the empirical fatigue shear capacity formula, which was also incorporated in the extended truss model by Ueda and Okamura (1982) as

$$\log_{10}\left(\frac{V_c}{V_{co}}\right) = -0.036 * (1 - r|r|) \cdot \log_{10} N \tag{10.4}$$

where $r = V_{c,min}/V_{c,max}$, $N$ is the number of fatigue cycles, and $V_{co}$ is defined as the static shear capacity of concrete. This formula and the FE computed results are compared as shown in Figure 10.10. The computed fatigue life and the focal point are roughly equivalent to the empirical formula given by

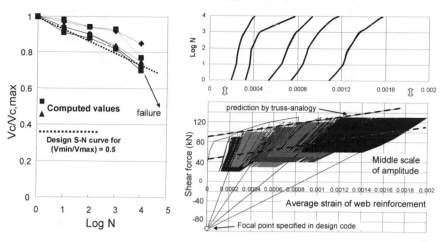

*Figure 10.10* Repeated shear force and stirrup strain relation under smaller amplitudes.

Equation 10.4, but in the case of small amplitudes just after diagonal cracking, the computed evolution of the stirrup mean strain tends to be slow. In fact, this tendency is also seen in the experiment but in the practical design it is simply ignored, to be on the safe side.

For further verification, non-web reinforced concrete beams were selected for computing the S-N curve of fatigue life under full and partial amplitudes. The aim was to have beams with the same rectangular section and detailing as those of the RC beam, but with the web steel removed. The rate of loading is 1.0 Hz, which is similar to the experiments. Figure 10.11 shows the computed S-N curve normalized by the computed static shear capacity. Generally, the analysis coincides with the empirical S-N formula with reasonable accuracy. The crack pattern after many cycles is shown in Figure 10.11a. It is clear that the failure mode is the typical shear accompanying diagonal cracking. The S-N diagram of full amplitude is computed for both slender and short beams (Figure 10.11b, c). The computed fatigue strength of the slender beam is close to the empirically derived S-N formula, while the short beam exhibits a slightly higher fatigue capacity than previously reported and the analysis matches the trend indicated by the facts (Higai 1978). Figure 10.11d shows fatigue life related to amplitude. According to the reduced amplitude, the logarithmic fatigue life is increased linearly in computation. The design formula and the computed one are close to each other.

## 10.2.3 The effect of shear transfer

It is a fact that the diagonal crack plane becomes smooth after many cycles. Repeated shear slip may cause local frictional failure at the contact point and crack roughness is erased. Thus, the loss of the shear transfer mechanism definitely amplifies a web deformation and it must be a part of fatigue

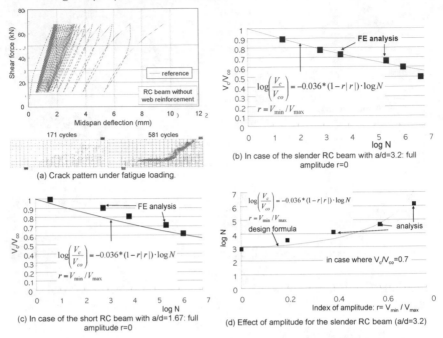

(a) Crack pattern under fatigue loading.

(b) In case of the slender RC beam with a/d=3.2: full amplitude r=0

(c) In case of the short RC beam with a/d=1.67: full amplitude r=0

(d) Effect of amplitude for the slender RC beam (a/d=3.2)

*Figure 10.11*  Finite element analysis of RC without web reinforcement under fatigue loading.

deterioration of RC beams. However, it is hard to quantitatively verify the effect of shear transfer deterioration solely with an experimental approach. Therefore, a numerical sensitivity analysis was conducted in which the factor for the shear slip accumulation was provisionally intensified one million times, as underlined in Table 10.1. The sensitivity analysis results are given in Figure 10.12.

The web strain becomes larger due to the rapid deterioration of shear transfer, and the pattern of unloading–reloading paths is greatly changed. The focal point of unloading–reloading paths is at the beginning of cycles. However, comparatively large plastic deformation of RC web reinforcement is generated under higher cycles and the fixed focal point is no longer relevant. This odd response caused by accelerated shear transfer fatigue means

*Table 10.1*  Sensitivity analysis for fatigue evolution of shear transfer

| Original modeling → | → Sensitivity analysis for trial |
|---|---|
| $\tau = X \cdot \tau_{or}(\varepsilon, \gamma)$ $X = 1 - \dfrac{1}{10}\log_{10}\left\{1 + \int |d(\gamma/\varepsilon)|\right\} \geq 0.1$ | $\tau = X \cdot \tau_{or}(\varepsilon, \gamma)$ $X = 1 - \dfrac{1}{10}\log_{10}\left\{1 + 10^{6}\int |d(\gamma/\varepsilon)|\right\} \geq 0.01$ |

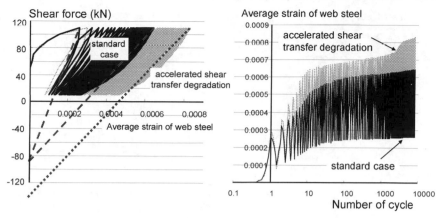

*Figure 10.12* Sensitivity fatigue analysis of RC beams for shear transfer.

that the fatigue non-linearity of RC in shear cannot be explained solely by the deterioration of rough crack shear transfer.

### 10.2.4 The effect of tension stiffness

Another source of fatigue deterioration is the degeneration of local bond or tension stiffness in a macroscopic point of view. Deterioration of the bond around the web reinforcement may allow further propagation of diagonal shear cracking and increase the crack width, which in turn may lead to reduced shear transfer along the cracks. Therefore, we have undertaken a sensitivity analysis by assuming fictitiously a great evolution of tension stiffness damage, as signaled in Table 10.2. Figure 10.13 shows the relation between shear force versus the strain of web reinforcement. It is not surprising that a high evolution of tension stiffening damage causes a large strain on the web steel. Here, the point of interest is the firm focal point like the fact before yielding.

This mechanism of structural fatigue rooted in tension stiffness has never been discussed, although static mechanistic investigation has been reported (Ueda *et al.* 1995). The bond of web reinforcement is thought to have a less direct effect on capacity. Unlike tension stiffness, shear transfer has been

*Table 10.2* Sensitivity analysis for fatigue evolution of tension stiffness

| Original modeling → | → Sensitivity analysis for trial |
|---|---|
| $Gd\varepsilon = 10^6 \cdot K^9 (1-K)^4 \cdot d\tilde{\varepsilon}$ | $Gd\varepsilon = \underline{10^{15}} \cdot K^9 (1-K)^4 \cdot d\tilde{\varepsilon}$ |
| $d\tilde{\varepsilon} = 0$, when $d\varepsilon \geq 0$ | $d\tilde{\varepsilon} = 0$, when $d\varepsilon \geq 0$ |
| $d\tilde{\varepsilon} = 9 \cdot \gamma^8 \cdot \left(\dfrac{d\varepsilon}{\varepsilon_o}\right)$, $\gamma = -\dfrac{\varepsilon - \varepsilon_{tp}}{\varepsilon_{max}}$, when $d\varepsilon < 0$ | $d\tilde{\varepsilon} = 9 \cdot \gamma^8 \cdot \left(\dfrac{d\varepsilon}{\varepsilon_o}\right)$, $\gamma = -\dfrac{\varepsilon - \varepsilon_{tp}}{\varepsilon_{max}}$, when $d\varepsilon < 0$ |

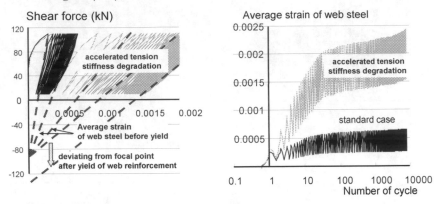

*Figure 10.13* Sensitivity fatigue analysis of RC beams for tension stiffness.

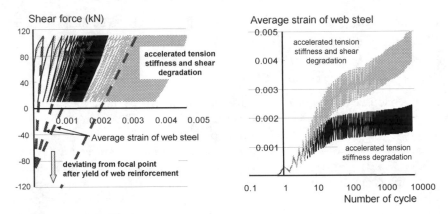

*Figure 10.14* Sensitivity fatigue analysis of RC beams for both tension and shear.

regarded as the critical mechanism by which shear is carried. But, when we integrate these two sorts of sensitivity analyses, tension stiffness fatigue cannot be left out of the discussion. Additional shear transfer decay is thought to accelerate the damage evolution, as shown in Figure 10.14. The influence of shear transfer on fatigue evolution is comparatively large where significant tension stress transfer deterioration has occurred across cracking. Thus, the combined effect should be the focus of future research.

### 10.2.5 Fatigue life of non-web reinforced RC beam

Another series of sensitivity analyses were conducted with the non-web reinforced beam (see Figure 10.11a). The standard material models lead to the fatigue capacity, which is almost consistent with the empirical formula as shown in Figure 10.11b. Even if the fatigue evolution of compression is

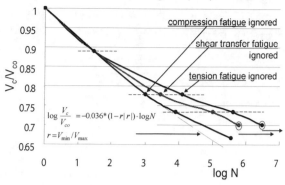

*Figure 10.15* Fatigue life sensitivity analysis of RC beams without web reinforcement.

forced to zero and other models remain unchanged, the S-N curve of RC is not much influenced, as shown in Figure 10.15.

But, if the tension fatigue model before and after cracking is the one left aside, computed fatigue life is substantially prolonged. This means that tension fatigue modeling plays a substantial role. The influence of shear transfer fatigue on structural fatigue life is also examined by assuming that the fatigue factor denoted by $X$ is unity all the time in Equation 10.2. The shear transfer fatigue modeling is also significant as shown in Figure 10.15, but its influence on structural capacity seems comparatively smaller than that of tension fatigue. A large amplitude or maximum shear produces low-cycle fatigue. In this case, the fatigue models of tension, compression and shear seem to play a minor role. In fact, the apparent cumulative damaging is caused chiefly by the time-dependent non-linear evolution already formulated in the original modeling. Analytically, the creep rupture may occur even when the shear force is kept constant as zero-amplitude according to time.

## 10.3 Corrosive cracking damage on fatigue life

Fatigue behavior is influenced by corrosive cracking formed along the main reinforcement. In this section, the effect of longitudinal cracking on fatigue strength is discussed. The corrosion gel formation and its mechanical effect are incorporated into the non-linear analysis (Chapter 9). The authors combined this procedure with the fatigue modeling of materials in order to examine the function of the direct path integral scheme, as shown in Figure 10.16.

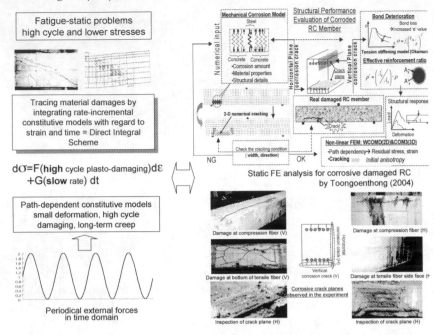

*Figure 10.16* Direct path-integral scheme for fatigue life simulation coupled with corrosive damage.

### 10.3.1 Corrosive cracking along main reinforcement: without shear reinforcement

Elevated shear capacity is experimentally known when corrosive cracking is induced along the main reinforcement within the shear span of an RC beam with no web reinforcement, and some predictive methods are proposed in Chapter 9. The increased shear capacity due to corrosive cracking is also shown in Figure 10.17, together with the numerical crack pattern at failure. It is clear that the shear load-carrying mechanism is definitely changed. The

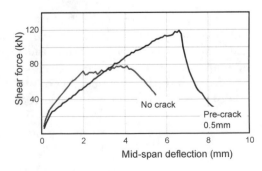

*Figure 10.17* Shear force versus displacement relation of corroded RC beam without stirrups.

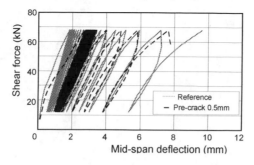

*Figure 10.18* Fatigue life of corroded RC beam without web reinforcement.

corrosion cracking may reduce the bond of main reinforcement and form the tied-arch action which results in increased static shear capacity. This phenomenon was also experienced and the analytical result follows this fact.

The behavior of a beam under fatigue loading was numerically investigated under the amplitude of 70% of the static strength of the non-damaged beam (no flange in Figure 10.1). The growing mid-span deflection and the induced shear cracking at each cycle are shown in Figure 10.18. The large initial deflection of the first cycle for the damaged beam is caused by pre-corrosive cracking. The numerical crack pattern is also shown for both sound and pre-damaged beams. The diagonal shear crack takes place subsequently for the pre-damaged beam, i.e., 581 cycles for the sound beam, and 25,000 cycles for the pre-damaged beam. The extended fatigue life of corroded RC beams without shear reinforcement is not quantitatively formulated in maintenance procedure but similar experience has been earned with slabs damaged by alkali aggregate reaction (ASR). ASR expansion may induce laminated cracking along member axes and it arrests shear crack propagation and prolongs fatigue life.

### 10.3.2 Corrosive cracking along tensile reinforcement at anchorage: with shear reinforcement

As shown in Figure 10.19, the monotonic loading analysis is compared between the sound beam and the one with corrosive pre-cracking in the anchorage zone, where the web reinforcement is placed analytically. The width of corrosive cracking is controlled to 0.5 mm by introducing the forced corrosion strain. Here, the strain localization of plain concrete is assumed normal to the main reinforcement. The monotonic loading also shows less influence on the load-carrying capacity and stiffness of the damaged beam compared to the referential one. The web reinforcement may successfully restrict the bond crack propagation at the anchorage zones. Although the initial longitudinal damage exists around anchorage, the failure along the corrosive crack plane is prevented by there being sufficient shear reinforcement. Thus, the truss shear resisting mechanism can function effectively.

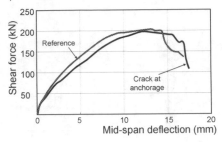

*Figure 10.19* Monotonic loading response of anchorage damaged RC beam.

The behavior of load under high repetition is numerically investigated under the amplitude of 100 kN with the loading range of 15–65% of maximum capacity. Figure 10.20 shows the accelerated damage initiated from the corrosive crack tip and the rapid evolution of average web strain can be seen at the member sections. The section Avg. 2 does not have a crack intersection at the beginning of loading and gives rise to a small deformation. In this case, the stable focal point is not identical in analysis, either.

The apparent reduction in concrete shear capacity was obtained with the increase in the number of loading cycles at section Avg. 1 near the corrosive cracking tip. To investigate the risk of stirrup rupture, the average web strain is converted to the induced stress in the web reinforcement. Figure 10.21

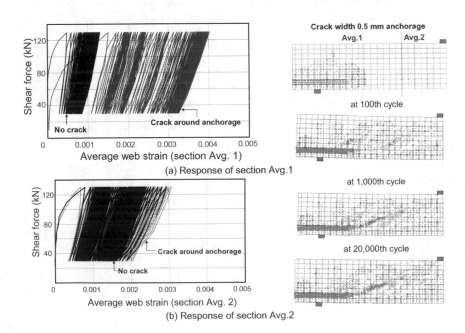

*Figure 10.20* Fatigue response of stirrups around the damaged anchorage.

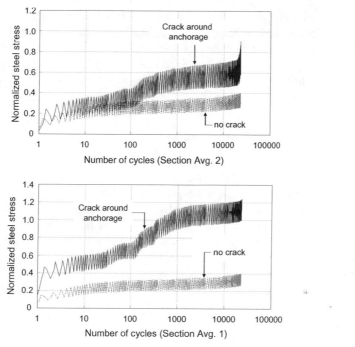

*Figure 10.21* Stress history for assessment of fatigue rupture of web reinforcement.

shows the response of cycle number and the stress level normalized by the yield strength inside the web reinforcement.

## 10.4 Fatigue life simulation of RC beams and verification: moving load passage

In this section, the behavior of RC beams under moving loads is discussed. The mechanistic behavioral understanding of RC members subjected to high reversal of stress is addressed and an S-N relation of RC beams under moving loads is proposed for practical use. Moreover, the authors try to identify some issues of importance on current fatigue design. Notably, the fatigue rupture of steel and reinforcement is disregarded in the analytical framework as the main focus is directed towards investigating the fatigue degradation of concrete.

Several investigations have been made on the moving-load fatigue behavior of RC slabs. Matsui (1987) and Perdikaris and Beim (1988) reported that the fatigue life of RC slabs is dramatically reduced under moving loads, where it is two or three orders lower than that of fixed pulsating loads. The phenomenon of the moving-load fatigue of RC beams is deemed to be different from that of RC slabs. Generally, the mechanism of moving-load fatigue in RC slabs is thought to be similar to that of fixed pulsating fatigue in RC beams, in which failure is observed to occur by diagonal cracking in the transverse direction to traffic. Unlike the case of fixed pulsating loads (Chang and Kesler 1958; Taylor 1959; Ueda and Okamura 1982), moving loads in RC beams may cause large reversals of stress–displacement. Thus, a better behavioral understanding of the fatigue response under this large stress reversal is of interest from a mechanistic point of view.

### 10.4.1 *Experimental investigation*

To examine the effect of moving loads on the shear fatigue mechanism of RC beams, an experiment on a simply supported beam without web reinforcement, of span length 2 m, is considered. The beam consists of both tension and compression reinforcements. Based on the mid-span loading, the shear span-to-depth ratio (a/d) is 3.2, with a tensile reinforcement ratio of 1.47%. The geometric, material and cross-sectional details for the beam are shown in Figure 10.22. As this chapter is focused mainly on examining the shear fatigue mechanism, the beam is designed to fail in shear mode. The nominal shear capacity of the concrete is 95 kN in the analysis and is almost the same as the prediction of shear capacity formulae by Niwa *et al.* (1987). The flexural yielding is estimated to occur at 220 kN of shear force, which is relatively higher than the shear capacity.

The beam is subjected to a step-wise moving load, in which loading is applied on predefined points, to observe the crack propagation at different positions of the load. In one passage, the beam is loaded at five different, equally spaced points, each with an offset distance of 25 cm. The first loading is applied at the mid-span of the beam. In each loading step a constant load of 140 kN is maintained, which is in the range 75–80% of the maximum shear capacity.

Figure 10.22 shows the loading set-up and crack pattern of the beam. The first loading, with amplitude of 140 kN, is applied statically with a loading rate of 0.2 mm/min. During this loading, only flexural cracks occurred. In the second loading, at an offset distance of 25 cm from the center,

Span length=2.0 m; a/d=3.2
Concrete: f'$_c$ =40 MPa
Reinforcement: f$_y$ =690 MPa;
ρ$_x$ =1.47 %

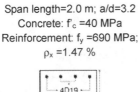

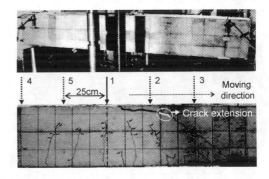

*Figure 10.22* Material and geometric details, loading set-up and crack pattern.

formation of diagonal cracking was observed. At the same time, the flexural cracks beneath the loading point were extended slightly. In the third loading point, neither new cracks nor appreciable extension of existing cracks were observed. Consequently, in the fourth loading point initiation of diagonal cracking was observed on the other half of the beam. During the fifth loading point, most of the flexural cracks extended and the diagonal crack formed during the second loading point was extended as shown in Figure 10.22. This indicates that, once a weak zone is formed, damage tends to accumulate on the previously damaged part due to other loading points. This effect is observed to be higher when the applied load is in the close neighborhood of the damaged zone. Lastly, the beam failed in the first loading point of the second passage, which is applied at the mid-span. The already extended diagonal crack propagated and penetrated into the compression zone. No crack-to-crack interaction mechanism was observed during the whole loading process. While this observation is reasonable for the case of beams, it does not necessarily imply that the same is true for RC slabs.

The applied load (140kN) corresponds to approximately 75–80% of the maximum capacity. According to the experimental reports by Chang and Kesler (1958), Higai (1978) and Sabry (1979), the life for the same load level under fixed loading (at the mid-span) is approximately 1,000 cycles. Analytical results reveal similar results. The computed fatigue life at different load levels, under fixed fatigue loading, is also presented in Section 10.4.2.

This experimental fact partly indicates that the fatigue life of RC beams as well as RC slabs may be significantly influenced under moving loads. At the considered load level nearly three orders of reduction in life can be observed as compared to the fixed pulsating load. Given this experimental fact, an analytical investigation is made using the multi-scale, fixed four-way crack modeling of concrete, based on the strain path and time-dependent fatigue constitutive models, to verify its applicability for the case of beams subjected to moving loads.

The direct path-integral scheme discussed in Section 10.1 is used to analyze the experimentally observed results for the step-wise moving load. Implicitly, the actual static capacity is assumed to be the same as the analytically obtained result. Accordingly, the experimental and analytical results are compared and the results are shown in Figure 10.23. The analytical result shows a relatively slight overestimation. In consideration of the sensitivity of fatigue with the load level, the prediction by the analysis is not far from reality. Figure 10.23 shows the maximum principal strain distribution based on the analytical result. It can clearly be seen that the experimentally observed crack pattern is well simulated by the analysis. Similar crack patterns were also observed in cut-out sections of pre-stressed bridge slabs subjected to a moving-type load (Nakatani 2002). Overall, the experimentally observed results are reasonably simulated by the analysis and this partly verifies the applicability of the analytical tool for the investigation of RC beams under moving loads.

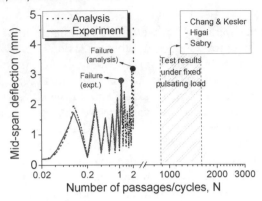

*Figure 10.23* Maximum principal strain distribution and progress of mid-span deflection with the number of passages.

## 10.4.2 Moving-load simulation and comparison with fixed-point loads

This section aims to analytically examine the behavior and fatigue life of RC beams under moving load on the basis of the multi-scale, fixed four-way crack modeling of concrete. For the purpose of investigation, a simply supported beam of span length 2 m is taken as a standard. Both beams with and without web reinforcement are examined. The shear span-to-depth ratio is 3.2, with a main reinforcement ratio of 1.47% and shear reinforcement ratio of 0.25%. The geometric, material and cross-sectional details for each are shown in Figure 10.24. The nominal shear capacity of the concrete is 99 kN. The shear capacity due to the web reinforcement, as estimated by the practical truss model, is 60 kN. The flexural yielding is estimated to occur at 220 kN of shear force, which is relatively higher than the shear capacities for each beam. In the analysis, the mesh model is generated by quadrilateral elements with eight nodes.

The beam is subjected to two types of loading, namely fixed pulsating and moving load. Fixed pulsating load is applied at the center of the beam with constant maximum load amplitude at a frequency of 1 Hz, until failure. Moving load is applied in such a way that the first three nodes on the left-most upper side of the beam are loaded in five incremental sub-steps. In the second step, the left-most of the loaded nodes is unloaded, while a new node, adjacent to the right side of the loaded joints, is loaded simultaneously with equal increments. In this manner a constant load is made to move to the right-most side, and similarly the procedure is repeated for each cycle until failure. The moving load is applied in an effective contact length of 10 cm, with a speed of 7.2 km/hr. The speed is equivalent to 1 Hz frequency of the fixed pulsating load. Thus, time-dependent effects for both types of loadings are similar. Each of the fatigue constitutive models discussed in Section 10.1 is time dependent, allowing loading speed effects to be taken into account automatically.

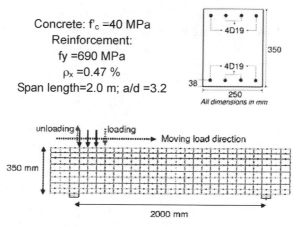

Concrete: f'$_c$ =40 MPa
Reinforcement:
fy =690 MPa
ρ$_x$ =0.47 %
Span length=2.0 m; a/d =3.2

*Figure 10.24* Moving load, geometric and material details.

Figure 10.25 shows typical responses of the progress of mid-span deflection with the number of cycles for both fixed pulsating and moving loads, at different stress levels. The amplitude of mid-span deflection for the same stress level is relatively large in the case of moving loads. In fact, a similar observation was reported in case of slabs (Matsui 1987). Here, the amplitude of the mid-span deflection for slabs is regarded as an indicator of the progress of damage. This may not necessarily be true for beams with shear failure mode, unlike the case of beams with bending failure mode.

Typical maximum principal strain contour plots for fixed pulsating and moving loads are indicated in Figure 10.25. The crack patterns for both types of loading are similar, except that the diagonal shear crack in the case of

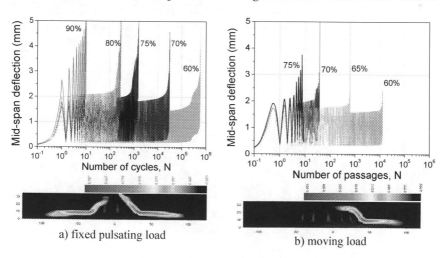

a) fixed pulsating load

b) moving load

*Figure 10.25* Load–deflection relation and maximum principal strain contour plot, beam without web bar.

moving loads runs parallel to the compression reinforcement before failure. This is quite similar to the experimentally observed crack pattern presented in Section 10.4.1. In addition, the flexural cracks in the case of moving loads are more extended to the compression zone and have no noticeable interaction with the diagonal shear crack. Based on the fatigue analysis, the life at different load levels is computed and the results are summarized in S-N diagram, normalized by the computed static capacity.

Figure 10.26 shows the S-N curves for the case of fixed pulsating and moving loads. The computed S-N curve for the fixed pulsating load almost coincides with the shear fatigue design equation proposed, based on extensive experimental results by Ueda and Okamura (1982). In contrast, the computed S-N curve for the moving-load case shows remarkable reduction in life, by nearly two or three orders, as compared to that of fixed fatigue loading. This is consistent with the experimentally obtained result under step-wise moving load (Section 10.4.1). The effect of a moving load is more pronounced at higher stress levels. This can be explained by the fact that the effect of stress reversal is predominant at higher stress levels. Interestingly, the order of fatigue life reduction is similar to that observed by Matsui (1987) for the case of RC slabs.

To investigate the effect of web reinforcement on the fatigue life of beams under moving load, a beam with web reinforcement is examined. The same details as for the case of the beam without web reinforcement are considered, except that the web reinforcement is 0.25%. As the main focus in this section is directed towards the fatigue degradation of concrete, which is predominant in the case of RC structures governed by shear, the authors disregard the fatigue rupture of web reinforcement. Indirectly, the reduction in shear resistance due to concrete is taken by the increase in the strain on the stirrups.

Figure 10.27 shows the experimental and analytical response of the beam with web reinforcement under static loading. The peak capacities for the experimental and analytical results are 331 kN and 339 kN, respectively,

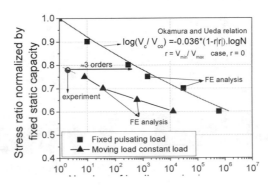

*Figure 10.26* S-N diagram for fixed pulsating and moving loads, beam without web reinforcement.

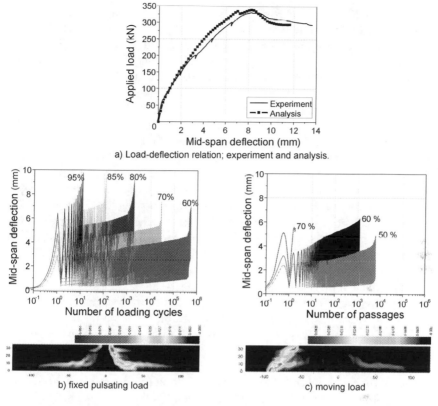

*Figure 10.27* Load–deflection relation and maximum principal strain contour plot, beam with web bar.

indicating close agreement. Prediction of the static capacity based on the practical truss model is around 308 kN. The difference may be attributed to the basic assumption of the truss model.

Figure 10.27b and c show the mid-span deflection in progress with the number of loading cycles and a typical maximum principal strain contour for the respective types of loading. As with the case of the beam without web reinforcement, the amplitude of mid-span deflection under moving load is higher. In contrast, the rate of increment of the amplitude is large for the beam with web reinforcement. Unlike the case of the beam without web reinforcement, the diagonal crack under the moving load occurs here near the support position. The difference is attributed to the variations of the ratio between applied shear force and shear capacity during movement of the load. When the load moves along the span of the beam, its behavior fluctuates from shallow to deep. In the case of the beam with web reinforcement, the critical shear location shifts towards the support position due to the invariability of the shear contribution by the web reinforcement with the movement of the load.

Figure 10.28 shows the S-N curves for both types of loading, and it can be seen that the computed curve for the fixed pulsating load is close to the shear fatigue design equation proposed by Ueda and Okamura (1982), with reasonable accuracy. The S-N curve for the moving load also shows a dramatic reduction in life, with a relatively pronounced effect as compared to that of the beam without web reinforcement. This difference may indicate the significance of shear transfer fatigue on shear cracks.

Overall, the above facts indicate that fatigue life in beams is reduced by two to three orders under moving loads as compared to fixed pulsating loads. Although the order of reduction is similar to that reported in the case of slabs (Matsui 1987), the mechanisms for the reduced life are not the same. The shear behavior of beams is strongly influenced by the shear span-to-depth ratio, unlike that of slabs. As a result, a constant moving load does not really represent a constant shear force-to-capacity ratio. In other words, if the moving load is applied in such a way that the shear force-to-capacity ratio is kept constant, the fatigue life may tend to be closer to that of the fixed pulsating load. Thanks to the advancements in computational mechanics, this type of loading can currently be simulated on an analytical basis, though practically it is not easy to conduct such an experiment. In the analysis, a constant shear force-to-shear capacity is maintained by varying the magnitude of the load with the movement away from the mid-span of the beam, according to the shear capacity relation proposed by Niwa *et al.* (1987).

Figure 10.29 shows the S-N curve for the constant shear ratio as compared to that of the fixed pulsating load. It is interesting to note that the fatigue life tends to approach to that of the fixed-fatigue loading at lower stress levels. Implicitly, the reduced fatigue life of beams under moving load is partly due to the combined effects of stress–displacement reversal and the variation of the shear force-to-capacity ratio at each section. In fact, the moving load induces not only a reversed stress path but also larger stress amplitude

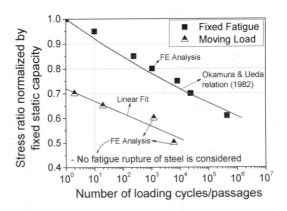

*Figure 10.28* S-N diagram for fixed pulsating and moving loads, beam with web reinforcement.

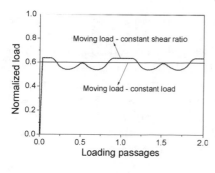 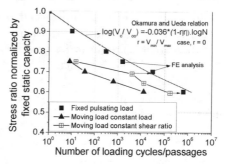

*Figure 10.29*   (a) Load profile for shear constant ratio, (b) S-N diagram for loading under constant shear ratio.

paths. Stress reversal is predominant mainly in the higher stress range near to the initiation of diagonal crack. On the other hand, at lower stress levels the fatigue life is governed by the combined effects of stress–displacement reversal and variation in the shear force-to-shear capacity ratio during the movement of the load.

The primary reasons for the reduced fatigue life of RC beams under moving load are the relatively large stress reversal caused by the movement of the load and the increase in the actual stress ratio caused by the movement of the load. The latter is caused by the variation of shear capacity with the a/d ratio. The relative influence of stress reversal becomes predominant with the initiation of diagonal cracks. With the movement of the load, the shear cracks are subjected to repeated displacement reversals. Shear transfer fatigue under displacement reversal is reported to cause a dramatic reduction in life, by nearly two to three orders, as compared to that without reversal (Gebreyouhannes *et al.* 2006). The accelerated reduction in shear transfer is accompanied by a further extension of the diagonal crack. This, in turn, creates high compressive stress on the uncracked portion of the section. As a result, the inclination of the crack becomes more horizontal and it runs parallel to the compression bar before penetrating the compression zone.

The mechanism of moving-load fatigue in the case of RC slabs is thought to be different as compared to that of RC beams. According to past investigations, the reduced life in slabs is due to the loss of punching shear capacity as a result of severe flexural cracking caused by the moving load (Matsui 1987). Stress reversal caused by the moving load is also a potential reason for the highly reduced life (Matsui 1987). Additional effects such as the loss of membrane forces could also be possible reasons.

In the case of slabs, extensive flexural cracks caused by moving loads may hinder the propagation of shear cracks in the longitudinal direction. The reversal of cyclic shear force along the extended flexural cracks largely reduces the stress transfer in the longitudinal direction. Hence, the slab acts no more as a single slab but rather as discrete beam elements. Consequently, the moving load acts in a similar manner as a fixed pulsating load for the beam–slab elements, so a diagonal shear failure plane appears only in the transverse direction normal to the traffic one.

Moving-load fatigue in RC beams is a two-dimensional problem in which the crack arrest mechanism in the longitudinal direction most likely does not prevail (Section 10.4.1). Moreover, unlike the fatigue life of slabs, that of RC beams is influenced by the variation of shear span-to-depth ratio caused by the movement of the load. However, according to current investigation, almost the same order of fatigue life reduction due to moving load, as in the case of RC slabs, is observed for RC beams, too. These facts suggest indirectly that the shear transfer fatigue could be the primary reason for the reduced fatigue life of RC slabs under moving loads. To reinforce the potential implications of this study for the case of slabs, further scrutiny is required.

### 10.4.3  Simplified fatigue life estimation

As explained in the previous sections, when a given beam is subjected to a moving load action, its behavior varies from shallow to deep beam. In other words, the beam may act as a shallow one when the load is close to the mid-span and will act as deep one when the load is close to the support, depending on the span length and effective depth of the beam. Thus, the apparent loading amplitude ratio (applied load normalized by central load static capacity) is not necessarily the governing load-to-capacity ratio, as the actual shear force-to-capacity ratio varies with the movement of the load. According to the equation for the shear design of beams by Niwa *et al.* (1987), the maximum stress ratio occurs at some distance away from the center. It is then evident that fixed fatigue at the point of the maximum shear stress ratio is more critical than at the mid-span.

This alternation of shear ratio during movement of the load is mentioned as one reason for the reduced fatigue life of RC beams under moving load with a shear failure mode. Accordingly, the shear fatigue life of RC beams could possibly be estimated based on the Japan Society of Civil Engineers (JSCE) standard specification formula proposed by Ueda and Okamura (1982). For a given beam, the maximum shear ratio can be expressed as a function of the apparent shear ratio and length to the effective depth ratio, as shown in Figure 10.30. Thus, the predicted S-N diagram based on the maximum shear ratio reasonably agrees with the result obtained by FE analysis, as shown in Figure 10.30. The difference is attributed to the accumulated damage caused by the loads away from the critical section. For practical purposes the difference can be covered by using an increased factor of safety, in the range 1.05–1.06 for the stress level.

### 10.4.4  Re-evaluation of current design

The idea that stress–displacement reversal plays an important role in the fatigue deterioration of RC members may indirectly imply that asymmetric-

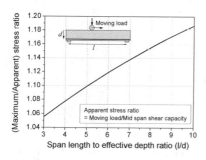 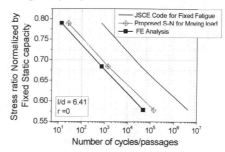

*Figure 10.30* S-N diagram based on simplified shear fatigue estimation method as compared to FE analysis.

type loadings could lead to more pronounced fatigue damage. Current fatigue assessment of RC members is based on fixed pulsating load at mid-span for beams or moving load at the central axis for slabs. However, these loading positions may not be the critical ones as far as fatigue is concerned. In fact, in practical situations, RC decks are subjected to traffic loading in a random manner. This random nature of loading may further aggravate the degree of fatigue damage.

Analytical investigation is conducted on beams with and without web reinforcement by applying an offset load alternating per cycle. First, a full cycle is applied at the center of the beam which is equivalent to one passage at the central axis in slabs. This is followed by the second and third full loading cycles of equal magnitude with an offset distance of 32.5 cm from the center, to the left and right of the beam, respectively. This asymmetric nature of loading, which will hereafter be referred to as *offset loading*, is supposed to create increased damage as compared to the fixed pulsating load. Figure 10.31 shows the computed S-N diagram for the offset loading as compared to fixed pulsating load. The life for the offset loading is observed to be lower by 1.0–1.5 orders for beams both with and without web reinforcement. The reduction in life is dependent on the offset distance and magnitude of the applied stress level. The effect is highly pronounced at higher stress levels due to the increased effect of shear displacement reversal. To investigate the effect of offset distance, loading at different values, $0.0l$, $0.1l$, $0.15l$ and $0.25l$, is examined.

Figure 10.32 shows the S-N diagram for each value of offset distance. There exists a critical distance from the center of the beam at which the fatigue life is largely reduced. This position is close to the mid-span of the beam, approximately at a distance of $0.1–0.15l$ from the center of the beam. If the offset distance is large (greater than $0.2l$), no pronounced effect to the fatigue life can be seen; rather, it tends to increase the fatigue life. This indirectly implies that loading points at a relatively far distance from the neighborhood of the critical position have less significance to fatigue life.

Overall, randomness of traffic loading may significantly influence the

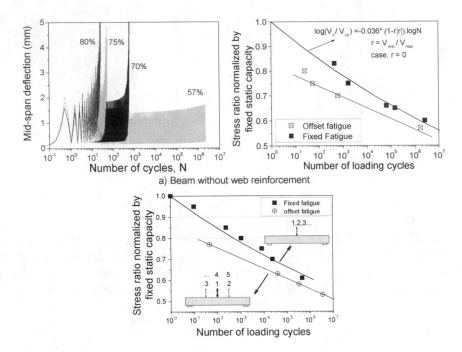

a) Beam without web reinforcement

*Figure 10.31* Effect of offset loading.

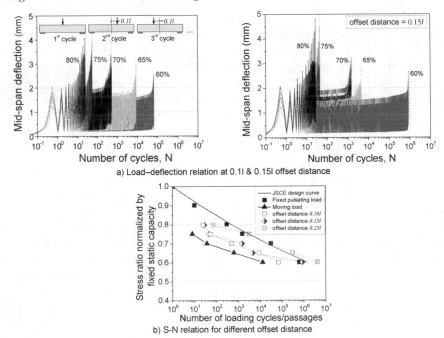

a) Load–deflection relation at 0.1l & 0.15l offset distance

b) S-N relation for different offset distance

*Figure 10.32* Effect of offset loading with loading distance.

fatigue life of RC structures. At material levels, fatigue damage becomes more critical if stress reversal is concerned. In the case of fixed fatigue loading, the stress path is single sided. However, in the case of moving and offset fatigue loading, the stress path varies in a reversed manner or a single-sided manner with larger amplitudes. According to material level experimental investigations, the relative damage for shear transfer under a reversed loading path is two or three orders higher than that of monotonic loading (Gebreyouhannes *et al.* 2006). Similarly, tension and compression fatigue are reduced by nearly two orders when stress reversal is concerned. It is also reported that fatigue life is highly influenced by stress range or amplitude (Cornelissen and Reinhardt 1984).

## 10.4.5 Effect of pre-induced vertical cracks

Concrete is an isotropic material with a relatively much higher compressive strength than its tensile strength. Its low tensile strength makes it susceptible to early-age cracking caused by excessive shrinkage and/or thermal effects. Thus cracks can easily be generated in RC members due to past loading and environmental attack. These cracks generally have a variety of widths and orientations and can influence the structural behavior of RC members. The influence of pre-cracks on the shear response of RC beams under static loading is reported by Pimanmas and Maekawa (2001d). Accordingly, RC beams with vertical pre-cracks show considerably higher shear capacity and reduced stiffness as compared to those which are not pre-cracked. The increase in shear capacity is attributed to the crack-to-crack interaction mechanism between the new and existing cracks, as illustrated in Figure 10.33. Given the mechanistic understanding of crack-to-crack interaction under static

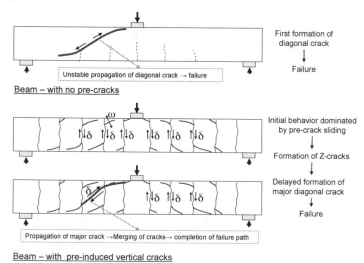

*Figure 10.33* Shear failure mechanism of pre-cracked beams.

loading, further investigation is required to check the viability and effect of the mechanism for cyclic loading paths.

The reduced stiffness of pre-cracked beams suggests that the amplitude of the mid-span deflection for pre-cracked beams is larger than that for sound ones, at least for the first few loading cycles. It has been reported that the amplitude of the mid-span deflection is an indicator for the progress of fatigue damage (Matsui 1987). Accordingly, it may be supposed that the fatigue life of pre-cracked beams is lower. On the other hand, it is also argued that the higher amplitude value does not necessarily indicate lower fatigue life, especially when the failure mode is in shear. Hence, it is of interest to gain an insight into how pre-existing cracks influence the fatigue behavior of RC members. The mechanics of shear anisotropy, which governs the activation or dormancy of cracks in an element with multi-cracks, is rooted in the analytical framework (Pimanmas and Maekawa 2001b). Thus, based on this mechanics rationale, the effect of pre-induced cracks on fatigue life is analytically investigated.

A beam with vertical pre-cracks is subjected to a fixed pulsating and moving-load fatigue. The pre-cracks are introduced by initially applying an axial force of 540 kN and then fully unloading the beam. The computed static capacity of the pre-cracked beam is observed to be higher by 11% as compared to the sound beam. The relative increase in capacity due to the presence of the pre-cracks is small as compared to that reported by Pimanmas and Maekawa (2001d), at 40–60%. This is because the applied axial load level is quite low as compared to the yield level of the reinforcement bars, resulting in small openings of the pre-cracks upon unloading.

Figure 10.34 and Figure 10.35 show typical fatigue response and the computed S-N diagram for both sound and pre-cracked beams under fixed pulsating and moving loads. The progress of mid-span deflection with the number of loading cycles at a load level of 60% is indicated for both types of loading. At this load level, the amplitude of mid-span deflection for the pre-cracked beam is higher by 20–30% than that of the sound beam. On the other hand, the fatigue life for the pre-cracked beam is longer by nearly one order as compared to that of the sound beam for fixed pulsating and moving loads. The increase is due to the fact that the formation of diagonal shear crack is delayed by the relaxation due to shear slip at the pre-cracks. At higher stress levels, the effect of pre-cracks is seen to adversely affect fatigue life because the majority of the anisotropic shear slip is mobilized during the first load cycle and no relaxation of stress concentration can be realized in subsequent cycles.

All in all, the fatigue behavior of pre-cracked beams as compared to sound ones is in close manifestation of their relative responses under static loading. The increased fatigue life due to the presence of vertical pre-cracks has important implications for the failure mechanism of slabs under moving load. In the case of RC slabs, vertical cracks transverse to the direction of traffic are created due to the combined effects of environmental factors and loading.

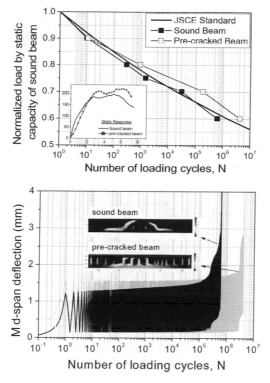

*Figure 10.34* (a) Progress of mid-span deflection b) S-N curves for pre-cracked and sound beam, fixed pulsating load.

The formation of these cracks may delay the formation of diagonal cracking in the longitudinal direction, thereby changing the failure mode from a radial type to a two-dimensional shear fracturing in the transverse direction.

## 10.5 Fatigue of RC slabs and three-dimensional verification

The fatigue life of RC bridge decks has long been a key issue in life-cycle design and maintenance. Maeda and Matsui (1984) and Matsui (1987) found a drastically shortened fatigue life of RC slabs, which was successfully verified through experiment, under traveling wheel loads. This pioneering work has greatly contributed to the fatigue design and codification of concrete bridges. Currently, new bridge deck structural systems, such as steel–concrete composites and non-metallic reinforcement, are being proposed. However, the performance of these new structures is mostly examined using full-size mock-ups in laboratories. There is, therefore, considerable demand for a versatile design concept. The authors presume that this is the right time to quantitatively clarify the mechanism of fatigue failure under both fixed and

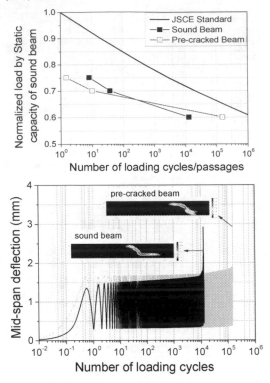

*Figure 10.35* (a) Progress of mid-span deflection, b) S-N curves for pre-cracked and sound beam under moving loads.

moving loads, with the shortened life resulting from traveling wheel-type loads being of particularly great interest.

In the field of bridge maintenance, seriously damaged RC slabs are sometimes discovered in service. There is demand for a quick and rational method of repair for such cases (Hasan *et al.* 1998; Oh *et al.* 2005). A method of retrofit design for damaged RC slabs is called for. Residual fatigue life and an effective repair design must be determined on a firm mechanistic basis to enable the assessment of remaining function. Otherwise, there is a danger of a non-strategic retrofit leading to costly maintenance – an outcome that has been experienced in real practice.

As material damage develops with three-dimensional scope in RC slabs under high-cycle load repetition, reduced degree-of-freedom elements such as Mindlin plates and three-dimensional fiber elements based on the in-plane hypothesis become useless for analysis purposes, because the failure mode is instead out-of-plane. This makes full three-dimensional treatment indispensable. In this section, the validated in-plane RC model incorporating creep and fatigue evolution (Section 10.4) will be extended into three-dimensional orthogonal space by means of the composition method

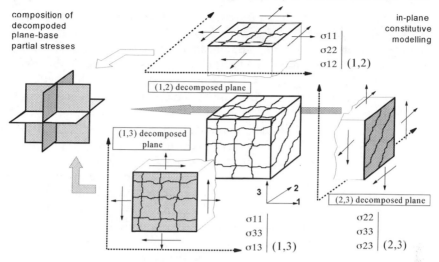

*Figure 10.36* Composition of sub-planes including non-orthogonal multi-directional cracks (Maekawa *et al.* 2003b).

proposed by Maekawa *et al.* (2003b), as summarized in Figure 10.36. This composition method is regarded as a simple extension of the multi-directional, non-orthogonal fixed-crack approach and has already been verified under low-cycle static and dynamic loads. In the past, RC columns subjected to mixed flexure, shear and torsion were targeted to investigate seismic performance. In the section that follows, the applicability of this approach to three-dimensional fatigue problems will be discussed. Whatever the complexity of the loading hysteresis, the multi-axial stress-carrying mechanism is formulated as a linear combination of one-dimensional sub-mechanisms representing the cracked concrete and reinforcement, as proposed by Collins and Vecchio (1982).

### 10.5.1 Failure mode under fixed and moving loads

The simply supported RC slab subjected to moving wheel-type loads described by Maeda and Matsui (1984) is used for the purpose of verification. The plane dimensions of the slab are 2 m × 3 m and it is 19 cm thick. Single layers of reinforcement consisting of D16 at 9 cm and D16 at 18 cm are provided on the tension and compression sides in the transverse (main) direction. The bar center-to-surface distance is 3 cm in both layers. The tension and compression reinforcement in the longitudinal direction (wheel-travel direction) consists of D13 at 25 cm and D13 at 40 cm, respectively. In this case, the bar center-to-surface distance is 4.45 cm. No web reinforcement is used. In computation, 30 MPa is assumed to be the compressive strength of the concrete and high-yield strength is assumed for the reinforcement (800 MPa) to prevent fatigue rupture of the steel. Analysis is carried out on a half-domain of the slab with the x-coordinate (load travel direction) defined as the axis of symmetry, as shown in Figure 10.37.

In the experiment, the two longitudinal edges were simply supported and transversely supported by elastic steel girders to replicate the greater dimensions of real bridge decks. In the analysis, all four peripheral edges are simply supported with free rotation and unbound horizontal translation, for simplicity. Actually, fatigue life is found to be computationally less sensitive to the effect of transverse boundary conditions, although deflection of the slab is much affected. The effect of transverse elastic support by steel girders will be discussed together with failure modes in the following section.

Two types of load, cyclic fixed-point loading and traveling wheel-type loading, are applied to allow for a comparison of failure mode and fatigue life, as shown in Figure 10.37, where the magnified displacement profile is drawn inverted for ease of understanding. As a wheel-type load (tire) is not a concentrated sharp line but rather a belt action with breadth, forces are applied simultaneously on six adjacent nodal points on the central strip of finite elements (see Figure 10.38), and the center of the resultant forces was gradually moved by shifting three sets of nodes on the loading FE strip. This results in a computed width of the wheel track of 17.3 cm.

The magnified direct path-integral is applied with the binary increasing magnification factor with each passage along the slab axis. Approximately 50 time-steps were implemented for a single load pass. About 3,000 steps in total were required for the whole simulation of one million passes. In order

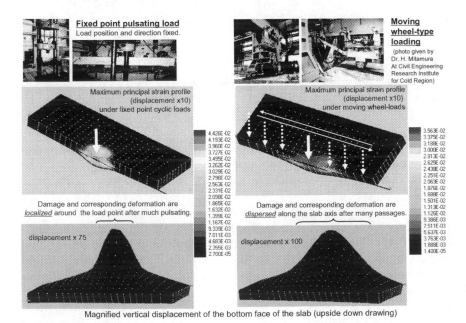

*Figure 10.37* Magnified deformations under different types of loading.

to accelerate the computation, a multi-frontal direct linear sparse matrix solution (Schenk and Gartner 2004) was applied. As a result, it takes only a couple of hours to complete the one million fatigue cycles. This may be faster than a real fatigue test. For each step, about 10 iterations were needed to reach equilibrium, at which unbalanced nodal forces are allowed to be no greater than 1% of the total applied forces. The total degree of freedom is about 4,200 for the discretized FE model (see Figure 10.37).

There was a great difference in failure modes with the two types of loading. Magnified details of the mode of failure are shown in Figure 10.38 for fixed-point loading. Several FE strip layers are selected and shown in the magnified displacement diagram along the track and transverse directions. The so-called punching shear mode of failure is identical with the conical-shaped plane of failure (Higai 1978; Kakuta and Fujita 1982; Graddy *et al.* 2002) under concentric pulsation. Localized deformation of the slab in both longitudinal and transverse directions takes place and distinct concentrated downward displacement is visible at the center of the slab.

In the case of moving loads, a diagonal shear failure plane also appears in the transverse direction normal to the wheel-track, but typical shear failure modes are not in evidence in the longitudinal direction of the slab (see Figure

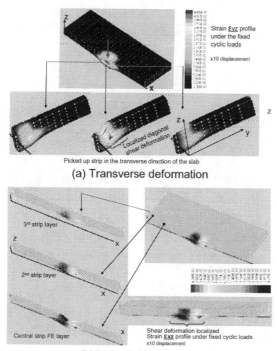

*Figure 10.38* Magnified deformation of RC slab under repeated fixed-point loading.

10.39a and Figure 10.39b). This computational simulation is very consistent with experimental observations. The Figures also show cross-sections obtained after the experiment by cutting the specimens.

Under moving loads, multi-directional flexural cracks occur over the whole domain of the RC slab. Diagonal shear cracks in the longitudinal direction are prevented by crack-to-crack interaction (Pimanmas and Maekawa 2001a). As a result, the load-carrying mechanism evolves from punching shear to semi in-plane load-carrying over the transverse direction. The three-dimensional direct path-integral scheme makes it possible to reproduce the mechanistic character of failure modes and deformation. Accordingly, the fatigue life is not necessarily invariant but can differ with loading and boundary conditions. This is discussed in the following section.

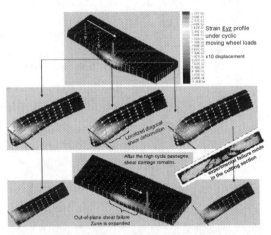

(a) Transverse deformation

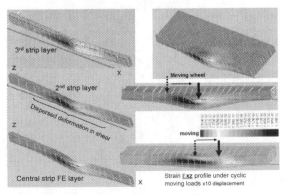

(b) Lateral deformation

*Figure 10.39* Magnified deformation of RC slab under traveling-wheel-type loading.

### 10.5.2 Computed fatigue life

Under high-cycle wheel load, deflection at the center of the slab gradually increases. The important issue is that the residual displacement also progresses, as shown in Figure 10.40, and the computational result agrees fairly well with reality. The drastically shortened fatigue life observed under this type of dynamic loading, as shown in Figure 10.41, is particularly noteworthy. Under fixed-point loading, the fatigue life diagram (S-N diagram) is quite similar to that of RC beams (Kakuta and Fujita 1982) but, with wheel-type loading, the slab life-cycle is considerably degraded (Maeda and Matsui 1984; Matsui 1987; Perdikaris and Beim 1988; Perdikaris *et al.* 1989).

Broadly speaking, life is reduced by 1/100–1/1000 in the computations, consistent with experimental observations. As explained by Maeda and Matsui (1984), the load-carrying mechanism may evolve as a result of crack-to-crack interaction and this results in a reduced area of diagonal shear fracture planes. This shift in the S-N diagram is well predicted by the full three dimensional fatigue analysis, in which load cycle at failure is judged

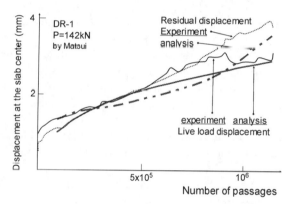

*Figure 10.40*  Deflection at the RC slab center.

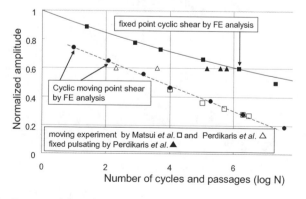

*Figure 10.41*  Computed S-N diagrams of RC slabs.

by monitoring the mode of out-of-plane deformation and deflection at the center of the slab. Fatigue shear failure is recognized as having occurred when a diagonal crack continues to expand but deformation of the surrounding block elements indicates recovery (shear localization).

### 10.5.3  Sensitivity of member life to microscopic fatigue damage

A series of sensitivity analyses were carried out on a non-web reinforced RC slab (see Section 10.5.2) to investigate the mechanism of shortened fatigue life under moving loads. Figure 10.42 shows deflection versus loading cycles when the amplitude is moderate (37% of static strength under fixed-point loading). This load level is not high enough to induce non-linear creep, so time-dependent deflection is negligible except just after loading. On the other hand, shear transfer fatigue along cracks and bond decay (tension stiffness) are predominant in both overall deflection and fatigue life. Concrete compression fatigue has comparatively little influence on member fatigue behavior. In fact, attention has been paid to the deterioration of shear transfer along crack planes in previous studies and some flakes and powder were reported to be removed from the crack planes. This sensitivity analysis confirms such observations quantitatively.

In the case of low-cycle fatigue under relatively higher amplitudes, the time dependency of material mechanical properties becomes comparatively larger because of the greater magnitude of sustained stresses during loading, as shown in Figure 10.43. In this case, the behavioral tendency is opposite to that when stress levels are lower. If the time-dependence of material properties is ignored, the computation forecasts shorter fatigue life because stress relaxation is not taken into account. Shear transfer decay under repeated

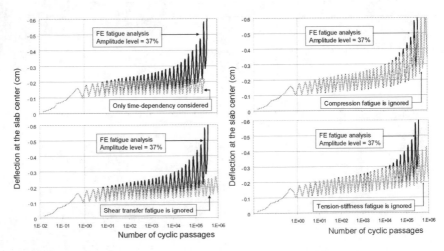

*Figure 10.42*  Fatigue life sensitivity analysis of RC beams without web reinforcement (higher cycle).

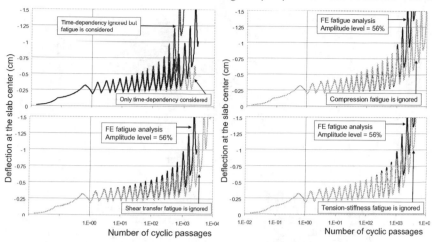

*Figure 10.43*  Fatigue life sensitivity analysis of RC beams without web reinforcement (lower cycle).

stress is a chief factor affecting RC slab life, as is the case in Figure 10.42. Furthermore, it should be noted that compression fatigue of the concrete cannot be ignored, either.

### 10.5.4 Sensitivity of shear transfer to fatigue life in water

Water exposure may affect the shear resistance of cracks in RC members. The authors investigated the effect of water on the shear degradation of cracked interfaces under two exposure conditions, namely when fully submerged and with water flowing downward through the cracked interface.

A specimen with a single crack is subjected to a reversed cyclic shear loading while simultaneously being exposed to water kept in the tank, as shown in Figure 10.44. Another specimen is subjected to the same loading condition under a dry case. The initial conditions are slightly different for both cases, with a 0.07 mm difference in initial crack width and 6 MPa in compressive strength. Bearing these variations in mind, the first cycle was applied statically under dry conditions to ensure the similarity of the static response, thus allowing comparisons to be made without bias. The dilation at the first peak loading is on average lower by 0.25 mm than the dry case. The differences between the specimens make direct comparison complicated. Nevertheless, the results point to better performance for the dry specimen.

Following the first cycle, the specimen was subjected to reversed cyclic loading under fully submerged conditions for up to 31 loading cycles. The general trend of the response is similar to that of the dry specimen in air. Further, the relative difference in propagation of shear slip between dry and submerged cases is small, with slightly higher values for the submerged case. It can be said that, although water has some accelerating effect on the damage

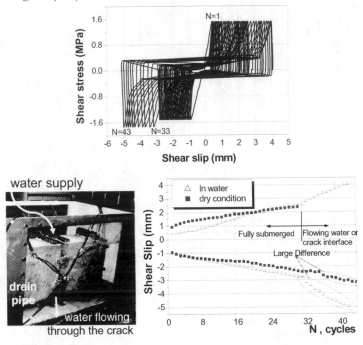

*Figure 10.44* Shear transfer response under fatigue loads in water.

rate when a specimen is submerged, the relative magnitude of the change is not large. Here, the water is kept within the tank and no direct water is supplied to the crack interface.

The experiment is continued by draining the water from the tank and water is directly poured into the crack interface with a downward flow, as shown in Figure 10.44. The specimen is further loaded for 11 cycles with an increased maximum shear load level of 1.70 MPa. This exposure condition greatly increases the probability of washing out crushed sludge particles due to the penetrating flow of water. The results are also indicated in Figure 10.44. A significant difference in degree of damage can be seen as compared to the test under dry conditions. If the results for the dry test are extrapolated, around 85 additional cycles are required to reach the same shear slip value exhibited by the case with downward water flow. This accelerated response in water can mainly be attributed to the washing out of crushed particles from the cracks. Further, water at the interface may act as a lubricant, thereby accelerating shear slip.

It can be concluded that water, in addition to reversed cyclic loading, can play a significant role in the shear deterioration of cracked interfaces. Hence, starting with the fatigue constitutive model for shear transfer in air (Maekawa *et al.* 2006a; Gebreyouhannes 2006), the authors proposed a simple computational model for transferred shear stress under repeated loading which takes into account the effect of water as

$$\tau = X.\tau_o(\delta,\omega)$$

$$X = 1 - \frac{1}{10}\log_{10}\left(1 + \int \left| d\left(\xi \cdot \xi_w \frac{\delta}{\omega}\right)\right|\right) \quad , \geq 0.1 \tag{10.5}$$

where $\tau$ is the transferred shear stress, $\tau_o$ is the original stress computed by the contact density model (Li and Maekawa 1987), $\delta$ and $\omega$ are crack shear displacement and opening, respectively, and $\xi$ and $\xi_w$ are acceleration factors to include the effects of loading pattern and water exposure.

Factor $\xi$ is taken to be 1.0 for single-sided fatigue and three orders higher for the reversed cyclic loading case (Gebreyouhannes 2006). The value of $\xi_w$ expresses the effect of water. As shown in Figure 10.44, it takes about 12 cycles to have the shear slip increment of 0.6 mm (from 2.4 to 3.0mm) in dry air, but just 4 cycles for 1.2 mm (from 2.8 to 4.0mm) in water. Broadly speaking, one cycle in water is equivalent to 10 cycles in air. Then, the authors adopted 10 as the value of $\xi_w$ for sensitivity analysis. The proposed model can tentatively be applied to combinations of dry and wet cycles, either with or without stress reversal, by taking definite integrals of the respective loading cycles.

In actual situations, the RC infrastructure is normally exposed to continuous wetting and ambient drying over long periods of time (Boothby and Laman 1999). Hence these combined exposure conditions should be studied in more detail so that a better simulation of reality can be constructed. Matsui (1987) investigated the effect of water on the fatigue life of RC decks submerged in water. Fatigue life was reported to be 1/250 less than that of a deck in dry conditions. Figure 10.45 offers an example simulation of a submerged RC slab whose details are the same as those in Section 10.5.1 and in which the wet-accelerated condition of shear transfer is assumed to be given by Equation 10.5.

The computed fatigue life is dramatically reduced, with about one order fewer cyclic passes. At higher amplitude, the reduction in fatigue life

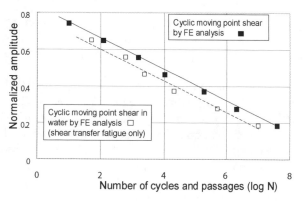

*Figure 10.45* Effect of water on the fatigue life of RC slabs: shear transfer.

is comparatively small because shear transfer has a minor effect in life, as discussed in the previous section. The results match the experiment in overall tendency, but on a quantitative basis the computation overestimates the structural fatigue life. This means that reduction of shear transfer due to wetting is an important factor in life reduction but it cannot solely explain slab performance. It has been reported that concrete on the bending compression side is also damaged seriously under coupled conditions of water with load repetition. The opening–closing of crack planes with water present may accelerate compression damage, too.

### 10.5.5 Effect of water on fatigue strength of concrete in compression

It is known that the compressive fatigue strength of concrete decreases under moist conditions and this characteristic is specified in some practical design codes. In fact, calcium hydroxide is released by concrete in water (Ozaki *et al.* 1995). Further, a great deal of cement sludge has been found between the pavement laid on bridges and the concrete slab deck. As the compression fatigue of concrete plays a substantial role under moving loads of medium intensity, leading to a life span of $10^3$–$10^4$ cycles (see Figure 10.43), the effect of water supplied as rainfall may be a deterioration of the life of the deck as well as shear transfer. To investigate this, a sensitivity analysis was carried out as follows.

The compressive fatigue strength of concrete under submerged conditions is known to be reduced by approximately 30% to around one million cycles, or the fatigue life is reduced to the order of 1/100 as compared to the fatigue life in air (Raithby and Galloway 1974; RILEM Committee 1984; Kaplan 1980). Broadly speaking, this implies that the rate of damage evolution is magnified 100-fold by water. Thus, to carry out sensitivity analysis related to the fatigue strength of concrete compression, the authors apply a magnified fatigue rate as

$$\lambda = 1000(1 - K_c) \cdot g \tag{10.6}$$

where the term $K_c$ influences the slope of the S-N curve in Equation 8.18. Thus, fatigue strength is altered so as to obtain a reduced life similar to the experimental evidence regardless of amplitude level, as shown in Figure 10.46.

Figure 10.47 shows the computed S-N diagram for the RC slab subjected to passes by wheels. The computed fatigue life is reduced by about 1/50–1/80, provided that fatigue deterioration of both shear transfer and concrete compression under submerged conditions is taken into account. This result is not so far from the experimental evidence (Matsui 1987; Nishibayashi *et al.* 1988). However, in reality, a fatigue life of about 1/250 has been reported experimentally in the laboratory. Thus, the computed fatigue life of RC slabs under water is somewhat overestimated.

In this sensitivity analysis, compression fatigue test results in water are

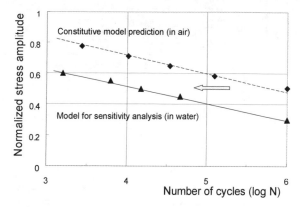

*Figure 10.46* Computed S-N diagram of concrete in compression under water.

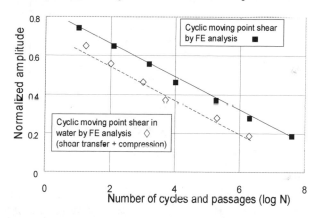

*Figure 10.47* Effect of water on the fatigue life of RC slabs: shear transfer and compression.

used. These specimens were well cured and it is thought that there was no clear crack on their surface. In fact, some cracks just above reinforcing bars were found at the very beginning of loading due to drying shrinkage and/ or initial defects and stress concentration close to the bars. Thus, water can penetrate into early-crack planes and it might induce a higher rate of deterioration. Initial defeats and their influences on material and structural fatigue lives will be discussed in the future study.

## 10.5.6 Effect of boundary conditions

It has been experimentally demonstrated that the fatigue life of slabs under traveling wheel-type loads is prolonged if displacement at the peripheral edge boundaries is inhibited. In order to qualitatively verify this effect, sensitivity analysis was carried out by setting up suitable fixed boundary conditions for displacement. The four sides of the referential slab were idealized as being

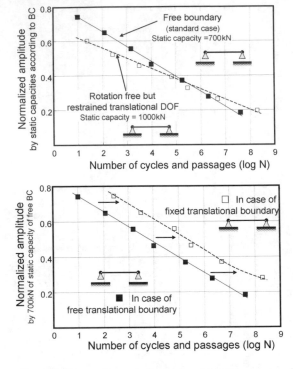

*Figure 10.48* Influence of boundary conditions on S-N diagram (analysis).

firmly fixed in the transverse direction, while a rotational degree of freedom was allowed. Figure 10.48 shows the computed S-N diagrams of the RC slab under these boundary conditions.

The number of fatigue punching shear failures increases by more than one order. The failure mode also changes, as shown in Figure 10.49. The zone of shear localization is seen to be more inclined than in the case of a free-translation boundary (see Figure 10.39). It can be understood that diagonal compression arises between the wheel track and the boundary support as a so-called arch action. Here, we should note that the static punching capacity also increases from 700 to 1000 kN under the fixed boundary conditions. Figure 10.48 shows the normalized S-N curve based on the fixed-point static capacity according to the assumed boundary conditions. The slope of S-N curve appears to be slightly changed, but a constant slope may be assumed for practical simplicity.

The damage in the longitudinal (travel) direction is not strongly localized but rather has a blunt appearance, similar to that shown in Figure 10.39. As a sensitivity check, a simply supported two-way RC slab, set up on the two edge lines in the direction of travel, was analyzed. The computed S-N diagram is similar to Figure 10.41 for traveling loads. Under traveling wheel loads, the diagonal crack planes do not propagate toward the track line but

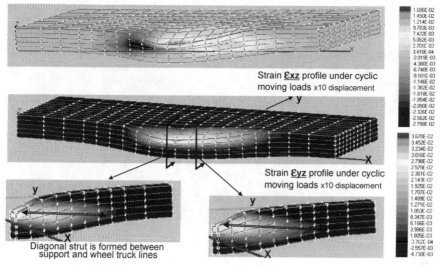

*Figure 10.49*   Fatigue shear failure mode with diagonal strut formation under fixed boundary.

spread in a restricted direction toward the main supports. In other words, the failure mode and corresponding fatigue capacity hardly change with transverse boundary conditions. This result qualitatively matches reality. As Matsui (1987) pointed out, the fatigue life can be characterized with respect to failure mode. It can be concluded that the performance of an RC slab under highly pulsating loads chiefly depends on the boundary conditions in the direction of travel.

## 10.6  Three-dimensional fatigue model of RC joint interface

The mechanistic non-linearity of concrete structures can be represented by combining continuum solids with dispersed cracks and discrete joints at points where distinct localization develops (Okamura and Maekawa 1991). In general, smeared-crack finite elements of two-dimensional or three-dimensional extent are allocated to structural members and the joint interfaces connect different members to take into account the localized deformation accompanying steel pullout, transverse shear slip and opening/closing of the joints, as shown in Figure 10.50. This section concentrates on the fatigue mechanics of RC joint interfaces and the computational simulation of structures under high-cycle load repetition.

Okamura and Maekawa (1991) raised the basic scheme of RC joint interface to deal with the pullout of reinforcing bars from footing for seismic analyses of shear walls. To meet the challenge, the strain–slip relation of embedded deformed bars and the shear transfer modeling along rough cracks were incorporated. Here, the integrated formula of the local bond–slip–strain

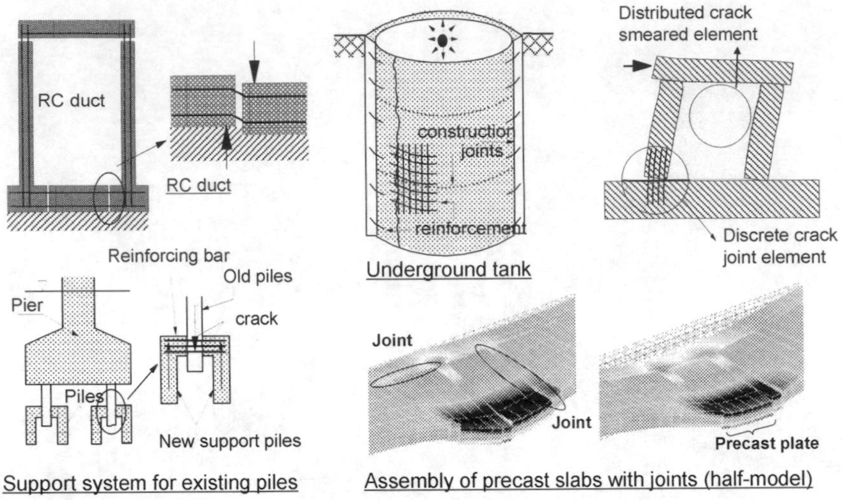

*Figure 10.50* Combination of smeared and discrete finite elements for structural concrete.

relation (Shima *et al.* 1987; Shin *et al.* 1988) was used with regard to the pullout slip of steel at the crack plane. In addition, the contact density model for rough crack interfaces by Li and Maekawa (1987) was built into the structural RC joint interface modeling. In view of the seismic performance of RC columns and shear walls, the basic scheme was fairly verified through systematically arranged experiments. Since then, much effort has been directed to developing constitutive modeling of steel pullout, shear transfer and dowel actions (Maekawa *et al.* 2003b) associated with local curvature of reinforcement (Maekawa and Qureshi 1996; Soltani and Maekawa 2007). In this section, the authors aim at extending these schemes to structural fatigue under repeated service loads as well as seismic ones.

In practice, structural performance needs to be examined, and direct integration of the space-averaged, path-dependent constitutive models is carried out for low-cycle seismic action as well as high-cycle fatigue, as shown in Figure 10.2. This general scheme was initially constructed for smeared-crack modeling (e.g., Collins and Vecchio 1982; Okamura and Maekawa 1991) with dispersed cracks in finite domains. In a similar way to its application to seismic analysis, it is expected that the scheme will also prove feasible for structural concrete with localized joint interfaces under fatigue loads. To verify this assumption, it is necessary to clarify the localized mechanics at the exact location of cracks or close to the interface. In particular, in this section on fatigue, the focus is on the bond pullout and dowel action of reinforcing bars at joints.

The fatigue simulation of structural concrete with interfaces is expected to be useful particularly for pre-cast structures with and without pre-stressing. In such structures, the source of non-linearity is chiefly the joint mechanism.

In this section, the effect of loading conditions is the main interest, so moving-wheel action is compared with the case of fixed-point loading, with attention directed primarily at the contrast with monolithically constructed RC slabs in which non-orthogonal multi-directional cracking is allowed.

The load-carrying mechanism of an RC joint interface consists of stress-transfer along concrete cracks or joint planes and the stress carried by reinforcement crossing the plane, as shown in Figure 10.51. When the reinforcement is arranged normal to the joint, we have

$$\sigma = \sigma_c + p \cdot \sigma_s$$
$$\tau = \tau_c + p \cdot \tau_s \qquad\qquad (10.7)$$

where $(\sigma, \tau)$ is the total stress applied to the unit plane of the joint, $(\sigma_c, \tau_c)$ is the stress transferred by interlock or frictional contact between the pair of joint surfaces, $p$ is the reinforcement ratio, $\sigma_s$ is the steel pullout stress, and $\tau_s$ is the dowel stress acting on the section of a single reinforcing bar. These stress components have been formulated under monotonic and cyclic action with a small number of repetitions in terms of joint opening, denoted by $\omega$, and shear slip along the joint, expressed by $\delta$. Opening $\omega$ is the section-averaged value having geometrical consistency with reinforcement pullout and is not exactly the same as the crack gap measured at the surface of the structure (Atimtay and Ferguson 1973), as shown in Figure 10.51. In this section, the formulation is extended to cover high-cycle fatigue.

### 10.6.1 High-cycle pullout of reinforcing bars

Shima *et al.* (1987) found a unique correlation between local bond stress, bond slip and the steel strain as a point-wise constitutive law. Their formula was integrated over the embedded reinforcement and a unique relation between the pullout slip of the reinforcement and the pullout force was mathematically and experimentally derived for a sufficiently long embedment length. This formula was extended to more general conditions with bond

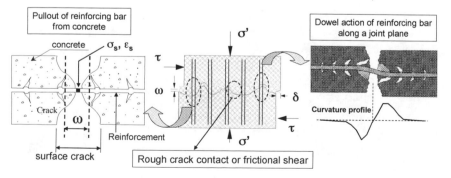

*Figure 10.51* Basic scheme for joint interface modeling.

deterioration zones close to joint planes (Okamura and Maekawa 1991; Maekawa *et al.* 2003b) as

$$\omega = 2S$$

$$s = \varepsilon_s \left( \beta + 3500\varepsilon_s \right)$$

(10.8)

$$s = \frac{S}{D} K_{fc}, \quad K_{fc} = \left( f'_c / 20 \right)^{2/3}$$

where $S$ is the pullout slip of the reinforcing bar, $s$ is the non-dimensional pullout slip, $f_c'$ is the concrete compressive strength in MPa, and $\beta$ is a non-dimensional factor related to the degree of bond deterioration close to the crack plane, with $\beta = 2$ for no deterioration in an idealized case (Shima *et al.* 1987) and $\beta = 6$ for normal bond deterioration of the RC joint interface. This factor has been adopted as the material constant for static and seismic analyses at low cycles. The symbol $\varepsilon_s$ represents the steel strain at the joint, as shown in Figure 10.51. Based on Equation 10.8, the cyclic hysteresis loop was also formulated by Shima *et al.* (1987). This formula is now codified for calculating the ductility of RC members in railway infrastructure.

However, the local bond gradually deteriorates with high cycling of the pullout forces and the strain of steel close to the free surface of the joint steadily increases, as shown in Figure 10.52 (Yamada *et al.* 1991). The pullout slip is equal to the integrated strain profile over the reinforcement. In general, the local bond corresponding to the gradient of the strain profile is much reduced near the free end of the embedded bar, but it is almost constant inside the concrete. Generally, bond deterioration proceeds close to the joint surface, while interior bonding hardly changes.

In Figure 10.53, the equivalent strain profiles are illustrated by dotted lines. These are shifted curves of the first loading cycle ($N = 1$) and lead to the pullout displacement (integral of strain profile) corresponding to each number of cycle ($N$). The shifted lines almost coincide with interior strain

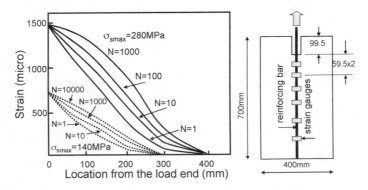

*Figure 10.52* Strain profiles along reinforcing bar under high cycle fatigue (Yamada *et al.* 1991).

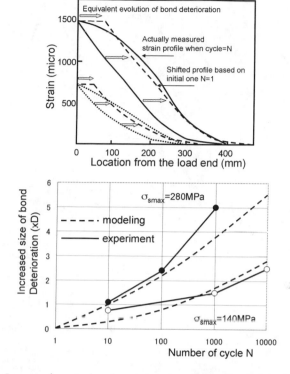

*Figure 10.53* Increased equivalent bond deterioration under fatigue loads.

profiles along reinforcement under any number of fatigue loading cycles. Then, the drift of the shifted curves can be defined as the equivalent degree of bond deterioration. This implies that the original bond pullout model can be upgraded simply by modifying the parameter related to the degree of bond deterioration, that is to say, $\beta$ in Equation 10.8 and Equation 10.9.

The inversely identified increased bond deterioration for different load levels is also illustrated in Figure 10.54. In general, the degree of damage evolution is nearly constant on the time–logarithmic scale. It is proportional to the stress applied at the extreme limit of the pullout force. The authors therefore propose the fatigue model of bond deterioration as

$$\beta = \frac{\beta_o}{z}, \ \beta_o = 6, \ z = 0.5 - 0.05 \log_{10} G_b < 1$$

$$G_b = \ln(10) \int \frac{m}{10^{m(s_y - \hat{s})}} d\hat{s}, \ m = 12 / s_y \tag{10.9}$$

where $s_y$ is the normalized yield pullout slip computed by Equation 10.8 with $\varepsilon_s$ = plastic yield strain, and $G_b$ is the damage index of bond deterioration.

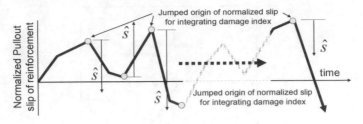

*Figure 10.54* Local pullout slip coordinates for integration of fatigue damage.

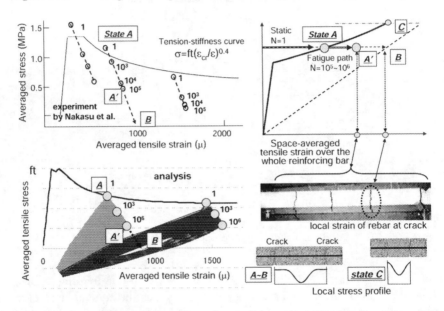

*Figure 10.55* Consistency with fatigue tension stiffness modeling (Nakasu and Iwatate 1996).

The differential unit denoted by ŝ expresses the fatigue amplitude and its local origin is set at the turning point of pullout, as shown in Figure 10.54. The consistency of the model with reality can readily be seen.

Application of the proposed model, as given by Equation 10.9, is limited to the steel pullout at a single crack or joint plane connected to a massive concrete solid. For the case of a dispersed cracked domain, the fatigue tension stiffness model is available (Maekawa *et al.* 2006a), as shown in Figure 10.55. As this space-averaged tension stiffness model incorporates delayed cracking due to fatigue and decay of the local bond, the two fatigue models cannot be compared directly. However, when loading is not that much above the cracking level (that is, when the crack spacing is large and the concrete between adjacent cracks can be assumed to remain stiff), the increased mean tensile strain predicted by the tension stiffness model closely correlates with the local pullout predicted by Equation 10.9.

Figure 10.55 shows the actual fatigue path from State A to State B, over which the averaged crack spacing is thought to be stable. In this experiment by Nakasu and Iwatate (1996), the average strain increment from A to A' is about 0.0002 after 100,000 cycles. As the experimental crack spacing is nearly 200 mm, the fatigue pullout from the concrete is approximately 200 mm × 0.0002 = 0.04 mm. As the maximum steel strain at the crack and the bar diameter are about 0.001 and 13 mm, respectively, a unit deterioration in the bond causes 0.001 × 13 mm = 0.013 mm of pullout. Then, the increased bond deterioration normalized by bar diameter is about 3 (0.04/0.013). Broadly speaking, this nearly matches the prediction also as shown in Figure 10.53. Thus, the smeared-crack approach in terms of tension stiffness (Maekawa *et al.* 2006a) and the discrete joint model presented in this book appear to be consistent in serviceability conditions.

## 10.6.2 High-cycle fatigue modeling of shear transfer

Stress transfer through a pair of joint planes in contact with each other depends greatly on the microscopic structure of the planes. Frictional contact corresponding to rather flat planes is the most fundamental means of transfer and this mechanical property is represented mainly by the coefficient of friction (Tassios and Vintzeleou 1987). In this case, the shear dilatancy associated with shear slip can be ignored and perfect plasticity can be assumed, as illustrated in Figure 10.56. As joint separation causes sudden loss of contact, the transferred shear is modeled to disappear in this situation. After re-contact, the transfer of shear stress begins again. This simple path-dependency is assumed for frictional contact planes in the structural analysis discussed in a later section. It is suitable for application to direct contact between different material solids (e.g., Lotfi and Shing 1994). The effect of fatigue will be represented by a reduced coefficient of friction. Further, Carol *et al.* (1997) presented a softening model of the joint interface to describe the transient state of cracking. Here, the interaction of softened crack shear and gap opening is taken into account. For simplicity in the structural analysis, a sudden jump from continuum to perfect separation is assumed. When we deal with rough crack planes, the contact density modeling discussed in Section 10.1.2

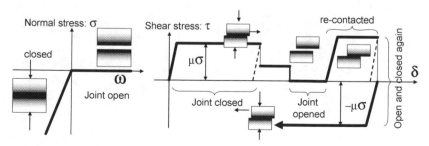

*Figure 10.56* Mechanism of frictional contact joint.

and Section 10.5.4 can be applied with the fatigue degradation factor and the computed transferred stresses is to be substituted into Equation 10.7.

### 10.6.3  High-cycle fatigue of dowel action

For normal rough cracks in concrete, the dowel action of reinforcing bars plays a minor role because of the significantly higher shear stiffness of crack joints with aggregate interlocking. For this reason, dowel action is generally ignored except where the reinforcement ratio is 6% or more. On the contrary, the dowel action of reinforcing bars becomes predominant if the interface planes are rather flat, such as in the case of construction joints and frictional interfaces between old and new concrete surfaces. For this reason, fatigue modeling of joints is required for structural simulation of pre-cast structures with lots of connection planes.

Monotonic and low cyclic models for dowel action have been proposed by Poli *et al.* (1992), Maekawa and Qureshi (1996) and Soltani and Maekawa (2007). These mechanistic models idealize the reinforcing bars as slender flexural beams of circular section that follow Euler–Kilchhof's in-plane hypothesis and are transversely supported on a fictitious bed. Dowel action is then represented by the shear force at the joint location. Maekawa and Qureshi (1996) proposed an intrinsic profile of curvature along the reinforcement and formulated the dowel force based on deformation compatibility in terms of pullout slip and transverse shear as

$$\tau_S = J(s, \delta) \tag{10.10}$$

where $s$ is the non-dimensional pullout slip defined by Equation 10.8.

The dowel strength of bars is associated with local yielding within the concrete except in the immediate vicinity of the joint, because the extreme fiber stress of a section within the concrete reaches a maximum due to the large induced curvature. The section at the joint plane achieves the maximum axial force but the local curvature is zero. This means that dowel strength depends greatly on the diameter of the reinforcing bars, as shown in Figure 10.57.

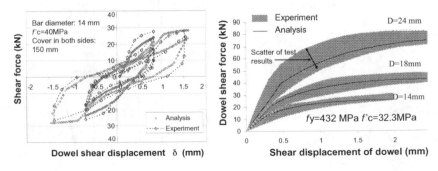

*Figure 10.57*  Low-cycle dowel shear transfer of reinforcing bar.

A non-fatigue simulation and an experimental verification under reversed cyclic loading (Vintzeleou and Tassios 1987; Soroushian *et al.* 1988) were performed by Soltani and Maekawa (2007), as shown in Figure 10.57.

In order to investigate the mechanism of dowel action fatigue deterioration, a repeated single-sided shear force was applied to the flat plane across which a single deformed bar of 13 mm diameter was arranged, as shown in Figure 10.58. First, a half-length of the push-off specimen (Mattock and Hawkins 1972) was cast into the formwork. All sides were made with smooth form surfaces. One day after casting, the form was stripped and the second half-length was immediately cast with additional forms attached to the already hardened concrete. This produced a flat shear plane and the applied shear force was carried only by the reinforcing bar. The compressive strength of the concrete was 40 MPa and the yield strength of the reinforcement was 395 MPa.

Three specimens with the same dimensions and materials were made and tested under different loading levels. Specimen A was used to confirm the static capacity for dowel action. Specimen B and Specimen C were subjected to high-cycle repetition of shear forces ranging up to 22.5 kN and 11.0 kN, respectively. The minimum load level was set as 10% of the maximum shear force in order to keep the specimen stable without relaxation of the set-up under high-cycle fatigue. Cyclic force was applied by means of a hydraulic servo-actuator and the specimen was laterally supported to prevent out-of-plane collapse. Joint kinematics was measured by the use of non-contact transducers and the local strain of the reinforcing bar was picked up by strain gauges, as shown in Figure 10.58. A pair of gauges was affixed to the bar surface upside down, so that both local curvature and mean axial strain could be obtained. The integral of local curvature with respect to location should coincide with the dowel shear slip displacement (Maekawa *et al.* 2003b).

The experimental shear force versus slip displacement relation for Specimen A is shown in Figure 10.59. After local yielding of the reinforcement (at about 30 kN), the dowel force gradually increased without showing a clear peak until the stroke capacity of the loading system was reached. The computed

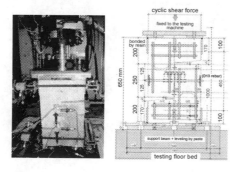

*Figure 10.58* Single-sided cyclic dowel shear experiment.

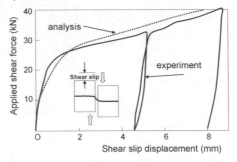

*Figure 10.59* Monotonic dowel shear force versus shear slip relation.

result matches the experiment fairly well, similarly to the verification by Maekawa *et al.* (2003b) and Soltani and Maekawa (2007).

Figure 10.60 indicates the relation between dowel shear force and transverse shear slip for Specimen B for various cycle totals. The shear slip increases as the number of cycles increases. The load level is high enough to induce considerable change in dowel shear, up to eight times the initial value. The residual dowel shear after release of the load also rises with the number of cycles. The incremental dowel shear is shown against the number of fatigue cycles in Figure 10.61. There exists a linear relation between incremental dowel shear and the logarithm of the number of fatigue cycles. This is similar

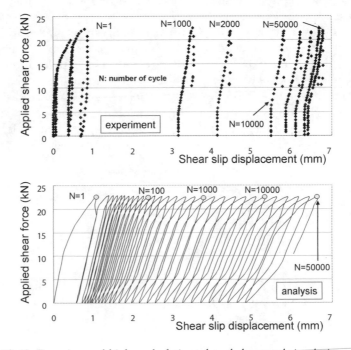

*Figure 10.60* Experimental high-cycle fatigue dowel shear and simulation.

to the S-N diagram for the fatigue life of materials. The experimental result for Specimen C is shown together with that for Specimen B in Figure 10.61. Although logarithmic linearity is clear in both cases, the rates of dowel shear evolution are very different.

Locally induced curvature at the location 2D from the joint plane is shown in Figure 10.62. The experimental curvature was calculated by assuming a linear profile of fiber strains. Here, there is a clear linear relation between curvature and the logarithmic number of cycles and the rate of curvature change is found to depend greatly on the magnitude of amplitude. As the steel reinforcing bars continued to work without fatigue rupture, the experimental data can be used to verify the fatigue model of dowel action.

The increase in local curvature and dowel shear slip is thought to result from reduced transverse stiffness of the concrete supporting the reinforcing bars. Further, it has been reported that the deterioration of concrete surrounding steel bars is not very localized, unlike the case of fatigue pullout of reinforcement. In consequence, when longitudinal cracking occurs in the cover concrete, the confining stiffness of the concrete acting on the steel bars is more or less homogeneously reduced and a large amount of dowel shear displacement is caused. Thus, the authors simply modify the original dowel action model described by $J$ in Equation 10.10 in regard to fatigue deterioration as

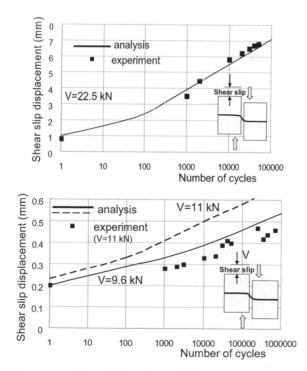

*Figure 10.61* Incremental dowel shear slip under high-cycle fatigue.

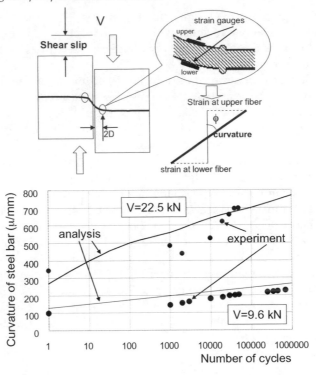

*Figure 10.62* Flexural curvature evolution of reinforcement associated with dowel shear.

$$\tau_S = J(s, \alpha \cdot \delta)$$

$$\alpha = 1.0 - 0.15\sqrt{\log_{10}\left(1 + G_d\right)}$$

$$G_d = \frac{10}{D} \cdot \int |d\delta|$$

(10.11)

where $\alpha$ is a global stiffness reduction factor related to the accumulated damage denoted by $G_d$ as above. Since the dowel shear slip is equal to the integrated local curvature over the whole steel domain, its factorization implies a reduction in concrete stiffness to restrain the flexural deformation. For the present, this proposal should be limited to applications where the fatigue is single sided, although there is no mathematical constraint on its application elsewhere.

The computed relation between dowel shear slip and force is shown in Figure 10.60 together with the experimental results. The incremental evolution of dowel shear is summarized in Figure 10.61 for different amplitudes. It appears that reasonable agreement is obtained between simulation results and reality at the higher shear force amplitude (22.5 kN), although the number of cases used for verification is limited. In the case of lower amplitude (11.0 kN), it seems that the predictive method gives rise to more rapid evolution of dowel fatigue than in reality (see Figure 10.61). Here, it should be noted that the dowel displacement at the first cycle (N = 1) is also overestimated a little, by about 10%. This inevitably results in a larger accumulated shear slip based on the definition of $G_d$ in Equation 10.11. In order to clarify sensitivity to this error, a slightly reduced force (9.6 kN) was applied in the numerical model so that the displacement at the first cycle matches the experimental value. Fatigue evolution now follows the experiment reasonably and closely. This means that the discrepancy at the lower amplitude can be attributed to a slight overestimation by the original model, yielding a ±10% discrepancy. The computed curvature, which is highly associated with dowel shear due to deformational compatibility, is also illustrated along with the experimental data in Figure 10.62. Reasonable consistency is evident.

### 10.6.4 Fatigue constitutive modeling of reinforcing bars

In the path-integral scheme of fatigue simulation (Maekawa *et al.* 2006a), the stress or deformational amplitude is not necessarily constant in the modeling of dowel action and bond pullout. In fact, the local stress and/or strain amplitude of structural concrete varies even though the amplitude of externally applied forces is constant. Thus, the fatigue strength model for the steel has to be re-formulated to make it consistent with any stress or strain path. The tensile fatigue strength of a reinforcing bar, denoted by $f_{sr}$ (MPa), is specified in the JSCE code (2002) as

$$f_{sr} = 190 \frac{10^a}{N^k} \left( 1 - \frac{\sigma_{sp}}{f_u} \right)$$

$$a = k_{0f} \left( 0.81 - 0.003\phi \right)$$

$$k = 0.12, \quad k_{0f} = 1.0$$

(10.12)

where $N$ is the number of cycles, and $\phi$ is the diameter of the reinforcing bar (mm). Symbol $\sigma_{sp}$ means the lower bound stress related to permanent loads and $f_u$ is the tensile strength (MPa). This formula derives from the S-N diagram of fatigue test data, which was generally obtained under constant-stress amplitude.

By following Palmgren–Miner's rule (the linear damage assumption, Palmgren 1924), this empirical formula can be converted into a discrete rate-type one so that the fatigue rupture of steel can be consistently monitored in the strain path-integral scheme as

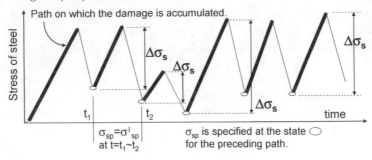

*Figure 10.63* Definition of stress amplitude and effective path for integrating steel
damage.

$$D_s = \sum \frac{1}{N} = \sum \left[ 190 \frac{10^\alpha}{\Delta\sigma_s} \left( 1 - \frac{\sigma_{sp}}{f_u} \right) \right]^{-1/k}$$

$$D_s = \frac{1}{190 \cdot 10^\alpha \cdot k} \int \left( 1 - \frac{\sigma_{sp}}{f_u} \right)^{-1/k} d\sigma_s$$

$$D_s = 1: \quad fatigue \;\; failure , \quad D_s < 1: \quad no \;\; failure$$

(10.13)

where $D_s$ represents the accumulated damage related to fatigue rupture of
the steel, and $\Delta\sigma_s$ is the stress amplitude as shown in Figure 10.63. When the
value of $D_s$ reaches unity, rupture of the reinforcement is inferred.

### 10.6.5 Accelerated direct-path integral

It has been shown that well-established methods of non-linear dynamic
analysis for RC structures can be extended to cover fatigue analysis in the
time–space domain by direct-path integration of constitutive models. The
fatigue effects of bond pullout, dowel action and shear transfer are simply
incorporated into the original models in terms of evolution derivatives, so
the resulting extended material models also cover highly non-linear behavior. Consequently, the fatigue analysis system shares the same scheme as
that for non-linear failure analysis under seismic loading. To accelerate the
computation, the logarithmic time derivative can be used in path integrating
the constitutive models. Analogous to Section 10.1.3, the damaging rates
with respect to deformational paths are simply integrated in the numerical
scheme as

$$G_b = \ln(10) \cdot \int \frac{\xi \cdot m}{10^{m(s_y - \hat{s})}} d\hat{s} \qquad \text{generalized bond pullout of Equation 10.9}$$

(10.14)

$$G_d = \frac{10}{D} \cdot \int \varsigma \cdot |d\delta| \quad \text{Ĺ generalized dowel action of Equation 10.11}$$

(10.15)

$$D_s = \frac{1}{190 \cdot 10^{\alpha} \cdot k} \int \varsigma \left(1 - \frac{\sigma_{sp}}{f_u}\right)^{-1/k} d\sigma_s \quad \begin{array}{l} \text{Ĺ generalized steel tension fatigue} \\ \text{of Equation 10.13} \end{array}$$

(10.16)

where $\varsigma$ is the integral accelerator (Maekawa *et al.* 2006a).

Material non-linearities in structures develop rapidly at the beginning of fatigue loadings, so the value of $\varsigma$ should be unity for accuracy. That is, the exact direct paths of forces and corresponding internal stresses should be tracked in real time. After many cycles, plasticity, damaging and fatigue development tend to be much reduced and the increment in non-linearity per cycle becomes exponentially small. At this stage, the value of $\varsigma$ can be larger for computational efficiency at each load step. A single cycle of computational load is equivalent to $\varsigma$–cycles. In Equations 10.14–10.16, time and strain path integrals only are magnified by $\varsigma$, but it should be noted that the terms of instantaneous non-linearity associated with evolution boundaries (plasticity and damage) are not magnified by $\varsigma$, but the path is strictly mapped out.

Figure 10.64 shows the computed displacement histories of an RC member with a discrete joint, as shown in Figure 10.65. Details of this member are given in the following section. The thin (green) line represents analysis in which direct integration of fatigue constitutive models was carried out up to 1,000 cycles. Here, the magnification factor was kept at unity over the whole loading path. The broad (black) line is the computed result based on the magnified integral. The factor $\varsigma$ for the first cycle is unity and the second cycle corresponds to $\varsigma=2$, the third to $\varsigma=5$, the fourth to $\varsigma=10$, and so on. Just a few substantial cycles are enough to simulate the material and structural

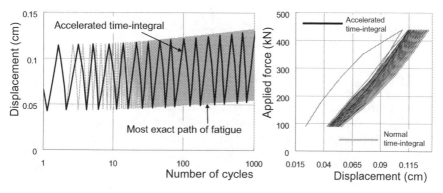

*Figure 10.64* Magnified direct time and path integral.

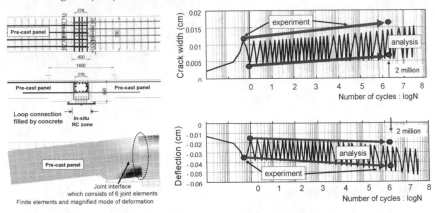

*Figure 10.65* RC joint interface subjected to combined bending and shear.

states with reasonable computational accuracy, as shown in Figure 10.64. The equivalence of the magnification method to the exact path-integral has already been examined for monolithic RC members (Maekawa *et al.* 2006a), so this demonstrates that a general analysis domain with or without a joint interface can be dealt with by the logarithmic time-integral method.

### 10.6.6 *Verification of joint interface*

In general, the reliability of non-linear modeling has to be crossways examined at different levels, that is, verification is necessary not only at the material level but also at the member or structural level. In this chapter, the fatigue behavior of RC beams with a single joint interface is simulated to check the accuracy of material modeling under non-uniform fields of stress and deformation. Second, the proposed finite element computation method is applied to more realistic structures whose non-linearity is chiefly rooted in joint interfaces. Here, slabs of PC–RC assembly under fatigue loading are selected for assessing the practical performance of the proposed method. For continuous zones of reinforced concrete domains, multi-directional non-orthogonal smeared-crack modeling, which has been verified under high-cycle fatigue loads (Maekawa *et al.* 2006a), is applied.

#### 10.6.6.1 *Joint interface connecting pre-cast RC panels*

A mixed structural member consisting of two pre-cast PC panels connected by RC joints is selected for experimental verification (Figure 10.65; Taira *et al.* 2007). This is a full-scale mock-up of part of the extended runway for Tokyo International Airport at Haneda. External loading sets up a combined flexure moment and shear force. Since this target for experimental verification consists of a statically determinate structure subjected to the constant amplitude of external forces, the resultant amplitude of these sectional forces becomes constant. However, the local stress conditions along the joint interface are not uniform but vary from tension to compression. Thus, this structure is appropriate for verifying the fatigue model in more general stress states. The joint is divided into six elements, as shown in Figure 10.65.

The junction between the PC panels at the RC connection core is not a rough crack plane but a construction joint. The surface was artificially processed by abrasion to produce some amount of friction at closure. The simulation was then carried out with frictional contact, as shown in Figure 10.65, and the frictional coefficient at closure was set equal to 0.5, according to past experience with concrete-to-concrete contact (Tassios and Vintzeleou 1987). The experimental and analytically obtained deflections and crack widths at the level of the tension reinforcement are shown in Figure 10.65 at one million and at two million cycles. The magnified time integration, similar to the case shown in Figure 10.64, is defined below.

The computationally predicted deflection matches the experimental measurements, but the crack width is slightly underestimated. The experimentally obtained crack width is a value measured at the surface of the concrete member, while the computational value is the average crack width at the level of the tension reinforcement (see Figure 10.51). In fact, the crack gap right on the surface of the reinforcing bars is known to be almost nil, and local crack width increases proportionally according to location, as illustrated in Figure 10.51. For this reason, a slight underestimation is thought to be reasonable in view of the mechanical model of bond pullout. In this experiment, the shear slip along the joint plane was not the same in experiment and analysis.

### 10.6.6.2 Pre-cast slab system under moving-wheel-type loads

The stress state of the mock-up specimen discussed in the previous section is, on the whole, two-dimensional. In this section, the verification is extended to more generic three-dimensional states. Figure 10.66 shows a laterally supported slab system consisting of PC pre-cast panels and RC block connectors.

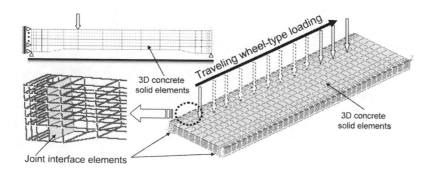

*Figure 10.66* Structural finite element modeling and boundary conditions.

The slab system is longitudinally supported through the joint interfaces linking the main body of the PC panels to the RC connectors, as shown in Figure 10.65. A traveling-wheel load is applied repeatedly to produce three-dimensional stress states accompanying principal stress rotation. This verification of the simulation scheme is focused on dowel action and bond pullout, so shear transfer along the joint plane is intentionally neglected by defining zero as the friction coefficient. Only normal compression is transferred at contact and it is assumed that no tension is transferred at all.

Figure 10.67 shows the simulation results, in which no external membrane

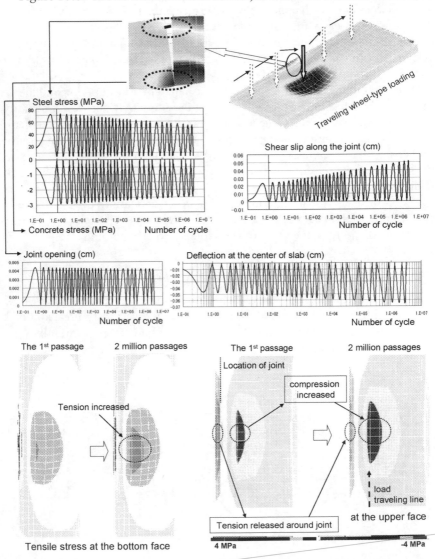

*Figure 10.67* Fatigue deformation mode and mechanism (dowel action only).

force is applied but out-of-plane bending and shear are produced. The computational traveling load is 800 kN and the corresponding deflection at the center of the slab, the tension reinforcement stress and the contact stress at the compressive extreme fiber are shown for the central joint plane. The magnified logarithmic time integral is the same value as adopted in the previous section. A gradual progressive deflection can be seen in the results and transverse shear slip is not insignificant, since shear transfer is unrealistically ignored so as to obtain a functional check of dowel action and steel pullout under fatigue action. Similar magnitudes of incremental shear slip at the joint and of structural deflection are obtained. Since the pre-cast panels are free from cracking due to pre-stressing numerically induced in terms of initial strain, the main source of non-linearity derives from the joint interfaces.

The maximum joint opening at the tension reinforcement is almost constant over the load history. A gradual decay of steel tensile stress is seen, as shown in Figure 10.67. This means that the embedded reinforcement exhibits reduced pullout stiffness. This is caused by evolving bond deterioration close to the joint plane. As a result, the bending moment shifts from joint areas to the pre-cast panels and the flexural tension at the bottom face of the slabs gradually increases. As the decay in joint stiffness reduces the subsequent risk of reinforcement fatigue failure at the joint, the initial stress amplitude of steel can be used for fatigue design in practice. Quite reasonably, this behavior appears to be the opposite of that exhibited by statically determinate structural systems, where stiffness decay resulting from fatigue directly causes shear deformation of beam webs (Ueda and Okamura 1982).

Figure 10.68 shows simulation results for the same panel system but with a constant membrane axial tension applied. This loading is analogous to the self-equilibrated sectional force induced by thermal action, assuming that the panel system is firmly fixed by external constraints. Under this severe loading condition, the joint plane is subjected to full tension at any one time and the reinforcement stress is larger than under free membrane action, as shown in Figure 10.67. This accelerates the fatigue of both dowel action and bond pullout. In fact, deflection is substantially increased, but the general tendency of overall behavior is similar to the case with no membrane tension. As the external tension is carried by the reinforcement, the tension reinforcement stress at the joint is elevated. At the same time, contact compression produced by flexure is consistently lower, as shown in Figure 10.68. In analysis, the embedment length of reinforcing bars is assumed to be sufficiently long, with loop anchorage as shown in Figure 10.65.

It is known that the fatigue life of RC slabs under moving-wheel loads is much less than the life under fixed-point pulsation. The mechanism of this difference in fatigue life has been investigated (Matsui 1987; Perdikaris and Beim 1988) and crack-to-crack interaction is thought to be one of main reasons for the difference (Maekawa *et al.* 2006a). In order to study the fatigue characteristics of the panel system under investigation here, a wheel-type

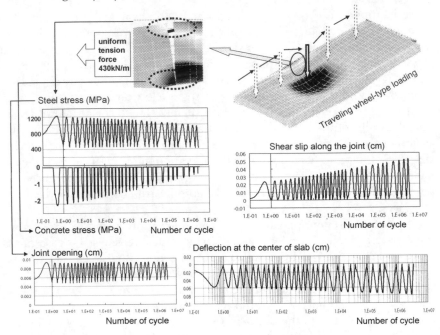

*Figure 10.68* Fatigue deformation mode and mechanism under membrane tension.

moving load is applied to the upper surface of the assembled slab system. Figure 10.69 compares the progress of deflection under fixed-point as well as moving-wheel loading. At the extreme end of fatigue life, a dramatic increment of deflection is experienced and no dynamic equilibrium solution can be found in the predictor–corrector method. There is a fair coincidence between the two. This is very different from the behavior of monolithic RC slabs. The main point is that crack location and the mode of joint deformation barely differ.

In fact, crack location and orientation do not change over the whole life of the structure, even though flexural tension somehow increases in the main pre-cast panels. That is, in contrast with a monolithic PC slab, the mode of deformation does not change. It can be said that this PC–RC mixed structural system has a central portion that is strengthened by pre-stressing and crack non-linearity is confined solely to the RC connection units. Since the non-linearity is structurally confined within a limited range, shear transfer decay provoked by principal stress rotation barely occurs. Here, the concrete that confines the reinforcement at joints is assumed to be rather massive, without spalling of the concrete cover or splitting cracks along the reinforcing steel. For more general cases, the dowel action model needs to be more generalized (Vintzeleou and Tassios 1987). As these fatigue analyses on wheel-type traveling loads are also time dependent, the delayed recovery of deflection

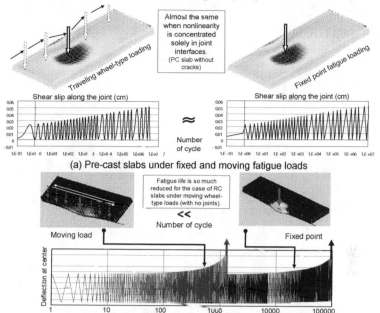

(a) Pre-cast slabs under fixed and moving fatigue loads

(b) Monolithically constructed slabs under fixed and moving fatigue loads (Maekawa *et al.* 2000)

*Figure 10.69* Effect of loading patterns and comparison with RC slabs with multi-directional cracking.

after the passage of wheel loads is consequentially computed. This effect was checked by changing the traveling speed of wheels, and a negligible influence was confirmed for these cases.

# References

## Main technical papers related to this book

*Journal of Advanced Concrete Technology (http://www.jstage.jst.go.jp/ browse/jact)*

Asamoto, S., Ishida, T. and Maekawa, K. (2006). "Time–dependent constitutive model of solidifying concrete based on thermodynamic state of moisture in fine pores." *JACT*, 4(2).

Asamoto, S., Ishida, T. and Maekawa, K. (2008). "Investigations into volumetric stability of aggregates and shrinkage of concrete as a composite." *JACT*, 6(1), 77–90.

El–Kashif, K. F. and Maekawa, K. (2004). "Time–dependent nonlinearity of compression softening in concrete." *JACT*, 2(2).

El–Kashif, K. F. and Maekawa, K. (2004). "Time–dependent post–peak softening of RC members in flexure." *JACT*, 2(3) 301–315.

Gebreyouhannes, E., Chijiwa, N., Fujiyama, C. Maekawa, K. (2008). "Shear fatigue simulation of RC beams subjected to fixed pulsating and moving loads." *JACT*, 6(1), 215–226.

Gebreyouhannes, E., Kishi, T. and Maekawa, K. (2008). "Shear fatigue response of cracked concrete interface." *JACT*, 6(3).

Ishida, T. and Li, C. (2008). "Modeling of carbonation based on thermo–hygro physics with strong coupling of mass transport and equilibrium in micro–pore structure of concrete." *JACT*, 6(2).

Ishida, T., Miyahara, S. and Maruya, T. (2008). "Chloride binding capacity of mortars made with various Portland cements and admixtures." *JACT*, 6(2).

Ishida, T., Maekawa, K. and Soltani, M. (2004). "Theoretically identified strong coupling of carbonation rate and thermodynamic moisture states in micro–pores of concrete." *JACT*, 2(2).

Maekawa, K. and El–Kashif, K. F. (2004). "Cyclic cumulative damaging of reinforced concrete in post–peak regions." *JACT*, 2(2).

Maekawa, K., Fukuura, N. and Soltani, M. (2008). "Path–dependent high cycle fatigue modeling of joint interfaces in structural concrete." *JACT*, 6(1), 227–242.

Maekawa, K., Gebreyouhannes, E., Mishima, T. and An, X. (2006). "Three–dimensional fatigue simulation of RC slabs under traveling wheel–type loads." *JACT*, 4(3) 445–457.

Maekawa, K., Ishida, T. and Kishi, T. (2003). "Multi–scale modeling of concrete performance: integrated material and structural mechanics (invited)." *JACT*, 1(2), 91–126.

Maekawa, K., Soltani, M., Ishida, T. and Itoyama, Y. (2006). "Time–dependent space–averaged constitutive modeling of cracked RC subjected to shrinkage and sustained loads." *JACT*, 4(1) 193–207.

Maekawa, K., Toongoenthong, K., Gebreyouhannes, E. and Kishi, K. (2006). "Direct path–integral scheme for fatigue simulation of reinforced concrete in shear." *JACT*, 4(1) 159–177.

Nakarai, K., Ishida, T. and Maekawa, K. (2006). "Modeling of calcium leaching from cement hydrates coupled with micro–pore formation." *JACT*, 4(3) 395–407.

Toongoenthong, K. and Maekawa, K. (2004). "Interaction of pre–induced damages along main reinforcement and diagonal shear in RC members." *JACT*, 2(3).

Toongoenthong, K. and Maekawa, K. (2005). "Multi–mechanical approach to structural performance assessment of corroded RC members in shear." *JACT*, 3(1) 107–122.

Toongoenthong, K. and Maekawa, K. (2005). "Computational performance assessment of damaged RC members with fractured stirrups." *JACT*, 3(1) 123–136.

Toongoenthong, K. and Maekawa, K. (2005). "Simulation of coupled corrosive product formation, migration into crack and propagation in reinforced concrete sections." *JACT*, 3(2), 253–265.

## Other journals

Hussain, R. and Ishida, T. (2008). "Enhanced multi–scale corrosion model for RC structures under severe coupled environmental actions of variable chloride and temperature conditions." *Materials and Structures* (under review).

Iqbal, P. and Ishida T. (2008). "Modeling of chloride transport coupled with enhanced moisture conductivity in concrete exposed to virtual marine environment " *Cement and Concrete Research* (under review).

Ishida, T., Iqbal, P. and Ho, T. (2008). "Modeling of chloride diffusivity coupled with non–linear binding capacity in sound and cracked concrete." *Cement and Concrete Research* (under review).

Ishida, T., Maekawa, K. and Kishi, T. (2007). "Enhanced modeling of moisture equilibrium and transport in cementitious materials under arbitrary temperature and relative humidity history." *Cement and Concrete Research*, 37(2).

Maekawa, K., Nakarai, K. and Ishida, T. (2007). "Chemo–physical and mechanical approach to performance assessment of structural concrete and soil foundation." *Transport Properties and Concrete Quality*, John Wiley & Sons.

Nakarai, K., Ishida, T. and Maekawa, K. (2006). "Multi–scale physicochemical modeling of soil: cementitious material interaction." *Soils and Foundations*, 46(5).

Nakarai, K., Ishida, T., Kishi, T. and Maekawa, K. (2007). "Enhanced thermodynamic analysis coupled with temperature–dependent microstructures of cement hydration." *Cement and Concrete Research*, 37(2) 139–150.

## Homepages on performance assessment software *DuCOM* and *COM3*

http://concrete.t.u–tokyo.ac.jp/en/demos/ducom/index.html
http://www.comse.co.jp/english/index.html
http://www.ducoms.com/

# General sources related to this book

Alca, N., Alexander, S. D. B. and MacGregor, J. G. (1997). "Effect of size on flexural behavior of high strength concrete beams." *Journal of ACI Structural*, 94(1), 59–67.

Al–Sulaimani, G. J., Kaleemullah, M., Basunbul, I. A. and Rasheeduzzafar (1990). "Influence of corrosion and cracking on bond behaviour and strength of reinforced concrete members." *Journal of ACI Structural*, 87(2), 220–231.

American Concrete Institute 222R–96. (2001). *Corrosion of Materials in Concrete.* ACI.

American Concrete Institute Committee 209. (2005). "Report on factors affecting shrinkage and creep of hardened concrete." ACI, 209.1R–05.

American Concrete Institute Committee 215 (1982). *Fatigue of Concrete Structures.* Shah, S. P. ed., ACI.

An, X., Maekawa, K. and Okamura, H. (1997). "Numerical simulation of size effect in shear strength of RC beams." *Journal of Materials, Concrete Structures and Pavement*, JSCE, V–35(564), 297–316.

Andrade, C., Alonso, C. and Molina, F. J. (1993). "Cover cracking as a function of bar corrosion; Part I: Experimental Test." *Materials and Structures*, 26, RILEM, 453–464.

Aoki, S., Miura, N., Takeda, N. and Sogo, S. (1994). "Strength development of high–strength concrete with belite high–content cement under high temperature rise." *Proceedings of JCI*, 16(1), 1317–1322.

Arai, M., Tamura, H., Sekiguchi, K., Umeda, H., Iwashimizu, T., Tatematsu, K., Ohashi, M., Urano, H., Kimura, Y., Akashi, T., Motoki, R., Yamasaki, J., Sotoya, Y., Imamoto, K., Annoura, T., Tohyama, S. and Fujisawa, I. (1999). (2000). "Study on influence of aggregate quality on drying shrinkage of concrete (Parts 1–8)." *Summaries of Technical Papers of Annual Meeting*, Architectural Institute of Japan, A–1 (materials and constructions), 741–746 (1999), 645–654 (2000) (in Japanese).

Arai, Y. (1984). *Chemistry of Cement Materials*, Dai–Nippon Tosho Publishing Co., Ltd., Tokyo.

Architecture Institute of Japan. (2003). *Shrinkage Cracking in Reinforced Concrete Structures: Mechanisms and Practice of Crack Control.*

Arthanari, S. and Yu, C. W. (1967). "Creep of concrete under uniaxial and biaxial stresses at elevated temperatures." *Magazine of Concrete Research*, 19(60), 149–156.

Asamoto, S. (2006). "Enhanced Multi–scale Constitutive Model of Solidifying Cementitious Composites and Application to Cracking Assessment of Concrete Structures." PhD Thesis, The University of Tokyo.

Asamoto, S. and Ishida, T. (2003). "Influence of liquid characteristic and its distribution in micro–pore on time–dependent mechanical behavior of concrete." *Advances in Cement and Concrete*, Proceedings of a conference held at Copper Mountain Colorado, 181–190.

Asamoto, S. and Ishida, T. (2005). "Coupled analysis of moisture state and mechanical behavior under arbitrary temperature conditions." *Proceedings of JCI*, 27(1), 451–456 (in Japanese).

Asamoto, S., Ishida, T. and Maekawa, K. (2006). "Time–dependent constitutive model of solidifying concrete based on thermodynamic state of moisture in fine pores." *Journal of Advanced Concrete Technology*, 4(2), 301–323.

ASTM C 876–91. (1999). *Standard Test Method for Half–Cell Potentials of Uncoated Reinforcing Steel in Concrete.* ASTM, USA.

ASTM G1–03. (2002). *Standard Test Practice for Preparing, Cleaning and Evaluating Corrosion Test Specimens.* ASTM, USA.

Atimtay, E. and Ferguson, P. M. (1973). "Early chloride corrosion of reinforced concrete: a test report." *Journal of ACI*, 70(9).

Atkinson, A. and Nickerson, A. K. (1984). "The diffusion of ions through water–saturated cement." *Journal of Materials Science*, 19, 3068–3078.

Auyeung, Y. B., Balaguru, P. and Chung, L. (2000). "Bond behavior of corroded reinforcement bars." *ACI Materials Journal*, 97(2), 214–220.

Award, M. E. and Hilsdorf, H. K. (1974). "Strength and deformation characteristics of plain concrete subjected to high repeated and sustained loads." ACI publication, SP 41-1, 1–13.

Ayano, K. and Sakata, K. (1997). "Concrete shrinkage strain under the actual atmosphere." *Proceedings of JCI*, 19(1), 709–714.

Badmann, R., Stockhausen, N. and Setzer, M. J. (1981). "The statistical thickness and the chemical potential of adsorbed water films." *J. Coll. Interf. Sci.*, 82(2), 534–542.

Bamforth, P. B. (1987). "The relationship between permeability coefficients of concrete using liquid and gas." *Magazine of Concrete Research*, 39, 3–11.

Barneyback, R. S. and Diamond, S. (1981). "Expression and analysis of pore fluids from hardened cement pastes and mortars." *Cement and Concrete Research*, 11, 279–285.

Bazant, Z. P. (1979). "Physical model for steel corrosion in concrete sea structures: theory." *Journal of Structural Division*, ASCE, 105(ST6), 34–45.

Bazant, Z. P. (1988). *Mathematical Modeling of Creep and Shrinkage of Concrete*. John Wiley & Sons, Inc.

Bazant, Z. P. and Panula, L. (1978a). "Practical prediction of time–dependent deformations of concrete Part I: Drying shrinkage, Part II: Basic creep." *Materials and Structures*, 11(65), 307–328.

Bazant, Z. P. and Panula, L. (1978b). "Practical prediction of time–dependent deformations of concrete Part III: Drying creep, Part IV: Temperature effect on basic creep." *Materials and Structures*, 11(66), 415–434.

Bazant, Z. P. and Panula, L. (1979). "Practical prediction of time–dependent deformations of concrete Part V: Temperature effect on drying creep." *Materials and Structures*, 12(69), 169–174.

Bazant, Z. P. and Prasannan S. (1989). "Solidification theory for concrete creep, I. Formulation, II. Verification and application." *Journal of Engineering Mechanics*, 115(8), 1691–1725.

Beaudoin, J. J., Ramachandran, V. S. and Feldman, R. F. (1990). "Interaction of chloride and C–S–H." *Cement and Concrete Research*, 20, 875–883.

Bentz, D. P. and Garboczi, E. J. (1991). "Percolation of phases in a three–dimensional cement paste microstructure model." *Cement and Concrete Research*, 21, 325–344.

Benveniste, Y. (1987). "A new approach to the application of Mori–Tanaka's theory in composite materials." *Mechanics of Materials*, 6, 147–157.

Berner, U. R. (1988). "Modelling the incongruent dissolution of hydrated cement minerals." *Radiochimica Acta*, 44/45, 387–393.

Berra, B. (2003). "Steel–concrete bond deterioration due to corrosion: finite–element analysis for different confinement levels." *Magazine of Concrete Research*, 3, June.

Boothby, T. E. and Laman, J. A. (1999). "Cumulative damage to bridge concrete deck slabs due to vehicle loading." *Journal of Bridge Engineering*, ASCE, 4(1), 80–82.

Breugel, K. (1991). "Simulation of Hydration and Formation of Structure in Hardening Cement–Based Materials." PhD Thesis, Delft Technological Institute.

Broomfield, J. P. (1997). "Corrosion of steel in concrete." Chapter 6, *Electrochemical Repair Techniques*, E & FN Spon, London.

Browne, R. D. (1967). "Properties of concrete in reactor vessels." *Proceedings of Conference on Prestressed Concrete Pressure Vessels*, group C, 131–151.

Browne, R. D. (1980). "Mechanisms of corrosion of steel in concrete in relation to design, inspection, and repair of offshore and coastal structures." *Performance of Concrete in Marine Environment*, ACI, SP–65, 169–204.

Brunauer, S., Emmet, P. H. and Teller, E. (1938). "Adsorption of gases in multimolecular layers." *Journal of American Chemical Society*, 60.

BS 8110. (1997). "Structural use of concrete." *Code of Practice for Design and Construction*.

Buil, M., Revertegat, E. and Oliver, J. (1992). "A model of the attack of pure water or under saturated lime solutions on cement." ASTM STP 1123, 227–241.

Bujadham, B. and Maekawa, K. (1992). "Qualitative studies on mechanisms of stress transfer across cracks in concrete." *Proceedings of JSCE*, 451(17), 265–275.

Byung, H. O., Bong, S. J. and Seong, C. L. (2004). "Chloride diffusion and corrosion initiation time of reinforced concrete structures." *Proceedings of the International Workshop on Microstructure and Durability*, Sapporo, Japan.

Cabrera, J. G. (1996). "Deterioration of concrete due to reinforcement steel corrosion." *Cement and Concrete Composites*, 18, 47–59.

Cabrera, J. G. and Ghoddoussi, P. (1992). "The effect of reinforcement corrosion on the strength of the steel concrete bond." *Proceedings of the International Conference on Bond in Concrete, Riga, Latvia*, 10/11–10/24.

Cano–Barrita, P. F. de J., Balcom, B. J., Bremner, T. W., MacMillan, M. B. and Langley, W. S. (2004). "Moisture distribution in drying ordinary and high performance concrete cured in a simulated hot dry climate." *Material and Structures*, 37, 522–531.

Carol, I., Prat, P. C. and Lopez, C. M. (1997). "Normal/shear cracking model: application to discrete crack analysis." *Journal of Engineering Mechanics*, 123(8), 765–773.

CEB. (1992). *Durable Concrete Structures, CEB Design Guide*, Thomas Telford, 3–7.

Cebeci, O. Z., Al–Noury, S. I. and Mirza, W. H. (1989). "Strength and drying shrinkage of masonry mortars in various temperature–humidity environments." *Cement and Concrete Research*, 19, 53–62.

Cervenka, V. and Cervenka, J. (1996). "Computer simulation as a design tool for concrete structures." *ICCE–96*, Bahrain.

Chali, A., Dilger, W. and Neville, A. M. (1969). "Time–dependent forces induced by settlement of supports in continuous reinforced concrete beams." *Journal of ACI*, November, 907–915.

Chang, T. S. and Kesler, C. E. (1958). "Fatigue behavior of reinforced concrete beams." *Journal of ACI*, 55(14), 245–254.

Chaube, R. P., Shimomura, T. and Maekawa, K. (1993). "Multiphase water movement in concrete as a multi–component system." *Proceedings of the 5th International ConCreep RILEM Symposium*, 139–144.

Chijiwa, N. (2006). "Time dependent simulation of reinforced concrete subjected to coupled mechanistic and environmental actions." *6th International PhD Symposium in Civil Engineering*, Zurich.

Chong, K. T., Gilbert, R. I. and Foster, S. J. (2004). "Modeling time–dependent cracking in reinforced concrete using bond–slip interface elements." *Computers and Concrete*, 1(2), 151–168.

Claeson, C. and Gylltoft, K. (2000). "Slender concrete column subjected to sustained and short–term eccentric loading." *Journal of ACI Structural*, 97(1), 45–52.

Collins, M. P. and Vecchio, F. (1982). "The response of reinforced concrete to in–plane shear and normal stresses." University of Toronto.

Connell, L. D. and Bell, P. R. F. (1993). "Modeling moisture movement in revegetating waste heaps: I. Development of a finite element model for liquid and vapor transport." *Water Resources Research*, 29 (5), 1435–1443.

Copeland, L. E. and Hayes, J. C. (1956). "Porosity of hardened Portland cement paste." *Journal of ACI*, 52–39, 633–640.

Cornelissen, H. A. W. and Reinhardt, H. W. (1984). "Uniaxial tensile fatigue failure of concrete under constant–amplitude and program loading." *Magazine of Concrete Research*, 36(129), 216–226.

Coronelli, D. (2002). "Corrosion cracking and bond strength modeling for corroded bars in reinforced concrete." *Journal of ACI Structural*, 99(3), 267–276.

Coronelli, D. and Gambarova, P. (2004). "Structural assessment of corroded reinforced concrete beams: modeling guidelines." *Journal of Structural Engineering*, ASCE, 130(8), 1214–1224.

Cowell, W. L. (1965). *Dynamic Tests of Concrete Reinforcing Steels*. Technical Reports R384, Naval Civil Engineering Laboratory, Port Hueneme, Calif.

Cusack, N. E. (1987). *The Physics of Structurally Disordered Matter*. Bristol edition.

Czernin, W. (1969). *Cement Chemistry for Construction Engineers*, translated by Kichiro Tokune, Giho–do.

Daimon, M., Abo–El–Enein, S. A., Hosaka, G., Goto, S. and Kondo, R. (1977). "Pore structure of calcium silicate hydrate in hydrated tricalcium silicate." *Journal of American Ceramic Society*, 60(3–4), 110–114.

Denny, A. J. (1996). *Principles and Prevention of Corrosion*, 2nd edition, New Jersey, Prentice Hall.

Dhakal, R. P. and Maekawa, K. (2002). "Modeling for post-yield buckling of reinforcement bar including buckling." *Journal of Structural Engineering*, ASCE, 128(9), 1139–1147.

Dhir, R. K., Hewlett, P. C. and Chan, Y. N. (1989). "Near surface characteristics of concrete: intrinsic permeability." *Magazine of Concrete Research*, 41(147), 87–97.

Dhir, R. K., Hewlett, P. C. and Dyer, T. D. (1996). "Influence of microstructure on the physical properties of self–curing concrete." *Journal of ACI*, 93(5), 465–471.

El–Bahy, A., Kunnath, S. K., Stone, W. C. and Taylor, A. W. (1999). "Cumulative seismic damage of circular bridge columns: benchmark and low–cycle fatigue tests." *Journal of ACI Structural*, 96(4), July–August.

El–Kashif, K. F. and Maekawa, K. (2004a). "Time–dependent nonlinearity of compression softening in concrete." *Journal of Advanced Concrete Technology*, Japan Concrete Institute, 2(2), 233–247.

El–Kashif, K. F. and Maekawa, K. (2004b). "Time–dependent post–peak softening of RC members in flexure." *Journal of Advanced Concrete Technology*, JCI, 2(3), 301–315.

England, G. L. and Illston, J. M. (1965). "Methods of computing stress in concrete from a history of measured strain Part 1." *Civil Engineering and Public Works Review*, 60 (4), 513–517.

England, G. L. and Illston, J. M. (1965). "Methods of computing stress in concrete from a history of measured strain Part 2." *Civil Engineering and Public Works Review*, 60 (5), 692–694.

England, G. L. and Illston, J. M. (1965). "Methods of computing stress in concrete from a history of measured strain Part 3." *Civil Engineering and Public Works Review*, 60 (6), 847–848.

England, G. L. and Ross, A. D. (1962). "Reinforced concrete under thermal gradients." *Magazine of Concrete Research*, 14(40), 5–12.

Enright, M. P. and Frangopol, D. M. (1998). "Service–life prediction of deteriorating concrete bridges." *Journal of Structural Engineering*, ASCE, 124(3), 309–317.

Esaki, F. and Ono, M. (2001). "Effect of loading rate on mechanical behavior of SRC shear walls." *Steel and Composite Structures*, 1(2), 201–212.

Espion, B. and Wastiels, J. (1989). "Creep and shrinkage tests carried out within the research program FRFC–FKFO 2.90001.80 on the behavior of partially prestressed concrete beams under long term sustained loading." Research report, Brussels Free University, Brussels, Belgium.

Espion, B. and Halleux, P. (1991). "Long term behavior of prestressed and partially presstressed concrete beams: experimental and numerical results – computer analysis of the effects of creep, shrinkage, and temperature changes on concrete structures." ACI Special Publication, 129, 19–38.

Famy, C., Scrivener, K. L., Atkinson, A. and Brough, A. R. (2002). "Effect of an early or a late heat treatment on the microstructure and composition of inner C–S–H products of Portland cement mortars." *Cement Concrete Research*, 32, 269–278.

Feldman, R. F. and Sereda, P. J. (1968). "A model for hydrated Portland cement paste as deduced from sorption–length change and mechanical properties." *Material Construction*, 1(6) 509–519.

Feldman, R. F. and Sereda, P. J. (1970). "A new model for hydrated cement and its practical implications." *Engineering Journal*, 53, 53–59.

Freiser, H. and Fernando, Q. (1963). *Ionic Equilibria in Analytical Chemistry*, John Wiley & Sons, Inc.

Frenaij, J. W. I. J., Walraven, J. C. and Reinhardt, H. W. (1988). "Time–dependent shear transfer in cracked reinforced concrete." Delft University Press.

Fujii, K. and Kondo, W. (1974). "Kinetics of the hydration of tricalcium silicate." *Journal of American Ceramic Society*, 57(11), 492–497.

Fujiwara T. (1984). "Change in length of aggregate due to drying." *Bulletin of the International Association of Engineering Geology*, 30, 225–227.

Funato, M., Hashimoto, M. and Kuramochi, S. (1991). "Method of quantitative analysis of silica gel generated by carbonation of cement hydrates." *Proceedings of Cement and Concrete*, 44, 252–257.

Gao, L. and Hsu, T. T. C. (1998). "Fatigue of concrete under uniaxial compression cyclic loading." *Journal of ACI Structural*, 95, 575–581.

Garboczi, E. J. (1990). "Permeability, diffusivity and microstructural parameters: a critical review." *Cement and Concrete Research*, 20, 591–601.

Gebreyouhannes, E. (2006). "Shear transfer of cracked concrete under fatigue loading." *Proceedings of 6th International PhD Symposium in Civil Engineering*, Zurich.

Gebreyouhannes, E., Kishi, T. and Maekawa, K. (2006). "Response of cracked concrete interfaces subjected to reversed cyclic shear loading and the effect of water." *Proceedings of the 10th East Asia–Pacific Conference on Structural Engineering and Construction*.

Gérard, B., Le Bellego, C. and Bernard, O. (2002). "Simplified modeling of calcium leaching of concrete in various environments." *Materials and Structures*, 35, 632–640.

Gilbert, R. I. (1999). "Deflection calculation for reinforced concrete structures: why we sometimes get it wrong." *Journal of ACI Structural*, 96(6), 1027–1032.

Gilbert, R. I. (2004). "Time–dependent cracking in reinforced concrete beams and slabs" *Proceedings of fib Symposium: The Challenge of Creativity*, Avignon, France, CD-6, 298–299.

Gilbert, R. I. and Smith, S. T. (2005). "Strain localization and its impact on the ductility of reinforced concrete slabs containing 500 MPa reinforcement." *Proceedings of 18th Aust. Conf. Mech. Structures and Materials, (ACMSM18)*, Deeks and Hao (eds), 2, 811–817.

Gjørv, O. E. and Sakai, K. (1995). "Testing of chloride diffusivity for concrete."

*Proceedings of the International Conference on Concrete Under Severe Conditions*, CONSEC95, 645–654.

Glass, G. K., Page, C. L. and Short, N. R. (1991). "Factors affecting the corrosion rate of steel in carbonated mortars." *Corrosion Science*, 32(12), 1283–1294.

Glass, G. K., Hassanein, N. M. and Buenfeld, N. R. (1997). "Neural network modeling of chloride binding." *Magazine of Concrete Research*, 49(181), 323–335.

Glasser, F. P., Kindness, A. and Stronach, S. A. (1999). "Stability and solubility relationships in AFm phases Part I. Chloride, sulfate and hydroxide", *Cement and Concrete Research*, 29, 861–866.

Glucklich, J. (1959). "Rheological behavior of hardened cement paste under low stress." *Journal of ACI*, 56 (23), 327–337.

Goto, S. and Roy, D. M. (1981). "The effect of W/C ratio and curing temperature on the permeability of hardened cement paste." *Cement and Concrete Research*, 11, 575–579.

Goto, T. and Uomoto, K. (1993). "Rate of alite hydration." *Annual meeting of ACI*, 47, 44–49.

Goto, Y. and Fujiwara, T. (1979). "Effect of aggregate on drying shrinkage of concrete." *Proceedings of JSCE*, 286, 125–137, (in Japanese).

Graddy, J. C., Kim, J., Whitt, J. H., Burns, N. H. and Klinger, R. E. (2002). "Punching–shear behavior of bridge decks under fatigue loading." *Journal of the ACI Structural*, 99(3), 257–266.

Gross, H. (1973). "On high–temperature creep of concrete." *2nd International Conference on Structural Mechanics Reactor Technology*, Berlin, H, H6/5.

Gross, H. (1975). "High–temperature creep of concrete." *Nuclear Engineering and Design*, 32, (1), 129–147.

Guo, X. H. and Gilbert, R. I. (2002). "An experimental study of reinforced concrete flat slabs under sustained service loads." *UNICIV Report* R–407.

Halamickova, P. *et al.* (1995). "Water permeability and chloride ion diffusion in Portland cement mortars: relationship to sand content and critical pore diameter." *Cement and Concrete Research*, 25(4), 790–802.

Hall, C. (1989). "Water sorptivity of mortars and concretes: a review." *Magazine of Concrete Research*, 41(147), 51–61.

Hamada, M. (1969). "Carbonation of concrete and corrosion of reinforcement." *Cement and Concrete*, 272, Japan Cement Association, 2–18.

Hanehara, S. and Tobiuchi, K. (1991). "Low–heat cement." *Cement and Concrete*, 535, 12–24.

Hannant, D. J. (1967). "Strain behavior of concrete up to 95°C under compressive stresses." *Proceedings of Conference on Prestressed Concrete Pressure Vessels*, group C, 177–191.

Hansen, T. and Nielsen, K. (1965). "Influence of aggregate properties on concrete shrinkage." *Journal of the ACI*, 62(7), 783–794.

Harada, S., Maekawa, K., Tsuji, Y. and Okamura, H. (1991). "Non–linear coupling analysis of heat conduction and temperature–dependent hydration of cement." *Concrete Library of JSCE*, 18.

Hasan, H. O., Cleary, D. B. and Ramirez, J. A. (1998). "Performance of concrete bridge decks and slabs reinforced with epoxy–coated steel under repeated loading." *Bond and Development of Reinforcement*, ACI, SP180–17, 391–404.

Hawkins, N. M. (1974). "Fatigue characteristics in bond and shear of reinforced concrete beams." *ACI publication* SP–41, 203–236.

Hearn, N., Detwiler, R. J. and Sframeli, C. (1994). "Water permeability and microstructure of three old concretes." *Cement and Concrete Research*, 24, 633–640.

Heffernan, P. J., Erki, M. and DuQuesnay, D. L. (2004). "Stress redistribution in cyclically loaded reinforced concrete beams." *Journal of the ACI Structural*, 101(2), 261–268.

Higai, T. (1978). "Fundamental study on shear failure of reinforced concrete beams." *Proceedings of JSCE*, No.279, 113–126.

Hillerborg, A. (1985). "A modified absorption theory." *Cement and Concrete Research*, 15, 809–816.

Hilsdorf, H. K. (1980). "Unveroffentlichte Versuche an der MPA Munchen." Private communication with Bazant, in Bazant, Z. P., Kim, J. and Panula, L. (1991). "Improved prediction model for time–dependent deformations of concrete. Part I: Shrinkage." *Materials and Structures*, 24(65), 327–345.

Hirose, T., Fujisawa, T., Nagayama, I., Yoshida, H. and Sasaki, T., (2001). "Design criteria for trapezoid–shaped CSG dams." *International Commission on Large Dams 2001 Workshop*, Dresden.

Hisasue, K. and Maekawa, K. (2005). "Time–dependent tension stiffness of reinforced concrete and effect of drying shrinkage." *Transaction of JSCE*, V, 555–556 (in Japanese).

Holmen, J. O. (1982). "Fatigue of concrete by constant and variable amplitude loading." *Fatigue of Concrete Structures Detroit, ACI* publication SP–75, 71–110.

Hori, M. (1962). "Strength of cementitious material based on surface energy." *Journal of the Ceramic Association*, Japan, 70, 54–59. (in Japanese).

Horiuchi, Z., Sugiyama, T., Tsuji, Y. and Hashimoto, C. (1998). "A study on the pore structure of concretes containing fly ash by electrical potential method." *Proceedings of JCI*, 20(1), 203–208 (in Japanese).

Houst, Y. F. and Wittmann, F. H. (1994). "Influence of porosity and water content on the diffusivity of $CO_2$ and $O_2$ through hydrated cement paste." *Cement and Concrete Research*, 24(6), 1165–1176.

Hsu, T. T. C. (1981) "Fatigue of plain concrete." *Journal of the ACI, Proceedings* 78(4), 292–305.

Hwan, O. B. (1991). "Fatigue life distributions of concrete for various stress levels." *ACI Journal Material*, 88(2), 122–128.

Igarashi, S. (2002). "Effect of ductile reinforcement covering surface of member." *Proceedings of JCI*, 24(2), 1273–1278.

Ikeda, S. and Uji, K. (1980). "Studies on the effect of bond on the shear behavior of reinforced concrete beams." *Journal of JSCE*, 293, 101–109.

Ikegami, M., Sato, H., Ichiba, T., Otsuki, N., Nishida, T., Terashi, M. and Oishi, K. (2004). "Simplified prediction method for degradation process of cement treated soil." *Proceedings of Japan Civil Engineering Society*, 59(3), 1073–1074 (in Japanese).

Imamoto, K. and Arai, M. (2005). "The influence of specific surface area of coarse aggregate on drying shrinkage of concrete." *Creep and Shrinkage and Durability Mechanics of Concrete and Other Quasi–brittle Materials (ConCreep.7)*, Nantes, France, 95–100.

Imamoto, K., Ishii, S. and Arai, M. (2006). "Drying shrinkage properties of concretes with several kinds of aggregate and the influence of specific surface areas of the aggregates." *Journal of Structural and Construction Engineering*, Architectural Institute of Japan, 606, 9–14 (in Japanese).

Imoto, H., Kurashige, R., Hironaga, M. and Yokozeki, K. (2004). "Leaching behavior of Portland cement hydrate in sodium chloride solution." *Proceedings of JCI*, 26(1), 903–908 (in Japanese).

Inoue, H., Sakai, E. and Daimon, M. (2002). "Analysis of hydration of calcium filler cement." *Proceedings of Cement and Concrete*, 56, 42–49.

Irawan, P. and Maekawa, K. (1994). "Three–dimensional analysis of strength and deformation of confined concrete columns." *Concrete Library of JSCE*, 24, 47–70.

Iriya, K., Hiramoto, M., Hattori, T. and Umehara, H. (1998). "Study on compressive creep in concrete at early age." *Proceedings of JSCE*, 599(V–40), 1–14.

Ishida, T. and Maekawa, K. (1999). "An integrated computational system for mass/energy generation, transport and mechanics of materials and structures." *Journal of JSCE*, 627(44).

Ishida, T. and Maekawa, K. (2000). "An integrated computational system for mass/energy generation, transport, and mechanics of materials and structures." *Concrete Library of JSCE*, 36, 129–144.

Ishida, T. and Maekawa, K. (2002). "Solidified cementitious material–structure model with coupled heat and moisture transport under arbitrary ambient conditions." *Proceedings of FIB Congress 2002*, Osaka.

Ishida, T., Chaube, R. P., Kishi, T. and Maekawa, K. (1997). "Analysis of autogenous and drying shrinkage of concrete based on its microstructure." *Proceedings of JSCE*, 578(V–37), 111–121 (in Japanese).

Ishida, T., Chaube, R. P., Kishi, T. and Maekawa, K. (1998). "Modeling of pore water content in concrete under generic drying–wetting conditions." *Concrete Library of JSCE*, 31, 275–288.

Ishida, T., Kishi, T. and Maekawa, K. (2007). "Enhanced modeling of moisture equilibrium and transport in cementitious materials under arbitrary temperature and relative humidity history." *Cement and Concrete Research*, 37, 565–578.

Israelachvili, J. N. (1985) *Intermolecular and Surface Forces with Applications to Colloidal and Biological Systems*. Academic Press, London.

Ito, H and Fujiki, Y. (1987). "About concrete blended blast furnace slag." *Proceedings of JSCE Symposium on Application of Blast Furnace Slag to Concrete*, 37–42.

Ito, K., Kishi, T. and Uomoto, T. (2002). "Microstructure of hardened cement paste formed in various curing temperatures." *Proceedings of JCI*, 24(1), 489–490.

Iwata, G. and Ishida, T. (2003). "Temperature sensitivity of moisture equilibrium in cementitious material." *Proceedings of JCI*, 25(1), 515–520 (in Japanese).

Izuno, K., Iemura, H., Yamada, Y. and Fujisawa, S. (1993). "Seismic damage assessment of RC structures using different hysteretic models." *Memories of The Faculty of Engineering*, Kyoto University, 55(1), 1–19.

Jansen, D. and Shah, S. P. (1997). "Effect of length on compressive strain softening of concrete." *Journal of Engineering Mechanics*, ASCE, 123(1), 25–35.

Jansen, D. C., Shah, S. P. and Rossow, E. C. (1995). "Stress–strain results of concrete from circumferential strain feedback control testing." *ACI Material Journal*, 92(4), 419–428.

Japan Cement Association. (2001). *Hardened Cement Paste Research Committee Report*, 292–301.

Japan Concrete Institute (1993). "The report of the JCI Committee on Carbonation." JCI (in Japanese).

Japan Society for Analytical Chemistry, The (2005). *Study Group on X–ray Analyses, Practical Side of Powder X–ray Analysis: Introduction to Rietveld Analysis*, 54–58.

Japan Society of Civil Engineers (1998a). "Long–term research on carbonation and corrosion of concrete using fly ash (Final report)." Tokyo, *Concrete Library of JSCE*, 64, 16 (in Japanese).

Japan Society of Civil Engineers (1998b). "Recommendation for design and construction of concrete using blast furnace slag." Tokyo, *Concrete Library of JSCE*, 63, 56–57 (in Japanese).

Japan Society of Civil Engineers (2000) "Creep and drying shrinkage of concrete II." JSCE (in Japanese).

Japan Society of Civil Engineers (2002). *Standard Specification for Design and Construction of Concrete Structures*. (English version of *Structural Performance Verification*).

Japan Society of Civil Engineers Concrete Committee (2005). "Interim report on deterioration of Tarui bridge" (in Japanese).

Japan Society of Civil Engineers. (2005). "Report on safety and serviceability of ASR damaged RC structures." Tokyo, *Concrete Library of JSCE*, 124.

John, J., Hiari, K. and Mihashi, H. (1990). "Influence of environmental moisture and temperature on carbonation of mortar." *Concrete Research and Technology*, JCI, 1(1), 85–94 (in Japanese).

Kakuta, Y. and Fujita Y. (1982). "Fatigue strength of reinforced concrete slabs failing by punching shear." *Proceedings of JSCE*, 317, 149–157.

Kaneko, R. and Ohgishi, S. (1991). "Direct–tension and splitting tensile fatigue characteristics on plain concrete." *Journal of Structural and Construction Engineering*, Architectural Institute of Japan, 419(1), 31–38.

Kaplan, S. A. (1980). "Factors affecting the relationship between rate of loading and measured compressive strength of concrete." *Magazine of Concrete Research*, 32(111), 79–88.

Kato, H., Miyagawa, T., Nakamura, A. and Doi, H. (2003). "Behaviors of chloride ions and effects of gypsum on corrosion of steel reinforcement in mortar using ground granulated blast furnace slag." *Proceedings of JSCE*, 746(61), 1–12.

Kato, E., Kato, Y. and Uomoto, T. (2005). "Development of simulation model of chloride ion transportation in cracked concrete." *Journal of Advanced Concrete Technology*, 3(1), 85–94.

Katz, A. J. and Thompson, A. H. (1987). "Prediction of rock electrical conductivity from mercury injection measurements." *Journal of Geophysical Research*, 92(B1), 599–607.

Kawakami, H. (1991). "Effect of type of aggregate on mechanical behavior of concrete." *Proceedings of JCI*, 13(1), 63–68.

Kawashima, K. (1992). "Dynamic strength and ductility of hollow circular cross–reinforced concrete bridge pier." *Proceedings of JCI*, 34–39 (in Japanese).

Kawasumi, M. (1992). "Determination of the instantaneous strain components in a creep prediction expression: relationships between static and pulse moduli of elasticity, pulse Poisson's ratios of cement pastes." *Abiko Research Laboratory Report*, Central Research Institute of Electric Power Industry, U92004 (in Japanese).

Keeton, J. R. (1965). "Study of creep in concrete." Technical reports R333–I, R333–II, R333–III, US Naval Civil Engineering Laboratory, Port Hueneme, California.

Khan, A. A., Cook, W. D. and Mitchell, D. (1997). "Creep, shrinkage and thermal strains in normal, medium and high–strength concretes during hydration." *ACI Journal Material*, 94(2), 156–163.

Kim S., Taguchi, S., Ohba, Y., Tsurumi, T., Sakai, E. and Daimon, M. (1995). "Carbonation reaction of calcium hydroxide and calcium silicate hydrates." Inorganic materials, 2(254), 18–25.

Kishi, T. and Maekawa, K. (1994). "Thermal and mechanical modeling of young concrete based on hydration process of multi–component cement minerals." *Proceedings of the International RILEM Symposium on Thermal Cracking in Concrete at Early Ages*, Munich, 11–18.

Kishi, T. and Maekawa, K. (1995). "Multi–component model for hydration heat of Portland cement", *Journal of Materials, Concrete Structures and Pavements*, JSCE, V–29 (526), 97–109.

Kishi, T. and Maekawa, K. (1996). "Multi–component model for hydration heat of blended cement with blast slag and fly ash." *Journal of Materials, Concrete Structures and Pavements*, JSCE, V–33 (550), 131–143.

Kishi, T. and Maekawa, K. (1997). "Multi–component model for hydration heating of blended cement with blast furnace slag and fly ash." *Concrete Library of JSCE*, 30, 125–139.

Kishi, T. and Dorjpalamyn, S. (1999). "Hydration heat modeling for cement with limestone powder", *IABSE Colloquium: Concrete Model Code for Asia*, Phuket, 133–138.

Kishi, T., Ozawa, K. and Maekawa, K. (1993a). "Multi–component model for hydration heat of concrete based on cement mineral compounds." *Proceedings of JCI*, 15(1), 1211–1216.

Kishi, T., Shimomura, T. and Maekawa, K. (1993b). "Thermal crack control design of high performance concrete." *Proceedings of Concrete 2000*, Dundee, 447–456.

Kishi, T., Ishida, T. and Maekawa, K. (2001). "Interaction of hydration heat of low water to cement ratio concrete with moisture equilibrium in solid micro–pore." *Journal of Materials, Concrete Structure and Pavements*, 690(53), 45–54.

Kishitani, K. (1963). *Durability of Reinforced Concrete*, Kajima Institute Publishing.

Kiyohara, C., Nagamatsu, S., Sato, Y. and Mihashi, H. (2002). "Study on modulus of elasticity and drying shrinkage of concrete with admixtures." *Proceedings of JCI*, 24(1), 339–344 (in Japanese).

Kiyohara, C., Nagamatsu, S., Sato, K. and Ueda, K. (1999). "Equation for predicting compressive Young's modulus of concrete." *Proceedings of JCI*, 21(2), 601–606.

Kjellsen, K. O., Detwiler, R. J. and Gjørv, O. E. (1990). "Pore structure of plain cement pastes hydrated at different temperature." *Cement Concrete Research*, 20(6), 927–933.

Kobayashi, K. (1991). "Carbonation of concrete." *Proceedings of JSCE*, 433(15), 1–14.

Kobayashi, K. and Syutto, K. (1986). "Diffusivity of oxygen in cementitious materials." *Concrete Engineering*, 24(12), 91–106.

Kodama, K. and Ishikawa, Y. (1982). "Study on fatigue characteristics of concrete." *Proceedings of JCI*, 4, 193–196.

Kokubu, M., Kobayashi, M., Okamura, H. and Yamamoto, Y. (1969). "Problems of lightweight aggregate concrete", *Concrete Library of JSCE*, 24, 1–13 (in Japanese).

Kokubu, M., Kobayashi, M., Okamura, H. and Yamamoto, Y. (1975). "Problems of lightweight aggregate concrete", *Selected papers of Masatane Kokubu Studies on Materials for Concrete Structures*, 315–343.

Komendant, J., Polivka, M. and Pirtz, D. (1976). "Study of concrete properties for prestressed concrete reactor vessels, final report–part II, Creep and strength characteristics of concrete at elevated temperatures." Report UCSESM 76–3 Prepared for General Atomic Company, Dept. Civil Engineering, University of California, Berkeley.

Komendant, J., Nicolayeff V., Polivka, M. and Pirtz, D. (1978). "Effect of temperature, stress level and age at loading on creep of sealed concrete." ACI Special Publication, 55, 55–81.

Kono, S., Bechtoula, H., Kaku, T. and Watanabe, F. (2002). "Damage assessment of RC columns subjected to axial load and bi–directional bending." *JCI Journal*, 24(2), 235–240 (in Japanese).

Kosuge, K. (1993). *Pozzlanic Reaction and Reaction Degree Measuring Method of Silica Fume*. Concrete technology series, Japan Society of Civil Engineers, 4, 1–6.

Kroone, B. and Crook, D. N. (1961). "Studies of pore size distribution in mortars." *Magazine of Concrete Research*, 13(39).

Kunii, T. and Hurusaki, S. (1980). *The Theory on Transfer Rate*, Bai–hu Kan Press.

L'Hermite, R. G., Mamillan, M. and Lefeve, C. (1965). "Nouveaux résultats de recherches sur la déformation de la rupture du béton." *Ann. Inst. Techn. Batiment Trav. Publics*, 207–208, 323–360 (in French).

L'Hermite, R. G., Mamillan, M. and Lefeve, C. (1968). "Further results of shrinkage and creep tests." *International Conference on the Structure of Concrete*, Cement and Concrete Association, 423–433.

Lee, H. S., Noguchi, T. and Tomosawa, F. (1998). "Fundamental study on evaluation of structural performance of reinforced concrete beam damaged by corrosion of longitudinal tensile main rebar by finite element method." *Journal of Structural Concrete Engineering*, AIJ, 506, 43–50.

Lettsrisakulrat, T., Watanabe, K., Matsuo, M. and Niwa, J. (2000). "Localization effects and fracture mechanism of concrete in compression." *Proceedings of JCI*, 22(3), 145–150.

Li, B. and Maekawa, K. (1987). "Contact density model for cracks in concrete." *IABSE Colloquium*, Delft, 51–62.

Li, B., Maekawa, K. and Okamura, H. (1989). "Contact density model for stress transfer across cracks in concrete." *Journal of the Faculty of Engineering, The University of Tokyo* (B), XL(1), 9–52.

Li, C. Q. (2003). "Life–cycle modeling of corrosion–affected concrete structures: propagation." *Journal of Structural Engineering*, ASCE, 129(6), 753–761.

Lide D. R. (2004). *CRC Handbook of Chemistry and Physics*, 85th ed., Boca Raton, London, New York, Washington, D.C.: CRC Press.

Liu, Y. and Weyers, R. E. (1998). "Modeling the time–to–corrosion cracking in chloride contaminated reinforced concrete structures." *ACI Journal Material*, ACI, 95(6), 675–681.

Lokhorst, S. J. and Breugel, K. (1997). "Simulation of the effect of geometrical changes of the microstructure on the deformational behavior of hardening concrete." *Cement and Concrete Research*, 27 (10), 1465–1479.

Lotfi, H. R. and Shing, P. B. (1994). "Interface model applied to fracture of masonry structures." *Journal of Structural Engineering*, 120(1), 63–80.

Lundgren, K. (2002). "Modeling the effect of corrosion on bond in reinforced concrete." *Magazine of Concrete Research*, 54(3), 165–173.

Luping, T. and Nilsson, L. O. (1992). "A study of the quantitative relationship between permeability and pore size distribution of hardened cement pastes." *Cement and Concrete Research*, 22, 541–550.

Mabrouk, R. T., Ishida, T. and Maekawa, K. (2002). "Unified solidification model of hardening concrete composite." *Control of Cracking in Early Age Concrete*, Mihashi and Wittmann (eds), Swets and Zeitlinger, Lisse, 57–66.

Maeda, K. (1989). "Study on a numerical analysis for carbonation of concrete." *Journal of Structural and Construction Engineering*, Transactions of AIJ, 402, 11–19.

Maeda, Y. and Matsui, S. (1984). "Fatigue of reinforced concrete slabs under trucking wheel load." *Proceedings of JCI*, 6, 221–224.

Maekawa, K. and Qureshi, J. (1996). "Computational model for reinforcing bar embedded in concrete under combined axial pullout and transverse displacement." *Proceedings of JSCE*, 538/V–31.

Maekawa, K. and Qureshi, J. (1997) "Stress transfer across interfaces in reinforced concrete due to aggregate interlock and dowel action." *Proceedings of JSCE*, 557/V–34, 159–172.

Maekawa, K. and Ishida, T. (2002). "Modeling of structural performances under coupled environmental and weather actions." *Materials and Structures*, 35, 591–602.

Maekawa, K. and El–Kashif, K. F. (2004). "Cyclic cumulative damaging of reinforced concrete in post–peak regions." *Journal of Advanced Concrete Technology*, 2(2), 257–271.

Maekawa, K., Chaube, R. and Kishi, T. (1996). "Coupled mass transport, hydration and structure formation: theory for durability design of concrete structures." *Integrated Design and Environmental Issues in Concrete Technology* (ed. Sakai), E & FN Spon, London.

Maekawa, K., Chaube, R. and Kishi, T. (1999). *Modeling of Concrete Performance,*

*Hydration, Microstructure Formation and Mass Transport*, E & FN Spon, London.

Maekawa, K., Kishi, T. and Ishida, T. (2003a). "Multi–scale modeling of concrete performance: integrated material and structural mechanics." *Journal of Advanced Concrete Technology*, 1(2), 91–126.

Maekawa, K., Pimanmas, A. and Okamura, H. (2003b). *Nonlinear Mechanics of Reinforced Concrete*. Spon Press, London.

Maekawa, K., Toongoenthong, K., Gebreyouhannes, E. and Kishi, K. (2006a), "Direct path–integral scheme for fatigue simulation of reinforced concrete in shear." *Journal of Advanced Concrete Technology*, 4(1), 159–177.

Maekawa, K., Soltani, M., Ishida, T. and Itoyama, Y. (2006b). "Time–dependent space–averaged constitutive modeling of cracked reinforced concrete subjected to shrinkage and sustained loads." *Journal of Advanced Concrete Technology*, 4(1), 193–207.

Maekawa, K., Gebreyouhannes, E., Mishima, T. and Xuehui, An. (2006c). "Three–dimensional fatigue simulation of RC slabs under traveling wheel–type loads." *Journal of Advanced Concrete Technology*, 4(3), 445–457.

Maekawa, K., Ishida, T. and Chijiwa, N. (2007). "Computational life–cycle assessment of structural concrete subjected to coupled severe environmental and mechanistic actions." *International Conference on Concrete Under Severe Environment*, CONSEC '07, France.

Maekawa, K., Fukuura, N. and Soltani, M. (2008). "Path–dependent high cycle fatigue modeling of joint interfaces in structural concrete." *Journal of Advanced Concrete Technology*, 6(1).

Malvar, L. J. (1998). "Review of static and dynamic properties of steel reinforcing bars." *Journal of ACI Material*, 95(5), 609–616.

Mander, J. B., Priestley, M. J. N. and Park, R. (1988). "Theoretical stress–strain model for confined concrete." *Journal of the Structural Division*, ASCE, 114, 1804–1826.

Mangat, P. S. and Elgarf, M. S. (1999). "Flexural strength of concrete beams with corroding reinforcement." *Journal of ACI Structural*, 96(1), 149–158.

Marti, P. (1985). "Basic tools of reinforced concrete beam design." *Journal of ACI, Proceedings*, 82(1), 45–56.

Martin–Perez, B. (2000). "A study of the effect of chloride binding on service life predictions." *Cement and Concrete Research*, 30, 1215–1223.

Maruya T. (1995). "Development of a Method of Analyzing the Movement of Chloride Ions in Concrete." PhD Thesis, The University of Tokyo.

Maruya, T., Matsuoka, Y. and Tangtermsirikul, S. (1992). "Simulation of chloride movement in hardened concrete." *Concrete Library of JSCE*, 20, 57–70.

Maruya, T., Tangtermsirikul, S. and Matsuoka, Y. (1998a). "Modeling of chloride ion movement in the surface layer of hardened concrete." *Concrete Library of JSCE*, 32, 69–84.

Maruya, T., Tangtermsirikul, S. and Matsuoka, Y. (1998b). "Modeling of movement of chloride ions in concrete surface layer." *Proceedings of JSCE*, 585(38), 79–95.

Matsui, S. (1987). "Fatigue strength of RC–slabs of highway bridge by wheel running machine and influence of water on fatigue." *Proceedings of JCI*, 9(2), 627–632.

Matsumoto, Y., Ueki, H., Yamasaki, T. and Murakami, M. (1998). "Model analysis on carbonation reaction with redissolution of $CaCO_3$." *Proceedings of JCI*, 20(2), 961–966.

Matsuo, T., Matsumura, T., Endo, T. and Tachibana, Y. (2002). "Numerical simulation of loading test for deteriorated RC box culvert." *Proceedings of JCI*, 24(2), 1297–1302.

Matsuo, T., Miyagawa, Y., Akiyama, T. and Iwamori, A. (2007). "Real–scale loading experiment of RC box culverts with corroded reinforcement." *Transactions of JSCE*, 62(5).

Matsuoka, R., Ono, M. and Esaki, F. (2001). "Cyclic effect and loading rate on the mechanical behavior of shear walls." *Proceedings of JCI*, 433–438 (in Japanese).

Matsushima, Y. (1969). "A study of hysteresis model from damage caused by Tokachi–Oki earthquake." *Summaries of Technical Papers of Annual Meeting*, AIJ, 587–588.

Matsuzato, H., Funato, M. and Yamazaki, Y. (1992). "Strength and microstructure of carbonated hardened cement." *Proceedings of Cement and Concrete*, 46, 592–597.

Mattock, A. H. and Hawkins, N. M. (1972). "Shear transfer in reinforced concrete: recent research.' *Journal of PCI*, 17(2), 55–75.

Mehta, P. K. and Manmohan C. (1980). "Pore size distribution and permeability of hardened cement paste." *7th International Congress on Chem. Cem.*, Paris, 3, 7–1/5.

Mihashi, H. and Numao, T. (1992). "Influence of temperature on water loss and shrinkage of hardened cement paste." *Proceedings of JCA*, 46, 702–707 (in Japanese).

Mishima, T., Yamada, K. and Maekawa, K. (1992). "Localized deformational behavior of a crack in RC plates subjected to reversed cyclic loads." *Proceedings of JSCE*, V–16(442), 161–70.

Miyagawa, T. (2003). "Safety evaluation of structures with ruptured reinforcing bars by ASR (intermediate report)." *Journal of JSCE*, 88, 83–84.

Molina, F. J., Alonso, C. and Andrade C. (1993). "Cover cracking as a function of rebar corrosion. Part 2: Numerical model." *Materials and Structures*, RILEM, 26, 532–548.

Morikawa, M., Seki, H. and Okamura, Y. (1988). "Basic study on cracking of concrete due to expansion by rebar corrosion." *Concrete Library of JSCE*, 11, June, 83–105.

Moriwake, A., Fukute, T. and Horiguchi, K. (1993). "Study on durability of massive concrete." *Proceedings of JCI*, 15(1), 859–864.

Mozer, J. D., Gerstle, K. H. and Tulin, L. G. (1970). "Time–dependent behavior of concrete beams." *Journal of Structural Division*, ASCE, 96(ST3), March, 597–612.

Murata, N. and Okada K. (1980). *Fresh Concrete Rheology and Concrete Elasticity and Creep*. Sankai–Do (in Japanese).

Murdock, J. and Kesler, C. (1958). "Effect of range of stress on fatigue strength of plain concrete beams." *Journal of ACI*, 55(12), 221–231.

Mutsuyoshi, K. and Machida, A. (1985). "Behavior of reinforced concrete members subjected to dynamic loading." *Journal of Materials, Concrete Structures and Pavements*, JSCE, 354/V–2, 81–90.

Nagataki, S. and Sato, R. (1983). "Deformation and crack analysis of RC members under sustained loads." *Proceedings of JCI*, 5(1), 461–464.

Nagataki, S., Ohga, H. and Saeki, T. (1987). "Analytical prediction of carbonation depth." *Annual Report of Cement Technology*, 41, 343–346.

Nakamura, H. and Higai, T. (1999). "Compressive fracture energy and fracture zone length of concrete." *Seminar on Post–peak Behavior of RC Structures Subjected to Seismic Loads*, Japan Concrete Institute, C51E(2), 259–272.

Nakarai, K. and Ishida, T. (2003). "Modeling of time–dependent material properties of cemented soil based on thermodynamic approach." *9th East Asia–Pacific Conference on Structural Engineering and Construction* (EASEC–9).

Nakarai, K., Ishida, T., Kishi, T. and Maekawa, K. (2005). "Enhanced modeling of microstructure formation of cement hydrates coupled with thermodynamics."

*Proceedings of Workshop on Cementitious Materials as Model Porous Media: Nanostructure and Transport Processes*, 147–151.

Nakarai, K., Ishida, T., Kishi, T. and Maekawa, K. (2007). "Enhanced thermodynamics analysis coupled with temperature–dependent microstructures of cement hydrates." *Cement and Concrete Research*, 37(2), 139–150.

Nakasu, M. and Iwatate, J. (1996). "Fatigue experiment on bond between concrete and reinforcement." *Transaction of JSCE*, V–426, 852–853.

Nakatani, S. (2002). "Experimental study on the fatigue durability of highway bridge slabs." National Institute for Land and Infrastructure Management, Ministry of Land, Infrastructure and Transport, Japan, 28.

Nasser, K. W. and Neville, A. M. (1965). "Creep of concrete at elevated temperatures." *Journal of ACI*, 62(87), 1567–1579.

National Institute of Natural Science, National Astronomical Observatory of Japan, (2002). "Chronological Scientific Table", Tokyo, Maruzen (in Japanese).

Neville, A. M. (1959). "Creep recovery of mortars made with different cements." *Journal of ACI*, 56(13), 167–174.

Neville, A. M. (1970). *Creep of Concrete*. North–Holland Publishing Co.

Neville, A. M. (1991). *Properties of Concrete*, Elsevier Science B. V.

Neville, A. M. (1995). *Properties of Concrete*. 4th edition, John Wiley & Sons, Inc.

Ngala, V. T. and Page, C. L. (1997). "Effects of carbonation on pore structure and diffusion properties of hydrated cement pastes." *Cement and Concrete Research*, 27(7), 995–1007.

Nilson, L. O., Massat, M. and Tang, L. (1994). "The effect of non–linear chloride penetration into concrete structures." *Concrete Durability*, 145–24, 469–486.

Nishi, T., Shimomura, T. and Sato, H. (1999). "Modeling of diffusion of vapor within cracked concrete." *Proceedings of JCI*, 21(2), 859–864.

Nishibayashi, S., Inoue, S. and Yamura, K. (1988). "Fatigue characteristics of reinforced concrete in water." *Concrete in Marine Environment*, ACI publication, SP 109–24, 543–561.

Nishida, N., Otsuki, M. A., Baccay and Hamamoto, J. (2005). *Temperature Dependency of Corrosion Rate of Steel Bars in Concrete Influenced by Material Segregation*. Tokyo Institute of Technology, Japan, Toa Corporation, Japan.

Nishida, T. (2005). "Influence of temperature on deterioration process of reinforced concrete members due to steel corrosion." PhD Thesis, Tokyo Institute of Technology, Japan.

Niwa, J., Yamada K., Yokozawa K. and Okamura, H. (1987). "Revaluation of the equation for shear strength of reinforced concrete beams without web reinforcement." *Concrete Library of JSCE*, 9, 65–84.

Numao, T. and Mihashi, H. (1991). "Moisture migration and shrinkage of hardened cement paste at elevated temperatures." *Transactions 11th International Conference on Structural Mechanics in Reactor Technology* (SMiRT), 37–42.

Nyame, B. K. and Illston, J. M. (1981). "Relationships between permeability and pore structure of hardened cement paste." *Magazine of Concrete Research*, 33(116), 139–146.

Oh, H., Sim, J. and Meyer, C. (2005). "Fatigue life of damaged bridge deck panels strengthened with carbon fiber sheets." *Journal of ACI Structural*, 102(1), 85–92.

Ohga, H. and Nagataki, S. (1988). "Prediction and evaluation of the depth of carbonation of concrete by accelerated test." *Proceedings of JSCE*, 390(8), 225–233.

Ohgishi, S., Ono, H. and Tanahashi, I. (1983). "Influence of surface energy on mechanical properties and swelling of building materials." *Journal of the Society of Materials Science*, Japan, 32(353), 67–83 (in Japanese).

Okada, K., Kobayashi, K. and Miyagawa, T. (1988). "Influence of longitudinal cracking due to reinforcement corrosion on characteristics of reinforced concrete members." *Journal of ACI Structural*, 85(2), 134–140.

Okamura, H. (2000). *Engineering of Reinforced Concrete*. Ichigaya Publishing Co. Ltd, 16 (in Japanese).

Okamura, H. and Higai, T. (1980). "Proposed design equation for shear strength of reinforced concrete beams without web reinforcement." *Proceedings of JSCE*, 300, 131–141.

Okamura, H. and Maekawa, K. (1991). *Nonlinear Analysis and Constitutive Models of Reinforced Concrete*. Giho–do Press, Tokyo.

Okamura, H., Farghaly, S. A. and Ueda, T. (1981). "Behaviors of reinforced concrete beams with stirrups failing in shear under fatigue loading." *Proceedings of JSCE*, 308, 109–122.

Ono, H. and Shimomura, T. (2005). "Role of aggregate in shrinkage of concrete." *Proceedings of JCI*, 27, (1), 457–462 (in Japanese).

Osada, M., Ueki, H., Yamasaki, T. and Murakami, M. (1997). "Simulation analysis on carbonation reaction of concrete members taking alkali element into consideration." *Proceedings of JCI*, 19(1), 793–798.

Osbaeck, B. (1992). "Prediction of cement properties from description of the hydration processes." *9th International Congress on the Chemistry of Cement*, 4, 504–510.

Ozaki, S., Sugata, N. and Mukaida, K. (1995). "Investigation on the reduction of fatigue strength of submerged concrete." *Proceedings of International Conference on Concrete Under Severe Conditions*, CONSEC, 2, 1694–1703.

Ozawa, K., Maekawa, K. and Okamura, H. (1992). "Development of high performance concrete." *Journal of The Faculty of Engineering*, The University of Tokyo (B), XLI(3), 381–439.

Page, C. L. and Vennesland, Ø. (1983). "Pore solution composition and chloride binding capacity of silica fume cement pastes." *Materials and Structures*, 16(1), 19–25.

Paillere, A. M., Buil, M. and Serrane, J. J. (1990). "Effect of fiber addition on the autogenous shrinkage of silica fume concrete." *Discussion, Author's Closure, ACI Journal Material*, 87(1), 82–83.

Pallewatta, T. M., Irawan, P. and Maekawa, K. (1995). "Effectiveness of laterally arranged reinforcement on the confinement of core concrete." *Journal of Materials, Concrete Structures and Pavements*, JSCE, V–28(520), 297–308.

Palmgren, A. G. (1924). "Die Lebensdauer von Kugellagern." *VDI–Zeitschrift des Vereines Deutscher Ingenieure*, 68(14), 339–341 (in German).

Pantazopoulou, S. J. and Papoulia, K. D. (2001). "Modeling cover–cracking due to reinforcement corrosion in RC structures." *Journal of Engineering Mechanics*, ASCE, 127(4), 342–351.

Papadakis, V. G., Vayenas, C. G. and Fardis, M. N. (1991a). "Physical and chemical characteristics affecting the durability of concrete." *ACI Journal Material*, 88(2), 186–196.

Papadakis, V. G., Vayenas, G. G. and Fardis, M. N. (1991b). "Fundamental modeling and experimental investigation of concrete carbonation." *ACI Journal Material*, 88(4), 363–373.

Park, K. B. and Noguchi, T. (2002). "Autogenous shrinkage of cement paste hydrated at different temperatures: influence of microstructure and relative humidity change." *Proceedings of 3rd International Reserach Seminar on Self–desiccation and Its Importance in Concrete Technology*, B. Persson and G. Fagerlund (eds), Lund, Sweden, 93–101.

Park, Y. J. and Ang, A. H. S. (1985). "Mechanistic seismic damage model for reinforced concrete." *Journal of Structural Engineering*, ASCE, 111(4), 722–739.

Perdikaris, P. C. and Beim, S. (1988). "RC bridge decks under pulsating and moving load." *Journal of Structural Division*, ASCE, 114(3), 591–607.

Perdikaris, P. C., Beim, S. R. and Bousias, S. N. (1989). "Slab continuity effect on ultimate and fatigue strength of reinforced concrete bridge deck models." *Journal of ACI Structural*, 86(4), 483–491.

Persson, B. (1997). "Moisture in concrete subjected to different kinds of curing." *Materials and Structures*, 30, 533–544.

Peschel, G. (1968). "The viscosity of thin water films between two quartz plates." *Matrx. et Constr.*, 1(N6), 529–534.

Petersen, E. E. (1958). "Diffusion in a pore of varying cross section." *A. I. Ch. E. Journal*, 4(3), 343–345.

Pickett, G. (1956). "Effect of aggregate on shrinkage of concrete and hypothesis concerning shrinkage." *Journal of ACI*, 52, 581–590.

Pimanmas, A. and Maekawa, K. (2001a). "Finite element analysis and behaviour of pre–cracked reinforced concrete members in shear." *Magazine of Concrete Research*, 53(4), 263–282.

Pimanmas, A. and Maekawa, K. (2001b). "Multi–directional fixed crack approach for highly anisotropic shear behavior in pre–cracked RC members." *Journal of Materials, Concrete Structures and Pavements*, JSCE, 50 (669), 293–307.

Pimanmas, A. and Maekawa, K. (2001c). "Control of crack localization and formation of failure path in RC members containing artificial crack device." *Journal of Materials, Concrete Structures and Pavements*, JSCE, 52 (683), 173–186.

Pimanmas, A. and Maekawa, K. (2001d), "Shear failure of RC members subjected to pre–cracks and combined axial tension and shear." *Journal of Materials, Concrete Structures and Pavements*, JSCE, 53 (690), 159–174.

Piron, D. (1991). *Electrochemistry of Corrosion*. NACE International.

Pirtz, D. (1968). "Creep characteristics of mass concrete for Dworshak Dam." Report 65–2 (Structural Engineering Laboratory), University of California, Berkeley.

Poli, S. D., Prisco, M. D. and Gambarova, P. G. (1992). "Shear response, deformations and subgrade stiffness of a dowel bar embedded in concrete." *Journal of ACI Structural*, 89(6), 665–675.

Powers, T. C. (1964). "The physical structure of Portland cement paste." *The Chemistry of Cement*, H. F. Taylor (ed.), Academic Press, New York, 391–416.

Powers, T. C. (1968a) "The thermodynamics of volume change and creep." *Matériaux et Constructions*, 1(6), 487–507.

Powers, T. C. (1968b). "Mechanisms of shrinkage and reversible creep of hardened cement paste." *International Conference on the Structure of Concrete*, Cement and Concrete Association, London, 319–344.

Powers, T. C. and Brownyard, T. L. (1947). "Studies of the physical properties of hardened Portland cement paste, Part 9: General summary of findings on the properties of hardened Portland cement paste." *Journal of American Concrete Institute*, 18(8), 972–992.

Pruijssers, A. F., Walraven, J. C. and Reinhardt, H. W. (1988). *Aggregate Interlock and Dowel Action Under Monotonic and Cyclic Loading*. Delft University Press.

Raithby, K. D. and Galloway, J. W. (1974). "Effect of moisture condition, age and rate of loading on fatigue of plain concrete." ACI Publication SP 41–1, 15–34.

Raja, R. H. (2006). "Coupling of the Effect of Chloride and Temperature on Corrosion of RC Structures Based on Thermodynamic Approach." Masters Thesis, The University of Tokyo.

Ramachandran, V. S., Feldman, R. F. and Beaudoin, J. J. (1981). *Concrete Science*. Heyden & Son Ltd.

Ravindrarajah, R. and Ong, K. (1987). "Corrosion of steel in concrete in relation to bar diameter and cover thickness." *Concrete Durability, Katherine and Bryant Mather International Conference*, ACI, SP–100, 1667–1677.

Regan, P. E. and Kennedy Reid, I. L. (2004). "Shear strength of RC beams with defective stirrup anchorages." *Magazine of Concrete Research*, 56(3), 159–166.

Reinhardt, H. W. and Gaber, K. (1990). "From pore size distribution to an equivalent pore size of cement mortar." *Materials and Structures*, 23(133), 3–15.

Reinhardt, H. W., Cornelissen, H. A. W. and Hordijk, D. A. (1986). "Tensile tests and failure analysis of concrete." *Journal of Structural Engineering*, ASCE, 112(11), 2462–2477.

Richart, F. E., Brandtezaeg, A. and Brown, R. L. (1928). *A Study of the Failure of Concrete Under Combined Compressive Stresses*. Bulletin 1985, University of Illinois, Urbana.

RILEM Committee 36–RDL (1984). "Long–term random dynamic loading of concrete structures." *Materials and Structures*, RILEM, 17(9), 1–28.

Rodriguez, J., Ortega, L. M. and Casal, J. (1997). "Load carrying capacity of concrete structures with corroded reinforcement." *Journal of Construction and Building Materials*, ASCE, 11(4), 239–248.

Roll, F. (1964). "Long–time creep recovery of highly stressed concrete cylinders." *Creep and Concrete*, ACI Special Publication, 9, 95–114.

Ross, A. D. (1958). "Creep of concrete under variable stress." *Journal of the ACI*, 53, 739–758.

Rostásy, F. S., Teichen, K. T. and Engelke, H. (1972). "Beitrag zur Klärungdes Zusammenhanges von Kriechen und Relaxation bei Normalbeton." *Amtlicheforschungs und Materialprufungsanstalt fur das Bauwesen*, Heft 139 (in German).

Rusch, H. (1960). "Research towards a general flexural theory for structural concrete." *Journal of the ACI*, 57(1), 1–27.

Rusch, H., Jumgwirth, D. and Hilsdorf, H. K. (1983). *Creep and Shrinkage: Their Effect on the Behavior of Concrete Structures*. New York: Springer–Verlag.

Ryshkewitch, E. (1953). "Composition and strength of porous sintered alumina and zirconia." *Journal of the American Ceramic Society*, 36, 65–68.

Sabry, A. Farghaly (1979). "Shear design of reinforced concrete beams for static and repeated loads." Dissertation for the degree of Doctor of Engineering, The University of Tokyo.

Saeki, T., Ohga, H. and Nagataki, S. (1991). "Mechanism of carbonation and prediction of carbonation process of concrete." *Concrete Library of JSCE*, 17, 23–36.

Saetta, A. V., Schrefler, B. A. and Vitaliani, R. V. (1993). "The carbonation of concrete and the mechanisms of moisture, heat and carbon dioxide flow through porous materials." *Cement and Concrete Research*, 23, 761–772.

Saetta, A. V., Schrefler, B. A. and Vitaliani, R. V. (1995). "2–D model for carbonation and moisture/heat flow in porous materials." *Cement and Concrete Research*, 25, 1703–1712.

Saito, H., *et al.* (1996). "Deterioration of cement hydrate by electrical acceleration test method." *Proceedings of JCI*, 1, 969–974.

Saito, M. and Imai, S. (1983). "Direct tensile fatigue of concrete by the use of friction grips." *Journal of the ACI*, September–October, 431–438.

Sakai, E. (1993). "The carbonation reaction." *Cement Chemistry Simplified*, Japan Cement Association, 105–112.

Sakai, E., Sakai, M., Asaga, K. and Daimon, M. (1998). "Phase composition model of cement hydration." *Annual Proceedings of Concrete Engineering*, 20(1), 101–106.

Sakai, E., Morioka, M., Xi, Z., Ohba, Y. and Daimon, M. (1999). "Carbonation reaction of hardened low heat Portland cement." *Journal of the Ceramic Society of Japan*, 107(6), 561–566.

Santhikumar, S. (1993). "Temperature Dependent Heat Generation Model for Mixed

Cement Concrete with Mutual Interactions Among Constituent Minerals." Masters Thesis, The University of Tokyo.

Sato, H., Yui, M. and Yoshikawa, H. (1995). "Diffusion behavior for Se and Zr in sodium–bentonite." *Proceedings of Materials Research Society Symposium*, Vol. 353, 269–276.

Sato, R., Ujike, I., Tezuka, M. and Yoshimoto, T. (1987). "Deformation and cracking of RC flexural members under sustained loads." *Proceedings of JCI*, 9(2), 217–222 (in Japanese).

Satoh, Y. *et al.* (2003). "Shear behavior of RC member with corroded shear and longitudinal reinforcing steels." *Proceedings of JCI*, 25(1), 821–826.

Schenk, O. and Gartner, K. (2004). "Solving unsymmetric sparse systems of linear equations with PARDISO." *Future Generation Computer Systems*, 20(3), 475–487.

Schiller, K. K. (1958). *Mechanical Properties of Non–Metallic Materials*, Butterworths, London, 35–50.

Schlaich, J. and Weischede, D. (1982). "Detailing of concrete structures." *Bulletin d'Information 150, Comité Euro–International du Béton*, Paris, 163.

Scott, B. D., Park, R. and Priestley, M. J. N. (1982). "Stress–strain behavior of concrete confined by overlapping hoops at low and high strain rate." *Journal of ACI*, 79(1), 13–27.

Sheikh, S. A. (1982). "A comparative study of confinement models." *Journal of ACI*, 79, 296–306.

Shima, H., Chou, L. and Okamura, H. (1987). "Micro and macro models for bond in reinforced concrete." *Journal of the Faculty of Engineering, The University of Tokyo (B)*, 39(2), 133–194.

Shima, H., Asaga, K., Daimon, M. and Goto, S. (1989). "Absorption rate of carbon dioxide in hardened cement." *Proceedings of Cement and Concrete*, 43, 406–411.

Shimomura, T. (1998). "Modeling of initial defect of concrete due to drying shrinkage." *Concrete under Severe Conditions 2, CONSEC 98*, 3, 2074–2083.

Shimomura, T. and Maekawa, K. (1997). "Analysis of the drying shrinkage behavior of concrete using a micromechanical model based on the micropore structure of concrete." *Magazine of Concrete Research*, 49(181), 303–322.

Shin, H., Maekawa, K. and Okamura, H. (1988). "Analytical approach of RC members subjected to reversed cyclic in–plane loading." *Proceedings of JCI Colloquium on Ductility of Concrete Structures and its Evaluation*, C2, 45–56.

Shioya, T., Higuchi, Y., Siokawa, H. and Takagisi, M. (2001). "Flexural shear experiments of RC beams with T–headed bars as stirrups." *Proceedings of JCI*, 23(3), 799–804.

Shirakawa, T., Shimazoe, Y., Aso, M. and Nagamatsu, S. (2001). "Investigation of carbonation mechanism using hardened cement paste." *Proceedings of JCI*, 23(2), 493–498.

Snowdon, L. and Edwards, A. (1962). "The moisture movement of natural aggregate and its effect on concrete." *Magazine of Concrete Research*, 14(41), 109–116.

Sogano, T. *et al.* (2001). "Numerical analysis on uneven settlement and seismic performance of Tokyo brick bridges." *Structural Engineering Design*, 17, JR–East, 96–109.

Soltani, M. and Maekawa, K. (2007). "Path–dependent mechanical model for deformed reinforcing bars at RC interface under coupled cyclic shear and pullout tension." *Engineering Structures*, Elsevier.

Someya, K., Daisoku, N., Tiong–Huan, W. and Nagataki, S. (1989). "Characteristics of binding of chloride ions in hardened cement pastes." *Annual Proceedings of Concrete Engineering*, 11(1), 603–608.

Song, C., Maekawa, K. and Okumara H. (1991). "Time and path–dependent uniaxial constitutive model of concrete." *Journal of The Faculty of Engineering, The University of Tokyo (B)*, XLI(1), 159–237.

Soroka, I. (1976). *Portland Cement Paste and Concrete*. The Macmillan Press Ltd, 114–125.

Soroushian, P., Obasaki, K., Baiyasi, M. I., El–Sweidan, B. and Choi, K. B. (1988). "Inelastic cyclic behavior of dowel bars." *Journal of ACI Structural*, 23–29.

Stanish, K., Hooton, R. D. and Pantazopoulou, S. J. (1999). "Corrosion effects on bond strength in reinforced concrete." *Journal of ACI Structural*, 96(6), 915–921.

Stauffer, D. and Aharony, A. (1992). *Introduction to Percolation Theory*, Taylor & Francis.

Stockl, S. (1981). "Versuche zum Einfluss der Belastungshohe auf das Kriechen von Beton." *Deutscher Auschuss fur Stahlbeton*, Heft 324 (in German).

Subramaniam, K. V., Popovics, J. S. and Shah, P. (2002). "Fatigue fracture of concrete subjected to biaxial stresses in the tensile C–T region." *Journal of Engineering Mechanics*, ASCE, 128(6), 668–676.

Suzuki, K., Nishikawa, N. and Hayashi, T. (1989a). "Carbonation of C–S–H with different Ca/Si ratios." *Proceedings of Cement and Concrete*, 43, 18–23.

Suzuki, Y., Harada, S., Maekawa, K. and Tsuji, Y. (1989b). "Evaluation of adiabatic temperature rise of concrete measured with the new testing apparatus." *Concrete Library of JSCE*, 13.

Suzuki, Y., Tsuji, Y., Maekawa, K. and Okamura, H. (1990a), "Qualification of heat evolution during hydration process of cement in concrete." *Proceedings of JSCE*, 414, 155–164.

Suzuki, Y., Tsuji, Y., Maekawa, K. and Okamura, H. (1990b). "Quantification of hydration–heat generation process of cement in concrete." *Concrete Library of JSCE*, 16.

Swatekititham, S. and Okamura, H. (2006). "Low chloride distribution in concrete structures near the seashore." *Proceedings of JSCE*, 62(1), 221–229.

Synder, K. A. *et al.* (1990). "Interfacial zone percolation in cement aggregate composites." 27, *Interfaces in Cementitious Composites*, ed. Maso J.C., London, E & FN Spon.

Tabata, M. and Maekawa, K. (1984). "Prediction model for plasticity and failure of concrete based on time dependence." *Proceedings of JCI*, 6(1), 269–272 (in Japanese).

Tachibana, Y., Kajikawa, Y. and Kawamura. M. (1990). "The mechanical behavior of RC beams damaged by corrosion of reinforcement." *Concrete Library of JSCE*, 14, March, 177–188.

Tae, S., Park, K. and Noguchi, T. (2001). "Autogeneous shrinkage and microstructure formation of cement–based materials." *Proceedings of JCI*, 23(2), 793–798 (in Japanese).

Tafel, J. (1906). *Zeitschrift für Eleckrochemie*. 12, 112–122 (in German).

Taira, Y., Suda, K., Aikawa, K. and Noguchi, T. (2007). "Cyclic fatigue loading experiment of connection joints in between pre–cast panels." *Transactions of JSCE*, V.

Takeda, T., Sozen, M. A. and Nielsen, N. N. (1970). "Reinforced concrete response TZO simulate earthquake." *Journal of Structural Division*, ASCE, 96(12), 2257–2573.

Takewaka, K. and Matsumoto, S. (1984) "Behaviors of reinforced concrete members deteriorated by corrosion of reinforcement." *Proceedings of JCI*, 6, 177–180.

Taniguchi, K., Kishi, T. and Ishida, T. (2000). "Moisture behavior during cement hydration." *Proceedings of JSCE*, 55(5), 510–511.

Tassios, T. P. and Vbintzeleou, E. N. (1987). "Concrete–to–concrete friction." *Journal of Structural Engineering*, ASCE, 113(4), 832–849.

Tatematsu, K., Arai, M., Iwashimizu, T., Kimura, Y., Urano, H., Imamoto, K. and Motoki, R. (2001). "Experimental study on drying shrinkage and pore size distribution of aggregate in Kansai area." *Journal of Structural and Construction Engineering*, Architectural Institute of Japan, 549, 1–6 (in Japanese).

Taylor, H. F. W. (1972). *The Chemistry of Cement*. 1, Academic Press ING, London.

Taylor, H. F. W. (1997). *Cement Chemistry*, 2nd edition, Thomas Telford, London, 231–237, 245.

Taylor, R. (1959). Discussion of a paper by Chang, T. S. and Kesler, C. E. "Fatigue behavior of reinforced concrete beams." *Journal of ACI*, 55(14), 1011–1015

Tazawa, E. and Miyazawa, S. (1994). "Influence of binder and mix proportion on autogenous shrinkage of cementitious materials." *Proceedings of JSCE*, 502(V–25), 43–52 (in Japanese).

Tazawa, E., Miyazawa, S., Sato, T. and Konishi, K. (1992). "Autogenous shrinkage of concrete." *Proceedings of JCI*, 14 (1), 561–566 (in Japanese).

Tepfers, R. (1979). "Tensile fatigue strength of plain concrete." *Journal of ACI*, August, 919–933.

Tepfers, R. (1982). "Fatigue of plain concrete subjected to stress reversals." *Fatigue of Concrete Structures Detroit*, ACI, publication SP–75, 343–372.

Tomosawa, F. (1974). "Cement hydration model." *Annual Report of JCA*, 28, 53–57.

Toongoenthong, K. and Maekawa, K. (2004a). "Interaction of pre–induced damages along main reinforcement and diagonal shear in RC members." *Journal of Advanced Concrete Technology*, 2(3), 431–443.

Toongoenthong, K. and Maekawa, K. (2004b). "Multi–mechanical approach to structural performance assessment of corroded RC members in shear." *Journal of Advanced Concrete Technology*, 3(1) 107–122.

Toongoenthong, K. and Maekawa, K. (2004c). "Computational performance assessment of damaged RC members with fractured stirrups." *Journal of Advanced Concrete Technology*, 3(1) 123–136.

Toongoenthong, K. and Maekawa, K. (2004d). "Simulation of coupled corrosive product formation, migration into crack and propagation in reinforced concrete sections." *Journal of Advanced Concrete Technology*, 3(2), 253–265.

Troxell, G. E., Raphael, J. E. and Davis, R. W. (1958). "Long–time creep and shrinkage tests of plain and reinforced concrete." *Proceedings of ASTM*, 58, 1101–1120.

Tsuda, M., Sato, R., Tottori, S. and Tezuka, M. (1995). "On the behaviors of PRC continuous beams subjected to sustained loads." *Proceedings of JCI*, 17(2), 697–702 (in Japanese).

Tsunomoto, M., Kajikawa, Y. and Kawamura, M. (1990). "Elasto–plastic analysis of expansive behaviour due to corrosion of reinforcement in concrete." *Concrete Library of JSCE*, 14, 189–199.

Tsutsumi, T., Yasuda, N., Matsushima, M. and Ohga, H. (1998). "Study on model of crack width due to corrosion products." *Journal of Materials, Concrete Structures and Pavements*, JSCE, 585 (V–38), 69–77.

Tuutti, K. (1982). *Corrosion of Steel in Concrete*. CBI Forskning/Research, Swedish Cement and Concrete Research Institute, 4, 468.

Uchida, K. and Sakakibara, H. (1987). "Formulation of the heat liberation rate of cement and prediction method of temperature rise based on cumulative heat liberation." *Concrete Library of JSCE*, 9.

Uchikawa, H. (1985). "Effect of blast furnace slag and fly ash on the diffusivity of alkali ions in hardened cement paste." *Cement and Concrete*, 460, 20–27 (in Japanese).

Uchikawa, H. (1986). "Effect of blending component on hydration and structure formation." *Proceedings of the 8th International Congress on the Chemistry of*

*Cement*, Rio de Janeiro.

Uchikawa, H., Uchida, S. and Hanehara, S. (1991). "Measuring method of pore structure in hardened cement paste." *Mortar and Concrete, il Cemento*, 88, 67–90.

Uchikawa, H., Sawaki, D. and Hanehara, S. (1993). "Effect of type of organic admixture and addition method on fluidity of fresh cement paste." *Proceedings of JCI Symposium on High-fluid Concrete*, 55–62.

Ueda, T. and Okamura, H. (1982). "Fatigue behavior of reinforced concrete beams under shear force." *IABSE Colloquium* (Lausanne), 415–422.

Ueda, T., Pantaratorn, N. and Sato, Y. (1995). "Finite element analysis on shear resisting mechanism of concrete beams with shear reinforcement." *Journal of Materials, Concrete Structures and Pavements*, 520/V–25, JSCE, 273–286.

Ueda, T., Zahran, M. and Kakuta, Y. (1999). "Shear fatigue behavior of steel–concrete sandwich beams." *Concrete Library of JSCE*, 33, 83–111.

Uematsu, K. and Kishi, T. (1997). "Effect of limestone powder on hydration heat process of cement." *Proceedings of the Annual Conference of JSCE*, 52(5), 180–181.

Ulm, F. J. and Coussy, O. (1996). "Strength growth as chemo–plastic hardening in early age concrete." *Journal of Engineering Mechanics*, ASCE, 122(12), 1123–1132.

Uomoto, T. (2000). *Non–Destructive Testing in Civil Engineering*. Elsevier, 671–678.

Uomoto, T. and Takada, Y. (1993). "Factors affecting concrete carbonation ratio." *Concrete Library of JSCE*, 21, 31–44.

Uomoto, T. and Ohshita, K. (1994). "A fundamental study on set–retardation of concrete due to superplasticizer." *Concrete Research and Technology*, JCI, 5(1), 119–129.

Upendra, J. C. (1964). "The effect of the elastic modulus of the aggregate on the elastic modulus, creep and creep recovery of concrete." *Magazine of Concrete Research*, 16 (48), 129–138.

Usui, T., Ishida, T., Nakarai, K. and Sakimura, T. (2005). "Calcium leaching analysis coupled with bentonite and surrounding soil." *Proceedings of JCI*, 27 (in Japanese).

Vaccaro, F. J., Rhoades, J. and Le, B. (1997). "The effect of temperature on VRLA reaction rates and the determination of battery state of charge. II. Fundamental considerations." *Telecommunications Energy Conference*. INTELEC 97, 19(19–23) 230–237.

Van Brakel, J. and Heertjes, P. M. (1974). "Analysis of diffusion in macroporous media in terms of a porosity, a tortuosity and a constrictivity factor." *International Journal of Heat and Mass Transfer*, 17, 1093–1103.

Van Mier, J.G.M. (1986). "Multiaxial strain–softening of concrete, Part I: Fracture." *Material and Structures*, RILEM, 19(111), 179–190.

Verbeck, G. J. (1958). "Carbonation of hydrated Portland cement." ASTM Special Technical Publication, 205, 17–36.

Viest, I. M., Elstner, R. C. and Hognestad, E. (1956). "Sustained load strength of eccentrically loaded short reinforced concrete column." *Journal of ACI*, 27(7), 727–755.

Vintzeleou, E. N. and Tassios, T. P. (1987). "Behavior of dowels under cyclic deformations." *Journal of ACI Structural*, 84, 18–30.

Volkwein, A. (1993). "The capillary suction of water into concrete and the abnormal viscosity of the pore water." *Cement and Concrete Research*, 23, 843–852.

Wallo, E. M., Yuan, R. L., Lott, J. L. and Kesler, C. E. (1965). *Sixth Progress Report on Prediction of Creep in Structural Concrete from Shot Time Tests*. T&AM Report 658, Department of Theoretical and Applied Mechanics, University of Illinois, Urbana.

Walraven, J. C. (1981). "Fundamental analysis of aggregate interlock." *Journal of Structural Division*, ASCE, 107(ST11), 2245–2270.

Walraven, J. C. and Reinhardt, H. W. (1981). "Theory and experiments on the mechanical behavior of crack in plain and reinforced concrete subjected to shear loading." *HERON*, 26(1A), 5–68.

Wanibushi, K., Nakaumura, A., Sakai, E., Osawa, S. and Daimon, M. (1999). "Effects of potassium chloride on hydration of ground granulated blast furnace slag." *Inorganic Materials*, 6, 207–212.

Washa, G. W. and Fluck, P. G. (1952). "Effect of compressive reinforcement on the plastic flow of reinforced concrete beams." *Journal of ACI*, 24(2), 89–108.

Watanabe, K. and Tateyama, M. (2002). "Shaking table tests on a new type bridge abutment with geogrid–reinforced cement treated backfill." *Proceedings of 7th International Conference on Geosynthetics*, Nice, France, Vol. 1.

Watanabe, K., Yokozeki, K., Otsuki, N. and Daimon, M. (2000). "Experimental investigation for analysis of calcium leaching of cementitious materials." *Proceedings of JCI*, 22(1), 217–222 (in Japanese).

Weiss, J., Kadir, G. and Shah, S. P. (2001). "Localization and size–dependent response of reinforced concrete beams." *Journal of ACI Structural*, 98(5), 686–695.

Welty, J. R., Wicks, C. E. and Wilson, R. E. (1969). *Fundamentals of Momentum, Heat, and Mass Transfer*, 3rd edition, John Wiley & Sons, Inc.

West, J. M. (1986). *Basic Corrosion and Oxidation*. 2nd edition. New York, John Wiley & Sons.

Westman, G. (1999). "Concrete Creep and Thermal Stresses." PhD Thesis, Division of Structural Engineering, Lulea University of Technology, 301.

Win Pa Pa (2004a). "Evaluation of Effect of Crack on Chloride Ions Penetration in Reinforced Concrete Structures." PhD Thesis, Saitama University, Saitama, Japan.

Win Pa Pa, Watanabe, M. and Machida, A. (2004b). "Penetration profile of chloride ion in cracked reinforced concrete." *Cement and Concrete Research*, 34, 1073–1079.

Wittmann, F. H. (1982). "Creep and shrinkage mechanisms." *Creep and Shrinkage in Concrete Structures*, Bazant and Wittmann (eds), John Wiley & Sons, Inc.

Wittmann, F. H., Bazant, Z. P., Alou, F. and Kim, J. K. (1987). "Statistics of shrinkage test data." *Cement, Concrete and Aggregates*, 9(2), 129–153.

Yamada, K., Shima, H. and Haraguchi, K. (1991). "Effects of cyclic loading and time on bond between steel bar and concrete." *Proceedings of JCI*, 12(2).

Yamamoto, T. and Hironaga, M. (2006). "Evaluation of leaching process of cementitious materials for engineered barrier system." *Proceedings of JCI*, 28(1), 713–718 (in Japanese).

Yamazaki, K. (1962). "Fundamental studies of the effects of mineral fines on the strength of concrete." *Transactions of JSCE*, 85, 15–46.

Yoda, A. (2002). "Neutralization of blast furnace slag cement concrete naturally exposed for 40 years and the effect of finishing materials." *Proceedings of Cement and Concrete*, Japan Cement Association, 56, 449–454.

Yoda, A. and Yokomuro, T. (1987). "Carbonation depth of concrete using different types of cement." *Proceedings of JCI*, 9(1), 327–332.

Yokozeki, K. (2004). "Long–term durability design of 1,000–year level on leaching of cement hydrates from concrete." PhD Thesis, Tokyo Institute of Technology (in Japanese).

Yokozeki, K., Motohashi, K., Okada, K. and Tsutsumi, T. (1997). "A rational model to predict the service life of RC structures in marine environment." *4th CANMET/ACI International Conference on Durability of Concrete*, SP 170–40, 777–798.

Yokozeki, K., Watanabe, K., Hayashi, D., Sakata, N. and Otsuki, N. (2003). "Modeling of ion diffusion coefficients in concrete considering hydration and temperature effects." *Concrete Library International*, 42, 105–119.

Yonekura, A. (1994). "Moisture dissipation and Creep." *Concrete Journal*, JCI, 32(9), 37–42 (in Japanese).

Yoon, S., Wang, K., Weiss, W. J. and Shah, S. P. (2000). "Interaction between loading, corrosion and serviceability of reinforced concrete." *ACI Journal Materials*, 97(6), 637–644.

York, G. P., Kennedy, T. W. and Perry, E. S. (1972). "Experimental investigation of creep in concrete subjected to multiaxial compressive stressed and elevated temperatures." Concrete for Nuclear Reactors, ACI Special Publication, 34, 647–700.

Yoshioka, Y. and Yonezawa, T. (1982) "Basic study about mechanical characteristics of reinforcement's corrosion." *Proceedings of the Annual Conference of the JSCE*, 271–272.

Yue, L. L. and Taerwe, L. (1992). "Creep recovery of plain concrete and its mathematical modeling." *Magazine of Concrete Research*, 44(161), 281–290.

Zhu, Y., Ishida, T. and Maekawa, K. (2004). "Multi–scale constitutive model of concrete based on thermodynamic states of moisture in micro–pores." *Proceedings of JSCE*, 760 (V–63), 241–260 (in Japanese).

# Index

# eBooks – at www.eBookstore.tandf.co.uk

## A library at your fingertips!

eBooks are electronic versions of printed books. You can store them on your PC/laptop or browse them online.

They have advantages for anyone needing rapid access to a wide variety of published, copyright information.

eBooks can help your research by enabling you to bookmark chapters, annotate text and use instant searches to find specific words or phrases. Several eBook files would fit on even a small laptop or PDA.

**NEW:** Save money by eSubscribing: cheap, online access to any eBook for as long as you need it.

## Annual subscription packages

We now offer special low-cost bulk subscriptions to packages of eBooks in certain subject areas. These are available to libraries or to individuals.

For more information please contact webmaster.ebooks@tandf.co.uk

We're continually developing the eBook concept, so keep up to date by visiting the website.

# www.eBookstore.tandf.co.uk

Printed in the United States
by Baker & Taylor Publisher Services